Interdisciplinary Biotechnological Advances

This series is an authoritative resource that explains recent advances and emerging directions in biotechnology, reflecting the forefront of research clearly and reliably, without excessive hype. Each volume is written by authors with excellent reputations and acknowledged expertise in the topic under discussion. The volumes span the entire field from an interdisciplinary perspective, covering everything from biotechnology principles and methods to applications in areas including genetic engineering, transgenic plants and animals, environmental problems, genomics, proteomics, diagnosis of disease, gene therapy, and biomedicine. The significance of these applications for the achievement of UN Sustainable Development Goals is highlighted. The series will be highly relevant for Master's and PhD students in Biotechnology, Nanochemistry, Biochemical Engineering, and Microbiology, medical students, academic and industrial researchers, agricultural scientists, farmers, clinicians, industry personnel, and entrepreneurs.

Sankhadip Bose • Amritesh Chandra Shukla
Mirza R. Baig • Sabyasachi Banerjee
Editors

Concepts in Pharmaceutical Biotechnology and Drug Development

Springer

Editors
Sankhadip Bose
School of Pharmacy
The Neotia University
West Bengal, West Bengal, India

Amritesh Chandra Shukla
Department of Botany
University of Lucknow
Lucknow, India

Mirza R. Baig
Department of Clinical Pharmacy and Pharmacotherapeutics
Dubai Pharmacy College for Girls
Dubai, United Arab Emirates

Sabyasachi Banerjee
Department of Pharmaceutical Chemistry
Gupta College of Technological Sciences
Asansol, West Bengal, India

ISSN 2730-7069 ISSN 2730-7077 (electronic)
Interdisciplinary Biotechnological Advances
ISBN 978-981-97-1147-5 ISBN 978-981-97-1148-2 (eBook)
https://doi.org/10.1007/978-981-97-1148-2

This Springer imprint is published by the registered company Springer Nature Singapore Pte Ltd.
The registered company address is: 152 Beach Road, #21-01/04 Gateway East, Singapore 189721, Singapore

Preface

The book *Concepts in Pharmaceutical Biotechnology and Drug Development* provides comprehensive coverage of the development of new pharmaceuticals and the enhancement of existing ones. There is a growing recognition among companies in various industrial sectors about the significance of sustainable development and the promising opportunities presented by biotechnology. The implementation of biotechnology has the potential to enhance the eco-friendliness of industrial processes and reduce both financial and operating expenses. Additionally, it has the potential to decrease the usage of raw materials and energy while also minimizing waste. The primary objective is to offer a comprehensive understanding of pharmaceutical biotechnology, including its underlying principles and practical applications from an industrial standpoint. The book provides a thorough overview of the ways in which biotechnology influences drug development, production, and regulation, and is a valuable resource for those seeking to enhance their understanding in this area. This book is designed to support educators in their teaching efforts and offers a reader-friendly exploration of the various stages involved in the development of new pharmaceuticals through biotechnology. This book is a valuable resource for individuals at various stages of their academic and professional careers, including undergraduates, graduates, pharmaceutical scientists, clinicians, and academic researchers. It provides convenient access to current practices in pharmaceutical biotechnology, making it particularly useful for those working in the interdisciplinary field of biochemistry, pharmacology, biopharmaceutics, and biotechnology. This book's concise and impartial content structure may also prove advantageous to corporate researchers.

Sarisa, West Bengal, India — Sankhadip Bose
Lucknow, India — Amritesh Chandra Shukla
Dubai, United Arab Emirates — Mirza R. Baig
Asansol, West Bengal, India — Sabyasachi Banerjee

Contents

Editors and Contributors

About the Editors

Sankhadip Bose (b. 05 October 1979) is a Professor and Head, School of Pharmacy, The Neotia University, West Bengal. He did his Master of Pharmacy in Pharmacognosy from K.L.E.S' College of Pharmacy, Hubli, Karnataka, and Ph.D. (Pharmaceutical Technology) from West Bengal University of Technology, Kolkata. He served as academic person for more than 17 years in Maulana Abul Kalam Azad University of Technology and guided more than 20 M. Pharm students in their research projects. Dr. Bose has successfully handled one Central Government Research Project from the Department of Science and Technology (DST), New Delhi, Govt. of India, in 2013. One Ph.D. has been awarded under the supervision of Dr. Bose, and presently another three Ph.D. projects are going on in Maulana Abul Kalam Azad University of Technology and two projects in The Neotia University under his supervision. He has published more than 50 research papers/articles in nationally and internationally reputed journals as well as more than 10 book chapters published by CRC Press, Springer Nature, Apple Academic Press, etc.

Dr. Bose has presented more than 70 research papers in different national and international conferences. Recently in 2019, he has presented one of his research articles in a SPER Sponsored International Conference in Kathmandu University, Nepal. He has delivered a speech in Upgradation Training Program of Registered Pharmacists organized by West Bengal Pharmacy Council in Burdwan Medical College in 2018 and also delivered another speech on "Phytochemical Screening for the Development of Herbal Medicine" in the 2nd National Conference of Society of Pharmaceutical Education and Technology (SPET) in Mehr Chand Polytechnic College in Jalandhar, Punjab, as a resource person. He was in the position of State Secretary (2016-2022) of Association of Pharmaceutical Teachers of India (APTI) and life member of Society of Pharmaceutical Education and Research (SPER). Presently, he is the CEC member of Association of Pharmaceutical Teachers of India (APTI).

Dr. Bose has been the Chairperson of Scientific Session of AICTE Sponsored National Conference on "Complex Diseases, Novel Therapeutics & Delivery Challenges," 2014, Gupta College of Technological Sciences, Asansol, West Bengal, India. He has become a member of many scientific committees of different national and international conferences. And at last, he has become the Executive Editorial Member and reviewer of many national and international journals.

Amritesh Chandra Shukla (b. 25 July 1973) did his D.Phil. and D.Sc. in Botany from the University of Allahabad, Allahabad and is an energetic Professor (Dept of Botany, University of Lucknow, Lucknow, India) and passionate researcher. He served as Reader/ Professor and Head of the Department of Horticulture, Aromatic and Medicinal Plants, Mizoram University, Aizawl (2007–16). He has been working on plant pharmaceuticals, natural products, drug development, and patenting, since 1995. He standardized various scientific methods viz. MSGIT, MDKT MS-97, and NCCLS (2002-03), successfully handled six major research projects, and is supervising one European Union Project (No. 8944160). He has developed commercial herbal drug formulations and holds patents in the USA, UK, Japan, and India. Besides this, he supervised master's (21) and Ph.D. thesis (10) and published 118 research papers/articles in nationally and internationally reputed journal as well as ten books (authored 04, edited 06) published by CRC Press (2019), Springer (2020), etc.

He serves as Editor, Associate Editor, and Member of the Editorial Board of many internationally reputed journals such as *American J of Food Technology*, *J of Pharmacology & Toxicology*, and *Research J of Medicinal Plants* and as reviewer for *Scientific Reports*, *Int. J. of Dermatology*, *European J. of Medicinal Plants*, *African J. of Agricultural Research*, *Int. J. of Agricultural Chemistry*, and *J of Herbs, Spices & Medicinal Plants*.

Prof. Shukla has vast overseas experience in initiating to cooperate R&D for medicinal and aromatic plants among India and several countries in Asia and the Pacific regions. He has been invited speaker in Indo-Brazilian Symposium (2020); delivered a lecture and chaired a session at ISHS-IV Asia Symposium on Quality Management in Postharvest Systems, Jeonju, Korea (2017); Member of the International Society for Horticultural Science, Belgium; invited scientist CAF, Gold Coast, Australia (2004); CAF, Fresno, USA (2000); 7th IWCSPP, Beijing, China (1998); and 4th EFPP, Berlin, Germany (1996), for his professional expertise and activities. He has also been a Visiting Professor at the University of Mauritius (2014); a Visiting Professor at the University of British Columbia, Canada (2013-14); a Member of the expert panel for promotion to Professor in the Department of Botany and Microbiology, College of Science, King Saud University, Saudi Arabia (2018); and a member of the Research Degrees and Scholarships Committee at Murdoch University, Perth, Australia (2019-21).

For his outstanding scientific contribution, Prof. Shukla has been honored Y.S. Murty Memorial Gold Medal; Prof. R.C. Pant Memorial Award; NESA—Scientist of the Year Award; ISC—Member of the Sectional Committee, Plant Sciences Sec of the 101st Indian Science Congress; and IPS Medal-98. He has been the Executive Councilor—Indian Botanical Society; Chairman—International Symposium on Sustainable Horticulture (ISSH-2016); Member—State Level Steering Committee for National Mission on Medicinal Plants, Government of Mizoram; Program Coordinator—Indian Academy of Sciences, Bangalore; Program Coordinator—CSIR-CIMAP Training; Convener—ASSOCHAM Conference; Course Director—RAMAPS; and Convener—NCNRM, India. Besides this, he is Fellow—Ethnobotanical Society;

Fellow—Society of Life Sciences; Fellow—Indian Botanical Society; Fellow—International Young Scientist Association; and Fellow of the Linnean Society of London (2021).

Mirza R. Baig is currently working as a Dean and Professor in Clinical Pharmacy in Dubai Pharmacy College for Girls. He did his Ph.D. in Clinical Pharmacy from University Science Malaysia (USM), Malaysia. He worked in Malaysia and India and has international teaching and research experience of about 20 years. He has been recognized as one of the top scientists in the world in the scientific index for 2022-23. He has one German patent 2022, in collaboration with King Abdul Aziz University, Saudi Arabia. He published five books as coauthor related to Clinical Pharmacy and recent one in Elsevier publication in June 2020 on Pharmacovigilance. He published about 80 research papers and articles in international journals with a recent paper published in *Inflammopharmacology*, a part of Springer Nature journal, 2023. He has been invited as a chairperson, speaker, and moderator in many international conferences and seminars. He serves on the editorial board of various international journals, including reviewer for a couple of Elsevier journals. He supervised numerous master's and Ph.D. students for their research projects. He also reviewed several master's and Ph.D. thesis for Indian and foreign universities as an external examiner. He received many awards in his career, such as Best Research Project 2019, in a conference held at the University of Sharjah, Best Professor in Clinical Pharmacy in Middle East Educational Awards 2018; Best oral presentation in DUPHAT Dubai 2017; Best Quality Poster in 2016 in DUPHAT Dubai; an Outstanding scientist by a research foundation in India 2016; Best oral presentation in Dubai 2015 & 2014; young scientist award in Malaysia in 2013; and Best presentation award in IPC, Chennai 2003, in India. His areas of research are pharmacoepidemiology, drug safety and efficacy, drug utilization studies, pharmacovigilance, and public health.

Sabyasachi Banerjee (b. 22 November 1992) is an Assistant Professor in the Department of Pharmaceutical Technology, Gupta College of Technological Sciences, Asansol, in Paschim Barddhaman District, West Bengal. He did his Master of Pharmacy in Quality Assurance and Pharma Regulatory Affairs from Gupta College of Technological Sciences, Asansol, West Bengal, and is pursuing a Ph.D. (Pharmaceutical Technology) on plant-based phytochemicals (isolation, characterization, preparing semi-synthetic derivatives, and so on) from Maulana Abul Kalam Azad University of Technology (formerly known as West Bengal University of Technology), Haringhata. He served as Assistant Professor for more than 7 years in Maulana Abul Kalam Azad University of Technology. Sabyasachi Banerjee also played a role as an editor in an edited book namely *Medicinal Spices and Herbs from India*, 2-volume set (Volume 1: *Introduction, Therapeutic Properties, and Commercial Products*/Volume 2: *Beneficial Health Effects of 20 Exotic Spices*), which is being published by Apple Academic Press (AAP) along with CRC Press (Taylor & Francis Group). He has already published more than 33 research papers/articles in internationally reputed journals like *Coordination Chemistry Reviews* (Elsevier, IF-20.6), *Pharmacology and Therapeutics* (Elsevier, IF-13.5), *Pharmacological Research* (Elsevier, IF-9.3), *Cells* (MDPI, IF-6), *Cancers* (MDPI, IF-5.2), *Marine Drugs* (MDPI, IF-5.4), *Life Sciences* (MDPI, IF-3.253), *Phytotherapy Research* (Wiley, IF-7.2), *Mini Reviews in Medicinal Chemistry* (Bentham, IF-3.8), *Current Drug Targets* (Bentham, IF-3.2), *Frontiers in Oncology* (Frontiers, IF-4.7), *Anti-Cancer Agents in Medicinal Chemistry* (Bentham, IF-2.8), *Russian Journal of General Chemistry* (Springer, IF-0.9), *Physiologia* (MDPI), etc. and currently around 5 are under process; as well as 13 book chapters published by Elsevier, CRC Press, Wiley & Sons, Springer Nature, Apple Academic Press, etc. along with 18 more accepted chapters. Apart from that he has presented in many national as well as international conferences. He is a permanent reviewer of the journals *Biosciences Biotechnology Research Asia* and *Journal of Pure and Applied Microbiology*, as well as he also serves as a reviewer for journals like *Frontiers in Pharmacology* (Frontiers, IF-5.6), *Phytomedicine*

(Elsevier, IF-7.9), *Medicinal Chemistry Research* (Springer, IF-2.6), *Phytotherapy Research* (Wiley & Sons, Inc., IF: 7.2), *European Journal of Medicinal Chemistry* (Springer, IF: 6.7), *European Journal of Medicinal Chemistry Reports* (Springer), etc. He has been a fellow life member of Scholars Academic and Scientific Society since 2021. He is also a fellow life member of Indian Chemical Society since August 2023.

Contributors

Dakshitha Akula Department of Biotechnology, School of Applied Science, REVA University Bangalore, Bangalore, Karnataka, India

T. Amala Moulana College of Paramedical Sciences, Malappuram, Kerala, India

T. A. Anboli School of Biotechnology, National Institute of Technology, Calicut, Kerala, India

Mohd Aslam Department of Chemistry, SRM Institute of Science and Technology, Delhi-NCR Campus, Modinagar, Ghaziabad, Uttar Pradesh, India

Department of Chemistry, Atma Ram Sanatan Dharma College, University of Delhi, New Delhi, India

Mirza R. Baig Department of Clinical Pharmacy and Pharmacotherapeutics, Dubai Pharmacy College for Girls, Dubai, United Arab Emirates

Asis Bala Division of Life Sciences, Institute of Advanced Study in Science and Technology (IASST), An Autonomous Institute Under Department of Science and Technology (Govt. of India) Vigyan Path, Guwahati, Assam, India

Sabyasachi Banerjee Department of Pharmaceutical Chemistry, Gupta College of Technological Sciences, Asansol, West Bengal, India

Santanu Banerjee Department of Pharmacology, Gupta College of Technological Sciences, Asansol, West Bengal, India

Subhasis Banerjee Department of Pharmaceutical Chemistry, Gupta College of Technological Sciences, Asansol, West Bengal, India

Sugato Banerjee Department of Pharmacology and Toxicology, National Institute of Pharmaceutical Education and Research, Kolkata, India

Sankhadip Bose School of Pharmacy, The Neotia University, Sarisa, West Bengal, India

Debprasad Chattopadhyay ICMR-National Institute of Traditional Medicine, Nehru Nagar, Belagavi, Karnataka, India

ICMR-National Institute of Traditional Medicine, Belagavi, India

School of Life Sciences, Swami Vivekananda University Barrackpore, Kolkata, India

Ankita Choudhury Silchar Medical College and Hospital, College of Pharmacy, Silchar, India

Avijit Choudhury Gupta College of Technological Sciences, Asansol, West Bengal, India

Khushboo Dasauni Department of Biotechnology, Kumaun University, Nainital, India

Avik Das Department of Pharmacology, Gupta College of Technological Sciences, Asansol, West Bengal, India

Rajeswar Das School of Pharmacy, The Neotia University, Sarisa, West Bengal, India

Susovan Das Acharya & BM Reddy College of Pharmacy, Bengaluru, India

Rabin Debnath ISF College of Pharmacy, Moga, Punjab, India

Sourav De Department of Pharmaceutical Technology, Eminent College of Pharmaceutical Technology, Kolkata, India

Gourab Dey Department of Pharmaceutical Technology, Eminent College of Pharmaceutical Technology, Kolkata, India

Arghya Kusum Dhar School of Pharmacy, The Neotia University, Sarisa, West Bengal, India

Subas Chandra Dinda School of Pharmacy, The Neotia University, Sarisa, West Bengal, India

Sumit Durgapal Department of Pharmaceutical Sciences, Kumaun University, Nainital, India

Shalesh Gangwar Department of Computer Science, Jamia Millia Islamia, New Delhi, India

Anirbita Ghosh Department of Pharmaceutical Technology, Jadavpur University, Kolkata, West Bengal, India

Nibir Ghosh Department of Pharmaceutical Technology, Eminent College of Pharmaceutical Technology, Kolkata, India

Somsubhra Ghosh School of Pharmacy, The Neotia University, Sarisa, West Bengal, India

Barun Das Gupta School of Natural Product Studies, Jadavpur University, Kolkata, West Bengal, India

Rajendra Gyawali Department of Pharmacy, Kathmandu University, Dhulikhel, Nepal

Pallab K. Haldar School of Natural Product Studies, Jadavpur University, Kolkata, West Bengal, India

Department of Pharmaceutical Technology, Jadavpur University, Kolkata, West Bengal, India

Abu Md Ashif Ikbal Department of Pharmaceutical Sciences, Drug Discovery Research Laboratory, Assam University (A Central University), Silchar, India

Sandipan Jana School of Pharmacy, The Neotia University, Sarisa, West Bengal, India

Department of Pharmaceutical Technology, Jadavpur University, Kolkata, West Bengal, India

Amit Kar Institute of Bioresources and Sustainable Development, Autonomous Institute Under Department of Biotechnology, Imphal, Manipur, India

Javed Khan Department of Chemistry, SRM Institute of Science and Technology, Delhi-NCR Campus, Modinagar, Ghaziabad, Uttar Pradesh, India

Department of Chemistry, Atma Ram Sanatan Dharma College, University of Delhi, New Delhi, India

Suma Krishnaswamy Phytazen Biosciences Pvt. Ltd, Bangalore, India

Sindhoora Lakshmi Department of Biotechnology, School of Applied Science, REVA University Bangalore, Bangalore, Karnataka, India

Subhash C. Mandal Department of Pharmaceutical Technology, Jadavpur University, Kolkata, India

Motlalepula G. Matsabisa Department of Pharmacology, University of the Free State, Bloemfontein, South Africa

Sujayita Mazumder School of Pharmacy, The Neotia University, Sarisa, West Bengal, India

Piyush Mohapatra Amity Institute of Biotechnology, Amity University, Noida, Uttar Pradesh, India

Nurul Hassan Mondal Dr. B. C. Roy College of Pharmacy and AHS, Dr Meghnad Saha Sarani, Durgapur, West Bengal, India

Aniruddha Mukherjee School of Pharmacy, The Neotia University, Sarisa, West Bengal, India

Pulok K. Mukherjee Institute of Bioresources and Sustainable Development, Autonomous Institute Under Department of Biotechnology, Imphal, Manipur, India

Khushbu Nailwal Department of Pharmaceutical Sciences, Kumaun University, Nainital, India

Tapan Kumar Nailwal Department of Biotechnology, Kumaun University, Nainital, India

Dinesh Kumar Lakshmi Narayanan Faculty of Pharmacy, AIMST University, Bedong, Kedah, Malaysia

Narahari Narayan Palei Amity Institute of Pharmacy, Amity University Lucknow Campus, Lucknow, Uttar Pradesh, India

Partha Palit Department of Pharmaceutical Sciences, Drug Discovery Research Laboratory, Assam University (A Central University), Silchar, India

Garima Pandey Department of Chemistry SRM Institute of Science and Technology, Delhi-NCR Campus, Modinagar, Ghaziabad, Uttar Pradesh, India

Bhaskara Nand Pant Department of Chemistry, Atma Ram Sanatan Dharma College, University of Delhi, New Delhi, India

Sompreeti Paul Department of Biotechnology, School of Applied Science, REVA University Bangalore, Bangalore, Karnataka, India

Monihar Pervin Department of Pharmaceutical Technology, Jadavpur University, Kolkata, West Bengal, India

Pallabita Rakshit Department of Pharmaceutical Technology, Jadavpur University, Kolkata, India

Anjali Rani Department of Chemistry, SRM Institute of Science and Technology, Delhi-NCR Campus, Modinagar, Ghaziabad, Uttar Pradesh, India

Department of Chemistry, Atma Ram Sanatan Dharma College, University of Delhi, New Delhi, India

Mrudula Rao Department of Biotechnology, School of Applied Science, REVA University Bangalore, Bangalore, Karnataka, India

Geethaa Sahgal Faculty of Pharmacy, AIMST University, Bedong, Kedah, Malaysia

Pooja Sankaranarayanan School of Biotechnology, National Institute of Technology, Calicut, Kerala, India

Neha Sharma Amity Institute of Virology and Immunology, Amity University Uttar Pradesh, Noida, India

Amritesh Chandra Shukla Department of Botany, University of Lucknow, Lucknow, India

T. V. Suchithra School of Biotechnology, National Institute of Technology, Calicut, Kerala, India

Jeevandran Sundarasekar Faculty of Applied Science, AIMST University, Bedong, Kedah, Malaysia

Aswani Thekkangil Department of Biotechnology, School of Applied Science, REVA University Bangalore, Bangalore, Karnataka, India

Sabu Thomas School of Chemical Sciences, Mahatma Gandhi University, Kottayam, India

Devinder Toor Amity Institute of Virology and Immunology, Amity University Uttar Pradesh, Noida, India

Ravichandran Veerasamy Faculty of Pharmacy, AIMST University, Bedong, Kedah, Malaysia

Seema Yadav Amity Institute of Pharmacy, Amity University Lucknow Campus, Lucknow, Uttar Pradesh, India

Part I
Sec A: Biotechnological Approaches in Formulation and Drug Design

Chapter 1
Agro-Wastes-Based Feedstock as a Source for Bionanomaterials Production: Outcomes and Challenges

Pooja Sankaranarayanan, T. A. Anboli, and T. V. Suchithra

Abstract Nanotechnology has notably impacted science and technology over recent years, especially in medicine, electronics, food safety, and energy. Using nanomaterials in agriculture has addressed several issues, such as crop protection, soil management, and post-harvest treatment. Literature reports suggest the use of nanotechnology in farming increases the nutrient uptake of soil. The positive outcomes of nanofertilizers, nanopesticides, and nanosensors led to further experimentation of nanomaterials in agriculture-based applications. Second-generation biofuel production from agro-wastes ignited the curiosity of developing bionanomaterials. The emergence of structured nanomaterials such as nanofibers, nanocellulose, and nanoporous membranes led to several applications in nano-based product synthesis. Nanocatalysis, nanosensors, smart nanoparticle delivery, and nanomaterials in nutrient uptake were some of the recent applications of nanotechnology in precision agriculture. Most of the synthesized bionanomaterials were found to have medicinal value, especially those that use zinc, iron, silver, and gold as precursors. The critical question lies in the potential toxicity of nanoparticles in the environment, which is still unanswered. In short, this chapter lists the possible sources for bionanomaterials production concerning their commercial applications. Despite having a greater scope in agriculture, bionanomaterials production from agro-wastes holds an upper hand due to its easy availability and reduction potential. If the toxicity of the nanomaterial is adequately addressed, the choice of agro-wastes as a precursor in developing bionanomaterials may improve the commercial viability of green synthesis.

Keywords Nanotechnology · Bionanomaterials · Agro-wastes · Nanocatalysis · Nanoparticles · Toxicity

---

P. Sankaranarayanan · T. A. Anboli · T. V. Suchithra (✉)
School of Biotechnology, National Institute of Technology, Calicut, Kerala, India
e-mail: drsuchithratv@nitc.ac.in

S. Bose et al. (eds.), *Concepts in Pharmaceutical Biotechnology and Drug Development*, Interdisciplinary Biotechnological Advances,
https://doi.org/10.1007/978-981-97-1148-2_1

1.1 Introduction

Nanomaterials have greater scientific significance owing to its smaller size, higher surface area, and unique morphological properties than conventional bulk particles. Nanoparticles (NPs) are mainly composed of three layers: surface, shell, and core. The method of synthesis decides the physical nature and application of nanomaterials. Methods used to prepare nanoparticles are mechanical grinding, laser ablation, electro-explosion, chemical vapor deposition, sol-gel synthesis, ultrasound-assisted synthesis, colloid nanoparticles synthesis, and green synthesis (Sasidharan et al. 2019). Out of all physical and chemical synthesis approaches, green synthesis has several advantages: low cost, simple and eco-friendly synthesis, and low energy requirements. As the name suggests, green synthesized nanoparticles get their "nano" form sculpted from plants' "green" elements, namely antioxidants, flavonoids, and phenolic compounds, which act as capping and reducing agents. There are various reviews on the use of plant precursors in developing nanoparticles with medical and non-medical applications. Some applications of nanoparticles are detailed in Table 1.1.

Advancement in the field of material sciences and a better understanding of remedial properties of plant-based nanoparticles resulted in improvised mechanisms of synthesis, thereby focusing on targeted applications (Zangeneh and Zangeneh 2020). Metallic nanoparticles like Gold (Au), Zinc (Zn), Silicon (Si), Silver (Ag), and Cerium (Ce) nanoparticles have several applications in the field of medicine, agriculture, textile, micro-electronics, food processing, and industry (Ngu et al. 2016). Simultaneously, carbon-based nanomaterials from plants have introduced the cost of practical fabrications to the electronics sectors in the form of fullerenes, carbon nanotubes, quantum dots, and graphene oxides (Jamdagni et al. 2018). The low-cost synthesis and power-saving nature of these fabrications paved the way to develop "nanoelectronics" and lifted nanotechnology to a greater height.

Using silicon and carbon nanoparticles from agricultural wastes as catalysts in the production of biofuels has reduced production costs by incorporating cheaper enzyme immobilization techniques in heterogenous catalysis (Hazmi et al. 2020). Nanocellulose from lignocellulosic biomass has raised the standards of fiber-based engineering applications. The richer content of fibrous materials in plant sources led to the development of many nanocomposites and nanodesigns, which can be used as nanofertilizers, nanosensors, nanopesticides, and even in improving soil fertility by increasing its nutrient content. Though nanobiomaterials possess fascinating applications and advantages, several issues leading to its toxic environmental impact were also expressed (Prasad et al. 2017).

Most of the produced nanomaterials are usually studied for its medicinal properties rather than its industrial outcome. Developing nano-based fertilizers, detergents, and sensors exposed the scope of agricultural and industrial applications in nanotechnology. For example, cost-efficient natural detergents from

Table 1.1 Applications of Bionanomaterials from agro-wastes

Nanomaterials	Mechanism/source	Average size	Shape	Outcomes	Applications	Reference
Chemical synthesis						
TiO_2 coconut shell nanocomposite	Calcination and chemical pretreatment methods for coconut shells	15 nm	–	Photocatalytic absorbent	Can be used as absorbent in removing carbamazepine	Khraisheh et al. (2013)
Nanoporous carbon materials	Multi-step chemical reduction process of rice husk using KOH, NaOH, $ZnCl_2$ followed by carbonization	–	–	Iodine and methylene blue adsorption property	Adsorbent materials in water purifiers	Shrestha et al. (2019)
Multi-walled carbon nanotubes—TiO_2-loaded nanocomposite	Modified sol-gel synthesis using tetrabutyl titanate	15 nm	Tube-like shape	Highest photodegradation efficiency in removal of methyl orange during photocatalytic experiment	High photocatalytic activity, potential in wastewater treatment	Wang et al. (2016)
Cellulose nanofibers	Chemical pretreatment of bamboo leaves, followed by ultra-sonication process	53.2 nm	–	Exhibit superior sound-absorption property	Can be used as insulating agents	Fan et al. (2018)
Biological synthesis						
CdS nanopesticides	Green synthesis from *Valoniopsis pachynema* algae	< 100 nm	Spherical	Anti-malarial activity against *Anopheles stephensi* and *A. sundaicus*	Nanopesticide	Sujitha et al. (2017)
Nanocapsules	Green synthesis from *Eucalyptus globulus* extract	380 nm	Spherical	Insecticidal activity against *Myzus persicae*	Nanoinsecticide	Khoshraftar et al. (2019)
Ag nanoparticles	Green synthesis from *Ligustrum lucidum* leaf extract	13 nm	Spherical	Exhibit synergistic action with fungicide, epoxiconazole	Fungistat	Huang et al. (2020)

(continued)

Table 1.1 (continued)

Nanomaterials	Mechanism/source	Average size	Shape	Outcomes	Applications	Reference
Fe_3O_4 Magnetic nanoparticles	Green synthesis from watermelon rinds	2–20 nm	Spherical	Excellent magnetic property and used as a catalyst in the preparation of 2-oxo-1,2,3,4-tetra hydropyrimidine	Catalyst	Prasad et al. (2016)
Nickel oxide nanocrystals	Green synthesis from *Nephelium lappaceum* L. (Rambutan) peel extract	104 nm	Face-centered crystalline lattice	Antibacterial activity against *Staphylococcus aureus* and *Escherichia coli*	Antimicrobial and free radical property	Yuvakkumar et al. (2014)
Chem-biosynthesis						
Ag nanoparticles	The extracellular matrix of the eggshell is used to reduce $AgNO_3$ to form AgNPs through chemical pretreatment followed by green synthesis	3.81 nm	Spherical shape with a porous nature	Reduces 4-nitrophenol to 4-aminophenol	Catalytic property	Liang et al. (2014)

corn cobs with fewer chemicals have gained attention due to its fascinating properties (Liu et al. 2020a). Similarly, bionanomaterials in pest control and dye degradation are not much surprising due to its inbuilt antimicrobial nature (Prasad et al. 2017). Ag and Au nanomaterials always synergize well in antimicrobial and photocatalytic applications (Liu et al. 2020b). Even though several works have been discussed on plant sources and their material synthesis, a detailed view of the source-based outcomes is essential. This chapter mainly focuses on the most fascinating nanomaterial outcomes from different agricultural sources and their applications in various fields to give an outlook in this sector. Certain limitations and misconceptions dealing with this technology have also been intimated.

1.2 Nanomaterials from Husk and Shell Waste

Applications of bionanomaterials synthesized from fruit, leaf, flower, and seed wastes were illustrated in Fig. 1.1.

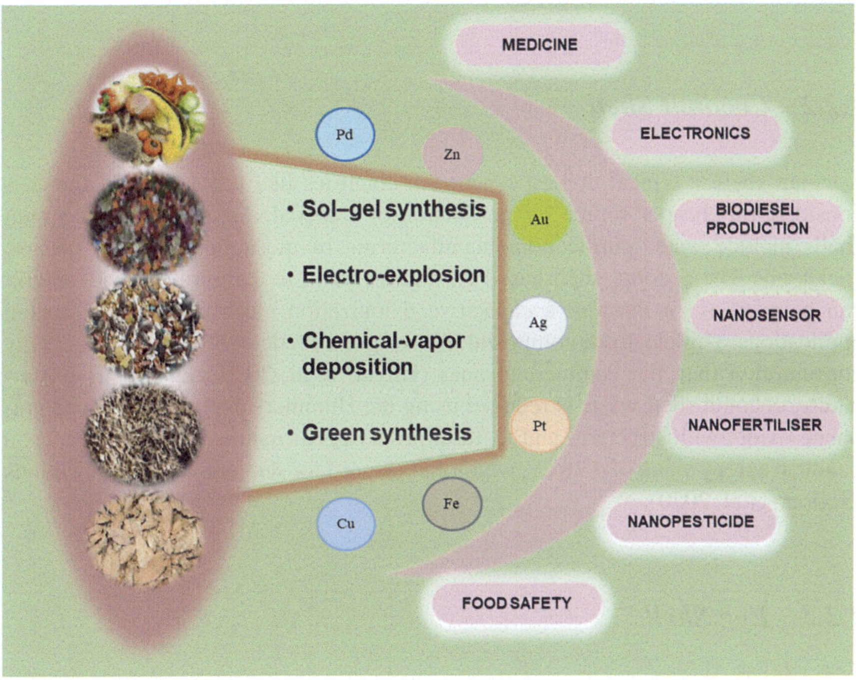

Fig. 1.1 Schematic representation of bionanomaterials synthesized from agro-wastes and their applications

1.2.1 Rice Husk

Rice husk (RH) is a biodegradable and fibrous-rich biomass generated during rice milling. It is rich in 50% cellulose, 25–30% lignin, 15–20% silica, and 10% moisture. RH has applications in dye degradation, cement processing, etc. Some recent bionanomaterials produced from RH and their industry-oriented outcomes are listed below. Carbon nanoparticles with an efficient quenching effect had been synthesized from RH using the thermal-assisted acid-carbonization method, which can be an alternative to existing fluorescent dyes due to its cheaper production costs and lesser toxicity (Ngu et al. 2016). Production of high-yield silicon dioxide nanoparticles from acid-treated RH ash in a two-step process yielded almost 93.4%. Also, it does not produce CO_2 or any other toxins, and the method is cost-effective due to using Na_2CO_3. The morphological and compositional characteristics of the produced SiO_2NPs were in par with the commercial silica, thereby expanding its applications in the field of microelectronics (Nayak and Datta 2020). RH ash is highly utilized in biodiesel and biomethane production due to its textural properties and thermal stability. For instance, a nano-bifunctional supermagnetic catalyst which is produced from RH char through K_2O doping wet impregnation method yielded a maximum of 98.6% of biodiesel from used cooking oil by using 4 wt% of catalyst, and it can be reused up to 5 times, without compromising its catalytic efficiency (Hazmi et al. 2020).

1.2.2 Coconut Shell

Coconut shell, a typical domestic waste in countries like Indonesia and India, is considered a cheaper source for production of charcoal. Applications of coconut shells include water purification, manufacturing of mosquito coils and incense sticks, etc. The coconut shell-based activated carbon in carbon nanotubes matrix can be used as an innovative capacitive deionization electrode for desalination applications. This electrode composite showed good cyclic stability with low energy consumption than the commercial ones (Huynh et al. 2020). Due to its fibrous nature, coconut shell waste is reduced using the Hummers method to produce graphene oxide, with structural and textural properties of natural graphite, and has a bandgap energy value of 4.38 eV, which can be used for semiconductor applications (Sujiono et al. 2020).

1.2.3 Egg Shell

Several studies reveal that the eggshell particles have 95% calcium carbonate and 5% inorganic components, making it a potential raw material for calcium-based nanoproducts. Calcium oxide nanomaterials from eggshell waste can be chemically synthesized using sol-gel method, which involves less temperature, alkaline additives and relatively cheaper than other chemical methods (Habte et al. 2019). Both nano-calcium oxide and calcium carbonate from eggshell have several applications in water treatment, automobile, and paper industry. Eggshell also has medicinal applications such as treatment of osteoporosis and in bone grafts. Calcium nitrate from eggshell wastes had been chemically treated to produce diopside powders, with which Poly (methyl methacrylate) was reinforced to make it a porous scaffold. This porous nanocomposite's mechanical and surface scaffolding assists in the biomedical field (Choudhary et al. 2020).

1.2.4 Walnut Shell and Husk

Walnut husks and shell pigments were believed to be a good resource for making fabric dyes and inks in Rome and Europe. Metal nanoparticles from various parts of the walnut tree have antimicrobial and antiradical applications. Iron oxide nanoparticles from walnut husk using a two-step co-precipitation method were investigated for their cytotoxic activity on mouse embryonic fibroblast cell lines and human colorectal adenocarcinoma cell lines. The cubic-shaped nanoparticles showed no toxicity on the cell lines up to 1000 μg/mL, making them suitable candidates for drug-related applications (Izadiyan et al. 2020). Walnut has excellent adsorbent capacity towards heavy metal ions, dyes, and organic compounds. Juglone dye extracted from walnut shell powder can be used as reducing and capping agents for the synthesis of Ag nanosols and nanodisks. Juglone has antimicrobial and antioxidant activity, enhancing the AgNPs' properties. Meanwhile, the addition of cetyl trimethyl ammonium bromide in nanoparticles resulted in micellar formation, hinting it as a potential surfactant (Zaheer 2019). The electrical potential of walnut shell is less explored, despite having high cellulose content. Liquefaction is used to obtain maximum cellulose from walnut shells, followed by the electrospinning process resulting in carbon nanofiber mats. Thinner micropore structure with higher specific surface area (408 m^2/g) was found to be a suitable electrode in lithium-ion batteries. Charge capacity of 150 mAh/g with excellent cyclic stability till 100 cycles proved it to be a sustainable electrode for batteries (Tao et al. 2017).

1.3 Bionanomaterials from Corn Cob

Metal nanoparticles and their antimicrobial activity have been closely examined for decades. Very recently in 2020, AgNPs from corn cob were found to be a potential antiparasitic agent against *Trypanosoma cruzi*. Corn cob is used as bioreductant along with $AgNO_3$ to produce nano xylan. Characterization of nanoparticles showed that nano xylan contains 81% xylan and 19% Ag. 100 μg/mL of nano xylan effectively eliminated 95% of parasite, which is considered a safer therapeutic alternative to benznidazole. In vivo study regarding the same is required to prove its therapeutic efficiency (Brito et al. 2020). Though eco-friendly products have a low yield, replacement of waste biomass conversion is fast and reliable. Replacing synthetic detergents with natural ones requires different approach to improve cleaning capacity. There was a recent study on a three-step cellulose nanospheres production process from corn cobs which can improve the cleaning capacity by the mechanism called Pickering emulsions. The nanospheres' life cycle assessment and biocompatibility studies proved it to be a recyclable detergent (Liu et al. 2020a). Use of organic fertilizers was found to impart higher nutrition content to the soil. Corn cob's biochar impregnated with macro (N, Mg, Ca, K, P) and micronutrients (Na, Zn, Fe) made into nanocomposites enhances plant growth by preserving nutrients for a long time by keeping the typical properties of slow releasing nanofertilizers such as water absorbance, salt index, retention and release abilities (Lateef et al. 2019).

1.4 Bionanomaterials from Fruit Waste

1.4.1 Citrus Peel Waste

The successful development in the green synthesis of nanoparticles is a great cause for significant hike in the applications of nanoparticles in the field of medicine. Au and AgNPs from fruit wastes have shown good stability at higher temperature and pressure. In addition to this, ZnONPs synthesized from fruit wastes have several applications in food packaging. A remodeled traditional approach can be taken by using an aqueous orange peel extract as a reducing agent to produce ZnONPs from Zinc acetate dihydrate. The minimal use of toxic chemicals and the catalytic capability of orange extract influence the microstructural stability of ZnONPs. Orange (*Citrus sinensis*) peel-based ZnONPs exhibit high bactericidal activity against *E.coli* and *S. aureus* at a concentration of 0.025 mg/mL, by interacting with the microbes via the cell membrane and damaging DNA (Doan Thi et al. 2020). Similar extraction technique is used to synthesize TiO2NPs using lemon (*Citrus limon*) peel aqueous extract and titania powder. Hesperidin present in lemon peel extract releases aglycone, which acts as capping and reducing agents for spherical TiO2NPs and it is found to be free from contamination, unlike commercial ones. The photolytic ability of TiO2NPs was 70% more efficient than previously reported ones (Nabi et al. 2020). Grapefruit (*Citrus paradisi)* peel-based distillate mixed with 1 mM $AgNO_3$ and reduced via the microwave-assisted sol-gel method for biosynthesizing AgNPs with an average size of 14.84 nm is found to have a broad-spectrum antibacterial potency against *E.coli, S. aureus,* and *Klebsiella pneumonia* with minimum inhibitory concentrations of 40, 20, and 40 μg/mL, respectively (Ayinde et al. 2019).

1.4.2 *Banana Peels*

Banana peel is a well-known agricultural waste for green synthesis due to its higher starch content. Au and Ag nanoparticles synthesized from banana peels have anti-cancer and antioxidant properties owing to their bioactive compounds. Au-dendrite nanocomposites derived from banana peel possess the ability to reduce chloroauric acid via hydrothermal routes (Liu et al. 2020b). It also has good biocompatibility, antimicrobial and anticancer potency. Besides, it has proven to have an efficient inhibitory effect on breast cancer cells at a concentration of 200 μg/mL, with the expense of a limited amount of chemicals. An in vivo study conducted in nude mice proved that support of infrared light in Au-dendrite nanomaterials suppressed tumor growth and migration (Liu et al. 2020b). Rich potassium content in banana peels can improve soil fertility. Shredded banana peel slurry mixed with potassium hydroxide was filtered, in which urea and citric acid were added and homogenized. The obtained nano mixture showed greater germination efficiency in the first planting week of tomato and fenugreek, thereby acting as a biostimulant in promoting seedlings growth performance (Hussein et al. 2019).

1.4.3 *Fruit Juices*

Nanomaterials obtained from fruit waste-mediated synthesis have various biological and engineering applications. CeO_2NPs preparation from cerium nitrate requires watermelon juice as a reducing agent. The simple solution combustion method followed by the calcination technique integrates the production of cubic-phase nanoceria. Photocatalytic experiments on synthesized CeO_2 nanomaterials showed degradation of about 98% methylene blue dye in the presence of UV irradiation. Antibacterial activity against *K. aerogenes* and *S. aureus* was also tested and maximum zone of inhibition was observed at 50 and 100 μL concentration respectively (Reddy Yadav et al. 2016). Solar-assisted ZnO nanoparticle was prepared using lime juice as a reducing and capping agent with zinc acetate dihydrate as a precursor. Citric acid content in lime juice reacts with zinc acetate to form zinc citrate, which further undergoes calcination to form zinc oxide. Solar energy-assisted green synthesis mechanism contributes to the crystalline nature of nanomaterials. But the presence of ascorbic acid in lime juice greatly defines the size of ZnO nanoparticle (Hinge and Pandit 2017). Studies regarding its biological efficiency and engineering applications were still under progress.

1.5 Bionanomaterials from Leaf Wastes

Green synthesis based on leaf extract marked the beginning of nano era in industrial applications. Impressive outcomes on energy conversion using leaf ash in heterogenous catalysis paved a way to extend its applications in biotechnology and its allied sectors. Bamboo ash, which contributes 70% of amorphous silica, has its applications as fillers, medical additives, carriers, and composites. One such outcome is the synthesis of SiO_2NPs from bamboo leaf ash via thermal combustion and alkaline extraction technique, making it a suitable filler for Polydimethylsiloxane (PDMS) membranes. Pure PDMS membranes vs PDMS incorporated with SiO_2 nanomaterials were analyzed for its pervaporation flux and thermal stability. A decrease in contact angle increases the selectivity of membranes which was studied using acetic acid-water mixture. The results proved that using Si-nanomaterials in PDMS decreases the contact angle, thereby improving its thermal stability and evaporation flux (Sethy et al. 2019). AgNPs and their antibacterial potency are well known for ages. AgNPs obtained from tomato (*Solanum lycopersicum)* leaves were entrapped in chitosan to transform them into Ag-chitosan nanoparticles through ion gelation. In vitro and in vivo evaluation of Ag-chitosan nanoparticles against *Ralstonia solanacearum,* a plant pathogen, was carried out. The presence of Ag and chitosan showed greater microbicide activity against pathogen within 48 hrs. In-depth understanding of Ag-chitosan nanoparticles in the environment as well as field study is required to analyze the potential use of nanocomposites as a targeted fertilizer. Zinc-based nanomaterials have unique structural and functional characteristics due to its vast availability in natural precursors (Santiago et al. 2019). *Nelumbo nucifera* (lotus) leaf extract was mixed with zinc nitrate in solution combustion approach to produce ZnONPs. The produced ZnO NPs, when fabricated with poly (3-hexylthiophene-2,5-diyl) (P3HT) in p-type organic field-effect transistor (OFET), shows prominent sensitivity for CO gas with field-effect mobility of 10^{-2} $cm^2/V/s$. The novelty of this approach lies in its eco-friendly synthesis and its capability to function in open air. Optimization of electrical parameters such as field-effect mobility, on and off current values enhances the performance of CO adsorption. The CO adsorption is dictated by the physical characteristics of active receptor layer (ZnO NPs) and conducting properties of P3HT (channel layer). The use of natural synthesized ZnO NPs in OFETs has greater significance over commercially available transistors due to their purity and reusability (Narayana et al. 2020). Such transistors can extend its application in real-time air quality examinations, medical diagnosis, and in nano forms of modulated electronics. Potential use of leaf ash as a catalyst enhancer for biofuel synthesis was also investigated, but its significance over other biological ash (rice husk ash) in terms of performance as well as productivity has to be reviewed.

1.6 Bionanomaterials from Flowers

Flower extract is one of the recent sources in green synthesis of nanomaterials. Ag nanomaterials obtained from plant extracts have shown antibacterial and antifungal activity. Very few reports of flower extract-based nanomaterials having potential applications in health care and industry were found. Aqueous extract of *Allamanda cathartica* (Golden trumpet) mixed with $AgNO_3$ solution incubated in the dark for

24 h and centrifuged. The obtained powder was analyzed for its physical characteristics. UV adsorption spectra within the range of 350–450 nm confirmed the formation of AgNPs, and they have significant antibacterial activity against *Salmonella typhimurium, S. aureus, E. coli*, and *K. pneumoniae* in comparison with standard antibiotic, streptomycin. Also, free radical scavenging assay of AgNPs indicated the enhancement of antioxidant activity from 40% to 88%, paving ways for several biomedical applications (Karunakaran et al. 2016). Synthesis of ZnO nanoparticle from *Nyctanthes arbor-tristis* (night flowering jasmine) using zinc acetate dihydrate resulted in crystalline-structured nanoparticles in the size range of 12–32 nm, proving to be an efficient antifungal agent against plant pathogens—*Alternaria alternata, Aspergillus niger, Botrytis cinerea, Fusarium oxysporum*, and *Penicillium expansum*. Highest minimum inhibitory concentration values were observed at 128 μg/mL for *Botrytis cinereal* and *Penicillium expansum* (Jamdagni et al. 2018). Metallic nanocomposites in the field of pharmacology is rarely studied. Spherical Au nanomaterials, with the average diameter range of 15–45 nm synthesized from flower extract (aqueous) of *Hibiscus sabdariffa*, had a potency in controlling myeloid leukemia effect in rodents. Induction of acute myeloid leukemia was done in 75 mice, which were divided into 6 subgroups—$HAucl_4$, *Hibiscus sabdariffa*, AuNPs, daunorubicin, untreated, and control. AuNPs enhanced the anti-inflammatory cytokines parameters as compared to untreated mice. The significant increase in antioxidant potential of the Au-treated mice is in par with that of daunorubicin-treated one, proving the potential of Au-NPs in the field of drug discovery (Zangeneh and Zangeneh 2020).

1.7 Bionanomaterials from Seed

Nanoproducts made out of seed extracts have several antimicrobial and anticytotoxic properties. Au and Ag nanomaterials have good antibacterial effect, like that of Zn and copper nanomaterials. A study was conducted on *Phoenix sylvestris* (silver date palm) seed extract as a reducing agent for shaping AgNPs from $AgNO_3$. The total phenolic and flavonoid content of the seed extract was assessed and free radical scavenging activity was tested with DPPH-scavenging assay and bactericidal activity analysis was carried out in Broth microdilution and disk diffusion methods. The higher amount of flavonoid content in the seed extract contributed to the higher antibacterial and antioxidant activity against *Acne vulgaris*, the acne-causing pathogen. Despite having high microbicidal activity, AgNPs as a cosmeceutical nanocomposite are not yet reported. Depending on their physical and morphological characteristics, AgNPs might be an effective tool in drug discovery (Qidwai et al. 2018). Unlike biological and fabrication applications certain nanomaterials from seed extracts have conducting applications. For example, *Caesalpinia bonducella* (fever nut) seed extracts were used to reduce Cu $(NO_3)_2{\cdot}3H_2O$ resulting in the formation of rice-shaped CuONPs. CuONPs were incorporated in paraffin-impregnated graphite electrode (Cu-PIG) to detect the presence of riboflavin. The stability of Cu-PIG stands strong and satisfactory for 120 days and makes it suitable to apply in real samples such as egg yolk, milk powder, and B-complex tablets. Recovery rates and detection sensitivity extend the nanofabricates into the field of electrochemical sensing (Sukumar et al. 2020).

1.8 Nanoproducts from Sugarcane Bagasse

Sugarcane bagasse, a biomass fraction obtained after extraction of sugarcane juice, is predominantly rich in cellulose, hemicellulose, and lignin. Almost 2–5% of the pulp and paper industry uses lignocellulosic biomass (sugarcane bagasse, straw) as its raw material. Silica from sugarcane bagasse can be modified with polyethylene imine (PEI) to produce magnetic paper. Silica-PEI and Cobalt ferrite nanoparticles were co-precipitated via the lumen loading process resulted in agglomeration due to their interaction with cellulose fibers to inculcate the magnetic ability. The magnetic property of the fibers was analyzed through its physical properties such as, the idea of replacing cobalt ferrite in place of magnetite results in higher magnetization, enhanced magnetic coercivity, tensile strength, hardness, and chemical stability. Toxicity analysis of these magnetic fibers must be overlooked to be a potential magnetographic printing paper (Piñeres et al. 2020). The sugarcane fibers' higher lignin and cellulose content contributes to the adsorption property. Using bioadsorbents in dye removal was previously done using rice husk ash and straw ash. Sugarcane bagasse was scrutinized for its adsorption properties and showed a maximum adsorption capacity of 9.41 mg/g on methylene blue dye in wastewater. Chemical reaction kinetic studies describe pseudo-second-order kinetics, and the process is endothermic, spontaneous, and feasible. High availability, excellent adsorption property, and eco-friendly properties of sugarcane bagasse help it to be a better alternative source in wastewater treatment (Siqueira et al. 2020).

1.9 Agro-Wastes in Improving Soil Fertility

Quality of plantation greatly depends on the cultivability and nutrient availability of the soil. Synthetic fertilizer cannot maintain the fertility of the soil due to its toxicity, which leads to less nutrient uptake by plants. To enhance optimal production, agricultural wastes should be used as a compost to increase the organic materials in soil, providing long-term soil fertility. The reusability of plant wastes in agriculture might make the food production unit self-sufficient. For example, the natural fertilizer prepared with mixture of sugarcane bagasse, cow manure, and filter mud yielded maximum saplings and showed better growth and higher stem weight compared to basic inorganic NPK fertilizers (Rahmad et al. 2020). Another field experiment study was conducted to investigate the effect of biochar, rice husk, and rice straw on maize yield. The combination of biochar with NPK fertilizer yielded 7.5 t/ha grain yield of maize and improved the flowering rate significantly. Soil fertility rates increased with the usage of higher proportion of agri-composting and lesser quantity of inorganic fertilizers. This outcome may be suitable for smallholder farmers to enhance the nutrient value of fragile lands with plantation wastes as an integrated soil management approach (Dzomeku et al. 2018).

1.10 Metallic Nanoparticles as Pesticides and Insecticides

Metallic nanoparticles and their role in bactericidal and fungicidal activities were well studied. But the synergistic activity of nanoparticles with pesticides/antibiotics was least evaluated. The usage of chemically synthesized AgNPs as pesticides and bactericides is commercially available (Rahmad et al. 2020). But the development of agricultural waste-based cidal agents was least focused. A nanocomposite developed by integrating AgNPs synthesized from *Ligustrum lucidum* (Glossy Privet) leaf extracts with epoxiconazole was evaluated for its antifungal property against *Setosphaeria turcica,* and its IC_{50} was found to be 170.20 μg/mL. Synthesized AgNPs when combined with fungicide epoxiconazole at a ratio of 9:1 can be an efficient fungistat. Further research regarding AgNPs properties such as surface charge, acid-base property, aggregation behavior and its fate on the environment must be carried out to make it a potent pesticide (Huang et al. 2020). Usage of copper and selenium nanoparticles against pests, insects, and larvae was extensively studied. An eco-friendly approach of using metal-contaminated soil-based copper nanoparticles against *Tribolium castaneum* excelled than those of synthetic pesticides (El-Saadony et al. 2020).

1.11 Limitations

Despite having several advantages of nanomaterials, there are debates regarding their adverse effects on the environment. Initially, nanoparticles in land have the potential to contaminate groundwater bodies and enter human body leading to pulmonary disorders. Nano TiO_2 and ZnO used in cosmetics enter human skin, causing skin allergies and organ toxicity. Though Au and AgNPs have efficient cytotoxic activities, some amount of cellular damage is inevitable. This can be overcome by the application of citrate-capped gold nanospheres, which have an effective cytotoxic effect, followed by its less toxic nature. Literature reviews suggest that many of the surfactants used in the manufacture of nanospheres are toxic (Ray et al. 2009). The choice of plants as precursors might be an alternative source for producing toxic nanomaterials. Meanwhile, there are two types of aggregation of nanomaterials in the environment—homo- and hetero-aggregation. Those that exist in water-dispersible colloidal form such as multi-walled carbon nanotubes have more surfactants, which get deposited in the soil for a long time, thereby affecting their stability.

In case of nanofertilizers, the biochemical composition of nanoparticles greatly impacts its fate on the environment. The use of "green" precursors was found to be less polluting, as they compost easily. Most chemically synthesized inorganic nanoparticles have toxic effects compared to organic nanoparticles. Since there are very few studies of nanomaterials in soil fertility, impact of nanomaterials on soil nutrition as well as with soil microbes is unclear. Progress in the development of tailored nanomaterials has a precise control over their efficiency. Bionanomaterials and their role in bioremediation need to be assessed in order to view their capability in integrated remediation. Meanwhile, nano-assisted farming requires experimentation to be performed under controlled conditions, with a variety of parameters optimized. Once the investigation regarding nano-agriculture is completed in small scale, field study has to be performed to evaluate its toxicity (Usman et al. 2020).

In recent times, the use of nanomaterials excelled in the field of medicine especially in case of sensing and drug delivery applications, marking its breakthrough in the management of several diseases. Still, several mysteries around the fate of nanomedicine in body's defence mechanism are not clear. But those therapies dealing with gene alterations and pre-natal stage treatments were not permitted, owing to their fatal outcomes. Similarly, workplace safety must be maintained while working with nanomaterials as they are too small and might enter patients' bodies, causing severe respiratory problems such as asthma and cold. Toxicity evaluation must be carried out to work with nano-fabricated devices (Odiba et al. 2017).

1.12 Significance of Using Bionanomaterials from Agro-Wastes

Nanomaterials produced from agro-wastes have lesser toxicity than that of the other green synthesis routes, mainly because of the organic compounds in the plant sources, such as proteins and flavonoids. Despite having lesser applications of bionanomaterials from agro-wastes in terms of electronics, food safety, and energy, the availability of literature on green synthesis makes it easier to produce nanocompounds at a cheaper rate (Odiba et al. 2017). The conversion of waste to wealth, especially in nano-applications, helps reroute the waste into raw materials, making it suitable for several applications. Agro-wastes-based bionanomaterials serve as a feedstock for soil fertility applications in nanosensors, food safety, bionanopesticides, biocatalysis, and biostimulant for flower, fruit, and seed production (Ray et al. 2009). Novel toxicity evaluation approaches for agro-based bionanoproducts must be extensively carried out to determine the futuristic prospects of using agro-wastes in nano-synthesis.

1.13 Conclusion

This chapter has highlighted the outcomes of agro-based bionanomaterials and their significance in the field of health care, electronics, agriculture, and engineering applications. Metallic nanobiomaterials have a greater demand in the near future due to their environmentally friendly synthesis and commercial viability. Furthermore, effective techniques have to be adapted in making the synthesis in simpler forms. The use of easily accessible raw materials such as agro-wastes, metal contaminated soils, and pre-processed industrial wastes might be a forthcoming energy source. Applications of bionanomaterials from agro-wastes are found to be suitable for precision agriculture in different forms such as nanofertilizers, nanoinsecticides, and nanopesticides. To conclude, scrutiny of plant-based bionanomaterials in day-to-day applications has to be assessed. More research revelations from the academic sector might add up more elite bionanomaterials for positive considerations in commercialization.

References

Ayinde WB, Gitari WM, Samie A (2019) Optimization of microwave-assisted synthesis of silver nanoparticle by Citrus paradisi peel and its application against pathogenic water strain. Green Chem Lett Rev 12:225–234. https://doi.org/10.1080/17518253.2019.1627427

Brito TK, Silva Viana RL, Gonçalves Moreno CJ et al (2020) Synthesis of silver nanoparticle employing corn cob Xylan as a reducing agent with anti-Trypanosoma cruzi activity. Int J Nanomedicine 15:965–979. https://doi.org/10.2147/IJN.S216386

Choudhary R, Venkatraman SK, Bulygina I, Senatov F, Kaloshkin S, Swamiappan S (2020) Designing of porous PMMA/diopside bone cement for non-load bearing applications. J Asian Ceramic Soc 8:1–11. https://doi.org/10.1080/21870764.2020.1793476

Doan Thi TU, Nguyen TT, Thi YD, Ta Thi KH, Phan BT, Pham KN (2020) Green synthesis of ZnO nanoparticles using orange fruit peel extract for antibacterial activities. RSC Adv 10:23899–23907. https://doi.org/10.1039/d0ra04926c

Dzomeku I, Illiasu O, Amegbor I (2018) Effects of biochar, rice husk and rice straw on productivity of maize (Zea mays L.) and sustainable soil fertility restoration. J Exp Agric Int 20:1–11. https://doi.org/10.9734/jeai/2018/39346

El-Saadony MT, Abd El-Hack ME, Taha AE, Fouda MMG, Ajarem JS, Maodaa SN, Allam AA, Elshaer N (2020) Ecofriendly synthesis and insecticidal application of copper nanoparticles against the storage pest *Tribolium castaneum*. Nanomaterials 10:587. https://doi.org/10.3390/nano10030587

Fan B, Yao Q, Wang C, Jin C, Wang H, Xiong Y, Li S, Sun Q (2018) Natural cellulose nanofiber extracted from cell wall of bamboo leaf and its derived multifunctional aerogel. Polym Compos 39:3869–3876. https://doi.org/10.1002/pc.24419

Habte L, Shiferaw N, Mulatu D, Thenepalli T, Chilakala R, Ahn JW (2019) Synthesis of nano-calcium oxide from waste eggshell by sol-gel method. Sustainability (Switzerland) 11:1–10. https://doi.org/10.3390/su11113196

Hazmi B, Rashid U, Taufiq-Yap YH, Ibrahim ML, Nehdi IA (2020) Supermagnetic nano-bifunctional catalyst from rice husk: synthesis, characterization and application for conversion of used cooking oil to biodiesel. Catalysts 10:225. https://doi.org/10.3390/catal10020225

Hinge SP, Pandit AB (2017) Solar-assisted synthesis of ZnO nanoparticles using lime juice: a green approach. Adv Nat Sci Nanosci Nanotechnol 8:045006. https://doi.org/10.1088/2043-6254/aa84e1

Huang W, Yan M, Duan H, Bi Y, Cheng X, Yu H (2020) Synergistic antifungal activity of green synthesized silver nanoparticles and epoxiconazole against Setosphaeria turcica. J Nanomater 2020:1–7. https://doi.org/10.1155/2020/9535432

Hussein HS, Shaarawy HH, Hussien NH, Hawash SI (2019) Preparation of nano-fertilizer blend from banana peels. Bull Natl Res Cent 43:1–9. https://doi.org/10.1186/s42269-019-0058-1

Huynh LTN, Pham TN, Nguyen TH et al (2020) Coconut shell-derived activated carbon and carbon nanotubes composite: a promising candidate for capacitive deionization electrode. Synth Met 265:116415. https://doi.org/10.1016/j.synthmet.2020.116415

Izadiyan Z, Shameli K, Miyake M, Hara H, Mohamad SEB, Kalantari K, Taib SHM, Rasouli E (2020) Cytotoxicity assay of plant-mediated synthesized iron oxide nanoparticles using Juglans regia green husk extract. Arab J Chem 13:2011–2023. https://doi.org/10.1016/j.arabjc.2018.02.019

Jamdagni P, Khatri P, Rana JS (2018) Green synthesis of zinc oxide nanoparticles using flower extract of Nyctanthes arbor-tristis and their antifungal activity. J King Saud Univ Sci 30:168–175. https://doi.org/10.1016/j.jksus.2016.10.002

Karunakaran G, Jagathambal M, Gusev A, Kolesnikov E, Mandal AR, Kuznetsov D (2016) Allamanda cathartica flower's aqueous extract-mediated green synthesis of silver nanoparticles with excellent antioxidant and antibacterial potential for biomedical application. MRS Commun 6:41–46. https://doi.org/10.1557/mrc.2016.2

Khoshraftar Z, Safekordi AA, Shamel A, Zaefizadeh M (2019) Synthesis of natural nanopesticides with the origin of Eucalyptus globulus extract for pest control. Green Chem Lett Rev 12:286–298. https://doi.org/10.1080/17518253.2019.1643930

Khraisheh M, Kim J, Campos L, Al-Muhtaseb AH, Walker GM, AlGhouti M (2013) Removal of carbamazepine from water by a novel TiO_2–coconut Shell powder/UV process: composite preparation and photocatalytic activity. Environ Eng Sci 30:515–526. https://doi.org/10.1089/ees.2012.0056

Lateef A, Nazir R, Jamil N, Alam S, Shah R, Khan MN, Saleem M, Rehman SU (2019) Synthesis and characterization of environmental friendly corncob biochar based nano-composite—a potential slow release nano-fertilizer for sustainable agriculture. Environ Nanotechnol Monit Manag 11:100212. https://doi.org/10.1016/j.enmm.2019.100212

Liang M, Su R, Qi W, Yu Y, Wang L, He Z (2014) Synthesis of well-dispersed ag nanoparticles on eggshell membrane for catalytic reduction of 4-nitrophenol. J Mater Sci 49:1639–1647. https://doi.org/10.1007/s10853-013-7847-y

Liu B, Li T, Wang W et al (2020a) Corncob cellulose nanosphere as an eco-friendly detergent. Nat Sustain 3:448–458. https://doi.org/10.1038/s41893-020-0501-1

Liu Y, Song X, Cao F, Li F, Wang M, Yang Y, Liu M, Liu A, Xin H, Wang X (2020b) Banana peel-derived dendrite-shaped au nanomaterials with dual inhibition toward tumor growth and migration. Int J Nanomedicine 15:2315–2322. https://doi.org/10.2147/IJN.S211076

Nabi G, Ain QU, Tahir MB, Nadeem Riaz K, Iqbal T, Rafique M, Hussain S, Raza W, Aslam I, Rizwan M (2020) Green synthesis of TiO2 nanoparticles using lemon peel extract: their optical and photocatalytic properties. Int J Environ Anal Chem 00:1–9. https://doi.org/10.1080/03067319.2020.1722816

Narayana A, Bhat SA, Fathima A, Lokesh SV, Surya SG, Yelamaggad CV (2020) Green and low-cost synthesis of zinc oxide nanoparticles and their application in transistor-based carbon monoxide sensing. RSC Adv 10:13532–13542. https://doi.org/10.1039/d0ra00478b

Nayak PP, Datta AK (2020) Synthesis of SiO2-Nanoparticles from Rice husk ash and its comparison with commercial amorphous silica through material characterization. Silicon 13:1209–1214. https://doi.org/10.1007/s12633-020-00509-y

Ngu PZZ, Chia SPP, Fong JFY, Ng SM (2016) Synthesis of carbon nanoparticles from waste rice husk used for the optical sensing of metal ions. New Carbon Mater 31:135–143. https://doi.org/10.1016/S1872-5805(16)60008-2

Odiba A, Ottah V, Ottah C, Anunobi O, Ukegbu C, Edeke A, Uroko R, Omeje K (2017) Therapeutic nanomedicine surmounts the limitations of pharmacotherapy. Open Med (Poland) 12:271–287. https://doi.org/10.1515/med-2017-0041

Piñeres OH, Salcedo EC, Herrera AP, Sánchez JH, Quintana GC (2020) Magnetic paper from sugarcane bagasse fibers modified with cobalt ferrite nanoparticles. Cellulose 27:3903–3918. https://doi.org/10.1007/s10570-020-03042-8

Prasad C, Gangadhara S, Venkateswarlu P (2016) Bio-inspired green synthesis of Fe3O4 magnetic nanoparticles using watermelon rinds and their catalytic activity. Appl Nanosci (Switzerland) 6:797–802. https://doi.org/10.1007/s13204-015-0485-8

Prasad R, Kumar M, Kumar V (2017) Nanotechnology: an agricultural paradigm. In: Nanotechnology: an agricultural paradigm. Springer, Cham, pp 1–372. https://doi.org/10.1007/978-981-10-4573-8

Qidwai A, Kumar R, Dikshit A (2018) Green synthesis of silver nanoparticles by seed of phoenix sylvestris L. and their role in the management of cosmetics embarrassment. Green Chem Lett Rev 11:176–188. https://doi.org/10.1080/17518253.2018.1445301

Rahmad M, Asrul L, Kuswinanti T, Musa Y (2020) The effect of sugarcane bagasse and filter mud compost fertilizer and manure application on the growth and production of sugarcane. Int J Sci Res Sci Technol 6:338–345. https://doi.org/10.32628/ijsrst196668

Ray PC, Hongtao Y, Fu PP (2009) Toxicity and environmental risks of nanomaterials: challenges and future needs. J Environ Sci Health C Environ Carcinog Ecotoxicol Rev 27:1–35

Reddy Yadav LS, Manjunath K, Archana B, Madhu C, Raja Naika H, Nagabhushana H, Kavitha C, Nagaraju G (2016) Fruit juice extract mediated synthesis of CeO2 nanoparticles for antibacterial and photocatalytic activities. Eur Phys J Plus 131:1. https://doi.org/10.1140/epjp/i2016-16154-y

Santiago TR, Bonatto CC, Rossato M, Lopes CAP, Lopes CAG, Mizubuti ESG, Silva LP (2019) Green synthesis of silver nanoparticles using tomato leaf extract and their entrapment in chitosan nanoparticles to control bacterial wilt. J Sci Food Agric 99:4248–4259. https://doi.org/10.1002/jsfa.9656

Sasidharan S, Raj S, Sonawane S, Sonawane S, Pinjari D, Pandit AB, Saudagar P (2019) Nanomaterial synthesis: chemical and biological route and applications. In: Nanomaterials synthesis: design, fabrication and applications. Elsevier, Amsterdam. https://doi.org/10.1016/B978-0-12-815751-0.00002-X

Sethy NK, Arif Z, Mishra PK, Kumar P (2019) Synthesis of SiO2 nanoparticle from bamboo leaf and its incorporation in PDMS membrane to enhance its separation properties. J Polym Eng 39:679–687. https://doi.org/10.1515/polyeng-2019-0120

Shrestha L, Thapa M, Shrestha R, Maji S, Pradhananga R, Ariga K (2019) Rice husk-derived high surface area nanoporous carbon materials with excellent iodine and methylene blue adsorption properties. C (Basel) 5:10. https://doi.org/10.3390/c5010010

Siqueira TCA, da Silva IZ, Rubio AJ, Bergamasco R, Gasparotto F, De Paccola EAS, Yamaguchi NU (2020) Sugarcane bagasse as an efficient biosorbent for methylene blue removal: kinetics, isotherms and thermodynamics. Int J Environ Res Public Health 17:526. https://doi.org/10.3390/ijerph17020526

Sujiono EH, Zurnansyah ZD, Dahlan MY, Amin BD, Samnur AJ (2020) Graphene oxide based coconut shell waste: synthesis by modified hummers method and characterization. Heliyon 6:e04568. https://doi.org/10.1016/j.heliyon.2020.e04568

Sujitha V, Murugan K, Dinesh D et al (2017) Green-synthesized CdS nano-pesticides: toxicity on young instars of malaria vectors and impact on enzymatic activities of the non-target mud crab Scylla serrata. Aquat Toxicol 188:100–108. https://doi.org/10.1016/j.aquatox.2017.04.015

Sukumar S, Rudrasenan A, Padmanabhan Nambiar D (2020) Green-synthesized rice-shaped copper oxide nanoparticles using Caesalpinia bonducella seed extract and their applications. ACS Omega 5:1040–1051. https://doi.org/10.1021/acsomega.9b02857

Tao L, Zheng Y, Zhang Y, Ma H, Di M, Zheng Z (2017) Liquefied walnut shell-derived carbon nanofibrous mats as highly efficient anode materials for lithium ion batteries. RSC Adv 7:27113–27120. https://doi.org/10.1039/c7ra02716h

Usman M, Farooq M, Wakeel A, Nawaz A, Cheema SA, Rehman HU, Ashraf I, Sanaullah M (2020) Nanotechnology in agriculture: current status, challenges and future opportunities. Sci Total Environ 721:137778. https://doi.org/10.1016/j.scitotenv.2020.137778

Wang Q, Wang Y, Duan B, Zhang M (2016) Modified sol-gel synthesis of carbon nanotubes supported titania composites with enhanced visible light induced photocatalytic activity. J Nanomater 2016:1–6. https://doi.org/10.1155/2016/3967156

Yuvakkumar R, Suresh J, Nathanael AJ, Sundrarajan M, Hong SI (2014) Rambutan (Nephelium lappaceum L.) peel extract assisted biomimetic synthesis of nickel oxide nanocrystals. Mater Lett 128:170–174. https://doi.org/10.1016/j.matlet.2014.04.112

Zaheer Z (2019) Eco-friendly walnut shell powder based facile fabrication of biogenic ag-nanodisks, and their interaction with bovine serum albumin. J Photochem Photobiol B 193:8–17. https://doi.org/10.1016/j.jphotobiol.2019.01.015

Zangeneh MM, Zangeneh A (2020) Novel green synthesis of Hibiscus sabdariffa flower extract conjugated gold nanoparticles with excellent anti-acute myeloid leukemia effect in comparison to daunorubicin in a leukemic rodent model. Appl Organomet Chem 34:1–13. https://doi.org/10.1002/aoc.5271

Chapter 2
Application of Biotechnology for Raw Material Analysis

Somsubhra Ghosh, Sujayita Mazumder, Rajeswar Das, and Sandipan Jana

Abstract Raw materials are the most important substance in the pharmaceutical and biotechnology industries nowadays. These are obtained and produced from different sources, by using several techniques. These are available in different forms and phases, so the processing, storage, and analytical process of the raw materials are very crucial and essential to maintain the highest standard of the finished product. Several instruments and analytical processes are available and used frequently in the industry for routine checkup of the raw materials. Conventional analytical techniques like spectroscopy, chromatography, etc. are mostly used among all other analytical techniques. Some the techniques are either time-consuming or very costly for routine analysis. Nowadays, several biotechnological techniques or processes are being used as alternatives to conventional techniques. Biotechnology is one of the emerging fields in the healthcare and pharmaceutical industry. It has several applications and uses, including patient care, treatment of several diseases, and even for diagnostic purposes. Apart from all those conventional uses, in the recent past, biotechnological techniques have also been used for analyzing and testing raw materials. A detailed discussion is provided systematically to highlight the application of biotechnology in the analysis of raw materials.

Keywords Biotechnology · Raw material · Analysis · 3D biological printing

S. Ghosh (✉) · S. Mazumder · R. Das · S. Jana
School of Pharmacy, The Neotia University, Sarisa, West Bengal, India
e-mail: somsubhra.ghosh@tnu.in

S. Bose et al. (eds.), *Concepts in Pharmaceutical Biotechnology and Drug Development*, Interdisciplinary Biotechnological Advances,
https://doi.org/10.1007/978-981-97-1148-2_2

2.1 Introduction

The rules and guidelines pertaining to biotechnology products do not provide a uniform definition of a raw material. Regulations make reference to elemental substances, constituents, and in-process materials. A raw material is defined as "starting materials, reagents, and solvents intended for the use in the production of intermediates or APIs" in the International Conference on Harmonisation (ICH) Q7. A monoclonal antibody used as a raw material in affinity chromatography is known as ICH Q6B. As a result, the producer must define a raw material's meaning according to its own standards on safety, quality, and adherence to the current Good Manufacturing Practices (cGMPs). A thorough assessment of the factors that must be taken into account for novel compounds is necessary to show that the raw material satisfies acceptable quality standards for use in the production of biotechnology products intended for human consumption. No specific threshold has been set for how much a raw material must adhere to cGMP guidelines (Noh et al. 2018). Numerous biotechnology and pharmaceutical enterprises, such as those affiliated with the Rx-360 Consortium, prioritize the oversight of supply quality, including raw materials. ICH Q9 provides guidance on incorporating quality risk management into materials management procedures. Since a biotechnology product may be manufactured using a range of raw materials, certain considerations are made when introducing a new raw material. Presently, the Chinese Pharmacopoeia (2020) uses two biotechnologies for the analysis of raw materials for animal and herbal medicine: polymerase chain reaction (PCR) and polymerase chain reaction-restriction fragment length polymorphism (PCR-RFLP) (Li et al. 2016).

2.2 Molecular Methods to Verify Raw Material Authenticity

Progress in sequencing technology, ongoing cost reductions in sequencing, collection of genetic data from a diverse range of species, and creation of multiple databases enable the advancement of molecular identification technology and the breeding of new varieties. Some DNA-based technologies that are capable of precisely identifying raw materials are barcoding, molecular markers, and fast detection.

Molecular markers: Molecular markers use DNA fragments carrying genetic information that differs between species to determine the principles governing gene organization and the manifestation of phenotypic features. They are resistant to differences in tissue and organ composition, changes in the external environment, and developmental phases. They are able to recognize distinct tissues, organs, and even individual cells in various stages of development. The markers have found widespread use in the study and identification of herbal medicines due to their stable qualities, broad dispersion, abundance, ease of selection, ease of operation, and speedy detection. For example, nucleotide sequences that have been inserted or deleted into the same gene in several individuals or related species are compared using insertion/deletion (InDel) markers to identify across species (Han et al. 2018).

Biotechnology-based rapid detection: These markers are widely utilized in the study and identification of herbal medicines because to their abundance, steady qualities, simplicity of selection, convenience of use, and speedy detection. For instance, nucleotide sequences that have been inserted or removed from the same gene in several individuals or related species are compared to identify individuals within a species using insertion/deletion (InDel) markers (Lin et al. 2016).

Chips: The method of identifying heavy metals, raw materials, and pesticide residues has been applied to chips. Gene chips integrate oligonucleotides or cDNA at a high density before affixing them to the surface of the supporting carrier in accordance with the specific array derived from nucleic acid molecule hybridization. Following the probe's hybridization with the carrier surface, a specialized apparatus with semiconductor sensors detects the biological signal. Synchronous recording of the content is then performed and the data is sent to the computer for examination. Ultimately, the carrier sample's gene phenotype and gene function may be determined. *Bupleurum chinense* can be authenticated by using gene chips that can differentiate it from other sources. The application of diversity arrays has been employed to differentiate between closely related species, including *Eucalyptus grandis*. For expeditious and precise screening of the protein component, the protein chip is employed for the purification and processing of supplementary material (Wang et al. 2018a). Furthermore, it exhibits preferential interactions with known molecules based on their molecular characteristics, gradually immobilizing specific known proteins onto a carrier. This technique may be applied to the identification of *Pheretima aspergillum* and *Chinemys reevesii*. By employing microfluidic chips, which facilitate the adaptable integration of diverse cell technologies on a controlled micro-platform, the essential steps of sample preparation, reaction, separation, and detection can be downsized onto a portable microchip. The manipulation of fluids in micro- and nanoscale environments is an additional capability of microfluidic chips. Pesticides and heavy metals can be found with this device, which enables techniques to regulate raw material safety. Microfluidics based on electrochemistry is one such technique (Wang et al. 2018b).

Using a competitive binding antibody approach for both antigen and labeling, immunoassay is based on the specific antigen-antibody reaction. The following immunoassay methods are used to identify toxic substances: gold immunochromatographic assay (GICA), chemiluminescent immunoassay (CLIA), enzyme-linked immunosorbent assay (ELISA), and fluorescence immunoassay (FIA). In an ELISA, antigens or antibodies are marked with enzymes that act as tracers. The enzyme catalyzes the development of color or light emission in the substrate to create a correlation between the substance to be tested's concentration and the system's degree of color development. Imidacloprid, neonicotinoid, and ethylene thiourea were evaluated using ELISA. Colloidal gold is used as the tracer marker in the Gold Immunochromatographic Assay (GICA) (Shi et al. 2020). Due to the elevated electron density of gold particles, microscopy enables the identification of dark-brown particles in proximity to the binding site of the protein labeled with gold. Red or pink dots are visible to the unaided eye when these markers build up significantly at the relevant ligands. For the purpose of finding pesticide residues like carbendazim, this method is frequently employed in conjunction with lateral flow strips (LFS). Profenofos is also detected by lateral flow immunoassay based on colloidal silver. The most used technique for finding pesticide residues is immunoassay. Nevertheless, the process of creating pesticide antibodies is challenging and expensive (Lei et al. 2022).

Process analysis technology (PAT) is a system designed to ensure the quality of the final product by organizing, evaluating, and controlling production through prompt measurement (i.e., while processing) of critical performance and quality characteristics of raw and in-process materials and processes (Songa and Okonkwo 2016). PAT refers to the integrated evaluation of chemical, physical, and biological qualities as "analysis" in a broad sense. Nonetheless, the predominant quality control methods applied in pharmaceutical production predominantly center on the analysis of physicochemical properties. Given the intricate nature of natural products characterized by complex mixtures, contemporary chromatographic and spectroscopic techniques, which predominantly employ a chemical approach, encounter significant challenges in the comprehensive identification of all components owing to their chemical complexity. Additionally, the existing data is insufficient to guarantee the clinical safety and efficacy of pharmaceuticals, as the correlation between the chemical information obtained through the chemical approach and the overall in vivo activity remains to be firmly established. Many compounds have low absorption in their spectra, which makes it challenging to detect them using chromatographic techniques (Peng et al. 2021a). This is especially true of some biological active components and bio-pollutants. The quality of synthetic medications and natural items has been regulated via the application of biotechnology. While biotechnological procedures are unable to furnish details on chemical components, they do offer insights into the direct bioactivity, clinical safety, and efficacy of biological products (Peng et al. 2021b).

Safety analysis: When a medication is used to treat a condition, two of its qualities that come into play are efficacy and safety. When taking medication, one must take into account the potential negative reactions (such as toxicity and side effects). An analysis of a medication's physicochemical features, pharmacokinetic traits including target and organ toxicity, and metabolic distributions of the drug are standard components of toxicity evaluation. Emerging new medications with increasing safety concerns include natural items and synthetic medicines. There is a need to build quick and precise drug safety analysis methodologies since the current approaches and techniques are not keeping up with the increasing demands (Fang et al. 2020).

Drug toxicity analysis: Animal models are used to assess and minimize the likelihood that people may be exposed to certain dangers by employing complete animals. Pharmacopoeia suggests evaluating drugs by contrasting their minimal lethality in pigeons with that of digitalis. Animal models for drug toxicity studies are often conventional animals, such as mice and rats. Though they are utilized in preclinical research, experimental animal models have low predictive power for drug metabolism and potential toxicity (Du et al. 2020).

Transgenic animal analysis techniques: Combining tagged protein-producing transgenic or gene-edited mice with optical biosensors might be a useful tool for quickly and extensively evaluating medication toxicity in preclinical studies as well as for studying the etiology and pathophysiology of several diseases (Zha et al. 2021). To track organelles, cells, or tissues throughout time for drug toxicity investigations, transgenic mouse models need to be developed. Human liver chimera models and mouse lines expressing the human leukocyte antigen (HLA) have been produced in the last 10 years because of the rapid growth of transgenic mice. Certain immune system alleles of humans are highly correlated with the reverse transcriptase inhibitor abacavir, which is used to treat HIV. Abacavir-induced hepatic damage was often observed in HLA mice given CpG-oligodeoxynucleotides. Although it is an expensive and time-consuming process, transgenic or gene-edited mice can aid in medication toxicity research (Mu et al. 2021).

2.3 High-Throughput Approaches for Drug Toxicity Biomarker Analysis

Toxicological biomarkers may be found by screening a variety of substances, such as proteins, DNA, or metabolites. This enables improved illness diagnosis and medication safety evaluations. High-throughput profiling tests hundreds of chemical changes concurrently, as opposed to examining one molecule at a time. Transcriptomics, also known as gene expression profiling, tracks variations in transcript levels and represents transcriptional and posttranscriptional control of gene expression (Ouyang et al. 2016). The primary constituents of vine tea, derived from the youthful leaves of *Ampelopsis grossedentata*, are predominantly flavonoids and polyphenolic compounds, which exhibit hepatotoxic effects in a dose-dependent manner. Exploring the toxicities associated with Herba Lysimachiae, commonly used for treating rheumatic arthralgia, involved proteome profiling. Utilizing DARTs-based proteomics, the investigation identified psoralen, a principal hepatotoxic component and blood entry agent in Fructus Psoraleae, as a potential target for toxicity. Metabolomic profiles are used in various biological processes to detect hormones and other signaling molecules, as well as compounds that function as intermediates and metabolic products. Significant liver damage has been linked to the well-known traditional Chinese herb heshouwu (*Polygonum multiflorum*) (Arora 2013).

2.4 3D Biological Printing of Organoids for Personalized Medicine Evaluation

The organoid offers a more physiologically appropriate model for research and therapeutic usage because it is a three-dimensional cell culture. Compared to traditional 2D cell cultures, it exhibits increased physiological relevance, self-organization, and self-renewal, providing new opportunities for preclinical drug testing and the development of therapeutic strategies. Organoids produced from patients offer several benefits when used as personalized tumor models. They could be produced simply and inexpensively to find and test novel anticancer medications. To enable more consistently replicable, dynamic, and resilient studies of organoids with the potential to enhance treatment decisions in personalized medicine, Schuster et al. developed an automated microfluidic platform for organoid culture, facilitating dynamic and combinatorial drug screening. 3DP, one of the most innovative advancements in pharmaceutical research, is now a powerful and innovative tool for tissue engineering, pharmaceutical analysis, sickness modeling, and the precise creation of customized dose forms (Wang et al. 2014). Using 3D computer data to generate customized drug carriers with the required size, shape, and structure can lead to the rapid, efficient, and cost-effective production of personalized medicine, various drug combinations, and sophisticated drug release patterns. Numerous 3DP methods, such as binder jetting, vat photopolymerization, pressure-assisted microsyringe, fused deposition modeling, inkjet printing, and selective laser sintering, have been developed as a result of research on pharmaceuticals and medicine. In the production of 5-fluorouracil-loaded tablets through digital light processing (3DP), Chen et al. employed acryl acid (AA) as a monomer, poly(ethylene glycol) dimethacrylate as a difunctional crosslinker, and acrylated hyperbranched polyester (AHBPE) as a multifunctional crosslinker. Their investigation revealed a direct correlation between the duration of printing and drug release with an increase in AA. Conversely, a rise in the quantity of AHBPE was associated with a decrease in both printing duration and drug release. Additionally, they suggested acrylated hyperbranched polyester (AHBPE) as a crosslinker for vat photopolymerization (3DP) of customized drug delivery (Kerwin 2008).

2.5 A Biotechnology Product's Life Cycle of a Raw Material

A quality agreement that allows the business to perform audits is frequently used by pharmaceutical product manufacturers to approve raw material vendors. In addition, if the raw material is crucial, a vendor will agree to notify of any changes made to the production process that could affect the raw material's quality. Specific raw material providers (like the food industry) are less inclined to hold the biotechnology company to stringent quality standards because of their much wider customer base. For manufacturers, there is also the expanding globalization of the raw material production sector. This is mirrored in supply chains, which span multiple global locations and are progressively more complex. It has long been believed that the raw material vendor's location marks the beginning of the raw material supply chain. An inherent premise of this paradigm is that the raw material's quality is appropriately controlled based on the vendor's quality data and extra testing conducted by the manufacturer. The Food and Drug Administration (FDA) and other regulatory bodies are showing a greater desire to tighten raw material controls, which should include more comprehensive raw material traceability (Wen et al. 2012).

2.6 Reducing the Risk of Introducing Raw Substances

A risk control plan should be put into place when risk has been recognized, examined, and assessed. Its goal is to lower risk to a level that can be managed. It is necessary to reduce unmanageable hazards. To reduce the risk associated with employing a vital raw material, for instance, appropriate internal testing can be done. A supplier's qualification may be revoked due to insufficient traceability of a raw material. The maker has an obligation to take all reasonable steps to keep acceptable risk at minimal levels after the risk has been acknowledged (Liu et al. 2010).

2.7 Future Directions

Implementing an upgraded methodology, as outlined in the ICH Q8, Q9, and Q10 guidances, begins with a risk-based approach to raw material management. Knowledge management is suggested by ICH Q10 Pharmaceutical Quality System guidelines. Furthermore, Quality by Design (QbD) ideas were incorporated into ICH Q8, which has been updated. A path toward designing a more resilient manufacturing process, which may involve creating a design space, can be given to the manufacturer by means of an upgraded development strategy. Regulatory bodies have embraced a QbD strategy as a twenty-first-century strategy for producing improved pharmaceutical goods. Many in the biotechnology industry are still working on putting this technique into practice, despite the fact that it supports initiatives taken by producers and authorities to improve the quality of the finished product (Thirumangalathu et al. 2009). A robust program for managing raw materials that employs comprehensive risk assessment and mitigation strategies may result in fewer deviations related to raw materials and/or changes in raw materials (Jameel et al. 2015). This can enable producers to stop any negative effects that raw material fluctuations may have on a biotechnology medicinal product's efficacy, safety, or quality (Tyagi et al. 2009).

2.8 Conclusion

Current Good Manufacturing Practices (cGMP) regulations require the biopharmaceutical industry to establish a system for the initial and recurring certification, approval, and selection of raw materials and their suppliers. For chemicals, natural products and synthetic medicines, bio-analytical technology is preferred. With unceasingly development and invention in the biology field, biotechnology may play a critical role in the quality control of natural products and synthetic medicinal. Next, companies ensure, through testing and acceptance programs, that the raw materials and supplies operation system provides rules so that it is capable of assigning use and function related to a "fit-for-use" or "fit-for-function." Any hazard related to raw material or stuff marked fit-for-function should be evaluated. Biopharmaceutical quality cannot be defined by a single set of nonsupervisory/compliance/quality criteria; in fact, a common standard component, i.e., swats or sugar, serves a wide range of functions with different criteria for fit. Some studies (e.g., USP, National Formulary) are restricted to the standardization of raw material identification. Now studies do not help much in the unique quality and safety, which is needed in biopharmaceuticals. Rather, nonsupervisory guidance states that the onus of determining the strictness with which GMPs must be followed and the level of supervision required to establish and preserve the excellent status of a transported raw material rests with the maker of pharmaceutical products. It appears that the guiding idea is that oversight should be proportionate to the risks associated with the particular material to its particular specified function and purpose, as determined by the maker of pharmaceutical products, taking into account factors such as material origin, derivate, force chain complexity, etc.

References

Arora M (2013) Cell culture media: a review. Mater Methods 3:175

Du M, Yang Q, Liu W, Ding Y, Chen H, Hua X et al (2020) Development of immunoassays with high sensitivity for detecting imidacloprid in environment and agro products using phage-borne peptides. Sci Total Environ 723:137909

Fang L, Liao X, Jia B, Shi L, Kang L, Zhou L et al (2020) Recent progress in immunosensors for pesticides. Biosens Bioelectron 164:12255

Han K, Wang M, Zhang L, Wang C (2018) Application of molecular methods in the identification of ingredients in Chinese herbal medicines. Molecules 23:2728

Jameel F, Hershenson S, Khan MA, Martin-Moe S (eds) (2015) Quality by design for biopharmaceutical drug product development. Springer

Kerwin BA (2008) Polysorbates 20 and 80 used in the formulation of protein biotherapeutics: structure and degradation pathways. J Pharm Sci 97:2924–2935

Lei P, Zhou Y, Zhao S, Dong C, Shuang S (2022) Carbon-supported X-manganate (XNi, Zn, and Cu) nanocomposites for sensitive electrochemical detection of trace heavy metal ions. J Hazard Mater 435:129036

Li JJ, Xiong C, Liu Y, Liang JS, Zhou XW (2016) Loop-mediated isothermal amplification (LAMP): emergence as an alternative technology for herbal medicine identification. Front Plant Sci 7:1956

Lin Y, Gritsenko D, Feng S, Teh YC, Lu X, Xu J (2016) Detection of heavy metal by paper-based microfluidics. Biosens Bioelectron 83:256–266

Liu W, Swift R, Torraca G, Nashed-Samuel Y, Wen Z-Q, Jiang Y, Vance A, Mire-Sluis A, Freund E, Davis J et al (2010) Root cause analysis of tungsten-induced protein aggregation in pre-filled syringes. PDA J Pharm Sci Technol 64:11–19

Mu X, Yao Y, Peng T, Wang D, Meng L, Wang Y et al (2021) Current status of organophosphorus pesticide residues in root and rhizome medicinal materials and research progress of rapid detection methods. Zhongguo Zhong Yao Za Zhi 46:5736–5743

Noh P, Kim WJ, Yang S, Moon BC (2018) Authentication of the herbal medicine Angelicae Dahuricae radix using an ITS sequence-based multiplex SCAR assay. Molecules 23:2134

Ouyang H, Shu Q, Wang W, Wang Z, Yang S, Wang L et al (2016) An ultra-facile and label-free immunoassay strategy for detection of copper (II) utilizing chemiluminescence self-enhancement of Cu (II)- ethylenediaminetetraacetate chelate. Biosens Bioelectron 85:157–163
Peng S, Yang S, Zhang X, Jia J, Chen Q, Lian Y et al (2021a) Analysis of imidacloprid residues in mango, cowpea and water samples based on portable molecular imprinting sensors. PLoS One 16:e0257042
Peng S, Wang A, Lian Y, Zhang X, Zeng B, Chen Q et al (2021b) Smartphone-based molecularly imprinted sensors for rapid detection of thiamethoxam residues and applications. PLoS One 16:e0258508
Shi R, Hu Z, Lu H, Liu L, Xu L, Liu Y et al (2020) Hierarchical nanostructuring array enhances mid-hybridization for accurate herbal identification via ITS2 DNA barcode. Anal Chem 92:2136–2144
Songa EA, Okonkwo JO (2016) Recent approaches to improving selectivity and sensitivity of enzyme-based biosensors for organophosphorus pesticides: a review. Talanta 155:289–304
Thirumangalathu R, Krishnan S, Ricci MS, Brems DN, Randolph TW, Carpenter JF (2009) Silicone oil- and agitation-induced aggregation of a monoclonal antibody in aqueous solution. J Pharm Sci 98:3167–3181
Tyagi AK, Randolph TW, Dong A, Maloney KM, Hitscherich C, Carpenter JF (2009) IgG particle formation during filling pump operation: a case study of heterogeneous nucleation on stainless steel nanoparticles. J Pharm Sci 98:94–104
Wang W, Ignatius AA, Thakkar SV (2014) Impact of residual impurities and contaminants on protein stability. J Pharm Sci 103:1315–1330
Wang B, Zhou Y, Qin G, Nie J (2018a) Identification of adulteration in Fritillariae cirrhosae Bulbus using multiplex ligation-dependent probe amplification technology. Chin J Pharm Anal 38:2104e9
Wang S, Lin B, Chen L, Li N, Xu J, Wang J et al (2018b) Branch-migration based fluorescent probe for highly sensitive detection of mercury. Anal Chem 90:11764–11769
Wen Z-Q, Torraca G, Masatani P, Sloey C, Phillips J (2012) Nondestructive detection of glass vial inner surface morphology with differential interference contrast microscopy. J Pharm Sci 101:1378–1384
Zha Y, Li Y, Hu P, Lu S, Ren H, Liu Z et al (2021) Duplex-specific nuclease-triggered fluorescence immunoassay based on dual-functionalized AuNP for acetochlor, metolachlor, and propisochlor. Anal Chem 93:13886–13892

Chapter 3
Application of Biotechnology in Pharmaceutical Manufacturing Control

Gourab Dey, Pallabita Rakshit, Nibir Ghosh, Sabyasachi Banerjee, Subhasis Banerjee, and Sourav De

Abstract Pharmaceutical biotechnology represents a rapidly evolving field wherein biotechnological principles are applied to diverse pharmaceutical applications. The integration and utilization of biotechnology in the pharmaceutical sector hold significant importance. The connection between biotechnological applications and a myriad of colors has spurred the need for a comprehensive classification system based on color. Advanced technologies and products have been innovated across various domains, including medicine (for the development of novel medicines, therapies, and biofuels), agriculture (in the creation of genetically modified plants and biological treatment), industrial biotechnology (for the production of chemicals, food, paper, and textiles), and environmental applications (such as biodiversity maintenance and bioremediation). Biotechnology plays a pivotal role in enabling the pharmaceutical industry to introduce novel products, processes, methods, and services while enhancing existing ones. Pharmaceutical analysis integrates biotechnological methods, which enhance drug quality assessment from raw materials to final products, offering insights into biological effects, efficacy, and safety. Various biotechnological approaches, including DNA barcoding, AI-enhanced hyperspectral imaging, and organoid modeling, have found applications in pharmaceutical analysis, providing comprehensive insights and intuitive results. This comprehensive chapter provides an overview of the application of biotechnology in the

G. Dey · N. Ghosh · S. De (✉)
Department of Pharmaceutical Technology, Eminent College of Pharmaceutical Technology, Kolkata, India

P. Rakshit
Department of Pharmaceutical Technology, Jadavpur University, Kolkata, India

S. Banerjee · S. Banerjee
Department of Pharmaceutical Chemistry, Gupta College of Technological Sciences, Asansol, West Bengal, India

S. Bose et al. (eds.), *Concepts in Pharmaceutical Biotechnology and Drug Development*, Interdisciplinary Biotechnological Advances,
https://doi.org/10.1007/978-981-97-1148-2_3

production of innovative pharmaceuticals. It delves into the influence of biotechnology on medical research and innovation across various medical domains. The healthcare sector offers an extensive array of biopharmaceutical products designed for therapeutic purposes. The review encompasses discussions on diverse categories of biotechnology-derived products, including gene therapy, monoclonal antibodies, vaccines, DNA fingerprinting, biopharmaceuticals, stem cell therapy, and pharmacogenomics. Additionally, the chapter explores the therapeutic applications associated with these biotechnology-based products.

Keywords Pharmaceutical technology · Stem cell therapy · Pharmacogenomics · Monoclonal antibody · DNA fingerprinting · Gene therapy

3.1 Introduction

The fields of chemical, physical, biological, and information technology form the foundation of pharmaceutical analysis. Drugs work through chemical components to produce their therapeutic effects, and the earliest analytical techniques to focus on this area were those based on chemicals. Drug quality can also be influenced by its physical characteristics, such as uniformity and crystal structure, which is why these attributes have drawn attention. However, pharmacists have not given biotechnology and cutting-edge information technology enough thought. Biotechnology plays a little function in pharmaceutical analysis. The search involved examining papers published in the *Web of Science* over the past 15 years (2007–2022) using the terms "pharmaceutical analysis" in conjunction with either "chemical analysis" or "biological analysis" (Fig. 3.1a). The amount of articles published every 5 years is displayed, which suggests that biological analysis has a limited function in pharmaceutical analysis while chemical analysis still holds a strong position. Certain identified index components may not accurately reflect the effectiveness or efficacy of natural products. This discrepancy is attributed to the intricate action mechanism of

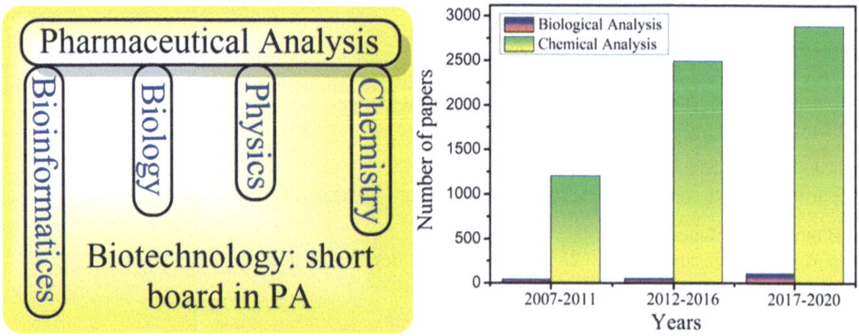

Fig. 3.1 Overview of current pharmaceutical analysis method

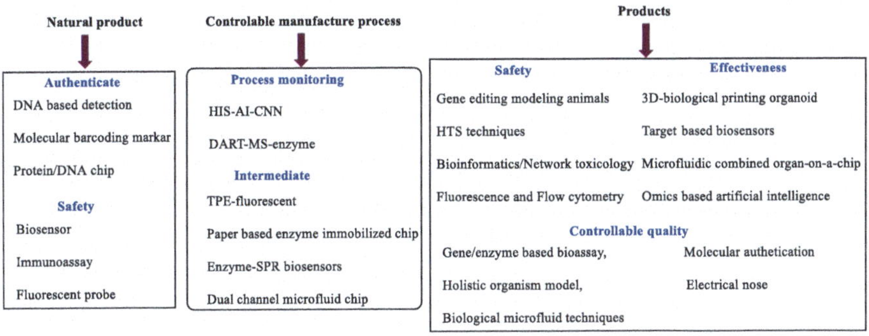

Fig. 3.2 Emerging biotechnologies and bioinformatics for pharmaceutical analysis

natural products, characterized by their complex composition. Despite the chemical and physical methods demonstrating high levels of sensitivity, accuracy, robustness, and throughput, they may not fully capture the nuanced effectiveness of these complex natural compounds (Fig. 3.1b). The study of biology forms the foundation of life sciences, especially medicine, and its field has evolved with the advancement of biotechnology (Zhang et al. 2020). Several biotechnological methods have been consistently employed in the pharmaceutical analysis process, spanning from raw materials to the production of finished products. These methodologies are preferred due to their efficacy in elucidating the complete biological effects, functions, or mechanisms of action of medications, thereby producing tangible and understandable results (Fig. 3.2). The primary assessment tools for examining the potency, security, and caliber of medications are biological detection procedures. This chapter delineates and investigates the applications of state-of-the-art biotechnologies in the realm of pharmaceutical analysis (Ramakrishnan et al. 2020).

3.2 Biotechnological Application in Pharmaceutical Manufacturing Control

Batch processing is usually employed in conventional pharmaceutical manufacture, and quality is assessed through laboratory testing on the samples that are gathered. To some extent, this traditional method has been successful in producing high-quality medication. The quality of recently launched products has improved, but the unacceptably high number of product recalls has raised worries about the quality of pharmaceutical items. Unacceptably high product recall frequencies are the reason why product quality continues. Natural products are renowned for their complex formulations, involving numerous compounds and ingredients, and exhibit considerable variability in quality among items from different producers or batches within the same manufacturing source. Consequently, monitoring the production processes of natural products poses a considerable challenge. To ensure the quality of the end product, the development of process analysis technology (PAT) has emerged as a method for strategically planning, evaluating, and controlling manufacturing. PAT involves the timely measurement (i.e., during processing) of crucial quality and performance attributes related to both raw and in process materials and processes (Zhang et al. 2020).

In PAT, the word "analysis" is widely understood to refer to combined chemical, physical, and biological properties (Ramakrishnan et al. 2020). Nonetheless, the majority of pharmaceutical manufacturing quality control methods concentrate on physicochemical property analysis. Since natural products are intricate mixtures, it is extremely difficult for existing chromatographic and spectroscopic methods that rely on the chemical approach to identify every component of these complicated mixtures due to their chemical complexity. Moreover, the absence of empirical support for the correlation between the chemical information provided by the chemical approach and overall in vivo activity raises concerns about the clinical safety and efficacy of medications. Chromatographic procedures, limited by their inability to detect certain substances, especially certain biologically active components and bio-pollutants, contribute to the insufficiency of available data. The quality of synthetic medications and natural items has been regulated through the application of biotechnology. Biotechnological techniques offer the advantage of furnishing insights into the immediate bioactivity, clinical safety, and efficacy of biological products, notwithstanding their inability to furnish information pertaining to chemical components.

3.3 Monitoring of Process

3.3.1 Chemical and Biological Coupling for Pharmaceutical Process

Few bioassays now available can meet the three assay requirements of speed, sensitivity, and multiplexing. Given the multi-component and multi-target nature of natural products, detection strategies with multiplexing capabilities are highly prized.

Consequently, scholars are exploring the integration of biological assays with chemical analysis. This amalgamation facilitates concurrent examinations of both biological and chemical aspects on a unified analytical platform, thereby decreasing instrument costs and variations. Recent developments in ambient ionization technology have made mass spectrometry (MS) a potentially inventive PAT tool in the production of natural products. Li and colleagues (Ramakrishnan et al. 2020) introduced an on-demand technique utilizing direct analysis in real-time mass spectrometry (DART-MS). This method assesses the biological activity of specific entities, namely thrombin and angiotensin-converting enzyme (ACE), employing a mass spectrometry probe as the substrate for the enzymes. Each probe has a distinct peptide sequence and a high-ionization-efficiency label made of 1-(2-pyrimidyl) piperazine. The methodology proficiently achieved the concurrent evaluation of the chemical composition and biological attributes of Danshen (*Salvia miltiorrhiza*) injection. The results from the detection process illustrated the capability of the platform to assess the quality of the drug across various dimensions, all under continuous real-time monitoring.

3.3.2 Quality Control for Pharmaceutical Process by Using Biological Soft Sensing

Because natural goods are diverse, there is variation in the raw materials and pharmaceutical products, which raises questions about the therapeutic consistency of these items. For this reason, multi-dimensional quality control is necessary to guarantee the security and effectiveness of these pharmaceutical items. In addition to being expensive, using various instruments for the quality analysis of various dimensions leads to differences between the equipment. A recent and well-liked technique in the monitoring industry is soft sensing. Thanks to this technology, it is possible to estimate a quality or activity in real time using a mathematical model and readily observable indications that are challenging to measure directly (Fraser et al. 2013). Consequently, soft sensing performs well when it comes to identifying quality indicators that are challenging to assess in industrial and pharmaceutical operations. Using the Shuxuening injection as an example, Zhong et al. (Li et al. 2019) employed a non-destructive method of simultaneously monitoring the injection's physical characteristics, chemical makeup, and biological activity by combining artificial intelligence (AI) with hyperspectral imaging (HSI) as a biological soft sensing technology. Drug component identification frequently necessitates the use of analytical tools like high-performance liquid chromatography (HPLC). Z-Gly-GlyArg-AMC acetate probe substrate and the DPPH-free radical scavenging assay are needed for antioxidant and anticoagulant actions. Image recognition methods are used to identify the injection's chromaticity and visible particles. The five quality indicators of total ginkgolides, total flavonol, anticoagulant activity, antioxidant activity, and visible particles were predicted by the HSI-AI technique. As a result, the platform may be utilized to monitor several quality attributes during the production process of pharmaceuticals. Under non-destructive circumstances, biological soft sensing methods in conjunction with HSI can simultaneously acquire many quality aspects of a sample.

3.4 Intermediate Product Analysis

Variations in natural products and the shifting of operational parameters in each process unit can cause variations in intermediate quality between batches in the pharmaceutical process, which can impact the quality of the finished product. Thus, quality control must be applied to the intermediates generated during the process.

3.4.1 Intermediate Product Analysis by Biological and Chemical Coupling

Uncertainty surrounds pharmaceutical items due to the variety of natural product sources and industrial settings. Every biological and chemical active ingredient that could have an impact on the potency and quality of medications is precisely and thoroughly identified. Shuxuening injection (SI) is primarily based on its antioxidant properties. To guarantee its quality, it is crucial to conduct both quantitative and qualitative evaluations of antioxidants in the pharmaceutical process. An online analytical approach called ABTSÍ (2,2-azinobis-(3-ethyl benzothiazoline-6-sulfonic acid))-CE was developed by Ma et al. (Li et al. 2019) to monitor different antioxidants during the injection preparation process. Initially, ABTSÍ was incorporated into the capillary and utilized for the concurrent identification and segregation of the primary antioxidants in the injection. The technique is a straightforward, dependable, and efficient instrument for the quantitative assessment of medications, as demonstrated by the experimental results, which SI validated. Certain indicators are unable to thoroughly assess TCM compositions with several components, hence ignoring the chemical complexity and several mechanisms that contribute to their pharmacological action. Wang et al. (Zhu et al. 2020) suggested a method for evaluating the quality of *Typha orientalis* by combining metabolomics and chemometrics with bioactive chemical comprehensive indicators based on thrombin. This study offers suggestions for developing techniques to assess a drug's chemical and biological qualities at the same time. This study offers suggestions for developing techniques to assess a drug's chemical and biological qualities at the same time (Table 3.1) (Fig. 3.3).

3.4.2 Biosensors for the Analysis of Intermediate Products

Small, integrated analytical devices known as biosensors are made up of biological elements in close proximity to physical sensors, which translate biological processes into quantifiable data. Through the utilization of biometric elements that transform data derived from the biological domain into chemical or physical signals, these sensors furnish precise quantitative or semi-quantitative analytical information (Zhu et al. 2020; Ma et al. 2018; Wang et al. 2020). In a research endeavor, the identification of bromocriptine was accomplished utilizing a label-free biosensor founded on surface plasmon resonance (SPR) technology (Andryukov et al. 2021). To ensure the precision and dependability of the findings, potential enzymes for the bromocriptine biosensor were first evaluated using a competitive inhibition experiment and a protein ligand docking simulation. Drug monitoring that is quick, real time, and label-free is possible with this enzyme SPR technique. The findings demonstrated that different enzymes and SPR technology might be used to monitor different medications. An alternative investigation proposed the utilization of hollow fiber-based

Table 3.1 Application of biotechnology in pharmaceutical manufacturing control

Category	Technique	Response	Applications	References
Biosensor for PAT	HF-AS	Lipase-binding ligand	Lotus leaf extracts	Li et al. (2019)
	Enzyme-based SPR	Acceptable therapeutic monitoring	Bromocriptine	Ma et al. (2018)
	TPE fluorescent probe	Monitor ACE activity	Tongmaiyangxin pill	Wang et al. (2020)
	Aptamer biosensor	Rapid detection of pathogenic *E. coli*	Licorice extracts	Jabbari et al. (2017)
	FRET fluorescent acceptor	Zearalenone and Partulin	Extracts of TCM	Tao et al. (2013)
Protein chip for target enzyme	PAD	Hypoglycemic and xanthine oxidase inhibitory activity	Mulberry, lotus, and salvia miltiorrhiza extracts	Wang et al. (2014), Zhang et al. (2022), Zhao et al. (2018)
	Dual channel microfluidic chip	Multiple biomarker assay	Qishen Yiqi pill	Gong et al. (2021)
Biological chemical coupling technology	ABTS+ CE	Major antioxidant in injection	Shuxuening injection	Guo et al. (2018)
	Metabolomics chemometrics	Bioactive chemical quality marker combination	Pollen of *T. orientalis*	Guo et al. (2019)
	DART-MS enzyme	Chemical component and biological activities	Danshen injection	Li et al. (2017)
Biological soft sensing technology	NIR machine learning	Chemical compounds and anti-inflammatory activity	Chrysanthemum extracts	Zhong et al. (2022)
	HIS-AI-CNN	Physical chemical and biological properties	Shuxuening injection	Ding et al. (2016)

affinity selection for identifying inhibitors within a medicinal plant extract (Bhavadharini et al. 2022). The initial step involved the adsorption of swine pancreas lipase onto the polypropylene hollow fiber surface to establish a resilient ligand-harvesting matrix. The known lipase was then used to optimize a number of binding capacity-related parameters, such as temperature, incubation duration, and enzyme concentration. Lotus leaf extracts containing lipase-binding ligands can be identified using the suggested technique. Tetraphenylethylene (TPE) is a common aggregation-induced emission-based fluorophore with a variety of uses in cell imaging and biosensors. TPE-based fluorescent probes were produced by Wang et al. (Vigneshvar et al. 2016) using two procedures. Two types of TPE were produced using normal

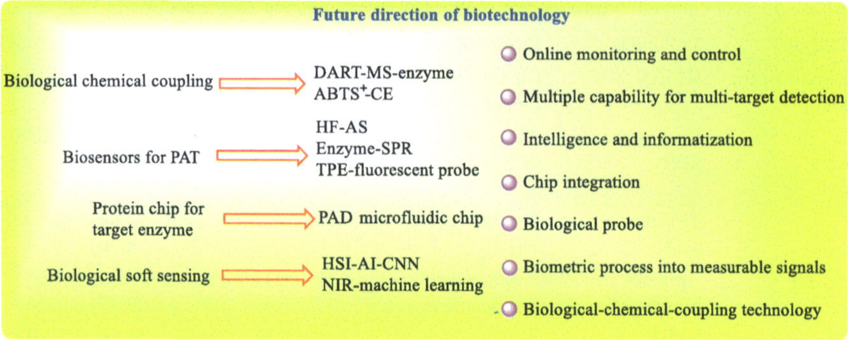

Fig. 3.3 Biotechnologies in pharmaceutical manufacturing process control and intermediate analysis

solid-phase synthesis and the McMurray reaction: TPE with the carboxylate group and TPE-SDKP. This assay functions as a method for monitoring angiotensin-converting enzyme (ACE) activity and assessing the effectiveness of ACE inhibitors sourced from extracts. Aptamers, single-stranded oligonucleotide molecules chosen through ligand evolution using an exponential enrichment mechanism, present numerous benefits, including notable specificity, robust stability, cost-effectiveness, ease of modification, and straightforward in vitro production. They present a viable alternative to the expensive antibody-based Enzyme-Linked Immunosorbent Assay (ELISA) and can serve as an alternative recognition element in biosensors alongside antibodies. Various sensing platforms utilizing aptamers have been developed for the detection of pollutants and cytotoxins in both food and pharmaceutical contexts. Cytotoxin and other contaminants have a significant negative impact on human health; hence their detection in the pharmaceutical process must be constantly monitored (Jabbari et al. 2017; Tao et al. 2013; Wang et al. 2014; Ahmadi et al. 2019; Caglayan and U¨stu¨ndag 2020; He et al. 2020; Jiang et al. 2020; Khan et al. 2018). In a recent study, patulin (PAT) and zearalenone (ZEN) were detected simultaneously by the use of fluorescent acceptors. In this work, fluorescent probes tagged FAM and CY3 on the aptamers of PAT and ZEN, respectively (Khan et al. 2020). Through pep stacking, both aptamers were adsorbed on the surface of graphene oxide (GO), causing an energy transfer known as fluorescence resonance between the fluorophore and GO. As a result, our detection platform detects TCM samples with good selectivity and reliability. The harm that pathogenic *E. coli* does to human health is becoming more and more noticeable (Mahmoud et al. 2019). In a recent study, pathogenic *E. coli* was quickly identified using licorice extracts and an aptamer-based electrochemical biosensor. *E. coli* was biotinylated after the sulfuration signal was initially detected in order to improve the interaction between the aptamer and the bacteria (Wu et al. 2018). When *E. coli* was present, a part of the biotinylated aptamer separated from the capture probe. Streptavidin-alkaline phosphatase was quantitatively bound by residual biotinylated aptamer probes. The outcomes demonstrated that bacteria in TCM and related disciplines can be quickly detected using the biosensor that was designed. The benefits of biosensors include excellent accuracy, quick analysis times, and strong specificity. The integration of computers and biosensors enables the automated collection and processing of data, leading to more accurate and scientifically robust outcomes and the establishment of automated detection systems. To fully harness the potential of integrated detection systems, there is a growing trend toward simultaneous coupling of chip technology and sensors.

3.4.3 Protein Chips in Intermediate Product Analysis

Paper-based microarrays are paper-based analytical instruments that have seen tremendous development in the last few years (Xing et al. 2020; Zhang et al. 2019, 2022; Zhao et al. 2018). They are inexpensive, simple to use, and capable of carrying out several chemical analyses. Furthermore, the high specific surface area of these paper devices makes it simple for molecules to bind and adsorb proteins. To lessen the pollution that experimental consumables produce, used paper devices can be burned. An excellent background signal for detecting colorimetric reactions is provided by the white paper. Paper-based microarrays are a type of analytical tool that have seen fast development in the last few years (Xing et al. 2020; Zhang et al. 2019, 2022; Zhao et al. 2018). They are inexpensive, simple to use, and capable of carrying out several chemical analyses. Furthermore, the high specific surface area of these paper devices facilitates easy protein adsorption and binding by molecules. To lessen the pollution that experimental consumables produce, used paper devices can be burned. An excellent background signal for detecting colorimetric reactions is provided by the white paper. In order to build paper-based microarrays and use them to identify active ingredients in TCM extracts, Gong et al. (Wang et al. 2019). presented an efficient omics method. Using 3D printing (3DP), customized paper containing polycaprolactone and chitosan was created, and a-glucosidase was rendered immobile on it. Using a paper-based analytical apparatus, the method was applied to investigate the hypoglycemic biological activity of lotus and mulberry extracts (PAD) (Gu et al. 2011; He et al. 2015). Another researcher examined the xanthine oxidase inhibitory activity of *S. miltiorrhiza* extracts after immobilizing xanthine oxidase on a PAD69. Therefore, natural products, intermediates, and preparations can be quickly tested for quality using paper-based microarrays (Mitchell et al. 2015). Biomarkers have been utilized as evaluative indicators of pharmaceutical quality, serving to appraise the biological attributes of formulations. A good biomarker should be clinically meaningful and should represent the drug's mode of action. In one study, ACEs and thrombin were used as quality biomarkers in a chip-based approach to assess the bio-consistency of natural products. For bioassays, a dual channel microfluidic chip was created (Ahmed et al. 2016), in which an enzymatic reaction takes place in one channel after the creation of an enzyme complex in the other. In addition to providing an effective surface for immobilizing enzymes, magnetic beads also act as a dependable, robust base for improved on-chip mixing and transport. Drug quality (or potency) is indicated by enzyme activity. Several batches of intermediates and preparations were evaluated in Qishen Yiqi Pills, and the findings indicated that the bioassay outperformed chromatographic fingerprinting in terms of discriminatory power for aberrant samples.

3.4.4 *Intermediate Product Analysis Through Biological Soft Sensing*

Although the correlation between spectrum and activity has often been disregarded, a predominant focus of research has been on elucidating the interconnection between chromatography/mass spectrometry and the biological activity of natural products. Using near-infrared reflectance spectroscopy (NIRS), chrysanthemum extracts' anti-inflammatory properties were discovered (Gong et al. 2021). NIRS was employed in this study together with information on anti-inflammatory activity. Through the utilization of the backpropagation artificial neural network, a sophisticated correlation between Q-markers and the overall anti-inflammatory efficacy has been established. In the context of drug quality control, the comprehensive near-infrared spectroscopy (NIRS) methodology can effectively quantify diverse components and predict the overall biological activity. However, the prediction cannot be made solely by network toxicology or even bioinformatics, and additional tests are needed to confirm the findings.

3.5 Conclusion

Due to their capacity to furnish insights into biological activity, clinical efficacy, and safety profiles of pharmaceuticals, biotechnological methodologies hold comparable importance to techniques reliant on physical and chemical analyses.

In certain situations, botanicals with similar chemical spectra may exhibit different biological activities when the bioactive constituents remain undetectable under specific analytical conditions. Conversely, botanicals with distinct chemical profiles may demonstrate comparable bioactivity if the phytochemicals responsible for the variation are biologically inert. These are just some of the advantages and disadvantages of chemical technology and biotechnology. In the future, biotechnology should be involved in every stage of the drug quality control process, in addition to chemical and physical investigations. Currently, synthetic and natural products receive less use of bioanalytical technologies than biological medications. This could be as a result of the fact that most scientists working in the field of pharmaceutical analysis have backgrounds in chemistry. Pharmaceutical analysis today ignores the value of bioassays and their applications. As basic biology continues to advance and bioanalytical techniques improve as a result, biotechnology can become increasingly important in the overall quality control of both synthetic and natural pharmaceuticals.

References

Ahmadi A, Danesh NM, Ramezani M, Alibolandia M, Lavaee P, Emrani AS et al (2019) A rapid and simple ratiometric fluorescent sensor for patulin detection based on a stabilized DNA duplex probe containing less amount of aptamer-involved base pairs. Talanta 204:641–646

Ahmed S, Bui M-PN, Abbas A (2016) Paper-based chemical and biological sensors: engineering aspects. Biosens Bioelectron 77:249–263

Andryukov BG, Lyapun IN, Matosova EV, Somova LM (2021) Biosensor technologies in medicine: from detection of biochemical markers to research into molecular targets (review). Sovrem Tekhnologii Med 12:70–83

Bhavadharini B, Kavimughil M, Malini B, Vallath A, Prajapati HK, Sunil CK (2022) Recent advances in biosensors for detection of chemical contaminants in food—a review. Food Anal Methods 15:1545–1564

Caglayan MO, U¨stu¨ndag Z (2020) Detection of zearalenone in an aptamer assay using attenuated internal reflection ellipsometry and its cereal sample applications. Food Chem Toxicol 136:111081

Ding G, Li B, Han Y, Liu A, Zhang J, Peng J et al (2016) A rapid integrated bioactivity evaluation system based on near-infrared spectroscopy for quality control of Flos Chrysanthemi. J Pharm Biomed Anal 131:391–399

Fraser K, Lane GA, Otter DE, Harrison SJ, Quek SY, Hemar Y et al (2013) Monitoring tea fermentation/manufacturing by direct analysis in real time (DART) mass spectrometry. Food Chem 141:2060–2065

Gong X, Shao J, Guo S, Pan J, Fan X (2021) Determination of inhibitory activity of salvia miltiorrhiza extracts on xanthine oxidase with a paper-based analytical device. J Pharm Anal 11:603–610

Gu Z, Zhao M, Sheng Y, Bentolila LA, Tang Y (2011) Detection of mercury ion by infrared fluorescent protein and its hydrogel-based paper assay. Anal Chem 83:2324–2329

Guo S, Shao J, Gong X (2018) Paper-based analytical devices prepared with polycaprolactone printing and their application in the activity determination of mulberry extracts. J Pharm Biomed Anal 161:28–34

Guo S, Lin X, Wang Y, Gong X (2019) Fabrication of paper-based enzyme immobilized microarray by 3D-printing technique for screening alpha-glucosidase inhibitors in mulberry leaves and lotus leaves. Chin Med 14:13

He Y, Wua WB, Fu JZ (2015) Rapid fabrication of paper-based microfluidic analytical devices with desktop stereolithography 3D printer. RSC Adv 5:2694–2701

He D, Wu Z, Cui B, Jin Z, Xu E (2020) A fluorometric method for aptamer-based simultaneous determination of two kinds of the fusarium mycotoxins zearalenone and fumonisin B1 making use of gold nanorods and upconversion nanoparticles. Mikrochim Acta 187:254

Jabbari S, Dabirmanesh B, Arab SS, Amanlou M, Daneshjou S, Gholami S et al (2017) A novel enzyme based SPR-biosensor to detect bromocriptine as an ergoline derivative drug. Sensor Actuator B Chem 240:519–527

Jiang M, Chen C, He J, Zhang H, Xu Z (2020) Fluorescence assay for three organophosphorus pesticides in agricultural products based on magnetic-assisted fluorescence labeling aptamer probe. Food Chem 307:125534

Khan IM, Zhao S, Niazi S, Mohsin A, Shoaib M, Duan N et al (2018) Silver nanoclusters based FRET aptasensor for sensitive and selective fluorescent detection of T-2 toxin. Sensor Actuator B Chem 277:328–335

Khan R, Sherazi TA, Catanante G, Rasheed S, Marty JL, Hayat A (2020) Switchable fluorescence sensor toward PAT via CA-MWCNTs quenched aptamer-tagged carboxy fluorescein. Food Chem 312:126048

Li ZH, Ai N, Yu LX, Qian ZZ, Cheng YY (2017) A multiple biomarker assay for quality assessment of botanical drugs using a versatile microfluidic chip. Sci Rep 7:12243

Li Z, Wang Y, Cheng Y (2019) Mass spectrometry-sensitive probes coupled with direct analysis in real time for simultaneous sensing of chemical and biological properties of botanical drugs. Anal Chem 91:9001–9009

Ma H, Li J, An M, Gao XM, Chang YX (2018) A powerful on line ABTS (þ)-CE-DAD method to screen and quantify major antioxidants for quality control of Shuxuening injection. Sci Rep 8:5441

Mahmoud M, Laufer S, Deigner HP (2019) Visual aptamer-based capillary assay for ethanolamine using magnetic particles and strand displacement. Mikrochim Acta 186:690

Mitchell HT, Noxon IC, Chaplan CA, Carlton SJ, Liu CH, Ganaja KA et al (2015) Reagent pencils: a new technique for solvent-free deposition of reagents onto paper-based microfluidic devices. Lab Chip 15:2213–2220

Ramakrishnan P, Loh WM, Gopinath SCB, Bonam SR, Fareez IM, Guad RM et al (2020) Selective phytochemicals targeting pancreatic stellate cells as new anti-fibrotic agents for chronic pancreatitis and pancreatic cancer. Acta Pharm Sin B 10:399–413

Tao Y, Zhang Y, Wang Y, Chen Y (2013) Hollow fiber based affinity selection combined with high performance liquid chromatography-mass spectroscopy for rapid screening lipase inhibitors from lotus leaf. Anal Chim Acta 785:75–81

Vigneshvar S, Sudhakumari CC, Senthilkumaran B, Prakash H (2016) Recent advances in biosensor technology for potential applications—an overview. Front Bioeng Biotechnol 4:11

Wang H, Huang Y, Zhao X, Gong W, Wang Y, Cheng Y (2014) A novel aggregation-induced emission based fluorescent probe for an angiotensin converting enzyme (ACE) assay and inhibitor screening. Chem Commun 50:15075–15078

Wang H, Zhao Y, Bie S, Suo T, Jia G, Liu B et al (2019) Development of an electrochemical biosensor for rapid and effective detection of pathogenic *Escherichia coli* in liquorice extract. Appl Sci 9:295

Wang X, Chen X, Li J, Evans OB, Wang H, Yang X et al (2020) Thrombin-based discovery strategy of bioactive-chemical quality marker combination for pollen of *Typha orientalis* by metabolomics coupled with chemometrics. Phytomedicine 75:153246

Wu S, Liu L, Duan N, Li Q, Zhou Y, Wang Z (2018) An aptamer-based lateral flow test strip for rapid detection of zearalenone in corn samples. J Agric Food Chem 66:1949–1954

Xing KY, Peng J, Shan S, Liu D, Huang YN, Lai W (2020) Green enzyme linked immunosorbent assay based on the single-stranded binding protein-assisted aptamer for the detection of mycotoxin. Anal Chem 92:8422–8426

Zhang M, Wang Y, Yuan S, Sun X, Huo B, Bai J et al (2019) Competitive fluorometric assay for the food toxin T-2 by using DNA-modified silver nanoclusters, aptamer-modified magnetic beads, and exponential isothermal amplification. Mikrochim Acta 186:219

Zhang Y, Lv X, Qu J, Zhang X, Zhang M, Gao H et al (2020) A systematic strategy for screening therapeutic constituents of Schisandra chinensis (Turcz.) Baill infiltrated blood-brain barrier oriented in lesions using ethanol and water extracts: a novel perspective for exploring chemical material basis of herb medicines. Acta Pharm Sin B 10:557–568

Zhang N, Li J, Liu B, Wang H, Zhang D, Li Z (2022) A facile "turn-on" fluorescent aptasensor for simultaneous detection of dual mycotoxins in traditional Chinese medicine based on graphene oxide and FRET. Toxicon 206:42–50

Zhao YW, Wang HX, Jia GC, Li Z (2018) Application of aptamer-based biosensor for rapid detection of pathogenic Escherichia coli. Sensors (Basel) 18:2518

Zhong Y, Ru C, Wang S, Li Z, Cheng Y (2022) An online, non-destructive method for simultaneously detecting chemical, biological, and physical properties of herbal injections using hyperspectral imaging with artificial intelligence. Spectrochim Acta A Mol Biomol Spectrosc 264:120250

Zhu X, Rehman KU, Wang B, Shahzad M (2020) Modern soft-sensing modeling methods for fermentation processes. Sensors 20:1771

Chapter 4
Applications of Biotechnology in Pharmaceutical Product Analysis

Mohd Aslam, Anjali Rani, Javed Khan, Bhaskara Nand Pant, and Garima Pandey

Abstract The pharmaceutical sector is critical to supplying safe and effective pharmaceuticals to people all around the world. Pharmaceutical product efficacy, safety, and quality are critical, needing stringent analytical methods to assure compliance with regulatory criteria. With its creative and adaptable approaches, biotechnology has revolutionized pharmaceutical product analysis by providing enhanced tools for a reliable and thorough evaluation. This chapter examines the various applications of biotechnology in pharmaceutical product analysis, focusing on how it has improved the identification, quantification, and characterization of active pharmaceutical ingredients, contaminants, and formulation components. DNA sequencing, proteomics, genomics, and bioinformatics techniques have allowed for the exact determination of constituents as well as the evaluation of their sources. Overall, the introduction of biotechnology to pharmaceutical product analysis has considerably improved the pharmaceutical sector's capacity to ensure reliability, safety, and functionality, instilling greater trust in healthcare practitioners and patients alike.

Keywords Biotechnology · Pharmaceutical analysis · Drug development · Biosensors

M. Aslam · A. Rani · J. Khan
Department of Chemistry, SRM Institute of Science and Technology, Delhi-NCR Campus, Modinagar, Ghaziabad, Uttar Pradesh, India

Department of Chemistry, Atma Ram Sanatan Dharma College, University of Delhi, New Delhi, India

B. N. Pant
Department of Chemistry, Atma Ram Sanatan Dharma College, University of Delhi, New Delhi, India

G. Pandey (✉)
Department of Chemistry, SRM Institute of Science and Technology, Delhi-NCR Campus, Modinagar, Ghaziabad, Uttar Pradesh, India

S. Bose et al. (eds.), *Concepts in Pharmaceutical Biotechnology and Drug Development*, Interdisciplinary Biotechnological Advances,
https://doi.org/10.1007/978-981-97-1148-2_4

Abbreviations

APIs	Active pharmaceutical ingredients
AQbD	Analytical quality by design
CE	Capillary electrophoresis
CMPs	Conjugated microporous polymers
ELISA	Enzyme-linked immunosorbent assay
GC	Gas chromatography
HPLC	High-performance liquid chromatography
MOFs	Metal-organic frameworks
MS	Mass spectrometry
PAT	Process analytical technology
PCR	Polymerase chain reaction
QbD	Quality by design
QC	Quality control
TLC	Thin-layer chromatography
WB	Western blotting

4.1 Introduction

Pharmaceutical chemical analysis is an essential area of study since medications have a significant impact on the well-being of people and animals. The pharmaceutical sector, biomedical research facilities, hospitals, and other industries have a significant demand for dependable, accurate, precise, and delicate analytical procedures. Analytical procedures are commonly used in the production of novel pharmaceuticals, degradation products, contaminants, biomarker determination, pharmacokinetic investigations, and diagnosis or analysis of therapeutic drugs in biological samples, among other things (Fanali et al. 2023; Klatte et al. 2017). Several analytical methods, which include supercritical fluid chromatography (SFC) (Gros et al. 2023; Cobo-Golpe et al. 2022), gas chromatography, high-performance liquid chromatography (HPLC) (Lo Faro et al. 2023; Lin et al. 2021; Zou et al. 2023; Aspromonte et al. 2019), and microfluidic techniques such as nano-liquid chromatography (nano-LC) (D'Orazio et al. 2019), capillary electrophoresis (CE) (Woźniakiewicz and Wietecha-Posłuszny 2023; Krait et al. 2021), and capillary electrochromatography (CEC) (Fanali et al. 2021; Chankvetadze et al. 2003), are now utilized in pharmaceutical companies and healthcare analysis. Bioanalysis is employed in the pharmaceutical sector throughout the initial phases of the discovery and development of drugs to give evidence to determine the metabolic and pharmacokinetic destiny of drug prospects. Bioanalysis is still employed in the preclinical and clinical stages of discovering drugs, and in certain circumstances, clinical therapeutic drug evaluation (Pandey et al. 2010). Parallel to this, the issue of pharmaceutical contamination has grown in prominence previously in the past. There are serious threats to human health and the ecosystem when there is persistent contamination from medicines in the water, food, and land in certain parts of the world (Patel et al. 2019; Rehman et al. 2015). The preliminary treatment of samples before analytical analysis is critical in the study of dietary, ecological, biological, and medicinal substances. Effective separation and accurate detection of complex and trace materials can be achieved using sample pretreatment methods (Wen et al. 2014). A number of substances exhibit excellent properties and potential use in separating and analysis of various pharmaceutical samples. Due to their functional diversity and porous structure, researchers have developed an interest in organic molecule-based framework porous materials. Different types of porous organic frameworks were documented, including conjugated microporous polymers (CMPs), covalent organic frameworks (COFs), hyper-crosslinked polymers (HCPs), metal organic frameworks (MOFs), and covalent triazine frameworks (CTFs) (Zhang et al. 2018). Because of its label-free and real-time detecting process, ease of use, and excellent detection, plasmonic-based nanosensors are among the most powerful tools for detecting and quantifying low amounts of molecular analytes. Plasmonic sensors are also useful for detecting a variety of other applications, including biosensors, environmental monitoring, medicines, and food control (Yesudasu et al. 2021; Tseng et al. 2017; Dissanayake et al. 2019; Muneer et al. 2020; Mauriz 2020).

For quantitative researchers, identifying and determining drugs, pollutants, compounds, and biomarkers in various matrices has become a challenging undertaking. Therefore, developing analytical tools to evaluate drugs, their derivatives, and biomarkers in biological samples is always necessary.

4.2 Importance of Biotechnology in Pharmaceutical Product Analysis

Biotechnology is critical in pharmaceutical product analysis, contributing to product safety, efficacy, and quality control of drugs. Here are some key reasons why biotechnology is relevant in this context.

4.2.1 Analytical Techniques

Biotechnology allows researchers access to cutting-edge analytical tools like genomics, proteomics, and metabolomics, which allow them to thoroughly study drug compounds at the molecular level. Pharmaceutical research has traditionally had an important role in the advancement of pharmaceutical development, directed by pharmacological and clinical research and supported by the field of chemistry. Active pharmaceutical ingredients (APIs) production and characterization, along with their evaluation to obtain provisional precautions and therapeutic efficacy information, must be carried out before drug candidates can be identified for further comprehensive research (Siddiqui 2017; Anon n.d.-a). Within the sector of pharmaceutical research, the analytical evaluation of large quantities of drug items, intermediary substances, medications, formulations of drugs, contaminants and compounds from degradation, and biological specimens comprising pharmaceuticals and metabolites of them is essential.

Titrimetric methods are being shown to be useful in kinetic evaluations, which can be then utilized for determining rates of reaction through the development of functional group evaluation procedures (Rahman et al. 2005; Basavaiah and Prameela 2003). Titrimetry was formerly used to estimate pharmacological product breakdown as well as drug concentrations (Matei et al. 2008). Different analytical techniques, including titrimetric, chromatographic, spectroscopic, and electrochemical techniques, are commonly employed to determine the quality and quantity of the drug product as well as the materials utilized during its synthesis (Beccaria and Cabooter 2020). Three main groups comprise the several techniques used to assess antioxidant capacity: spectrometry, electrochemical tests, and chromatography (Munteanu and Apetrei 2021).

4.2.2 Quality Control

Biotechnology is instrumental in establishing rigorous quality control standards for pharmaceutical products. Quality control (QC) measures are essential in a pharmaceutical manufacturing control plan to ensure product quality (Deidda et al. 2018). Companies submit QC procedures to regulatory agencies in each country where a pharmaceutical item is supplied to assist in market licencing procedures (Åsberg et al. 2016). The quality by testing concept, commonly known as trial-and-error, is a conventional strategy for developing analytical methods. Such a technique surely does not facilitate additional modifications that may be required due to a lack of understanding of the possible effect on method performance (Rozet et al. 2013). Capillary or chip-based electrophoretic techniques used for separation have essentially superseded traditional slab-gel SDS-PAGE methods utilized for biological impurity and purity testing. In QC labs, image capillary isoelectric focusing techniques have mostly superseded gel IEF techniques (Rogers et al. 2018). The quality data of TCM preparations is restricted by the capabilities of current quality traceability technology, making them unsuitable for use in production or sales. One of the key methods for quality assurance and evaluation of herbal medicines is the use of chemical fingerprints, sometimes referred to as chemical markers (Liu et al. 2018). The definitions of Analytical Quality by Design (AQbD) and Quality by Design (QbD) are comparable when it comes to manufacturing processes. As a result, AQbD also consists of four primary phases: (a) calculating the necessary analytical method performances (ATP), (b) identifying pertinent procedure parameters and quality standards (the screening stage), (c) defining the analytical DS (the

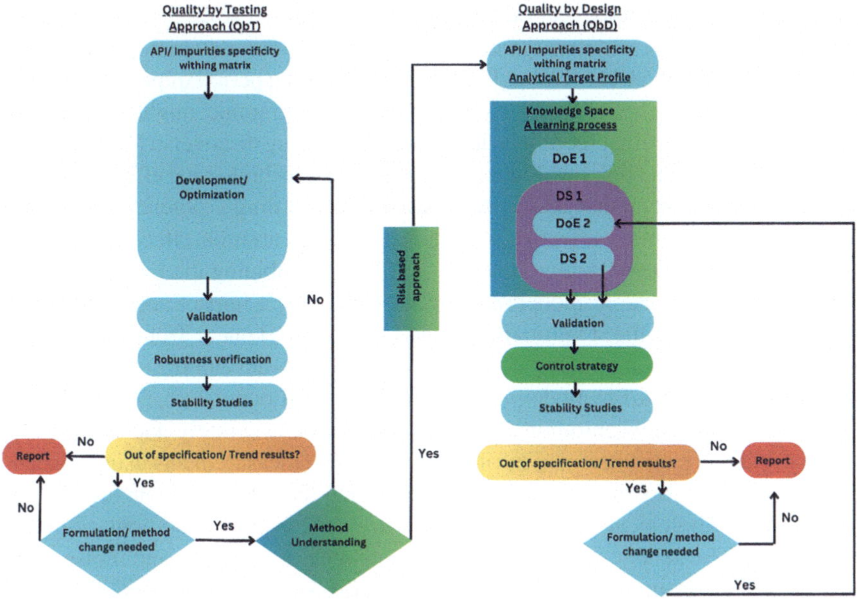

Fig. 4.1 Comparison of Quality by Testing (QbT) approach (Left) and Quality by Design (QbD) approach (Right) (Dispas et al. 2018)

robust optimization) allows for method optimization and risk assessment, (d) application of a control technique to ensure the method is improved continuously (Fig. 4.1) (Dispas et al. 2018).

In addition to the digital revolution, modelling approaches for process understanding, prediction, and consistency have received more attention as a result of the FDA and EMA's support of Quality by Design (QbD) and Process Analytical Technology (PAT) concepts. As a result, these paradigms allowed for a big deal of flexibility for selection and use cases of the quantitative methodologies (Narayanan et al. 2020). The most effective method for quality control in routine procedures as well as research projects is metabolomics. To track the metabolites present, untargeted metabolomic fingerprint can be utilized and statistical control charts can be generated by multivariate statistical analysis to confirm that the composition is maintained throughout the production process (Mattoli et al. 2023).

4.2.3 *Pharmacokinetics and Pharmacodynamics*

Biotechnology aids in studying the pharmacokinetics (the manner in which drugs are absorbed, metabolized, distributed, and excreted) and pharmacodynamics (manner in which drugs show interactions within the body) of pharmaceuticals (Chowdhury et al. 2021). Studies on the metabolism and pharmacokinetics of potential medications have become a significant aspect of drug discovery and development programmes, and they typically begin concurrently with biological activity screening (Üëá et al. 2011). Throughout the early learning period of preclinical drug discovery, a comprehensive and rigorous PK/PD programme might serve as an intermediate connecting preclinical development and drug discovery (Meibohm and Derendorf 2002). The link involving a drug's dose and its biological effects is examined by pharmacodynamic analysis. To ascertain the drug's potency, maximum response, and minimum effective concentration, concentration-effect curves may need to be established (Crommelin et al. 2020). By combining the two, pharmacokinetic/pharmacodynamic modelling forecasts the drug's impact according to its pharmacokinetic profile. Optimizing dosage tactics can benefit from this (Sarkar et al. 2022). There are multiple potential benefits to developing and applying mechanism-based PK/PD modelling in different animal populations: (a) It characterizes the dosage concentration effectiveness relation accurately and thoroughly, (b) it provides effectiveness and intrinsic performance measurements using concentrations instead of doses, (c) it enables research into the role of influencing physiological and pathological factors, in addition to resistance in conditions (Felmlee et al. 2012). The genetic variations of drug-metabolizing enzymes and transporters have a significant impact on the pharmacokinetics of natural products. This has garnered increasing attention in recent years, but despite the wide range and widespread applications of natural products worldwide, there is still a deficiency of knowledge in this area (Rao et al. 2019). The volume of distribution (Vd) and clearance (CL) are the two most significant PK parameters for pharmaceuticals. A proportionality constant known as the Vd is described as the relation among the observed plasma concentration and the dosage, or amount of drug supplied. The volume of blood or plasma from which a certain medication is entirely eliminated per unit of time (hours or minutes) is known as CL, which is a measure of drug elimination (van den Anker et al. 2018).

4.2.4 Drug Discovery and Development

Biotechnology techniques such as recombinant DNA technology and gene expression profiling are used to identify prospective therapeutic targets, find novel medication candidates, and improve their properties. The Human Genome Project has so far had a substantial effect on new drug discovery (Boulnois 2000). The development of microarray technology, which allows for the simultaneous study of the expression of tens of thousands of genes, as well as the increasing sophistication of proteome analysis, has created numerous opportunities for those working in drug discovery (Lander 1999).

Currently, biosensors are employed in assay development, validation, lead optimization, target identification, excretion, absorption, metabolism, distribution, and toxicity (Grimm et al. 2004). Proteomics is becoming increasingly significant in the target identification and validation stages of the drug discovery process.

Nanotechnology innovations are being used to improve drug delivery, drug discovery, and pharmaceutical manufacturing (Jain 2009).

Medical biotechnology is critical in today's climate for understanding the molecular factors that cause health problems, enhancing diagnostic processes, and generating tailored treatments. This approach enables disease therapy with an intriguing new pharmacological module that targets a milieu of causes (Anwar et al. 2022). Gram staining can distinguish among both kinds of bacteria and continues to be utilized in clinical specimens that include cerebrospinal fluid or biopsies for early bacterial infection detection since it is a simpler and faster way of identifying bacteria compared to the cultivation of cells (Lokko et al. 2018). However, due to the precision of such surgeries made possible by robots, they have been expanded to meet the needs of more people. It allows an experienced surgeon that is a recognized specialist in the discipline to operate on an individual in a different region of the world (Anwar et al. 2022). Traditionally, medicinal chemistry has taken advantage of any opportunity to contribute to the identification of new therapeutic molecules. Although medicinal chemistry as a profession has improved optimization into a process free of substantial unidentified barriers, the task of lead identification has yet to reach a comparable level of maturity (Venuti 1989; Anon n.d.-b).

4.3 Biotechnological Techniques in Pharmaceutical Product Analysis

Biotechnological techniques have revolutionized pharmaceutical product analysis, providing powerful tools for researchers and analysts to assess the safety, efficacy, and quality of drugs and biopharmaceuticals. Here are some of the most important biotechnological approaches used in pharmaceutical product analysis.

4.3.1 Polymerase Chain Reaction

Recent advancements in methodology have made it possible to enhance an organism's genetic abnormalities in DNA or RNA in vitro using polymerase chain reaction (PCR), perhaps eliminating the need for culture. Primer annealing, primer extension, and denaturation of double-stranded DNA comprise the fundamental three-step cycle process of PCR (Schochetman et al. 1988). Certain DNA fragments of interest are surrounded by two oligonucleotide primers that combine to adjacent strands in the polymerase chain reaction (PCR) in culture technique for enzymatic synthesis of specific segments of DNA (Erlich 1989). Through a cyclic process called PCR, practically any sequence of nucleic acids found in a complex substance can be processed, producing a large number of identical copies that are simple to assess (Kubista et al. 2006).

Recent improvements in PCR amplification methods have influenced crucial characteristics such as misincorporation rate, specificity (target versus non-target amplification), and PCR product maximal length (Mfuh et al. 2019). The number of amplification items is evaluated by many PCR techniques, oligonucleotide probe typing, or direct sequencing; therefore infrequent mis-incorporated nucleotides are not found (Erlich et al. 1985).

Additionally, it has been claimed that novel PCR-based techniques improve the assay's sensitivity and specificity. Furthermore, real-time Reverse Transcriptase-PCR is more sensitive than traditional RT-PCR assay, which can be very helpful in early infection diagnosis (Shen et al. 2020).

4.3.2 Enzyme-Linked Immunosorbent Assay (ELISA)

ELISAs are immunoassays through which an enzyme acts as the "signal generator" or "reporter" while one reactant remains stationary on a solid surface. Enzyme-linked immunosorbent assays (ELISA) and solid-phase immunoassays technically simplified antigen quantification and antibody detection (Butler 2000).

An indirect ELISA to identify particular antibodies and a direct competitive ELISA to identify soluble antigens are two of the six unique ELISA methods that demonstrate the diversity of ELISAs. Soluble antigens can be detected using an antibody-sandwich ELISA, specific antibodies can be found using a double antibody-sandwich ELISA, and cell-surface antigens and antibodies specific to surface antigens can be found using two cellular ELISAs (Hornbeck 1992). The antigen that has coated the well is attached by a primary detection antibody with a tagged enzyme in a direct ELISA (Schots et al. 1988). Sandwich ELISA is used to analyse more complicated solutions. The capacity of a configuration to fit an antigen between two antibodies is where the test gets its name (Katsurada et al. 2007). A competitive ELISA differs from sandwich, direct or indirect ELISAs in that it employs the method of efficient interactions (Libeau et al. 1995). Any antigen present is attached to the primary antibody when it comes into contact to an unpurified sample. Complexes of antigen and antibody are formed in proportion to the number of antigens in the material being examined (Starr and Tessier 2019). Lesser number of antibodies are able to attach to the antigen in greater number of antigen-antibody complexes that develop (Sittampalam et al. 1996).

4.3.3 Mass Spectrometry (MS)

Proteomics is the methodical analysis of every protein present in a given tissue or cell, with mass spectrometry at the heart of most proteomic techniques. For a long time, mass spectrometry was limited to tiny and thermostable compounds due to a lack of adequate procedures for softly ionizing and transferring ionized molecules from the condensed phase to the gas phase without significant fragmentation (Domon and Aebersold 2006). Mass Spectrometry is unique among analytical techniques utilized for an astonishingly wide range of applications by analysing the molecule's inherent mass property with exceptionally high accuracy (Mann et al. 2001).

Currently, there are three key applications of MS in proteomics. MS is the preferred technique for characterizing and controlling the quality of recombinant proteins and other macromolecules, which is a significant task in the field of biotechnology (Glish and Vachet 2003). It is also widely utilized in protein identification, whether in traditional biochemical research or large-scale proteomic ones (Zubarev and Makarov 2013). Finally, because MS assesses a protein's molecular weight, it is the method of choice for detecting and characterizing posttranslational modifications and has the potential to identify any covalent change that alters the mass of a protein (Cooks et al. 2006).

4.3.4 High-Performance Liquid Chromatography

Pharmaceutical analysis often entails conducting trials to determine how much of a medication is present in a pharmaceutical formulation of therapeutic substances. This method needs to be tried out extensively, and even after it is established and proven, there may be situations where it still needs to be tried out again (Ganorkar and Shirkhedkar 2017). Multidimensional column chromatography (also known as column switching, multiphase, multicolumn, or coupled column chromatography) is an effective method for separating multicomponent materials. The approach has been used in thin-layer chromatography (TLC) for many years, although in a somewhat different way (Milroy et al. 2011).

Chemicals dissolved in solution can be separated and quantified using a particular chromatography technique. Based on the mechanism of action, chromatography systems are frequently classified into four types: adsorption, partition, ion-exchange, and size exclusion (Kupiec 2004). When it comes to bulk medications and pharmaceutical formulation quality control, this method is crucial. It encompasses various tasks such as analysing active pharmaceutical ingredients (API), characterizing impurities, figuring out enantiomeric purity, and discovering breakdown of substances to evaluate product stability. Additionally, the byproducts of drugs in biological specimens are identified using it (Yabré et al. 2018).

4.3.5 Western Blotting

Northern and Southern blotting gave rise to Western blotting (WB), commonly referred to as immunoblotting or protein blotting (Southern 2004; Alwine et al. 1977). The nature of the gel, molecular mass of proteins being transferred, and the membrane being utilized are all crucial factors in ensuring the effective migration of proteins from a gel to a solid membrane support (Kurien and Scofield 2006).

Researching the regulatory molecular mechanisms underlying protein turnover, energy metabolism, and chronic physiological adaptations can benefit greatly from the extensive range of applications provided by the Western blot (WB) (Bass et al. 2017). A usual approach for identifying proteins and subsequent modifications on proteins is the western blot, which can provide numeric or indirect information on the target protein in both straightforward and complicated biological materials (Vallejo-Illarramendi et al. 2013). One popular method for analysing proteins is western blotting (He and Herr 2010). Purification and concentration of proteins determined by the extracts of cell or tissue are usually the first steps in this multistep process, protein size separation on a nitrocellulose or polyvinylidene fluoride (PVDF) and polyacrylamide gel containing sodium dodecyl sulphate (SDS) membrane-mediated mobilization protein isolation, obstructing membrane proteins without specificity, conjugation of a secondary antibody with a labelled fluorescent or chemiluminescent molecule, identification of a signal indicating the binding of an antigen or antibody densitometry analysis using software to identify relevant protein bands (Mishra et al. 2017; Murphy and Lamb 2013).

4.4 Applications in Drug Development and Quality Control

Biotechnology plays a crucial role in the pharmaceutical sector for both drug development and quality control. Here are some examples of major applications in both domains. The development of quick and precise drug safety analysis methodologies is necessary since current approaches and techniques are unable to keep up with the demands of the industry (Chen et al. 2022). This section provides an overview of the development of biological approaches used to analyse medication safety (Fig. 4.2).

4.4.1 Target Identification and Validation

Target validation and identification are the initial crucial steps in the therapeutic research process. Discovering new targets usually proteins, whose modification could halt or reverse the course of a disease, is the aim of target discovery (Howbrook et al. 2003). Nucleic acid microarrays have significantly improved our knowledge of both normal and abnormal fields of biochemistry, which in turn has influenced the choice of targets for medication development (Vernell et al. 2003; Wang et al. 2004).

It enables intuitive and accessible analysis of the data that links targets and diseases, while also giving methods for the inquiry of specific illness concept (Koscielny et al. 2017). Relevant information linking a target to a disease is stored on the Target Evaluation Services for all potential human targets, including proteins and RNA molecules (Malone et al. 2010).

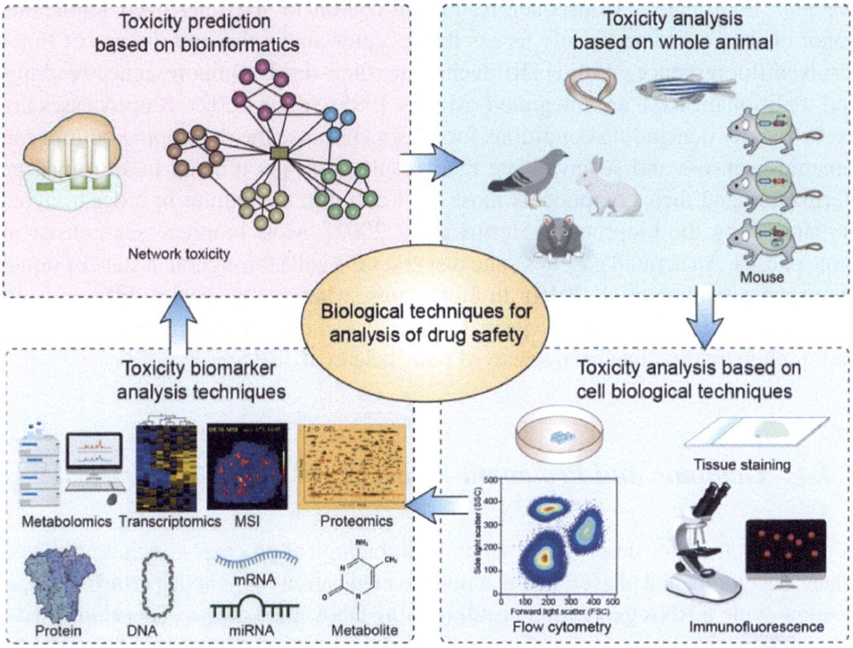

Fig. 4.2 A summary for drug assessment by biological methods (Chen et al. 2022)

4.4.2 Structural Biology

The earliest documented success stories reported in the early 1990s with structural biology as a therapeutic research technique date back to the mid-1970s (Beddell et al. 1976; Erickson et al. 1990). Nowadays, many corporate pharmaceutical development initiatives depend heavily on structural biology and SBDD, and these areas of study are also the focus of research for numerous university groups (Van Duin et al. 2003). This is the stage for the medication approach to research where SBDD has most potential to make a difference, via influencing the main components' modification in chemicals (Scapin 2006). Known as in-cell or in situ structure commitment, the viewing of cellular frameworks in their native surroundings is still one of structural biology's long-term objectives. Indeed, the viability of this strategy is demonstrated by recent developments in mass spectrometry, chemical crosslinking, data processing, cryo-electron tomography, and data processing (Cramer 2021).

4.4.3 Bioprocess Monitoring

HTRF (Homogeneous Time-Resolved Fluorescence) technology is based on the principle of fluorescence resonance energy transfer (FRET) where non-radiative energy is transferred from a donor fluorophore (rare-earth lanthanides) to an acceptor fluorophore. Consequently, the utilization of the long-lived lanthanide donor enables us to accurately assess the acceptor signal through the use of time-resolved fluorescence. With HTRF technique, time-resolved fluorescence readings and traditional FRET are integrated (Mathis 1995; Selvin 2002). Bioprocesses are exceptionally demanding conditions for sensors because the developing culture can penetrate sensors and so invalidate their results. Sensors utilized in situ must be sterilizable and their components must not leach into the culture in order to avoid contaminating the bioprocess (Harms et al. 2002). Most bioprocesses consist of three phases. An aerated gaseous state disperses the cells throughout a state of liquid atmosphere (Bluma et al. 2010). In bioreactors, where cells convert different substrates into new biomass and expected substances, the main goal of sensor creation and implementation has been achieved (Biechele et al. 2015).

4.4.4 Genomic and Proteomic Analysis

A potent method for describing alterations in biological processes, including illness states, developmental phases, and reactions to medications or genetic perturbations, is genome-wide mRNA expression profiling using DNA microarrays (Tian et al. 2004). Complying with morphology, degree of differentiation, and metastasis presence or absence, traditional pathological classification methods pose challenges in measuring the complex genetic and proteomic changes that drive cancer progression and cause substantial clinical heterogeneity among individual tumours and patients (Wulfkuhle et al. 2004). Among the many areas of pharmaceutical research and development where gene expression approaches may be helpful are identification of target, drug enhancement, and mechanism of action analysis. One method is for obtaining genes with unknown functions and study them with microarrays (Cunningham 2000). Drug discovery has access to a greater number of targets for investigation thanks to the application of genomics in the identification and validation of druggable genes. Genomic analysis of targets has been made possible by developments in antisense technology, genetically modified specimens, and short interfering RNA (siRNA), which mimics natural RNA interference (RNAi) (Kiriiri et al. 2020).

4.4.5 Bioassays

Bioassays are tests which identify genetic action for the extraction of substances derived from a living organism. Although clinical trials and studies using whole animals may be included in this category, they are not since they are generally used to study the efficacy and toxicity of drugs that have already been shown to have some activity in vitro (Houghton 2000). According to Eli Lilly & Company's Assay Guidance manual, a group of substances known as a biological test, or bioassay, which produces a detectable signal, can be used to quantify a biological process. When analysing a medicine, the biological activity or activities of one or more components are quantified by calculating their ability to produce the desired biological activity in terms of units on an in-vitro culture of living cells or on test organisms in the wild (Indrayanto et al. 2021). "Biological methods are described for the assay of certain substances and preparations whose potencies cannot be adequately assured by chemical or physical analysis," according to European Pharmacopeia 10 (Zhang et al. 2023). The objective of finding out how a medicine affects a particular kind of living thing or biological test system is known as a biological assay, or bioassay. Direct or indirect tactics might be used to carry out bioassays (Rosso 2010).

4.4.6 *Stability Testing*

Analysis of pharmaceutical product stability is necessary to guarantee the safety, effectiveness, and quality of a medication composition in an intricate set of procedures requiring a substantial financial investment, a substantial time commitment, and scientific expertise; one can ensure the scientific and commercial success of a pharmaceutical product by having a thorough understanding of the drug development process and the various tasks and benchmarks that make up an all-encompassing development plan (Singh and Bakshi 2000). The most important procedures carried out during the development phase are tests for stability and drug evaluations, and requirements must be met in order to confirm the authenticity, strength, and purity of the ingredients and the final goods. The ability of a certain composition to remain within its chemical, physical, biological, toxicological, secure, and instructive parameters in a given container or closure system is known as product stability (Kommanaboyina and Rhodes 1999). In order to determine the best storage conditions, determine a pharmaceutical drug's shelf life, and suggest labelling guidelines, stability analysis evaluates the impact of outside variables on the calibre of the material or item (Annison 2011).

4.5 Biosensors

Analytical instruments that convert biological reactions into electrical signals are called biosensors. Biosensors need to be extremely precise, unaffected by environmental factors like temperature and pH, and reusable (Mehrotra 2016). Studies in engineering, biology, and chemistry across multiple disciplines have been required for the creation of biological detectors, including materials, converting devices, immobilization techniques, etc. Three categories are distinguished by the mechanics underlying the biosensor substances: biocatalytic, microbe-based, bio-affinity, etc. (Sharma et al. 2003). Biosensors can be classified into various types, including DNA biosensors, enzyme-based, tissue-based biosensors, immunosensors, thermal or piezoelectric biosensors. Hicks and Updike (1967) reported the first-ever enzyme-based detector. Van der Walls forces, covalent and ionic bonding are examples for immobilization techniques that have been used to construct enzyme biosensors (Wang 2008; Akyilmaz et al. 2010). Organelle-based sensors were developed using mitochondria, chloroplasts, membranes, and microsomes. The biosensor exhibited remarkable stability; nevertheless, its detection time was prolonged and its specificity was diminished (Grattieri and Minteer 2018). Specific antibodies that adhere to pathogens or infections, or that interact with body's defence system, are based on immunosensors because of their high affinity for certain antigens (Gizeli and Lowe n.d.). The fundamental principle of DNA biosensors is the ability of a single-strand polynucleotides to identify and bind to its counterpart strand in a sample, because the interaction occurs when the two polynucleotide segments create stable hydrogen interactions with one another (Kavita 2017). The aforementioned biosensor materials are combined with a physical transducer to create thermal or calorimetric biosensors (Ramanathan and Danielsson 2001). Piezoelectric biosensors come in a couple of varieties: quartz crystal microbalances and surface acoustic wave devices. The quantification of variations in a piezoelectric crystal's frequency of resonance, which result from mass shifts inside the crystal framework, is their foundation (Skládal 2016). The successful application of biosensors in illness detection, prevention, rehabilitation, patient health monitoring, and human health management has significantly advanced various aspects of healthcare. Additionally, infections, viruses, and bacteria can be found using biosensors. Without naturally extracting blood from the human body, these sensors identify substances (Naresh and Lee 2021). The biosensor, as previously said, is composed of various segments or parts, such as the transducer, bioreceptor, analyte, electronics, and reader display. Regarding the outcomes that were showing, the biosensor's display reading device is often coupled with related communication or electronic gadgets. Agricultural studies, forensic testing, biomedical testing, health monitoring, trash tracking and monitoring, and diagnostic and mental patient management are among the situations where biosensors are used (Altug et al. 2022). In the field of medicine, glucose biosensors have typically been used to treat diabetes. Biosensors not only provide medical care and therapy but also enable quicker illness diagnosis and patient status monitoring (Carpenter et al. 2018).

4.5.1 Applications of Biosensors

Compared to conventional methods, biosensors offer more stability and sensitivity, which is why they are employed across numerous industries, including the culinary, pharmaceutical, and marine sectors. The manufacturing industry, where medical, healthcare, and clinical services are the main industries, is where biosensors have found their greatest uses (Fig. 4.3).

It seems beneficial that biosensors have been designed in response to the need for straightforward, instantaneous, cost-effective, and selective processes (Scognamiglio et al. 2014). Using cobalt phthalocyanine-based enzymatic biosensors, Ghasemi et al. investigated the monitoring of beer ageing (Ghasemi-Varnamkhasti et al.

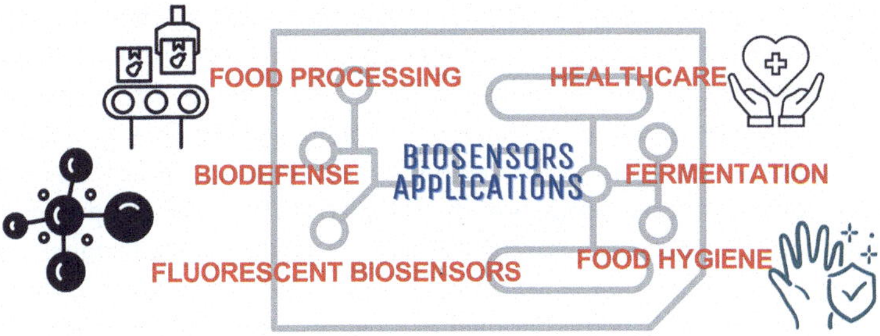

Fig. 4.3 Applications of biosensors in various fields

2012). Multichannel biosensors, which evaluate the electrophysiological functioning of the gustatory epithelial cells, are a more effective way to combine lipid coatings with electrochemical approaches as biosensors for rapid and specific sweetener assessment (Luong et al. 1997). When glucose was administered, the taste epithelium biosensor produced signals that were faint but had favourable waveforms, whereas sucrose sustained signals with negative spikes (Terry et al. 2005).

By keeping an eye on the existence of the process's byproducts, biosensors monitor the presence of antibodies, enzymes, biomass, and products. Biosensors are highly selective, inexpensive, easily automated, and have straightforward instrumentation, allowing them to precisely regulate the fermentation industry and yield repeatable results (Mao et al. 2008). Many important variables are essential in the complex procedure of fermentation, most of which are challenging to monitor in real time. The bio-enzymatic method of producing glucose is combined with the use of glucose biosensors, which are successfully managing the fermentation and saccharification seminars in factories (Yoo and Lee 2010).

Food quality includes look, aroma, chemical makeup, crispness, flavour, composition, etc. (Scognamiglio et al. 2014). Accurate nutrition monitoring and quick screening for chemical and biological contaminants are essential for ensuring the nutritiousness as well as quality of foodstuff. Biosensors are utilized in the detection of both general and specific toxic metals due to their ability to react with only the hazardous portions of metal ions (Vigneshvar et al. 2016). Using biological sensors in healthcare is growing rapidly. In clinical settings, type 2 diabetes is frequently diagnosed with glucose biosensors, a condition that requires precise blood glucose control (Chen et al. 2013). In order to diagnose infectious diseases, biosensors are frequently used in medicine. Research is underway to develop a possible biosensor technology for pathogen detection, antibiotic susceptibility testing, and urinary tract infection (UTI) diagnosis (Haleem et al. 2021). Additional biosensor applications include the following: the impact of oxazaborolidines on immobilized fructosyltransferase in dental diseases; the microfluidic impedance assay for regulating endothelin-induced cardiac hypertrophy; the quantitative assessment of cardiac markers in undiluted serum; the assay for histone deacetylase (HDAC) inhibitors from resonance energy transfer; and the biochip for quick and accurate detection (Alhadrami 2018). Biosensors are used in medical research to identify diseases as soon as it is practical to do so, thereby extending the scope of patient care. In the study of cancer, electrochemical biosensors and biomarkers are used to facilitate more rapid and accurate detection (Haleem et al. 2021). It has the ability to recognize harmful diseases and find bio-recognition elements like enzymes, antibodies, or biomolecules. Many biomarkers, including cardiac troponin I (cTnI), interleukins, myoglobin, natriuretic peptides of the B type (BNP), C-reactive protein (CRP), and interferons, have been recognized for use as a result of advancements in non-technology and medicine (Jin et al. 2020). Despite having a predictive value separate from the more established traditional risk factors, only a small number of these biomarkers have become essential diagnostic tools in the medical industry (Nabaei et al. 2018).

Research on cancer and the development of new therapeutics both use fluorescent biosensors as diagnostic tools. One or more fluorescent probes (produced chemically, enzymatically, or biologically) can be deposited onto tiny scaffolds known as fluorescent biosensors via a binding site (VanEngelenburg and Palmer 2008). When working with complex solutions, luminous biosensors can identify ions, substances, and biological markers of proteins and show the target's presence, movement, or condition (serum, cell secretions) (Ma et al. 2016). In pharmaceutical research programmes, luminous biosensors are utilized for lead optimization, post-screening result analysis, and high-throughput, extensive testing techniques to identify medications (Giuliano and Taylor 1998). In intravital imaging, image-guided surgery, assessing disease development and outcome of treatment, and early biomarker identification in scientific diagnostic procedures, luminescent biological sensors are effectively employed (Morris 2013).

Biosensors can be used for military objectives in the case of a biological assault. Such biosensors are primarily designed to identify, in almost real time, organisms known as biowarfare agents (BWAs), which include viruses, toxins, and bacteria (both vegetative and spore-forming) (Metzgar et al. 2013).

4.6 Drug Toxicity Analysis

When contemplating a drug's usage in illness therapy, two characteristics of its properties to evaluate are efficacy and safety. When using medications, it is vital to consider the undesirable reactions (such as toxicity and side reactions). Evaluations of drug biochemical patterns of distribution, physical, chemical, and pharmacological aspects, such as genetic damage, target toxicity, and organ harmful effects, are frequently included in toxicity assessments.

Genetically engineered or modified genetic mice producing labelled biomolecules and biosensors with optical properties exhibit promise in preclinical investigations for disease aetiology, pathogenesis, and quick large-scale drug toxicity screening. It is critical in drug toxicity studies to have accessible transgenic mice models for imaging organelles, cells, and even tissues throughout time. Over the past decade, there has been significant evolution in genetically modified mice specimens and lines, such as human liver chimeric animals and those bearing the human leukocyte antigen (HLA), illustrating notable advancements in this field (Bissig et al. 2018; Susukida et al. 2020). Conventional animal models with lengthy lifespans, such as rats and mice, are costly. As a result, alternatives to mammals for toxicological research are required. Hepatotoxicity has been seen in the seed of *Psoralea corylifolia*, also known as Fructus Psoralea, and its possible hepatotoxic pathways were studied in the zebrafish model. In another research using zebrafish embryos, the cardiovascular damage caused by four common macrolides—azithromycin, clarithromycin, tilmicosin, and tylosin—was investigated. A classic nootropic medication called *Bacopa monnieri* (Brahmi) was also investigated for its mitochondrial damage in zebrafish (Zhang et al. 2021; Yan et al. 2019; Author 2018). Drug toxicity alters or damages cell morphology, which is phenotypically classified according to anatomical criteria. Dyeing of tissues was utilized in traditional toxicological assessments for pathological screening; however it cannot be used to examine organelle destruction or uncover the processes of endogenous molecular control produced by toxic medicines (Karlsson and Hanrieder 2017). Because it can identify and locate antigens in a variety of tissues and cells, immunofluorescence has become an essential tool employed for assessment in numerous disorders that are infectious (Glancy 1897). According to their ability to absorb light or fluoresce, cells in a cell solution are counted, analysed, and sorted using flow cytometry. It makes it possible to assess the biological, chemical, and physical characteristics of individual cells in suspension, including their size, shape, and internal complexity (Nolan et al. 1999). The study of toxicogenomic, a branch of toxicology, uses genetic sequencing methods to look the consequences of chemicals regarding people's health and the environment, including both biological and therapeutic substances. It includes identifying pathogen-associated organisms and dangerous biomarkers. Toxicogenomics, with its new genomic and high-throughput techniques, offers unparalleled potential for enhancing regulatory decision-making and assessing risks (Liu et al. 2019). The delicate leaves of *Ampelopsis grossedentata* are used to make grapevine tea, which is mainly made up of flavonoids and polyphenolic chemicals that have varying degrees of hepatotoxicity. Gene expression characterization of vine tea indicated 34 potentially harmful components and 57 potentially hepatotoxic targets (Pan et al. 2020). Meanwhile, proteome profiling assesses the concentration of protein and evaluates gene regulation, translation, and posttranslational stability of protein (Rubinstein 2006).

4.7 Future Perspective

Advances in genomes, proteomics, and metabolomics will provide more extensive insights into the molecular makeup of medicinal products in the future. This will improve our ability to detect and quantify contaminants as well as assess product quality. Biotechnology is going to keep playing a significant part towards the evolution of personalized medicine. To determine patient-specific doses and treatment regimens, tailored therapies necessitate precise analytical methods. Future studies will concentrate on developing environmentally friendly biotechnological approaches for pharmaceutical analysis that reduce waste and conserve resources.

4.8 Conclusion

The use of biotechnology into pharmaceutical product analysis has transformed the profession by giving revolutionary tools and methodologies for ensuring pharmaceutical safety, efficacy, and quality. This chapter has emphasized some of the most important biotechnology applications in this setting, such as the utilization of bioprocesses, molecular biology, and genomics technologies. These applications have not only enhanced the accuracy and efficiency of pharmaceutical analysis, but they have also helped to discover novel drug candidates and personalized medicine. The pharmaceutical sector can now manage complicated challenges such as identifying contaminants, characterizing biomolecules, and assuring batch-to-batch uniformity thanks to advances in biotechnology. Furthermore, through expediting target selection, high-throughput screening, and lead optimization, it has been essential in accelerating the development of new drugs.

References

Akyilmaz E, Yorganci E, Asav E (2010) Do copper ions activate tyrosinase enzyme? A biosensor model for the solution. Bioelectrochemistry 78:155–160

Alhadrami HA (2018) Biosensors: classifications, medical applications, and future prospective. Biotechnol Appl Biochem 65:497–508

Altug H, Oh SH, Maier SA, Homola J (2022) Advances and applications of nanophotonic biosensors. Nat Nanotechnol 17:5–16

Alwine JC, Kemp DJ, Stark GR (1977) Method for detection of specific RNAs in agarose gels by transfer to diazobenzyloxymethyl-paper and hybridization with DNA probes. Proc Natl Acad Sci USA 74:5350–5354

Annison H (2011) Book review: book review. Criminol Crim Justice 11:277–278

Anon (n.d.-a) analytical 2.pdf

Anon (n.d.-b) med chem.pdf

Anwar M, Rather RA, Farooq Z (2022) Fundamentals and advances in medical biotechnology. https://doi.org/10.1007/978-3-030-98554-7

Åsberg D, Nilsson M, Olsson S et al (2016) A quality control method enhancement concept—continual improvement of regulatory approved QC methods. J Pharm Biomed Anal 129:273–281

Aspromonte J, Wolfs K, Adams E (2019) Current application and potential use of GC × GC in the pharmaceutical and biomedical field. J Pharm Biomed Anal 176:112817

Author T (2018) Mitochondrial toxicity submitted for consideration for the anniversary of toxicological sciences invited by Dr. Kendall Wallace Joel N. Meyer. https://doi.org/10.1093/toxsci/kfy008/4798828

Basavaiah K, Prameela HC (2003) Two simple methods for the estimation of albendazole and its dosage forms using chloramine-T. Farmaco 58:527–534

Bass JJ, Wilkinson DJ, Rankin D, Phillips BE, Szewczyk NJ, Smith K, Atherton PJ (2017) An overview of technical considerations for Western blotting applications to physiological research. Scand J Med Sci Sport 27:4–25
Beccaria M, Cabooter D (2020) Current developments in LC-MS for pharmaceutical analysis. Analyst 145:1129–1157
Beddell CR, Goodford PJ, Norrington FE, Wilkinson S, Wootton R (1976) Compounds designed to fit a site of known structure in human haemoglobin. Br J Pharmacol 57:201–209
Biechele P, Busse C, Solle D, Scheper T, Reardon K (2015) Sensor systems for bioprocess monitoring. Eng Life Sci 15:469–488
Bissig KD, Han W, Barzi M, Kovalchuk N, Ding L, Fan X, Pankowicz FP, Zhang QY, Ding X (2018) P450-humanized and human liver chimeric mouse models for studying xenobiotic metabolism and toxicity. Drug Metab Dispos 46:1734–1744
Bluma A, Höpfner T, Lindner P, Rehbock C, Beutel S, Riechers D, Hitzmann B, Scheper T (2010) In-situ imaging sensors for bioprocess monitoring: state of the art. Anal Bioanal Chem 398:2429–2438
Boulnois GJ (2000) Drug discovery in the new millennium: the pivotal role of biotechnology. Trends Biotechnol 18:31–33
Butler JE (2000) Enzyme-linked immunosorbent assay. J Immunoass 21:165–209
Carpenter AC, Paulsen IT, Williams TC (2018) Blueprints for biosensors: design, limitations, and applications. Genes (Basel) 9:375. https://doi.org/10.3390/genes9080375
Chankvetadze B, Kartozia I, Yamamoto C, Okamoto Y, Blaschke G (2003) Comparative study on the application of capillary liquid chromatography and capillary electrochromatography for investigation of enantiomeric purity of the contraceptive drug levonorgestrel. J Pharm Biomed Anal 30:1897–1906
Chen C, Xie Q, Yang D, Xiao H, Fu Y, Tan Y, Yao S (2013) Recent advances in electrochemical glucose biosensors: a review. RSC Adv 3:4473–4491
Chen S, Li Z, Zhang S, Zhou Y, Xiao X, Cui P, Xu B, Zhao Q, Kong S, Dai Y (2022) Emerging biotechnology applications in natural product and synthetic pharmaceutical analyses. Acta Pharm Sin B 12:4075–4097
Chowdhury EA, Meno-Tetang G, Chang HY, Wu S, Huang HW, Jamier T, Chandran J, Shah DK (2021) Current progress and limitations of AAV mediated delivery of protein therapeutic genes and the importance of developing quantitative pharmacokinetic/pharmacodynamic (PK/PD) models. Adv Drug Deliv Rev 170:214–237
Cobo-Golpe M, Ramil M, Cela R, Rodríguez I (2022) Supercritical fluid chromatography time-of-flight mass spectrometry enantiomeric determination of basic drugs in sewage samples. J Chromatogr A 1673:463088
Cooks RG, Ouyang Z, Takats Z, Wiseman JM (2006) Ambient mass spectrometry. Science 311:1566–1570
Cramer P (2021) AlphaFold2 and the future of structural biology. Nat Struct Mol Biol 28:704–705
Crommelin DJA, Mastrobattista E, Hawe A, Hoogendoorn KH, Jiskoot W (2020) Shifting paradigms revisited: biotechnology and the pharmaceutical sciences. J Pharm Sci 109:30–43
Cunningham MJ (2000) Genomics and proteomics: the new millennium of drug discovery and development. J Pharmacol Toxicol Methods 44:291–300
D'Orazio G, Fanali C, Fanali S, Gentili A, Chankvetadze B (2019) Comparative study on enantiomer resolving ability of amylose tris(3-chloro-5-methylphenylcarbamate) covalently immobilized onto silica in nano-liquid chromatography and capillary electrochromatography. J Chromatogr A 1606:460425
Deidda R, Orlandini S, Hubert P, Hubert C (2018) Risk-based approach for method development in pharmaceutical quality control context: a critical review. J Pharm Biomed Anal 161:110–121
Dispas A, Avohou HT, Lebrun P, Hubert P, Hubert C (2018) 'Quality by design' approach for the analysis of impurities in pharmaceutical drug products and drug substances. Trends Anal Chem 101:24–33

Dissanayake NM, Arachchilage JS, Samuels TA, Obare SO (2019) Highly sensitive plasmonic metal nanoparticle-based sensors for the detection of organophosphorus pesticides. Talanta 200:218–227
Domon B, Aebersold R (2006) Mass spectrometry and protein analysis. Science 312:212–217
Erickson J, Neidhart DJ, VanDrie J et al (1990) Design, activity, and 2.8 Å crystal structure of a C2 symmetric inhibitor complexed to HIV-1 protease. Science 249:527–533
Erlich HA (1989) Polymerase chain reaction. J Clin Immunol 9:437–447
Erlich A, Gelfand D, Sninsky JJ (1985) Recent
Fanali C, Della PS, Fanali S (2021) Capillary electrochromatography applied to drug analysis. J Chromatogr Open 1:100015
Fanali C, Della Posta S, Gentili A, Chankvetadze B, Fanali S (2023) Recent developments in electromigration techniques related to pharmaceutical and biomedical analysis—a review. J Pharm Biomed Anal 235:115647
Felmlee MA, Morris ME, Mager DE (2012) Mechanism-based pharmacodynamic modeling. Methods Mol Biol 929:583–600
Ganorkar SB, Shirkhedkar AA (2017) Design of experiments in liquid chromatography (HPLC) analysis of pharmaceuticals: analytics, applications, implications and future prospects. Rev Anal Chem 36(3):20160025. https://doi.org/10.1515/revac-2016-0025
Ghasemi-Varnamkhasti M, Rodríguez-Méndez ML, Mohtasebi SS, Apetrei C, Lozano J, Ahmadi H, Razavi SH, Antonio de Saja J (2012) Monitoring the aging of beers using a bioelectronic tongue. Food Control 25:216–224
Giuliano KA, Taylor DL (1998) Fluorescent-protein biosensors: new tools for drug discovery. Trends Biotechnol 16:135–140
Gizeli E, Lowe CR (n.d.) Immunosensors
Glancy M (1897) Chapter 26. Cary grant, mak a hollywood legend, vol 2021, pp 358–375
Glish GL, Vachet RW (2003) The basics of mass spectrometry in the twenty-first century. Nat Rev Drug Discov 2:140–150
Grattieri M, Minteer SD (2018) Self-powered biosensors. ACS Sens 3:44–53
Grimm J, Perez JM, Josephson L, Weissleder R (2004) Novel nanosensors for rapid analysis of telomerase activity. Cancer Res 64:639–643
Gros Q, Réset L, Duval J, Horie S, Toyota Y, Funada Y, Hayakawa Y, Lesellier E, West C (2023) Defining a generic column set for achiral supercritical fluid chromatography applied to pharmaceuticals or natural products. J Chromatogr A 1687:463667
Haleem A, Javaid M, Singh RP, Suman R, Rab S (2021) Biosensors applications in medical field: a brief review. Sens Int 2:100100
Harms P, Kostov Y, Rao G (2002) Bioprocess monitoring. Curr Opin Biotechnol 13:124–127
He M, Herr AE (2010) Polyacrylamide gel photopatterning enables automated protein Immunoblotting in a two-dimensional microdevice. J Am Chem Soc 132:2512–2513
Hornbeck P (1992) Enzyme-linked immunosorbent assays. Curr Protoc Immunol 1:2.1.1–2.1.22
Houghton PJ (2000) Use of small scale bioassays in the discovery of novel drugs from natural sources. Phyther Res 14:419–423
Howbrook DN, Van Der Valk AM, O'Shaughnessy MC, Sarker DK, Baker SC, Lloyd AW (2003) Developments in microarray technologies. Drug Discov Today 8:642–651
Indrayanto G, Putra GS, Suhud F (2021) Validation of in-vitro bioassay methods: application in herbal drug research. In: Profiles of drug substances, excipients and related methodology, 1st edn. https://doi.org/10.1016/bs.podrm.2020.07.005
Jain KK (2009) The role of nanobiotechnology in drug discovery. Adv Exp Med Biol 655:37–43
Jin X, Liu C, Xu T, Su L, Zhang X (2020) Biosensors and bioelectronics artificial intelligence biosensors: challenges and prospects. Biosens Bioelectron 165:112412
Karlsson O, Hanrieder J (2017) Imaging mass spectrometry in drug development and toxicology. Arch Toxicol 91:2283–2294
Katsurada A, Hagiwara Y, Miyashita K, Satou R, Miyata K, Ohashi N, Navar LG, Kobori H (2007) Novel sandwich ELISA for human angiotensinogen. Am J Physiol Renal Physiol 293:956–961

Kavita VJ (2017) DNA biosensors-a review. J Bioeng Biomed Sci 07(2):222. https://doi.org/10.4172/2155-9538.1000222
Kiriiri GK, Njogu PM, Mwangi AN (2020) Exploring different approaches to improve the success of drug discovery and development projects: a review. Futur J Pharm Sci 6:1. https://doi.org/10.1186/s43094-020-00047-9
Klatte S, Schaefer HC, Hempel M (2017) Pharmaceuticals in the environment—a short review on options to minimize the exposure of humans, animals and ecosystems. Sustain Chem Pharm 5:61–66
Kommanaboyina B, Rhodes CT (1999) Trends in stability testing, with emphasis on stability during distribution and storage. Drug Dev Ind Pharm 25:857–868
Koscielny G, An P, Carvalho-Silva D et al (2017) Open targets: a platform for therapeutic target identification and validation. Nucleic Acids Res 45:D985–D994
Krait S, Konjaria ML, Scriba GKE (2021) Advances of capillary electrophoresis enantioseparations in pharmaceutical analysis (2017–2020). Electrophoresis 42:1709–1725
Kubista M, Andrade JM, Bengtsson M et al (2006) The real-time polymerase chain reaction. Mol Asp Med 27:95–125
Kupiec T (2004) Quality-control analytical methods: high-performance liquid chromatography. Int J Pharm Compd 8:223–227
Kurien BT, Scofield RH (2006) Western blotting. Methods 38:283–293
Lander ES (1999) Array of hope. Nat Genet 21:4
Libeau G, Prehaud C, Lancelot R, Colas F, Guerre L, Bishop DHL, Diallo A (1995) Development of a competitive ELISA for detecting antibodies to the peste des petits ruminants virus using a recombinant nucleoprotein. Res Vet Sci 58:50–55
Lin Z, Tai HC, Petruzzella E, Tang F, Ye YK, Chadwick J, Ding W, He BL, Miller SA (2021) A generic liquid chromatography-mass spectrometry method for monitoring bis(pinacolato)diboron mutagenic impurity in pharmaceutical compounds. J Chromatogr Open 1:100009
Liu C, Guo D-A, Liu L (2018) Quality transitivity and traceability system of herbal medicine products based on quality markers. Phytomedicine 44:247–257
Liu Z, Huang R, Roberts R, Tong W (2019) Toxicogenomics: a 2020 vision. Trends Pharmacol Sci 40:92–103
Lo Faro AF, Berardinelli D, Sprega G, Tini A, Carlier J, Farkas T, Busardò FP, Chankvetadze B (2023) Development of an enantioselective high-performance liquid chromatography-tandem mass spectrometry method for the quantitative determination of methorphan and its O-demethylated metabolite in human blood and its application to post-mortem samples. J Pharm Biomed Anal 230:115384. https://doi.org/10.1016/j.jpba.2023.115384
Lokko Y, Heijde M, Schebesta K, Scholtès P, Van Montagu M, Giacca M (2018) Biotechnology and the bioeconomy—towards inclusive and sustainable industrial development. New Biotechnol 40:5–10
Luong JHT, Bouvrette P, Male KB (1997) Developments and applications of biosensors in food analysis. Trends Biotechnol 15:369–377
Ma F, Li Y, Tang B, Zhang CY (2016) Fluorescent biosensors based on single-molecule counting. Acc Chem Res 49:1722–1730
Malone J, Holloway E, Adamusiak T, Kapushesky M, Zheng J, Kolesnikov N, Zhukova A, Brazma A, Parkinson H (2010) Modeling sample variables with an experimental factor ontology. Bioinformatics 26:1112–1118
Mann M, Hendrickson RC, Pandey A (2001) Analysis of proteins and proteomes by mass spectrometry. Ann Rev Biochem 70:437–473
Mao XL, Wu J, Bin YY (2008) Application of electrochemical biosensors in fermentation. Fenxi Huaxue/Chin J Anal Chem 36:1749–1755
Matei N, Birghila S, Popescu V, Dobrinas S, Soceanu A, Oprea C, Magearu V (2008) Kinetic study of vitamin C degradation from pharmaceutical products. Rom Rep Phys 53:343–351
Mathis G (1995) Probing molecular interactions with homogeneous techniques based on rare earth cryptates and fluorescence energy transfer. Clin Chem 41:1391–1397

Mattoli L, Gianni M, Burico M (2023) Mass spectrometry-based metabolomic analysis as a tool for quality control of natural complex products. Mass Spectrom Rev 42:1358–1396
Mauriz E (2020) Recent progress in plasmonic biosensing schemes for virus detection. Sensors (Switzerland) 20:1–27
Mehrotra P (2016) Biosensors and their applications—a review. J Oral Biol Craniofac Res 6:153–159
Meibohm B, Derendorf H (2002) Pharmacokinetic/pharmacodynamic studies in drug product development. J Pharm Sci 91:18–31
Metzgar D, Sampath R, Rounds MA, Ecker DJ (2013) The value and validation of broad spectrum biosensors for diagnosis and biodefense. Virulence 4:752–758
Mfuh KO, Achonduh-Atijegbe OA, Bekindaka ON, Esemu LF, Mbakop CD, Gandhi K, Leke RGF, Taylor DW, Nerurkar VR (2019) A comparison of thick-film microscopy, rapid diagnostic test, and polymerase chain reaction for accurate diagnosis of plasmodium falciparum malaria. Malar J 18:1–8
Milroy CS, Stevenson PG, Mnatasakyan M, Shalliker RA (2011) Multidimensional high-performance liquid chromatography. Hyphenated Altern Methods Detect Chromatogr 18:251–285
Mishra M, Tiwari S, Gomes AV (2017) Protein purification and analysis: next generation western blotting techniques. Expert Rev Proteomics 14:1037–1053
Morris MC (2013) Fluorescent biosensors—probing protein kinase function in cancer and drug discovery. Biochim Biophys Acta 1834:1387–1395
Muneer S, Sarfo DK, Ayoko GA, Islam N, Izake EL (2020) Gold-deposited nickel foam as recyclable plasmonic sensor for therapeutic drug monitoring in blood by surface-enhanced Raman spectroscopy. Nano 10:1–12
Munteanu IG, Apetrei C (2021) Analytical methods used in determining antioxidant activity: a review. Int J Mol Sci 22:3380. https://doi.org/10.3390/ijms22073380
Murphy RM, Lamb GD (2013) Important considerations for protein analyses using antibody based techniques: down-sizing Western blotting up-sizes outcomes. J Physiol 591:5823–5831
Nabaei V, Chandrawati R, Heidari H (2018) Magnetic biosensors: modelling and simulation. Biosens Bioelectron 103:69–86
Narayanan H, Luna MF, von Stosch M, Cruz Bournazou MN, Polotti G, Morbidelli M, Butté A, Sokolov M (2020) Bioprocessing in the digital age: the role of process models. Biotechnol J 15:e1900172. https://doi.org/10.1002/biot.201900172
Naresh V, Lee N (2021) A review on biosensors and recent development of nanostructured materials-enabled biosensors. Sensors (Switzerland) 21:1–35
Nolan JP, Lauer S, Prossnitz ER, Sklar LA (1999) Flow cytometry: a versatile tool for all phases of drug discovery. Drug Discov Today 4:173–180
Pan X, Zhou J, Chen Y, Xie X, Rao C, Liang J, Zhang Y (2020) Classification, hepatotoxic mechanisms, and targets of the risk ingredients in traditional Chinese medicine-induced liver injury. Toxicol Lett 323:48–56
Pandey S, Pandey P, Tiwari G, Tiwari R (2010) Bioanalysis in drug discovery and development. Pharm Methods 1:14–24
Patel M, Kumar R, Kishor K, Mlsna T, Pittman CU, Mohan D (2019) Pharmaceuticals of emerging concern in aquatic systems: chemistry, occurrence, effects, and removal methods. Chem Rev 119:3510–3673
Rahman N, Anwar N, Kashif M (2005) Application of π-acceptors to the spectrophotometric determination of lisinopril in commercial dosage forms. Farmaco 60:605–611
Ramanathan K, Danielsson B (2001) Principles and applications of thermal biosensors. Biosens Bioelectron 16:417–423
Rao T, Tan Z, Peng J, Guo Y, Chen Y, Zhou H, Ouyang D (2019) The pharmacogenetics of natural products: a pharmacokinetic and pharmacodynamic perspective. Pharmacol Res 146:104283
Rehman MSU, Rashid N, Ashfaq M, Saif A, Ahmad N, Han JI (2015) Global risk of pharmaceutical contamination from highly populated developing countries. Chemosphere 138:1045–1055

Rogers RS, Abernathy M, Richardson DD, Rouse JC, Sperry JB, Swann P, Wypych J, Yu C, Zang L, Deshpande R (2018) A view on the importance of "multi-attribute method" for measuring purity of biopharmaceuticals and improving overall control strategy. AAPS J 20:7. https://doi.org/10.1208/s12248-017-0168-3

Rosso A (2010) Statistical analysis of experimental designs applied to biological assays, pp 1–42

Rozet E, Lebrun P, Hubert P, Debrus B, Boulanger B (2013) Design spaces for analytical methods. Trends Anal Chem 42:157–167

Rubinstein AL (2006) Zebrafish assays for drug toxicity screening. Expert Opin Drug Metab Toxicol 2:231–240

Sarkar T, Bharadwaj KK, Salauddin M, Pati S, Chakraborty R (2022) Phytochemical characterization, antioxidant, anti-inflammatory, anti-diabetic properties, molecular docking, pharmacokinetic profiling, and network pharmacology analysis of the major phytoconstituents of raw and differently dried Mangifera indica (Himsagar cultivar). Appl Biochem Biotechnol 194:950–987

Scapin G (2006) Structural biology and drug discovery. Curr Pharm Des 12:2087–2097

Schochetman G, Ou C-Y, Wanda K (1988) Polymerase chain reaction. J Infect Dis 158:1154–1157; https://www.jstor.org/stable/30137034

Schots A, Van der Leede BJ, De Jongh E, Egberts E (1988) A method for the determination of antibody affinity using a direct ELISA. J Immunol Methods 109:225–233

Scognamiglio V, Arduini F, Palleschi G, Rea G (2014) Trends in analytical chemistry biosensing technology for sustainable food safety. Trends Anal Chem 62:1–10

Selvin PR (2002) Principles and biophysical applications of lanthanide-based probes. Annu Rev Biophys Biomol Struct 31:275–302

Sharma SK, Sehgal N, Kumar A (2003) Biomolecules for development of biosensors and their applications. Curr Appl Phys 3:307–316

Shen M, Zhou Y, Ye J, AA AAL-M, Kang Y, Zeng S, Cai S (2020) Recent advances and perspectives of nucleic acid detection for coronavirus. J Pharm Anal 10:97–101

Siddiqui MR (2017) Analytical techniques in pharmaceutical analysis: a review. Arab J Chem 10:S1409–S1421

Singh S, Bakshi M (2000) Stress test to determine inherent stability of drugs. Pharm Technol 4:1–14

Sittampalam GS, Smith WC, Miyakawa TW, Smith DR, McMorris C (1996) Application of experimental design techniques to optimize a competitive ELISA. J Immunol Methods 190:151–161

Skládal P (2016) Piezoelectric biosensors. Trends Anal Chem 79:127–133

Southern EM (2004) Detection of specific sequences among DNA fragments separated by gel electrophoresis. J Mol Biol 98:503–517

Starr CG, Tessier PM (2019) Selecting and engineering monoclonal antibodies with drug-like specificity. Curr Opin Biotechnol 60:119–127

Susukida T, Aoki S, Shirayanagi T, Yamada Y, Kuwahara S, Ito K (2020) HLA transgenic mice: application in reproducing idiosyncratic drug toxicity. Drug Metab Rev 52:540–567

Terry LA, White SF, Tigwell LJ (2005) The application of biosensors to fresh produce and the wider food industry. J Agric Food Chem 53:1309–1316

Tian Q, Stepaniants SB, Mao M et al (2004) Integrated genomic and proteomic analyses of gene expression in mammalian cells. Mol Cell Proteomics 3:960–969

Tseng SY, Li SY, Yi SY, Sun AY, Gao DY, Wan D (2017) Food quality monitor: paper-based plasmonic sensors prepared through reversal nanoimprinting for rapid detection of biogenic amine odorants. ACS Appl Mater Interfaces 9:17306–17316

Üëá Ü, Ìè Ì, Ìè È, Öêå Ü (2011) No Title 11:165–178

Vallejo-Illarramendi A, Marciano DK, Reichardt LF (2013) A novel method that improves sensitivity of protein detection in PAGE and Western blot. Electrophoresis 34:1148–1150

van den Anker J, Reed MD, Allegaert K, Kearns GL (2018) Developmental changes in pharmacokinetics and pharmacodynamics. J Clin Pharmacol 58:S10–S25

Van Duin M, Woolson H, Mallinson D, Black D (2003) Genomics in target and drug discovery. Biochem Soc Trans 31:429–432

VanEngelenburg SB, Palmer AE (2008) Fluorescent biosensors of protein function. Curr Opin Chem Biol 12:60–65
Venuti MC (1989) Chapter 31. The impact of biotechnology
Vernell R, Helin K, Müller H (2003) Identification of target genes of the p16INK4A-pRB-E2F pathway. J Biol Chem 278:46124–46137
Vigneshvar S, Sudhakumari CC, Senthilkumaran B, Prakash H (2016) Recent advances in biosensor technology for potential applications—an overview. Front Bioeng Biotechnol 4:1–9
Wang J (2008) Electrochemical glucose biosensors. Chem Rev 108(2):814
Wang S, Sim TB, Kim YS, Chang YT (2004) Tools for target identification and validation. Curr Opin Chem Biol 8:371–377
Wen Y, Chen L, Li J, Liu D, Chen L (2014) Recent advances in solid-phase sorbents for sample preparation prior to chromatographic analysis. Trends Anal Chem 59:26–41
Woźniakiewicz A, Wietecha-Posłuszny R (2023) Determination of carbamazepine and its main metabolite in human hair by capillary electrophoresis and liquid chromatography techniques, both coupled with mass spectrometry. J Pharm Biomed Anal Open 1:100009
Wulfkuhle J, Espina V, Liotta L, Petricoin E (2004) Genomic and proteomic technologies for individualisation and improvement of cancer treatment. Eur J Cancer 40:2623–2632
Yabré M, Ferey L, Somé IT, Gaudin K (2018) Greening reversed-phase liquid chromatography methods using alternative solvents for pharmaceutical analysis. Molecules 23:1065. https://doi.org/10.3390/molecules23051065
Yan Z, Huang X, Xie Y, Song M, Zhu K, Ding S (2019) Macrolides induce severe cardiotoxicity and developmental toxicity in zebrafish embryos. Sci Total Environ 649:1414–1421
Yesudasu V, Pradhan HS, Pandya RJ (2021) Recent progress in surface plasmon resonance based sensors: a comprehensive review. Heliyon 7:e06321
Yoo EH, Lee SY (2010) Glucose biosensors: an overview of use in clinical practice. Sensors 10:4558–4576
Zhang S, Yang Q, Wang C, Luo X, Kim J, Wang Z, Yamauchi Y (2018) Porous organic frameworks: advanced materials in analytical chemistry. Adv Sci (Weinh) 5:1801116. https://doi.org/10.1002/advs.201801116
Zhang C, Qian DD, Yu T, Yang H, Li P, Li HJ (2021) Multi-parametric cellular imaging coupled with multi-component quantitative profiling for screening of hepatotoxic equivalent markers from Psoraleae fructus. Phytomedicine 93:153518
Zhang X, Wu X, He Q, Wang J, Mao Q, Liang Z, Xu M (2023) Research progress on substitution of in vivo method(s) by in vitro method(s) for human vaccine potency assays. Expert Rev Vaccines 22:270–277
Zou L, Guo X, McElderry JD (2023) Platform headspace gas chromatography method for high-throughput determination of residual solvents in pharmaceutical materials. J Pharm Biomed Anal 229:115349
Zubarev RA, Makarov A (2013) Orbitrap mass spectrometry. Anal Chem 85:5288–5296

Chapter 5
Development of Nanoparticles: Recent Developments and Future Prospects

Ravichandran Veerasamy

Abstract The biotechnological approach to nanoparticle development constitutes a remarkable convergence between biology, chemistry, and engineering, holding both recent successes and bright potential for the future. This method uses biological systems and processes to create, modify, and use nanoparticles for a variety of purposes. Recent advances in the production of biotechnological nanoparticles have taken advantage of organic processes in cells, plants, and even bacteria. These organisms are used in biosynthesis to produce nanoparticles with specific properties. For example, one can engineer bacteria and fungi to use their enzymatic activity to convert metal ions into nanoparticles. Compared to conventional chemical processes, this strategy is environmentally friendly, sustainable, and often produces more stable and biocompatible nanoparticles. Nanobiotechnology has made significant progress in the field of biomedical applications. Targeted drug delivery, imaging, and diagnostics are made possible by nanoparticles designed to interact with biological systems. To reduce side effects and increase treatment effectiveness, functionalized nanoparticles can deliver therapeutic drugs directly to the diseased area. In addition, nanoparticle image contrast facilitates early disease identification. In addition, scientists are working on nanoparticle-based biosensors that can identify specific chemicals involved in disease and transform point-of-care diagnostics. However, standardization, scalability, and ethical issues remain. There are engineering challenges to ensure reliable and reproducible nanoparticle fabrication using biological systems. As with any new technology, it is important to carefully consider the ethical implications of using living organisms to produce nanoparticles as well as any potential impacts on the environment. Interdisciplinary collaboration and thoughtful innovation will be key to maximizing the potential of biotech

R. Veerasamy (✉)
Faculty of Pharmacy, AIMST University, Bedong, Kedah, Malaysia
e-mail: ravichandran_v@aimst.edu.my

S. Bose et al. (eds.), *Concepts in Pharmaceutical Biotechnology and Drug Development*, Interdisciplinary Biotechnological Advances,
https://doi.org/10.1007/978-981-97-1148-2_5

nanoparticles as the field evolves. This chapter addresses potential future perspectives on this rapidly evolving topic while providing an in-depth review of current developments in biosynthesis, characterization, and application of nanoparticles.

Keywords Nanoparticles · Recent developments · Characterization · Biotechnology · Biomedical applications

5.1 Introduction

The size of nanoparticles is on the nanoscale, with a range of 1 to 100 nanometers (nm). Metals, metal oxides, semiconductors, polymers, and even biological components can be used to create them. Because of their small size, nanoparticles have different features and behaviors that distinguish them from bulk materials and microparticles. Nanoparticles have at least one dimension on the nanometer scale, which is approximately 1000 times smaller than the width of a human hair. Because of their small size and high surface area-to-volume ratio, they are extremely reactive and capable of unusual interactions with other materials. Quantum effects become relevant at the nanoscale. Because of their small size, nanoparticles can show quantum confinement, in which the behavior of electrons is limited. The optical, electrical, and magnetic properties could therefore change. The primary surface properties of nanoparticles stem from their huge surface area. Surface atoms or molecules can behave differently than bulk material atoms or molecules, influencing reactivity, adsorption, and catalytic characteristics. Nanoparticles can outperform their bulk counterparts in terms of mechanical qualities. Nanoparticles, for example, may have improved hardness and tensile strength (Altammar 2023).

Metal nanoparticles, which have optical and catalytic capabilities, include gold, silver, and platinum nanoparticles. They have numerous applications in electronics, catalysis, and biomedicine. Iron oxide and titanium dioxide nanoparticles are two examples of metal oxide nanoparticles that have applications in magnetic materials, catalysts, and sunscreen. Quantum dots are semiconductor nanoparticles with customizable electronic properties that are extensively utilized in displays, imaging, and solar cells. Polymeric Nanoparticles: Made of organic polymers, these nanoparticles are used in drug delivery systems, coatings, and nanocomposites (Alhalili 2023).

5.2 Synthesis of Nanoparticles

Chemical reduction, sol-gel procedures, physical vapor deposition, and biological approaches can all be used to create nanoparticles. These technologies enable exact control of size, shape, and content, allowing nanoparticles to be engineered for specific uses.

5.2.1 Chemical Reduction

Chemically reduced nanoparticles are a frequent and flexible category of nanoparticles. The employment of reducing chemicals to transform metal ions in a solution into metal nanoparticles is a method of chemical reduction. These techniques provide exact control of the nanoparticles' size, shape, and composition. The use of metal salts as precursors is usually the first step in chemical reduction processes. Common options include silver nitrate ($AgNO_3$), gold chloride ($HAuCl_4$), or other metal precursors, depending on the type of nanoparticle that is needed. There is a reduction agent applied to the metal precursor solution. The metal ions are reduced to their elemental state by the reducing agent through a chemical process. Common reducing agents include hydrazine, citrate ions, and sodium borohydride ($NaBH_4$). It is possible to modify the size, shape, and properties of the resulting nanoparticles by varying the reaction temperature, reaction duration, and agent concentration. It is possible to alter these factors to tailor nanoparticles for certain uses (Szczyglewska et al. 2023).

5.2.2 Sol-Gel Procedures

By using a chemical process to convert a precursor sol or solution into a solid gel, then drying and calcining the gel to create nanoparticles, sol-gel processes produce nanoparticles. With tight control over size, shape, and content, this method is widely utilized to generate a wide spectrum of nanoparticles and nanocomposites. The process begins with a precursor sol, which is a colloidal suspension comprising precursor compounds of metal or metal oxide. This sol is created by dissolving metal salts or alkoxides in a solvent. A controlled chemical process, commonly involving hydrolysis and condensation, results in the precursor sol forming a gel, which is a three-dimensional web of interconnected nanoparticles. This gel is made up of a continuous liquid phase and a three-dimensional network of solid nanoparticles. During the drying process, the gel's liquid component is removed. There are several methods for completing this process, such as freeze drying, supercritical fluid drying, and air drying. Then calcinated the dry gel at high temperature (typically above 400 °C), the gel is converted into a solid nanoparticle structure during. This entails removing organic components and crystallizing or transforming the nanoparticles into their final form (Bokov et al. 2021).

Sol-gel methods allow for fine control of nanoparticle size, shape, and content. This feature is very useful for customizing materials for specific uses. Typically, the process yields high-purity nanoparticles and nanocomposites. Sol-gel methods can produce a variety of materials, including oxide and non-oxide nanoparticles. In the case of silica nanoparticles, they are biocompatible and can be used in medical and biological applications (Sopan Mahato et al. 2023).

5.2.3 Physical Vapor Deposition

The process of producing physical vapor deposition (PVD) nanoparticles involves depositing vapor-phase material onto a substrate. PVD is a method that uses sputtering, evaporation, and chemical vapor deposition (CVD). PVD has various disadvantages in addition to its benefits. In PVD, the material to be deposited into nanoparticles is obtained from a material source. This source can be a heated crucible for evaporation, a liquid or solid precursor for CVD, or a solid target for sputtering. The material source is exposed to an energy source, which may be thermal heating, plasma, or laser irradiation, depending on the PVD process. The material transitions into the vapor phase as a result of this energy source. The material that has vaporized is directed towards a substrate, which may be a wafer, a solid surface, or another suitable medium. As the vapor condenses on the substrate, nanoparticles are created. The vaporized material may nucleate and form nanoparticles on the substrate. The PVD technology determines the specific mechanisms of growth, and variables like temperature, pressure, and deposition rate can be changed to control the process (Dobrzański et al. 2015; Smits et al. 2014; Vaseghi and Nematollahzadeh 2020).

5.2.4 Biological Approaches

In biological nanoparticle production, live organisms including bacteria, fungi, and plants as well as biological substances like enzymes and peptides are used to create nanoparticles. While there are many benefits to this sustainable and eco-friendly approach, there are some disadvantages as well. The precise method of biologically synthesizing nanoparticles can vary depending on the biological agent employed. Select an appropriate biological agent, such as bacteria (e.g., *E. coli*), fungi (e.g., *Aspergillus niger*), plant extracts, or biological compounds (e.g., enzymes) (Ghosh et al. 2021; Karunakaran et al. 2023; Pandit et al. 2022). If utilizing microorganisms, culture them in a suitable medium under controlled conditions. Alternatively, plant extracts or pure biological substances can be used. Introduce the desired metal ions (e.g., silver, gold) as precursor compounds into the culture medium or solution containing the biological agent (Duran and Seabra 2018; Hasan et al. 2018; Kumar et al. 2021). Through a bio-reduction process, the biological agent converts metal ions into metallic nanoparticles.

Larger nanoparticles are eventually produced by the nucleated nanoparticles' continued development and fusion. Reaction time and the concentration of precursor components are two examples of variables that can be used to influence the size and shape of the nanoparticles. After being produced, the nanoparticles are cleaned to get rid of extra biological material and other pollutants, and they are separated from the biological medium. The generated nanoparticles are characterized using a variety of techniques to ascertain their dimensions, composition, and properties (Nguyen et al. 2023).

5.2.5 *Polymeric Nanoparticle*

Synthesis of polymeric nanoparticles involves creating nanoscale particles from large molecules called polymers, which are composed of repeating subunits. These nanoparticles can be applied to coatings, nanocomposites, and medicine delivery, among other things. Depending on the polymer, the desired properties, and the intended usage, many processes can be employed to make polymeric nanoparticles. Choose a polymer that meets the requirements of the planned use. When creating nanoparticles, poly(lactic-co-glycolic acid) (PLGA), polyethylene glycol (PEG), chitosan, and other polymers are frequently used. Use a solvent that is compatible with the chosen polymer to dissolve the polymer. This results in a polymer solution, and additional chemicals or surfactants may be needed to regulate the nanoparticles' characteristics. Following methods like centrifugation, evaporation, or dialysis, the solvent is eliminated, leaving the polymer nanoparticles floating in a liquid. In certain situations, cross-linking agents or other chemical treatments may be utilized to enhance the stability or functionality of nanoparticles (Han et al. 2018; Zielińska et al. 2020).

5.3 Characterization of Nanoparticles

Nanotechnology involves a thorough analysis and understanding of the physical, chemical, and structural properties of particles that are typically between 1 and 100 nanometers in size. Because of their size, structure, and composition, nanoparticles exhibit unique behaviors that require characterization techniques to fully comprehend. High-resolution imaging and size determination are possible with electron microscopy (including transmission and scanning electron microscopy), while X-ray diffraction provides crystallographic information. Optical and chemical properties can be analyzed with spectroscopic techniques like UV-Vis, FTIR, and Raman spectroscopy. Moreover, surface charge and particle size distribution can be ascertained using zeta potential and dynamic light scattering investigations, while surface area and porosity are commonly assessed through BET (Brunauer-Emmett-Teller) examination. Magnetic, thermal, and electrical properties are examined using specialized techniques; surface characterization techniques such as atomic force microscopy and X-ray photoelectron spectroscopy help to analyze the surface topography and composition. The magnetic behavior and electron spin characteristics of nanoparticles are investigated using EPR spectroscopy. Designing nanoparticles for a range of uses, such as drug administration, catalysis, materials science, and electronics, requires an understanding of these characteristics. Moreover, nanoparticle characterization is a fundamental aspect of nanoscience research and development since it ensures the security and utility of nanomaterials across a wide range of fields. These characterization techniques are essential for assessing and optimizing the characteristics of nanoparticles to fit particular uses, such as energy conversion, environmental remediation, and medical and materials research. Combining these approaches allows researchers to fully comprehend the behavior of nanoparticles and maximize their capabilities (Jagadeesh et al. 2023; Ravichandran et al. 2022).

5.4 Challenges in the Synthesis of Nanoparticles (Han et al. 2019; Khandel and Shahi 2018; Patil and Chandrasekaran 2020; Ying et al. 2022)

The production of nanoparticles is a critical subject in nanoscience and nanotechnology, with significant potential benefits across a wide range of sectors. It does, however, present significant hurdles that researchers must solve in order to make nanoparticles with the appropriate qualities. Some of the major issues in the synthesis of nanoparticles are as follows: Attaining precise control over the dimensions and morphology of nanoparticles can pose challenges, as numerous synthesis methods yield a broad spectrum of sizes and forms. The uniformity of nanoparticles is crucial in specific applications, such as catalysis and drug administration. Many synthesis processes work well in the lab but are difficult to scale up for commercial output. For practical applications, developing cost-effective and scalable synthesis techniques is critical. It might be difficult to consistently reproduce the same nanoparticles with precise qualities, which can cause problems with quality control and product reliability. Variation in synthesis conditions and contaminants can have an effect on repeatability. Because of van der Waals forces or electrostatic interactions, nanoparticles tend to agglomerate, which might impact their properties and uses. It is difficult to stabilize nanoparticles and prevent agglomeration. Nanoparticle production frequently entails the use of poisonous or hazardous substances. It is critical to ensure the safety of both researchers and the environment. Furthermore, determining the possible toxicity of nanoparticles is a continuing challenge.

The ecological consequences of nanoparticle synthesis, encompassing the generation of waste products and the utilization of energy and resources, play a crucial role in the promotion of sustainable nanotechnology. Because of their small size and unique features, characterizing nanoparticles is difficult. Developing appropriate tools for precisely characterizing nanoparticles is a continuous challenge. The surface characteristics of nanoparticles influence their behavior and usefulness significantly. It might be difficult to create appropriate surface changes and coatings for specialized purposes. Some advanced nanoparticle synthesis methods are expensive and necessitate specialized equipment, limiting access to academics and enterprises with minimal resources. As nanoparticles find applications in a variety of industries, worries about their safety, regulation, and ethical use continue to emerge, posing difficulties in negotiating these complex issues.

5.5 Applications of Nanoparticles

Nanoparticles are utilized in medication delivery systems, as therapeutic agent carriers, and for imaging in diagnostics. They can target specific cells or tissues, increasing therapy efficacy and precision. Transistors and memory devices are two examples of electronic components made with nanoparticles. Displays make use of quantum dots. Nanoparticles act as efficient catalysts in chemical reactions, allowing for the creation of a wide range of chemical compounds. In materials research, nanoparticles are used to enhance material properties like coating durability and composite strength. In environmental remediation, they can be applied to wastewater treatment and to eliminate pollutants from soil and water (Baetke et al. 2015; Singh et al. 2019; Stark et al. 2015; Zazo et al. 2016).

5.6 Recent Developments in Microbes-Mediated Synthesis of Nanoparticles

Scientists have been investigating a wide variety of microorganisms, including uncommon strains and extremophiles, in order to generate nanoparticles. The bacteria have the capacity to generate specialized nanoparticles. Extremophiles, which thrive in severe conditions such as high salinity or temperatures, are capable of generating nanoparticles. An understanding of the mechanics driving the synthesis of microbial nanoparticles has enabled more exact control over their size, shape, and characteristics. To tailor nanoparticles for specific applications, researchers have focused on the genetic and enzymatic aspects of biogenic production.

Making biocompatible nanoparticles for biological uses is becoming more and more common. Using microbial manufacturing techniques, low-cytotoxic nanoparticles have been produced that have potential uses in medication delivery, medical imaging, and other healthcare fields. There have been attempts to increase the production of microbial nanoparticles for application in the industry. To satisfy the demands of large-scale nanoparticle production for commercial usage, researchers have been focusing on optimizing production processes such as bioreactors and fermentation systems (Patra et al. 2018; Yusuf et al. 2023).

The use of microbes in the manufacture of multimetallic nanoparticles has received attention. These nanoparticles, made up of various metals, have unique features for catalysis, sensing, and other applications. The controlled construction of these complex structures by microbes has shown promise. Several companies and startups have developed with the intention of commercializing microbially produced nanoparticles and nanoparticle-based goods. Antimicrobial coatings, nanocatalysts, and nanomaterials for diverse sectors are examples of this. Microbial production of nanoparticles has been investigated for environmental applications such as heavy metal and pollutant removal from water and soil. This eco-friendly strategy is in line with sustainability and green technologies (Arora et al. 2020; Basavegowda and Baek 2021).

Researchers have been exploring with microbial synthesis to create nanoparticle morphologies other than spheres and rods. These new forms could find use in nanoelectronics and catalysis. Some studies have concentrated on the utilization of microbial consortia, which entail the interaction of different microorganisms, to improve nanoparticle synthesis. Consortia in nanoparticle production can be tailored to function synergistically. Microbial synthesis has also been used to produce nanoparticles for bioremediation applications such as eliminating toxins from damaged environments (Table 5.1).

Table 5.1 Current studies on the microbial-facilitated production of nanoparticles

Microorganism	Nanoparticle type	Applications	Reference
Escherichia coli	Ag	Antimicrobial	Saeed et al. (2020)
Mycobacterium sp.	Au	Anticancer	Camas et al. (2018)
Bacillus subtilis	Fe_3O_4	Antimicrobial	Sundaram et al. (2012)
Serratia nematodiphila	Zinc sulfide	Antibacterial	Malarkodi and Annadurai (2013)
Lactobacillus sporogenes	ZnO	Antimicrobial	Mishra et al. (2013)
Gordonia amicalis	Ag	Antioxidant, scavenging	Sowani et al. (2016)
Morchella esculenta	Au	Antimicrobial and cytotoxicity	Acay (2021)
Cladosporium perangustum	Ag	Antioxidant, anticancer, and nano-toxicological study	Govindappa et al. (2020)
Trichoderma harzianum	Au	Antibacterial	Tripathi et al. (2018)
Thermophilic fungus *Humicola* sp.	CeO_2	Antibacterial	Khan and Ahmad (2013)
Penicillium chrysogenum	Pt	Cytotoxicity	Subramaniyan et al. (2018)
Penicillium janthinellum DJP06	Au	Anticancer	Pareek et al. (2020)
Trichoderma longibrachiatum	Ag	Antifungal against phyto-pathogenic fungi	Elamawi et al. (2018)
Nocardiopsis sp. MBRC-1	Ag	Antimicrobial	Manivasagan et al. (2013)
Marine endophytic actinomycetes	Cu	Antibacterial	Rasool and Hemalatha (2017)
Rhodococcus sp. (Actinomycete)	Au	Antimicrobial	Ahmad et al. (2003)
Streptomyces sp. JAR1	Ag	Antimicrobial	Chauhan (2013)
Streptomyces sp. LK3	Ag	Acaricidal activity against *Rhipicephalus microplus* and *Haemaphysalis bispinosa*	Karthik et al. (2014)

Table 5.1 (continued)

Microorganism	Nanoparticle type	Applications	Reference
Nostoc linckia	Ag	Antibacterial	Vanlalveni et al. (2018)
Amphiroa rigida	Ag	Antibacterial, cytotoxicity, and larvicidal	Gopu et al. (2021)
Galaxaura elongate	Au	Antibacterial	Abdel-Raouf et al. (2017)
Cystoseira baccata	Au	Anticancer	González-Ballesteros et al. (2017)
Spirulina platensis	Pd	Adsorbent	Sayadi et al. (2018)
Ulva armoricana sp.	Ag	Bactericidal	Massironi et al. (2019)
Chlorella ellipsoidea	Ag	Photophysical, catalytic, and antibacterial	Borah et al. (2020)

5.7 Future Prospects of Microbes-Mediated Synthesis of Nanoparticles (Arora et al. 2020; Koul et al. 2021; Qureshi et al. 2021; Yang et al. 2022a)

The future prospects of microbial nanoparticle production are excellent, with applications ranging from healthcare to environmental remediation to materials research and beyond. Here are some prospects for this field in the future:

Customized Nanoparticle Properties:

- Tailored nanoparticles: Researchers may create microorganisms that produce nanoparticles with precisely controlled characteristics including size, shape, and surface modifications through genetic engineering and bioprocess optimization.
- Application-specific nanoparticles: Tailored nanoparticles can be created for a range of uses, such as electronics, catalysis, medicine delivery, and more.

Advanced Biomedical Applications:

- Microbially synthesized nanoparticles can be engineered to selectively transport medicinal compounds to specific cells or tissues, thereby minimizing undesirable side effects and enhancing the effectiveness of treatment.
- Imaging and diagnostics: Biocompatible nanoparticles can serve as contrast agents in sophisticated imaging techniques like MRI and CT scans to identify and track diseases at an early stage.

Production of Green Nanomaterials:

- Sustainable nanomaterials: Microbial synthesis aligns with green and sustainable technology trends. The procedure is environmentally safe, uses less harmful chemicals, and saves energy.

- Renewable nanomaterials: Microorganisms can be used to manufacture nanomaterials from renewable resources, so helping to develop sustainable and circular economies.

Environmental Remediation:

- Heavy metal removal: Nanoparticles generated by microbes can be employed to efficiently remove heavy metals and pollutants from water and soil, assisting in environmental cleanup efforts.
- Pollution control: These nanoparticles have the potential to be used to create novel materials for air and water purification systems, resulting in less pollution and cleaner ecosystems.

Nanoelectronics and Sensing:

- Novel Nanoelectronic Devices: Nanoparticles generated by microbes with unique features can be integrated into advanced nanoelectronic devices such as sensors and memory devices.
- Improved Sensing: These nanoparticles have the potential to make sensors more sensitive and selective for environmental monitoring, healthcare, and security applications.

Multimetallic and Multifunctional Nanoparticles:

- Advanced Catalysis: Microbial synthesis can be employed to build complex multimetallic nanoparticles with enhanced catalytic characteristics, enhancing chemical reaction efficiency and generating sustainable energy.
- Multifunctional Materials: Nanoparticles with a variety of qualities can be created and used in multifunctional materials for a variety of industries, including aerospace and electronics.

Emerging Microbial Strains and Engineering:

- Custom Microbial Strains: Further study may lead to the identification and engineering of microorganisms with better nanoparticle synthesis capabilities, allowing nanoparticle attributes to be fine-tuned.
- Synthetic Biology Approaches: Synthetic biology advances can be used to manipulate microbes for specific nanoparticle manufacturing goals.

Bioremediation and Biomineralization:

- Improved Bioremediation: Microbially generated nanoparticles can be employed to boost bioremediation processes in contaminated environments by improving pollutant biological breakdown.
- Biomineralization: Through biomineralization processes, research in this area may lead to the production of novel materials for construction, building, and repair.

Industrial Scale-Up and Commercialization:

- Large-Scale Production: As microbial nanoparticle synthesis technologies advance, they are likely to find greater commercial-scale usage across a variety of industries.

- Innovative Products: Coatings, filters, and functional nanomaterials based on microbially generated nanoparticles may be developed in the future.

Regulatory Structures and Safety Measures:

Ongoing research will concentrate on developing safety criteria and regulatory frameworks for the manufacture and implementation of microbially produced nanoparticles, assuring their safe usage in a variety of industries and applications.

The future of microbial nanoparticle synthesis holds considerable promise for environmentally benign and sustainable nanotechnology solutions. It is expected to continue making substantial contributions in sectors ranging from healthcare and materials research to environmental protection and industrial manufacturing. The practical application of nanoparticles produced by microorganisms is expected to expand as research in this area progresses.

5.8 Recent Development in Plant-Mediated Synthesis of Nanoparticles (Munir et al. 2021; Nasrollahzadeh et al. 2019; Rani et al. 2023)

Using plant materials like leaves, stems, roots, and extracts, plant-mediated nanoparticle synthesis (also called "green synthesis") creates nanoparticles in an eco-friendly and sustainable way. Nanoparticles are created through the process of extracting metal ions from solutions in this method. The technology is not only ecologically clean and sustainable, but it also provides unique options for designing nanoparticles for specific applications. Researchers continue to investigate different plant sources and green synthesis optimization methodologies, broadening the potential for sustainable nanotechnology.

The green synthesis of nanoparticles typically follows these steps:

- Select a plant source based on the desired properties of the nanoparticles. Various plant species, including medicinal herbs, algae, and trees, have been used.
- A fine powder is made by cleaning, drying, and grinding the chosen plant material. Immersing plant material in an appropriate solvent is one method for preparing plant extracts.
- Combine the botanical substance or extract with a solution that contains metal ions, typically metal salts such as silver nitrate or gold chloride.
- The plant material or extract contains reducing agents such as phytochemicals, enzymes, and proteins, which are responsible for reducing metal ions and forming nanoparticles. This decrease triggers the commencement of nucleation and growth mechanisms.
- Nanoparticles are formed and can be isolated using methods like centrifugation, filtration, or ultracentrifugation.

Recent advances in the field of plant-mediated nanoparticle synthesis have demonstrated great progress in leveraging plants' inherent ability to generate nanoparticles with a variety of applications. Diverse Plant Sources: Scientists have broadened the plant species employed in nanoparticle manufacturing. Various plant sources, including marine plants, medicinal herbs, and agricultural crops, have been studied for their ability to synthesize nanoparticles, in addition to traditional plants like Aloe vera and green tea. Plant-mediated synthesis is a valuable method for producing biocompatible nanoparticles. Current studies have focused on creating biocompatible nanoparticles that can be utilized in medication administration, cancer treatment, and tissue regeneration. The ability to interact harmoniously with biological systems is of utmost importance in these applications. Green Synthesis: Scientists are increasingly dedicated to creating ecologically conscious and sustainable ways for synthesizing nanoparticles using plants as a basis. These green synthesis methods reduce the use of harmful chemicals while also lowering the environmental effect of nanoparticle creation. Targeted Delivery: Technological advances have resulted in the development of nanoparticles with greater targeting capabilities. Surface modification and biofunctionalization of plant-mediated nanoparticles allow for more precise drug delivery and therapeutic efficiency.

Enhanced characterization: Advances in analytical techniques have enhanced the characterization of plant-mediated nanoparticles. Advanced microscopy, spectroscopic, and analytical tools will be used to provide extensive insights into the structural and functional features of these nanoparticles. Phytochemical Control: Scientists are investigating the utilization of plant extracts rich in specific phytochemicals to control the dimensions, structure, and properties of artificially produced nanoparticles. This enables the tailoring of nanoparticles for diverse applications. Plant-mediated nanoparticles have demonstrated promise in combating antibiotic-resistant microorganisms. Recent studies have investigated the antibacterial properties of these substances for various medical and environmental uses. Plant-mediated nanoparticles have found use in catalysis, where their unique features improve reaction speeds and selectivity. They are also being researched for energy-related applications like solar cells and fuel cells. Commercialization and Industrial Scale-Up: Efforts are currently underway to commercialize plant-mediated nanoparticles and scale up their production for industrial usage, with an emphasis on lowering production costs and assuring quality consistency (Table 5.2).

These new findings underscore the growing interest in plant-mediated nanoparticle synthesis as a sustainable and adaptable technique with applications ranging from healthcare and agriculture to environmental remediation and energy technology. The topic has enormous promise for innovative solutions to different global concerns as researchers continue to investigate new plant sources, enhance synthesis methods, and broaden the variety of applications.

Table 5.2 Recent research in plant-mediated synthesis of nanoparticles

Plant name	Nanoparticle type	Applications	Reference
Amaranthus spinosus	Silver	Antibacterial	Preenanka and Sebastian (2022)
Cycas pschannae	ZnO	Antibacterial	Sudha et al. (2020)
Cyrtrandroemia nicobarica	ZnO	Antioxidant	Sudha et al. (2021a)
Knema andamanica	ZnO	Antibacterial	Sudha et al. (2021b)
Leea asiatica	ZnO	Antioxidant	Ali et al. (2021a)
Leea grandifolia	ZnO	Antibacterial	Ali et al. (2021b)
Manilkara littoralis	ZnO	Antioxidant	Ali et al. (2022)
Durio zibethinus seed	Ag	Antimicrobial, cytotoxic, photocatalytic	Sumitha et al. (2018)
Artocarpus altilis	Ag	Antimicrobial, antioxidant	Ravichandran et al. (2016)
Durio zibethinus	ZnO	Antimicrobial, antioxidant, cytotoxic, photocatalytic	Ravichandran et al. (2020)
Durio zibethinus rind	Ag	Antimicrobial, antioxidant	Sumitha et al. (2019)
Parkia speciosa	Ag	Antimicrobial, antioxidant, photocatalytic	Ravichandran et al. (2019)

5.9 Future Prospects of Plant-Mediated Synthesis of Nanoparticles (Rani et al. 2023; Ankamwar et al. 2020; Shyam et al. 2021)

Plant-mediated nanoparticle synthesis has a bright future, with continuous research and development pointing to a wide range of possible uses and advances. Here are some of the most promising future prospects for this environmentally friendly and sustainable method:

Customized Nanoparticle Properties:

- Tailored nanoparticles: Researchers will continue to investigate the utilization of various plant sources, extracts, and optimization methodologies to tailor nanoparticles with precise attributes for specific purposes, such as size, shape, and surface changes.

Biomedical and Healthcare Applications:

- Targeted medication delivery: Because plant-mediated nanoparticles are biocompatible, they can be used for targeted drug delivery, improving treatment efficacy and decreasing side effects.
- Theranostics: Theranostics, or the dual role of nanoparticles in diagnosis and therapy, will continue to improve in areas such as cancer.

Environmental Remediation and Water Purification:

- Effective pollutant removal: Plant-mediated nanoparticles have the ability to remove heavy metals, pollutants, and toxins from water and soil, so contributing to environmental cleanup and long-term water purification.
- Improved air quality: These nanoparticles could be used in air filtration technology to combat air pollution.

Nanoelectronics and Sensing:

- High-performance sensors: Future research will concentrate on combining plant-derived nanoparticles into highly sensitive and selective sensors for use in healthcare, environmental monitoring, and security.
- Low-power, energy-efficient electronics: Plant-mediated nanoparticles can help to produce low-power, energy-efficient electronic devices.

Multimetallic and Multifunctional Nanoparticles:

- Advanced catalysis: Researchers will investigate the utilization of plant-mediated synthesis to produce complex multimetallic nanoparticles with improved catalytic capabilities, which will have an impact on fields such as sustainable energy production and chemical processes.
- Multifunctional materials: Multifunctional nanoparticles will be incorporated into materials with a wide range of uses, such as aerospace, electronics, and sophisticated coatings.

Novel Plant Sources and Phytochemicals:

- New plant sources will be investigated: Researchers will continue to explore diverse plant species in order to uncover unique phytochemicals with nanoparticle production potential.
- Plant-mediated synthesis can be integrated into sustainable agriculture and agro-waste use strategies, benefiting both agriculture and nanotechnology.

Standardization and Large-Scale Production:

- Industry acceptance: If the technique advances, plant-mediated synthesis may see increased application in a variety of industries, if standardized methods for large-scale manufacturing are developed.
- Green nanotechnology: The method follows the tenets of green nanotechnology, with a focus on sustainability and environmental friendliness.

The future of plant-mediated nanoparticle synthesis is defined by ongoing research, innovation, and an expanding array of applications in a variety of sectors. This method provides a sustainable and environmentally benign approach to nanotechnology and is predicted to play an important role in tackling many healthcare, environmental, and materials science concerns.

5.10 Recent Development in Synthesis of Nanoparticles Using Biopolymers (Alavi and Rai 2019; El-Sherbiny and Salih 2018; Vijayakumar et al. 2019)

The utilization of biopolymers and biomolecules for the production of nanoparticles is an environmentally friendly and sustainable approach that has gained popularity due to its biocompatibility and minimal ecological impact. Biopolymers and biomolecules serve as both reducing and stabilizing agents throughout the nanoparticle synthesis process (Table 5.3).

These examples demonstrate the adaptability and biocompatibility of nanoparticles made from biopolymers and biomolecules. This green synthesis method opens up a plethora of opportunities for designing nanoparticles for specific uses in healthcare, materials research, environmental remediation, and other fields. Researchers are constantly looking for new biopolymers and biomolecules to help progress this sector.

Table 5.3 Recent research in synthesis of nanoparticles using biopolymers and biomolecules

Biopolymers	Method	Applications	Reference
Alginate	Calcium ion regulation of alginate gel formation	Using doxorubicin as a model drug, the drug-loading capacity was evaluated	Rajaonarivony et al. (1993)
Sodium alginate and BSA	Emulsion solidification	Determining the kinetic characteristics of the metabolism of 5-FU sodium alginate-125I BSA nanoparticles	Yi (1999)
Alginate and chitosan	Alginate is pre-gelated ionotropically with calcium chloride, and then it complexes with chitosan		Sarmento et al. (2006)
Alginate	Modified coacervation or ionotropic gelation	Monitor the formation of complexes between polyelectrolytes with opposite charges as carriers for insulin nanoparticles	Motwani et al. (2007)
Chitosan	Complex coacervation technique	Mucoadhesive nanoparticulate carrier system optimization for sustained medication administration in the eyes	Mansouri et al. (2004)
Chitosan	Ionic crosslinking of chitosan solution with TPP	Encapsulate nucleic acids (DNA) and increase the effectiveness of both in vitro and in vivo transfection	Sharma et al. (2007)
Chitosan–alginate	Water-in-oil reverse microemulsion	Put plasmid DNA in an encapsulated form to effectively distribute genes for transfection	Sharma et al. (2007)

Polymer-based nanoparticle synthesis has advanced significantly in recent years, resulting in versatile and adjustable nanoparticles with a wide range of uses. Polymer Nanoparticles for Drug Delivery: Polymer-based nanoparticles, such as polymeric micelles and dendrimers, have gained popularity as drug delivery vehicles. Recent advancements have focused on increasing drug-loading capacities, controlling release, and improving the stability and targeting of drug-loaded polymer nanoparticles. Multifunctional Polymer Nanoparticles: Researchers are working to create multifunctional polymer nanoparticles by combining different components, such as imaging agents, targeting ligands, and medicinal payloads, into a single nanoparticle system. These nanoparticles are intended for theranostic applications, allowing both diagnosis and treatment at the same time. Responsive Polymer Nanoparticles: Smart polymers that can react to environmental factors such as temperature, pH, or light have been integrated into nanoparticle design. These responsive polymer nanoparticles allow for regulated medication release, on-demand drug delivery, and tailored therapy in response to certain body conditions.

The usage of biodegradable and biocompatible polymers in nanoparticle manufacturing, such as PLGA and chitosan, has grown. These polymers are useful for medicinal applications since they are low in toxicity. Polymer chemistry and bioconjugation techniques advancements have enabled the production of targeted polymer nanoparticles. These nanoparticles have the ability to recognize specific biomarkers on cells or tissues, which improves their precision in medication delivery and imaging applications. Polymer nanoparticles are rapidly being investigated for applications in personalized medicine. To enhance treatment outcomes, researchers are designing nanoparticles suited to an individual patient's specific needs, taking genetic and molecular profiles into account. Polymer nanoparticles have found use in environmental remediation, such as contaminant removal from water and soil. They are capable of effectively capturing and sequestering toxins, hence contributing to cleaner and safer ecosystems. Recent improvements in polymerization techniques, such as controlled/living polymerization methods and polymer nanoencapsulation, have enhanced control over the size, shape, and characteristics of polymer nanoparticles (Makadia and Siegel 2011; Yang et al. 2022b).

5.11 Future Prospects of Synthesis of Nanoparticles Using Biopolymers (Begines et al. 2020; Vodyashkin et al. 2022)

The prospects for manufacturing nanoparticles utilizing biopolymers and biomolecules are bright, with several fascinating advances and applications on the horizon. Here are some significant potential prospects for this environmentally friendly and sustainable approach:

Tailored Nanoparticle Design:

- The progress in biopolymer and biomolecule-mediated synthesis will provide precise manipulation of the dimensions, configuration, and attributes of nanoparticles, facilitating customized designs for particular uses.

Biomedicine and Healthcare in Applications:

- Targeted drug delivery systems will benefit greatly from the use of biopolymer and biomolecule-stabilized nanoparticles. These nanoparticles will enhance treatment effectiveness and minimize the occurrence of negative effects.
- Theranostics: The development of nanoparticles for simultaneous diagnosis and therapy (theranostics) will continue, allowing for individualized medicine solutions.

Environmental Remediation and Green Technologies:

- Improved Environmental Cleanup: Nanoparticles made from biopolymers and biomolecules will be used to efficiently remove pollutants such as heavy metals from water and soil.
- Sustainable Water Treatment: Biopolymer-stabilized nanoparticles will be used in green technologies for sustainable water purification and wastewater treatment.

Advanced Sensors and Diagnostics:

- Biopolymer and biomolecule-stabilized nanoparticles will be integrated into sensors and diagnostic devices with higher sensitivity and selectivity for applications in healthcare and environmental monitoring.

Multifunctional Nanoparticles:

- Advanced Catalysis: Researchers will investigate the application of these nanoparticles in catalytic reactions, which will contribute to increased efficiency and sustainability in chemical processes.
- Multifunctional Materials: The integration of versatile nanoparticles into materials for many purposes, such as electronics and catalysis, will increasingly prevail.

Emerging Biopolymers and Biomolecules:

- New Sources and Materials: Researchers will continue to look into new biopolymers and biomolecules in order to broaden the spectrum of materials accessible for nanoparticle manufacturing.
- Sustainable Agriculture: Biopolymers derived from agricultural waste and renewable sources will help to establish sustainable agricultural methods.

The use of biopolymers and biomolecules in nanoparticle production has the potential to address significant concerns in healthcare, the environment, and materials science in the future. As researchers continue to improve and optimize this green synthesis approach, it is anticipated to become a cornerstone of sustainable nanotechnology, addressing a wide range of global concerns.

5.12 Summary

Researchers have advanced microbial nanoparticle manufacturing by looking for particular properties in extremophiles and uncommon strains. For medication delivery and medical uses, biocompatible nanoparticles with minimal cytotoxicity are created. A focus is on scaling up microbial nanoparticle synthesis for industrial usage, particularly multimetallic nanoparticles. Microbial synthesis contributes to environmental applications by eliminating heavy metals and contaminants in a sustainable manner. Tailoring characteristics, biomedical applications, nanoelectronics, and complex multimetallic architectures are all possibilities for microbial nanoparticle manufacturing in the future. Plant-mediated nanoparticle synthesis is a green approach that uses a variety of plant sources and has shown promise in medication delivery, catalysis, and energy applications. Current research is geared at commercialization and industrial scale-up. The green technique of biopolymer and biomolecule-mediated nanoparticle production yields versatile nanoparticles with applications in medication administration, diagnostics, and environmental cleanup. Tailor-made nanoparticle creation, breakthroughs in medicinal applications, environmental remediation, and investigation of developing biopolymers for sustainable agriculture are all possibilities for the future. In conclusion, these approaches to sustainable nanoparticle synthesis have a wide range of applications, ranging from healthcare to environmental issues, with ongoing research focusing on large-scale production and commercialization.

References

Abdel-Raouf N, Al-Enazi NM, Ibraheem IBM (2017) Green biosynthesis of gold nanoparticles using Galaxaura elongata and characterization of their antibacterial activity. Arab J Chem 10:S3029–S3039

Acay H (2021) Utilization of *Morchella esculenta* -mediated green synthesis golden nanoparticles in biomedicine applications. Prep Biochem Biotechnol 51:127–136

Ahmad A, Senapati S, Khan MI, Kumar R, Ramani R, Srinivas V et al (2003) Intracellular synthesis of gold nanoparticles by a novel alkalotolerant actinomycete, *Rhodococcus* species. Nanotechnology 14:824–828

Alavi M, Rai M (2019) Recent progress in nanoformulations of silver nanoparticles with cellulose, chitosan, and alginic acid biopolymers for antibacterial applications. Appl Microbiol Biotechnol 103:8669–8676

Alhalili Z (2023) Metal oxides nanoparticles: general structural description, chemical, physical, and biological synthesis methods, role in pesticides and heavy metal removal through wastewater treatment. Molecules 28:3086

Ali S, Sudha KG, Karunakaran G, Kowsalya M, Kolesnikov E, Rajeshkumar MP (2021a) Green synthesis of stable antioxidant, anticancer and photocatalytic activity of zinc oxide nanorods from Leea asiatica leaf. J Biotechnol 329:65–79

Ali S, Sudha KG, Karunakaran G, Kowsalya M, Kolesnikov E, Gorshenkov MV et al (2021b) Novel *Leea grandifolia* leaves mediated synthesis of ZnO nanorods for photocatalytic and anticancer applications. Appl Organ Chem 35:e6239

Ali S, Govindaraj Sudha K, Karunakaran G, Kowsalya M, Kolesnikov E, Gorshenkov MV et al (2022) Anticancer and photocatalytic activities of zinc oxide nanorods synthesized from Manilkara littoralis leaf extract. Mater Chem Phys 277:125541

Altammar KA (2023) A review on nanoparticles: characteristics, synthesis, applications, and challenges. Front Microbiol 14:1155622

Ankamwar B, Kirtiwar S, Shukla AC (2020) Plant-mediated green synthesis of nanoparticles. In: Patra JK, Shukla AC, Das G (eds) Advances in pharmaceutical biotechnology. Springer, Singapore, pp 221–234. https://doi.org/10.1007/978-981-15-2195-9_17; [cited 2023 Dec 4]

Arora N, Thangavelu K, Karanikolos GN (2020) Bimetallic nanoparticles for antimicrobial applications. Front Chem 8:412

Baetke SC, Lammers T, Kiessling F (2015) Applications of nanoparticles for diagnosis and therapy of cancer. Br J Radiol 88:20150207

Basavegowda N, Baek K-H (2021) Multimetallic nanoparticles as alternative antimicrobial agents: challenges and perspectives. Molecules 26:912

Begines B, Ortiz T, Pérez-Aranda M, Martínez G, Merinero M, Argüelles-Arias F et al (2020) Polymeric nanoparticles for drug delivery: recent developments and future prospects. Nano 10:1403

Bokov D, Turki Jalil A, Chupradit S, Suksatan W, Javed Ansari M, Shewael IH et al (2021) Nanomaterial by sol-gel method: synthesis and application. In: Wang Z (ed) Advances in materials science and engineering, pp 1–21

Borah D, Das N, Das N, Bhattacharjee A, Sarmah P, Ghosh K et al (2020) Alga-mediated facile green synthesis of silver nanoparticles: photophysical, catalytic and antibacterial activity. Appl Organ Chem 34:e5597

Camas M, Sazak Camas A, Kyeremeh K (2018) Extracellular synthesis and characterization of gold nanoparticles using mycobacterium sp. BRS2A-AR2 isolated from the aerial roots of the Ghanaian mangrove plant, Rhizophora racemosa. Indian J Microbiol 58:214–221

Chauhan R (2013) A biological approach to synthesis of silver nanoparticles with Streptomyces sp JAR1 and its antimicrobial activity. Sci Pharm 81:607–621

Dobrzański LA, Gołombek K, Lukaszkowicz K (2015) Physical vapor deposition in manufacturing. In: AYC N (ed) Handbook of manufacturing engineering and technology. Springer, London, pp 2719–2754. https://doi.org/10.1007/978-1-4471-4670-4_29; [cited 2023 Dec 4]

Duran N, Seabra AB (2018) Biogenic synthesized Ag/Au nanoparticles: production, characterization, and applications. Curr Nanosci 14:82–94

Elamawi RM, Al-Harbi RE, Hendi AA (2018) Biosynthesis and characterization of silver nanoparticles using Trichoderma longibrachiatum and their effect on phytopathogenic fungi. Egypt J Biol Pest Control 28:28

El-Sherbiny IM, Salih E (2018) Green synthesis of metallic nanoparticles using biopolymers and plant extracts. In: Kanchi S, Ahmed S (eds) Green metal nanoparticles, 1st edn. Wiley, pp 293–319. https://doi.org/10.1002/9781119418900.ch10; [cited 2023 Dec 4]

Ghosh S, Ahmad R, Zeyaullah M, Khare SK (2021) Microbial nano-factories: synthesis and biomedical applications. Front Chem 9:626834

González-Ballesteros N, Prado-López S, Rodríguez-González JB, Lastra M, Rodríguez-Argüelles MC (2017) Green synthesis of gold nanoparticles using brown algae Cystoseira baccata: its activity in colon cancer cells. Colloids Surf B: Biointerfaces 153:190–198

Gopu M, Kumar P, Selvankumar T, Senthilkumar B, Sudhakar C, Govarthanan M et al (2021) Green biomimetic silver nanoparticles utilizing the red algae Amphiroa rigida and its potent antibacterial, cytotoxicity and larvicidal efficiency. Bioprocess Biosyst Eng 44:217–223

Govindappa M, Lavanya M, Aishwarya P, Pai K, Lunked P, Hemashekhar B et al (2020) Synthesis and characterization of endophytic fungi, Cladosporium perangustum mediated silver nanoparticles and their antioxidant, anticancer and nano-toxicological study. Bio Nano Sci 10:928–941

Han J, Zhao D, Li D, Wang X, Jin Z, Zhao K (2018) Polymer-based nanomaterials and applications for vaccines and drugs. Polymers 10:31

Han X, Xu K, Taratula O, Farsad K (2019) Applications of nanoparticles in biomedical imaging. Nanoscale 11:799–819
Hasan M, Ullah I, Zulfiqar H, Naeem K, Iqbal A, Gul H et al (2018) Biological entities as chemical reactors for synthesis of nanomaterials: progress, challenges and future perspective. Mater Today Chem 8:13–28
Jagadeesh P, Rangappa SM, Siengchin S (2023) Advanced characterization techniques for nanostructured materials in biomedical applications. Adv Ind Eng Polym Res 1:S2542504823000180
Karthik L, Kumar G, Kirthi AV, Rahuman AA, Bhaskara Rao KV (2014) Streptomyces sp. LK3 mediated synthesis of silver nanoparticles and its biomedical application. Bioprocess Biosyst Eng 37:261–267
Karunakaran G, Sudha KG, Ali S, Cho E-B (2023) Biosynthesis of nanoparticles from various biological sources and its biomedical applications. Molecules 28:4527
Khan SA, Ahmad A (2013) Fungus mediated synthesis of biomedically important cerium oxide nanoparticles. Mater Res Bull 48:4134–4138
Khandel P, Shahi SK (2018) Mycogenic nanoparticles and their bio-prospective applications: current status and future challenges. J Nanostructure Chem 8:369–391
Koul B, Poonia AK, Yadav D, Jin J-O (2021) Microbe-mediated biosynthesis of nanoparticles: applications and future prospects. Biomol Ther 11:886
Kumar JA, Krithiga T, Manigandan S, Sathish S, Renita AA, Prakash P et al (2021) A focus to green synthesis of metal/metal based oxide nanoparticles: various mechanisms and applications towards ecological approach. J Clean Prod 324:129198
Makadia HK, Siegel SJ (2011) Poly lactic-co-glycolic acid (PLGA) as biodegradable controlled drug delivery carrier. Polymers 3:1377–1397
Malarkodi C, Annadurai G (2013) A novel biological approach on extracellular synthesis and characterization of semiconductor zinc sulfide nanoparticles. Appl Nanosci 3:389–395
Manivasagan P, Venkatesan J, Senthilkumar K, Sivakumar K, Kim S-K (2013) Biosynthesis, antimicrobial and cytotoxic effect of silver nanoparticles using a novel *Nocardiopsis* sp. MBRC-1. Biomed Res Int 2013:1–9
Mansouri S, Lavigne P, Corsi K, Benderdour M, Beaumont E, Fernandes JC (2004) Chitosan-DNA nanoparticles as non-viral vectors in gene therapy: strategies to improve transfection efficacy. Eur J Pharm Biopharm 57:1–8
Massironi A, Morelli A, Grassi L, Puppi D, Braccini S, Maisetta G et al (2019) Ulvan as novel reducing and stabilizing agent from renewable algal biomass: application to green synthesis of silver nanoparticles. Carbohydr Polym 203:310–321
Mishra M, Singh Paliwal J, Singh SK, Selvarajan E, Subathradevi C, Mohanasrinivasan V (2013) Studies on the inhibitory activity of biologically synthesized and characterized zinc oxide nanoparticles using Lactobacillus sporogenes against. J Pure Appl Microbiol 7:1263–1268
Motwani S, Chopra S, Talegaonkar S, Kohli K, Ahmad F, Khar R (2007) Chitosan–sodium alginate nanoparticles as submicroscopic reservoirs for ocular delivery: formulation, optimisation and in vitro characterisation. Eur J Pharm Biopharm 1:S0939641107003086
Munir H, Bilal M, Mulla SI, Abbas Khan H, Iqbal HMN (2021) Plant-mediated green synthesis of nanoparticles. In: Inamuddin BR, Ahamed MI, Khan A (eds) Advances in green synthesis. Springer International Publishing, Cham, pp 75–89. https://doi.org/10.1007/978-3-030-67884-5_4; [cited 2023 Dec 4]
Nasrollahzadeh M, Mahmoudi-Gom Yek S, Motahharifar N, Ghafori GM (2019) Recent developments in the plant-mediated green synthesis of Ag-based nanoparticles for environmental and catalytic applications. Chem Rec 19:2436–2479
Nguyen NPU, Dang NT, Doan L, Nguyen TTH (2023) Synthesis of silver nanoparticles: from conventional to 'modern' methods—a review. PRO 11:2617
Pandit C, Roy A, Ghotekar S, Khusro A, Islam MN, Emran TB et al (2022) Biological agents for synthesis of nanoparticles and their applications. J King Saud Univ Sci 34:101869
Pareek V, Bhargava A, Panwar J (2020) Biomimetic approach for multifarious synthesis of nanoparticles using metal tolerant fungi: a mechanistic perspective. Mater Sci Eng B 262:114771

Patil S, Chandrasekaran R (2020) Biogenic nanoparticles: a comprehensive perspective in synthesis, characterization, application and its challenges. J Genet Eng Biotechnol 18:67

Patra JK, Das G, Fraceto LF, Campos EVR, Rodriguez-Torres MDP, Acosta-Torres LS et al (2018) Nano based drug delivery systems: recent developments and future prospects. J Nanobiotechnol 16:71

Preenanka R, Sebastian D (2022) Characterization of green synthesized antibacterial silver nanoparticles from Amaranthus spinosus L. extract. Bio Nano Sci 12:502–511

Qureshi A, Blaisi NI, Abbas AAO, Khan NA, Rehman S (2021) Prospectus and development of microbes mediated synthesis of nanoparticles. In: Ansari MA, Rehman S (eds) Microbial nanotechnology: green synthesis and applications. Springer, Singapore, pp 1–15. https://doi.org/10.1007/978-981-16-1923-6_1; [cited 2023 Dec 4]

Rajaonarivony M, Vauthier C, Couarraze G, Puisieux F, Couvreur P (1993) Development of a new drug carrier made from alginate. J Pharm Sci 82:912–917

Rani N, Singh P, Kumar S, Kumar P, Bhankar V, Kumar K (2023) Plant-mediated synthesis of nanoparticles and their applications: a review. Mater Res Bull 163:112233

Rasool U, Hemalatha S (2017) Marine endophytic actinomycetes assisted synthesis of copper nanoparticles (CuNPs): characterization and antibacterial efficacy against human pathogens. Mater Lett 194:176–180

Ravichandran V, Vasanthi S, Shalini S, Ali Shah SA, Harish R (2016) Green synthesis of silver nanoparticles using Atrocarpus altilis leaf extract and the study of their antimicrobial and antioxidant activity. Mater Lett 180:264–267

Ravichandran V, Vasanthi S, Shalini S, Shah SAA, Tripathy M, Paliwal N (2019) Green synthesis, characterization, antibacterial, antioxidant and photocatalytic activity of Parkia speciosa leaves extract mediated silver nanoparticles. Results Phys 15:102565

Ravichandran V, Sumitha S, Ning CY, Xian OY, Kiew YU, Paliwal N et al (2020) Durian waste mediated green synthesis of zinc oxide nanoparticles and evaluation of their antibacterial, antioxidant, cytotoxicity and photocatalytic activity. Green Chem Lett Rev 13:102–116

Ravichandran V, Rohini K, Roy A, Rajeshkumar S (2022) Microbial mediated synthesis, characterisation and application of selenium nanoparticles. In: Savita SA, Jain R, Pati PK (eds) Mycology: current and future developments. BENTHAM SCIENCE PUBLISHERS, pp 62–102; [cited 2023 Dec 4]. https://www.eurekaselect.com/node/206938

Saeed S, Iqbal A, Ashraf MA (2020) Bacterial-mediated synthesis of silver nanoparticles and their significant effect against pathogens. Environ Sci Pollut Res 27:37347–37356

Sarmento B, Ferreira D, Veiga F, Ribeiro A (2006) Characterization of insulin-loaded alginate nanoparticles produced by ionotropic pre-gelation through DSC and FTIR studies. Carbohydr Polym 66:1–7

Sayadi MH, Salmani N, Heidari A, Rezaei MR (2018) Bio-synthesis of palladium nanoparticle using Spirulina platensis alga extract and its application as adsorbent. Surf Interfaces 10:136–143

Sharma BV, Ali M, Baboota S, Ali J (2007) Preparation and characterization of chitosan nanoparticles for nose to brain delivery of a cholinesterase inhibitor. Indian J Pharm Sci 69:712–713

Shyam A, Chandran SS, George B (2021) Plant mediated synthesis of AgNPs and its applications: an overview. Inorg Nano-Met Chem 51:1646–1662

Singh S, Kumar V, Romero R, Sharma K, Singh J (2019) Applications of nanoparticles in wastewater treatment. In: Prasad R, Kumar V, Kumar M, Choudhary D (eds) Nanobiotechnology in bioformulations. Springer International Publishing, Cham, pp 395–418. https://doi.org/10.1007/978-3-030-17061-5_17; [cited 2023 Dec 4]

Smits K, Grigorjeva L, Millers D, Kundzins K, Ignatans R, Grabis J et al (2014) Luminescence properties of zirconia nanocrystals prepared by solar physical vapor deposition. Opt Mater 37:251–256

Sopan Mahato S, Mahata D, Panda S, Mahata S (2023) Perspective chapter: sol-gel science and technology in context of nanomaterials—recent advances. In: Pal Singh J, Shankar Acharya S,

Kumar S, Kumar Dixit S (eds) Sol-gel method—recent advances. IntechOpen; [cited 2023 Dec 4]. https://www.intechopen.com/chapters/86894

Sowani H, Mohite P, Munot H, Shouche Y, Bapat T, Kumar AR et al (2016) Green synthesis of gold and silver nanoparticles by an actinomycete Gordonia amicalis HS-11: mechanistic aspects and biological application. Process Biochem 51:374–383

Stark WJ, Stoessel PR, Wohlleben W, Hafner A (2015) Industrial applications of nanoparticles. Chem Soc Rev 44:5793–5805

Subramaniyan SA, Sheet S, Vinothkannan M, Yoo DJ, Lee YS, Belal SA et al (2018) One-pot facile synthesis of Pt nanoparticles using cultural filtrate of microgravity simulated grown *P chrysogenum* and their activity on bacteria and cancer cells. J Nanosci Nanotechnol 18:3110–3125

Sudha KG, Ali S, Karunakaran G, Kowsalya M, Kolesnikov E, Rajeshkumar MP (2020) Eco-friendly synthesis of ZnO nanorods using *Cycas pschannae* plant extract with excellent photocatalytic, antioxidant, and anticancer nanomedicine for lung cancer treatment. Appl Organ Chem 34:e5511

Sudha KG, Ali S, Karunakaran G, Kowsalya M, Kolesnikov E, Gorshenkov MV et al (2021a) Cyrtrandroemia nicobarica-synthesized ZnO NRs: a new tool in cancer treatment. JOM 73:364–372

Sudha KG, Ali S, Karunakaran G, Kowsalya M, Kolesnikov E, Gorshenkov MV et al (2021b) An eco-friendly production of ZnO NRs using Knema andamanica (Warb) extracts for photocatalytic and anticancer applications. Inorg Chem Commun 134:109030

Sumitha S, Vasanthi S, Shalini S, Chinni SV, Gopinath SCB, Anbu P et al (2018) Phyto-mediated photo catalysed green synthesis of silver nanoparticles using Durio zibethinus seed extract: antimicrobial and cytotoxic activity and photocatalytic applications. Molecules 23:3311

Sumitha S, Vasanthi S, Shalini S, Chinni S, Gopinath S, Kathiresan S et al (2019) Durio zibethinus rind extract mediated green synthesis of silver nanoparticles: characterization and biomedical applications. Phcog Mag 15:52

Sundaram PA, Augustine R, Kannan M (2012) Extracellular biosynthesis of iron oxide nanoparticles by Bacillus subtilis strains isolated from Rhizosphere soil. Biotechnol Bioprocss Eng 17:835–840

Szczyglewska P, Feliczak-Guzik A, Nowak I (2023) Nanotechnology–general aspects: a chemical reduction approach to the synthesis of nanoparticles. Molecules 28:4932

Tripathi RM, Shrivastav BR, Shrivastav A (2018) Antibacterial and catalytic activity of biogenic gold nanoparticles synthesised by *Trichoderma harzianum*. IET Nanobiotechnol 12:509–513

Vanlalveni C, Rajkumari K, Biswas A, Adhikari PP, Lalfakzuala R, Rokhum L (2018) Green synthesis of silver nanoparticles using Nostoc linckia and its antimicrobial activity: a novel biological approach. Bio Nano Sci 8:624–631

Vaseghi Z, Nematollahzadeh A (2020) Nanomaterials: types, synthesis, and characterization. In: Srivastava N, Srivastava M, Mishra PK, Gupta VK (eds) Green synthesis of nanomaterials for bioenergy applications, 1st edn. Wiley, pp 23–82. https://doi.org/10.1002/9781119576785.ch2; [cited 2023 Dec 4]

Vijayakumar V, Samal SK, Mohanty S, Nayak SK (2019) Recent advancements in biopolymer and metal nanoparticle-based materials in diabetic wound healing management. Int J Biol Macromol 122:137–148

Vodyashkin AA, Kezimana P, Vetcher AA, Stanishevskiy YM (2022) Biopolymeric nanoparticles–multifunctional materials of the future. Polymers 14:2287

Yang Y, Waterhouse GIN, Chen Y, Sun-Waterhouse D, Li D (2022a) Microbial-enabled green biosynthesis of nanomaterials: current status and future prospects. Biotechnol Adv 55:107914

Yang F, Cabe M, Nowak HA, Langert KA (2022b) Chitosan/poly(lactic-co-glycolic)acid nanoparticle formulations with finely-tuned size distributions for enhanced Mucoadhesion. Pharmaceutics 14:95

Yi Y-M (1999) Preparation and distribution of 5-fluorouracil ^{125}I sodium alginate-bovine serum albumin nanoparticles. World J Gastroenterol 5:57

Ying S, Guan Z, Ofoegbu PC, Clubb P, Rico C, He F et al (2022) Green synthesis of nanoparticles: current developments and limitations. Environ Technol Innov 26:102336
Yusuf A, Almotairy ARZ, Henidi H, Alshehri OY, Aldughaim MS (2023) Nanoparticles as drug delivery systems: a review of the implication of nanoparticles' physicochemical properties on responses in biological systems. Polymers 15:1596
Zazo H, Colino CI, Lanao JM (2016) Current applications of nanoparticles in infectious diseases. J Control Release 224:86–102
Zielińska A, Carreiró F, Oliveira AM, Neves A, Pires B, Venkatesh DN et al (2020) Polymeric nanoparticles: production, characterization, toxicology and ecotoxicology. Molecules 25:3731

Chapter 6
Production of Biopharmaceuticals on Genetically Modified Organisms

Sandipan Jana, Barun Das Gupta, Amit Kar, Anirbita Ghosh, Monihar Pervin, Asis Bala, Motlalepula G. Matsabisa, Rajeswar Das, Somsubhra Ghosh, Pulok K. Mukherjee, and Pallab K. Haldar

Abstract The production of biopharmaceuticals using genetically modified organisms represents a transformative paradigm in the pharmaceutical industry. This abstract explores the advancements, challenges, and implications of employing genetically modified organisms for the synthesis of biopharmaceuticals. The genetic modification of organisms, ranging from bacteria and yeast to mammalian cells, enables the engineered expression of therapeutic proteins, antibodies, and vaccines. This review delves into the molecular techniques employed in genetically modified organisms to optimize protein production, the scalability of bioprocessing, and the regulatory landscape governing the use of genetically engineered platforms. Additionally, the chapter highlights the potential of genetically modified organisms to revolutionize the production of complex biologics and discusses the ethical and safety considerations associated with their widespread application in biopharmaceutical manufacturing. The synthesis of biopharmaceuticals on genetically modified organisms not only enhances production efficiency but also contributes to the expanding landscape of precision medicine, heralding a new era in therapeutic innovation and healthcare delivery.

S. Jana (✉)
School of Pharmacy, The Neotia University, Sarisa, West Bengal, India

Department of Pharmaceutical Technology, Jadavpur University, Kolkata, West Bengal, India

B. D. Gupta
School of Natural Product Studies, Jadavpur University, Kolkata, West Bengal, India

A. Kar · P. K. Mukherjee
Institute of Bioresources and Sustainable Development, Autonomous Institute Under Department of Biotechnology, Imphal, Manipur, India
e-mail: director.ibsd@nic.in

S. Bose et al. (eds.), *Concepts in Pharmaceutical Biotechnology and Drug Development*, Interdisciplinary Biotechnological Advances,
https://doi.org/10.1007/978-981-97-1148-2_6

Keywords Genetically modified organisms · Biopharmaceuticals · Upstream processing · Downstream processing

6.1 Biopharmaceuticals: An Introduction

In the past, the term "Biopharmaceuticals" referred to medications produced through biotechnological procedures utilizing molecular biology techniques. Today, these medications are widely used in various areas of medicine and are considered to be highly effective therapy options for numerous illnesses including cancer and metabolic disorders. Biopharmaceuticals encompass any beneficial medication produced using genetically modified organisms, microbes, or compounds derived from biological processing (Kesik-Brodacka 2018). The production of these medications involves various biotechnology processes such as genetic engineering, recombinant human technology, gene transfer, and antibody production (Walsh 2000). The first therapeutic protein and anti-diphtheria serum was introduced in 1894 by Farbwerke Hoechst to combat a catastrophic diphtheria epidemic in Europe (Strohl and Knight 2009). Later developments included the production of heterologous insulin refined from pigs and cows and sold under the brand name "Iletin" and the cloning and expression of human insulin, somatostatin, and also growth hormone (HGH) in *Escherichia coli* in the late 1970s, which led to the creation of the first recombinant human insulin (Bhatia and Goli 2018; Vecchio et al. 2018).

A. Ghosh · M. Pervin
Department of Pharmaceutical Technology, Jadavpur University, Kolkata, West Bengal, India

A. Bala
Division of Life Sciences, Institute of Advanced Study in Science and Technology (IASST), An Autonomous Institute Under Department of Science and Technology (Govt. of India) Vigyan Path, Guwahati, Assam, India
e-mail: asisbala_ju@yhaoo.co.in

M. G. Matsabisa
Department of Pharmacology, University of the Free State, Bloemfontein, South Africa
e-mail: motlalepula.matsabisa@mrc.ac.za

R. Das · S. Ghosh
School of Pharmacy, The Neotia University, Sarisa, West Bengal, India
e-mail: rajeswar.das@tnu.in; somsubhra.ghosh@tnu.in

P. K. Haldar
Department of Pharmaceutical Technology, Jadavpur University, Kolkata, West Bengal, India

School of Natural Product Studies, Jadavpur University, Kolkata, West Bengal, India

6.2 Production Processes

There are two stages of biopharmaceutical production—upstream processing and downstream processing.

6.2.1 Upstream Processing

Living cells such as those taken from patients that need to be genetically modified are used to produce biopharmaceutical drug substances using viral vectors. To create an ideal environment for cell development, it is important to choose the best cell line, culture medium, growth factors, and process optimization. Some crucial factors to take into account are the regulation of temperature, pH, oxygen supply, and the sanitation of supplies and equipment utilized in the upkeep of an environment free of microorganisms. The procedure involves identifying and isolating the gene that produces the same protein expression in a healthy organism and selecting an appropriate host cell type for genetic engineering. Strict control and cGMP compliance over the handling of these living cells are critical to the outcome of the upstream production, which includes the amount of desired biopharmaceuticals produced, the type and quantity of biomolecular structural variants formed, and the kind and also the amount of specific impurities that need to be managed during purification (Jozala et al. 2016; Gronemeyer et al. 2014).

6.2.2 Downstream Processing

Downstream processing is a crucial part of the purification of biological products. It involves three steps: (1) removing most contaminants through purification, (2) eliminating particular impurities and undesirable target biomolecule forms that might have developed during isolation and purification, and (3) polishing. Centrifugation, filtration, sedimentation, and flotation are the primary techniques used to separate the products. However, continuous downstream processing has been known to increase process complexity rather than efficiency in purification. Nevertheless, the benefits of high output in continuous working units outweigh this drawback (Mehta 2019).

The current and most commonly used purification methods include continuous centrifugation, which involves retrieving the product from the extracellular media, cleaning and recovering it from inclusion bodies, clarifying cell homogenates, and harvesting precipitated protein. Filtration is another method used to eliminate byproducts and concentrate culture media containing the desired expressed biomolecules. Cell lysis by physical, chemical, and enzymatic methods is yet another method used for the production of biopharmaceuticals like recombinant proteins such as interferon-β, interleukin-2, protein C, and proinsulin, which are found to be produced as inclusion bodies inside the cells (Mehta 2019).

6.3 Bacteria in Biopharmaceuticals Production

Bacteria can grow quickly on low-cost substrates, and researchers have extensively studied the method of producing recombinant proteins on a large scale in bioreactors. Microorganisms are a useful tool in producing recombinant proteins due to their metabolic power, diversity, and ability to adapt to various environmental conditions. The most commonly used microorganism is *Escherichia coli*, which has advanced genetics and a vast range of mutant host strains and cloning vectors. *E. coli's* high growth rate, basic media requirement, ease of handling, high yield, and economic viability make it the most suitable for large-scale biopharmaceutical production, and *E. coli* can also produce recombinant proteins up to 80% of its dry matter (Sobolewska-Ruta and Zaleski 2019; Pham et al. 2019).

The incidence of diabetes is increasing globally, which is leading to an increase in demand for recombinant insulin therapies. For many years, insulin derived from cows and pigs was used to treat diabetes (Jozala et al. 2016). However, this was not the best option because many people experienced adverse reactions such as serum sickness. Genetic engineering made it possible to transfer genes across different biological species with ease and paved the way for efficient insulin manufacturing. Downstream processing of recombinant human insulin/analogue manufacture from *E. coli* inclusion bodies has been carried out. Recombinant human insulin derived from DNA was first introduced by Eli Lilly & Company in 1982 and by Novo Nordisk in 1988 (Alyas et al. 2021). Humulin, the first of several recombinant insulin to be licensed for use in general medicine, is made by Eli Lilly and Genentech. Recombinant insulin is still mainly produced in *E. coli*, where inclusion bodies are often used to express human insulin and its analogues. To generate sufficient amounts of proinsulin, recombinant *E. coli* is used. *E. coli* is transformed with proinsulin producing plasmids to create this recombinant protein. The plasmid includes proinsulin coding genes and a kanamycin mono sulphate resistance gene, which prolongs its survival rate (Siew and Zhang 2021). Next, the transformed cells are cultured on tryptic soy broth supplemented with kanamycin mono sulphate, an antibiotic. *E. coli* cells that have not undergone transformation are eliminated by kanamycin mono sulphate. To maximize yields during product synthesis, the culture and fermentation conditions are closely monitored. Biologically active human insulin is produced by post-translational processing and thorough purification of the resultant medicinal product (Alyas et al. 2021).

The *Pseudomonas* genus is one of the most complex bacterial genera that researchers have investigated, showing a great deal of metabolic and physiological variety. A thorough investigation of *Pseudomonas* genomes and proteomics has led to a greater understanding of the system and molecular level of *Pseudomonas* synthetic processes. Due to its property of a vast array of genetic manipulation, fast growth rate, and the ability to secrete proteins, *Pseudomonas fluorescens* has emerged as an expression system for antibiotics production. To produce G-CSF, the human g-csf gene was cloned into a periplasmic expression vector. The created plasmids were then inserted into the *P. fluorescens* chassis strain, which has a specific phenotype for soluble expression of G-CSF (Wang et al. 2020).

Recombinant *Bacillus brevis* was used to secrete about 100 mg L^{-1} of a Fab against human urokinase-type plasminogen activator. Besides *Bacillus brevis*, *Bacillus choshinensis*, and *B. megaterium* were employed to produce antibody fragments with high binding affinity to antigens, which also come under the category of biopharmaceuticals. Growth factors like the human epithelial growth factor (hEGF), used to treat diseases like cancer, are produced from recombinant *Bacillus brevis* HPD31 and *B. subtilis*. Interferon alpha, beta, and gamma are produced by several recombinant *B. subtilis* strains cultured on a complex medium (Lakowitz et al. 2018).

6.4 Role of Fungi in Biopharmaceuticals

With the first successful report (1978) on the modification of *Saccharomyces cerevisiae* by plasmid DNA, the age of genetic engineering in yeasts was launched. It is used in the production of vaccinations against the human papillomavirus (HPV) and human serum albumin, as well as therapeutic proteins, peptides, and antibodies. The hepatitis B vaccine being one of the recombinant vaccines had been produced from the *S. cerevisiae* (yeast) unlike the production process of conventional vaccines which was time consuming and had the risk of pathogenicity (Kulagina et al. 2021). Also, subunit vaccines and virus-like particle (VLP)-based vaccines prefer using yeast as the heterologous expression system in their process of production. β-D-glucose polysaccharides, or β-glucans, are present in the cell walls of yeasts, and their immunomodulatory and adjuvant properties make them ideal for use in WYB vaccines. The β-glucans have an affinity to bind to specific receptors of immune cells like the dectin-1 receptor, the complement receptor 3 (CR3), and toll-like-receptors (TLRs). This property of binding to these specific receptors renders β-glucans as anticancer agents. Cytotoxic T cell responses against tumours and enhancement of haemopoiesis also proved β-glucans as anticancer agent in clinical trials. Human pathogen contamination poses a risk to the labour-intensive, expensive, and time-consuming process of purifying therapeutic human proteins from blood or tissue extracts and due to the manufacture of human therapeutic proteins in genetically modified yeast, bacteria, and also cultured mammalian cells, a variety of recombinant therapeutic proteins are now available for patients, resolving these concerns. All manufacturing methods, meanwhile, have their limitations. Yeast and bacteria, for example, are unable to synthesis complex human proteins including genuine post-translational modifications. Although only "simple" proteins are currently produced in yeast and bacteria, efforts are underway to humanise these non-mammal systems (Roohvand et al. 2017).

Pichia pastoris (*Komagataella* species) is yeast that uses methanol as a carbon source. It is the first known methylotrophic yeast that may be used to produce heterologous proteins utilising the methanol-inducible, endogenous AOX1 promoter.

Also, this species possess comparatively lesser degree of hypermannosylation so there will be lesser problems of post-translational modifications in their products. These features make it to be a well-enough option to be genetically manipulated in order to use it as a host to produce biopharmaceuticals (Ahmad et al. 2014). *P. pastoris* was found to secrete biologically active human growth hormone (HGH) efficiently, according to a patent application. In addition to HGH, *P. pastoris* is also used in producing human calcitonin (hCT) and human alpha-fetoprotein (AFP), peptide hormones that control blood levels of calcium and phosphate. AFP, or alpha foetal protein, is a plasma protein that is produced by the liver during foetal development and is used as a tumour marker in cancer therapy.

Genetically modified *Y. lipolytica* strains have found their application in biopharmaceuticals production, specifically in enzyme replacement therapy, and AzurRx Biopharma, Inc. (Brooklyn, NY, USA/Langlade, France), INRA and also Mayoly Spindler (Chatou, France) together developed lipase enzyme from these strains which were found to overexpress LIP2. Pancreatic insufficiency has been treated with the help of this recombinant lipase enzyme. OXY2810, a recombinant human acid α-glucosidase produced from *Y. lipolytica*, is used to treat glycogen build up, that is, Pompe disease. A GCase decorated with M6P glycans is being developed to treat neuronopathic Gaucher disease, which is caused by glucocerebroside buildup in neuronal cells. Recombinant glucocerebrosidase (GCase) is undergoing preclinical testing for the treatment of Parkinson's disease (Madzak 2021).

6.5 Transgenic Animals as Recombination Protein Expression Medium

Transgenic animals prove to be quite a good platform for biopharmaceutical production due to the following factors: economically viable, appreciable productivity of synthesized proteins, and quality of products. Animal bioreactors can create recombinant proteins that are structurally close to their original forms owing to the evolutionary link that exists between humans and different mammal species. Numerous animal species, such as mice, cattle, pigs, goats, sheep, and rabbits, have been employed as bioreactors (Bertolini et al. 2016). The expression method of complex recombinant proteins in milk is thought to be the best option currently available for the production of biopharmaceuticals when it comes to animal platforms. The production of a targeted protein in milk can be directed towards the mammary gland by using specific promoters for the gland, such as αs1-casein and whey acid protein (WAP). The technologies related to the purification of recombinant proteins produced from milk have dramatically advanced, enabling the completion of several specialised studies for the synthesis of biopharmaceuticals in transgenic animal milk (Chaible et al. 2010). The mammary gland's expression system is the sole animal platform that can produce recombinant proteins for approved commerce for therapeutic use.

For example, in 2006 the European Medicines Agency (EMA) approved the marketing of the first recombinant antithrombin drug, Atryn, produced from transgenic animals in European Union countries. Goats and rabbits collaborated to approve Ruconest~ (Pharming) which is a recombinant human C1 esterase inhibitor protein (C1INH), another biopharmaceutical derived from milk which was approved by European Medicines Agency in 2012 and FDA in 2014 (Bertolini et al. 2016). The animal platform has been proven to be a very reliable method for producing biopharmaceuticals through testing, approval, and growing convergence.

The production of biopharmaceuticals through animal transgenesis is still being developed. While significant progress has been made, convincing advancements must be made to guarantee efficiency in lowering transgenesis costs and allow the process to be implemented on an industrial scale. Additional advancements in the field are provided by novel molecular genetic techniques in conjunction with recent developments in genome annotation and sequencing, in vitro embryo creation, and reproductive technologies like nuclear transfer. Transgenic animals have a number of advantages over cultured cells, chief among them being the simplicity with which a huge number of useful proteins may be created and extracted. This is a result of both the biology of the producing cell and the "low tech" character of the fundamental production method, which uses animals in the field rather than intricate fermentation containers (Shakweer et al. 2023).

Important sources of production of biopharmaceuticals

Sources	Description	Reference
Milk	• The expression method of complex recombinant proteins in milk is thought to be the finest option currently available for the production of biopharmaceuticals when it comes to animal platforms • The mammary gland shows great potential to synthesize proteins and other biochemicals for baby nourishment during lactation. Milk contains a complex and wealthy combination of proteins, fats, and carbohydrates. The primary milk protein is casein, which comes in five different forms in mice, four in sheep and also cows, and two in humans. The milk of transgenic animals can express a vast array of foreign proteins • Human α antitrypsin has been successfully expressed in sheep to levels of up to 35 g/L using the ovine BLG promoter. Complex multi-chain proteins like fibrinogen and transgene combinations intended to enhance the lactating mammary glands innate capacity for protein processing are examples of the work that has been done in this area • In 2006, the European Medicines Agency (EMA) approved the marketing of the first recombinant antithrombin drug produced from goat milk, Atryn, in the European Union countries. Recombinant proteins such as human IGF-1, hGH, human lysozyme, human lactoferrin, human parathyroid hormone, and anti-clotting factors such as protein C, Antithrombin III, factor VII, factor VIII, factor IX, fibrinogen, and tissue plasminogen activator have also been successfully expressed in transgenic animals	Niemann et al. (2012), Bertolini et al. (2016)

Sources	Description	Reference
Urine	• The urinary system has some advantages over the mammary gland when it comes to producing certain proteins. Urine is formed by animals well before the recombinant protein and lactation can mature more effectively. Membrane-associated proteins called uroplakins are only expressed in the evolved uroepithelium of the bladder and urethra • Scientists have successfully directed human granulocyte macrophage-colony stimulating factor (hG-CSF) and human growth hormone (hGH) to express in mice using the mouse uroplakin II gene promoter. However, a major challenge with this process of production is the lower capacity for synthesis in the kidney and bladder compared to the mammary gland	Niemann et al. (2012), Bertolini et al. (2016)
Blood	• Transgenic animal blood is a growing source of human polyclonal antibodies due to the development of recombinant human proteins. Improved immune systems in cattle, pigs, and also rabbits are being developed • These animals can be vaccinated with a broad range of antigens to produce a plentiful supply of human polyclonal antibodies. These antibodies have significant advantages over monoclonal antibodies and can be useful in passive immunotherapy. It is possible to express proteins using animal blood as a bodily fluid • Examples of recombinant proteins produced in the blood of transgenic animals include human a1-antitrypsin in the serum of transgenic rabbits and transchromosomic cattle producing human polyclonal antibodies	Niemann et al. (2012), Bertolini et al. (2016)
Seminal fluid	• Semen is a bodily fluid produced by male sex glands in significant amounts (200–300 mL each time) • Studies suggest that porcine seminal fluid is a rich source of bioactive proteins. This is because the protein secretion process is exocrine, meaning it doesn't harm the animal's health • Seminal plasma can be a great source for producing bioactive proteins due to its high capacity for protein processing, stability, and purification	Bertolini et al. (2016)

6.6 Transgenes in Production of Genetically Modified Organisms

Pronuclear microinjection-introduced transgenes are intended to operate upon insertion into various locations within the host DNA. The number of copies of the transgenic, the place of insertion into the host genome, and the regulatory elements—that is, the sequences that control transcription, RNA processing, and translation—all affect how the transgene expresses itself. Before a desired gene can be successfully cloned and put into the organism, it must go through a number of changes. To enable gene expression, a promoter sequence is inserted. The transgenes must guide the profuse production and specific protein transport to the intended organ while preserving the health and welfare of the animal. They also need to gradually merge into the host genome and be transferred in a Mendelian fashion (Meyer 1995; Kaundal et al. 2014).

It could be required to modify codons to match those that are most commonly utilised by the host if the protein originates from progressively distant species. Additionally, if the protein is not normally released, the generating cell may be directed to secrete if a signal peptide is inserted at the N-terminus. A lot of genes are too big to fit neatly into plasmid cloning vectors. A "minigene" comprising two or one introns and 3′ or 5′ flanking regions believed to contain +ve regulatory elements is frequently created by combining cDNA and genomic sequences. Nevertheless, cryptic splice sites can occasionally result in abnormal RNA splicing when mixing and matching in this manner. The placement of the transgene within the host genome has a significant role in its ability to be expressed. The closeness of activation sequences, silencers, promoters, and enhancers that are found naturally may also affect the transgenic expression pattern and intensity. Sometimes flanking the transgenic construct with insulator elements improves the level, consistency, and also specificity of transgene expression. Research has demonstrated that several types of insulators can keep surrounding natural enhancers and other regulatory components from influencing transgenic promoters. They can also stop heterochromatin effects from extending into the transgene (Niemann et al. 2012).

6.7 Challenges Faced in Production Due to Post-Translational Modifications in Some Organisms

6.7.1 Glycosylation

One of the main obstacles to yeasts producing viable human therapeutic proteins is thought to be proper post-translational modification like glycosylation. The low number of authorised biopharmaceuticals from plant, insect, and yeast platforms is mostly due to unfavourable glycosylation. Biopharmaceuticals that need highly precise and intricate N-glycans similar to those found in humans must therefore be created using mammalian expression platforms (Roohvand et al. 2017). Recombinant human (rh) glycoproteins differ from their endogenous counterparts due to glycosylation, which is species and cell-specific as well as culture-dependent which affects the accuracy of manufacturing methods. Proper glycosylation is also crucial for mAbs, the primary product class of biopharmaceuticals. The majority of monoclonal antibodies (mAbs) belong to the IgG class and have a glycosylation site in the Fc region at amino acid position asparagine (Asn)–297; however a small fraction of mAbs have an additional site in the Fab region. Cell engineering focuses a lot of attention on the sugar moieties since they frequently directly affect pharmacokinetics, pharmacodynamics, and immunogenicity. The degree of sialylation and mannosylation can have an effect on the proper functioning of the mAb (Amann et al. 2019).

6.7.2 Carboxylation

Biopharmaceuticals requiring carboxylation must be expressed on platforms derived from multicellular species, ideally from mammalian hosts, as carboxylation activity is only documented inside multicellular organisms, such as mammals and Drosophila (Amann et al. 2019).

6.7.3 Tyrosine Sulfation

Appropriate tyrosine sulfation appears to affect biological activity and binding in addition to modulating protein stability and also biopharmaceuticals requiring sulfotyrosine, like recombinant factor VIII (rhFVIII), are produced in mammalian cell systems because tyrosine sulfation cannot be carried out by *E. coli* or other prokaryotic expression platforms (Ezban et al. 2014).

6.8 Conclusion

Importance for biopharmaceuticals is rising due to increasing incidence of chronic diseases. Treatment for autoimmune diseases, genetic problems, and chronic illnesses like cancer has been transformed by biopharmaceuticals. Genetically modified organisms, from microorganisms to transgenic mammals, are characterized to aid in the development of biopharmaceuticals. They are substantially rich in pharmacological and pharmacokinetic properties. Genetic engineering has proved to be a boon in the pharmaceutical field. Despite few limitations in the production process, biopharmaceuticals are being successfully manufactured defying all odds. Therefore, more and more research should be carried out to contribute to one of the most pivotal industries in the healthcare arena.

References

Ahmad M, Hirz M, Pichler H, Schwab H (2014) Protein expression in Pichia pastoris: recent achievements and perspectives for heterologous protein production. Appl Microbiol Biotechnol 98:5301–5317

Alyas J, Rafiq A, Amir H, Khan SU, Sultana T, Ali A, Ahmad A (2021) Human insulin: history, recent advances, and expression systems for mass production. Biomed Res Ther 8(9):4540–4561

Amann T, Schmieder V, Faustrup Kildegaard H, Borth N, Andersen MR (2019) Genetic engineering approaches to improve posttranslational modification of biopharmaceuticals in different production platforms. Biotechnol Bioeng 116(10):2778–2796

Bertolini LR, Meade H, Lazzarotto CR, Martins LT, Tavares KC, Bertolini M, Murray JD (2016) The transgenic animal platform for biopharmaceutical production. Transgenic Res 25:329–343

Bhatia S, Goli D (2018) History, scope and development of biotechnology. In: Introduction to pharmaceutical biotechnology, vol 1, pp 1–61

Chaible LM, Corat MA, Abdelhay E, Dagli ML (2010) Genetically modified animals for use in research and biotechnology. Genet Mol Res 9(3):1469–1482

Ezban M, Vad K, Kjalke M (2014) Turoctocog alfa (NovoEight®)–from design to clinical proof of concept. Eur J Haematol 93(5):369–376

Gronemeyer P, Ditz R, Strube J (2014) Trends in upstream and downstream process development for antibody manufacturing. Bioengineering 1(4):188–212

Jozala AF, Geraldes DC, Tundisi LL, Feitosa VDA, Breyer CA, Cardoso SL, Pessoa A Jr (2016) Biopharmaceuticals from microorganisms: from production to purification. Braz J Microbiol 47:51–63

Kaundal R, Yang Y, Nottoli T, Chi T (2014) Towards the molecular mechanisms of transgenerational epigenetic inheritance: insights from transgenic mice. In: Transgenerational epigenetics. Academic Press, pp 75–85

Kesik-Brodacka M (2018) Progress in biopharmaceutical development. Biotechnol Appl Biochem 65(3):306–322

Kulagina N, Besseau S, Godon C, Goldman GH, Papon N, Courdavault V (2021) Yeasts as biopharmaceutical production platforms. Front Fungal Biol 2:733492

Lakowitz A, Godard T, Biedendieck R, Krull R (2018) Mini review: recombinant production of tailored bio-pharmaceuticals in different bacillus strains and future perspectives. Eur J Pharm Biopharm 126:27–39

Madzak C (2021) Yarrowia lipolytica strains and their biotechnological applications: how natural biodiversity and metabolic engineering could contribute to cell factories improvement. J Fungi 7(7):548

Mehta A (2019) Downstream processing for biopharmaceuticals recovery. In: Pharmaceuticals from microbes: the bioengineering perspective, pp 163–190

Meyer P (1995) Understanding and controlling transgene expression. Trends Biotechnol 13(9):332–337

Niemann H, Kind A, Schnieke A (2012) Production of biopharmaceuticals in transgenic animals. In: Pharmaceutical biotechnology: drug discovery and clinical applications, pp 71–111

Pham JV, Yilma MA, Feliz A, Majid MT, Maffetone N, Walker JR, Yoon YJ (2019) A review of the microbial production of bioactive natural products and biologics. Front Microbiol 10:1404

Roohvand F, Shokri M, Abdollahpour-Alitappeh M, Ehsani P (2017) Biomedical applications of yeast-a patent view, part one: yeasts as workhorses for the production of therapeutics and vaccines. Expert Opin Ther Pat 27(8):929–951

Shakweer WME, Krivoruchko AY, Dessouki SM, Khattab AA (2023) A review of transgenic animal techniques and their applications. J Genet Engineer Biotechnol 21(1):1–14

Siew YY, Zhang W (2021) Downstream processing of recombinant human insulin and its analogues production from E. coli inclusion bodies. Bioresour Bioprocess 8(1):1–27

Sobolewska-Ruta A, Zaleski P (2019) Cell banks preparation in biopharmaceuticals production. Post Mikrob Adv Microbiol 58(1):87–100

Strohl WR, Knight DM (2009) Discovery and development of biopharmaceuticals: current issues. Curr Opin Biotechnol 20(6):668–672

Vecchio I, Tornali C, Bragazzi NL, Martini M (2018) The discovery of insulin: an important milestone in the history of medicine. Front Endocrinol 9:613

Walsh G (2000) Biopharmaceutical benchmarks. Nat Biotechnol 18(8):831–833

Wang S, Cui J, Bilal M, Hu H, Wang W, Zhang X (2020) Pseudomonas spp. as cell factories (MCFs) for value-added products: from rational design to industrial applications. Crit Rev Biotechnol 40(8):1232–1249

Chapter 7
Drug Delivery in Biotechnology: Present and Future

Seema Yadav, Narahari Narayan Palei, Subas Chandra Dinda, and Arghya Kusum Dhar

Abstract Nanobiotechnology is the integration of nanotechnology with biotechnology, with a specific emphasis on the utilisation of nanoscale techniques in biological systems. Nanobiotechnology significantly contributes to the field of drug delivery by utilising nanoscale materials, such as nanoparticles and liposomes, to enhance the effectiveness of medications. Nanoparticles possess a reduced size and altered properties that provide accurate drug localisation, regulated release, and enhanced bioavailability. This offers distinctive resolutions to challenges encountered in conventional drug administration techniques. Herein we highlight the significance of achieving optimal drug delivery and subsequently explore the potential of biotechnology to address these challenges. We explore contemporary methodologies, including the utilisation of nanotechnology for targeted drug delivery. Additionally, we discuss intelligent drug delivery systems, including those that are triggered by specific inputs or guided by artificial intelligence. The development of biodegradable implants, the use of artificial intelligence for personalised treatment, and the creation of microbes for the administration of drugs are also discussed. The chapter includes coverage of cutting-edge technology such as 3D printing for the customisation of medication forms and continuous monitoring devices, highlighting the important ethical inquiries and safety hazards linked to these novel methods. Finally, it provides a concise overview of the key progress made and anticipated breakthroughs in drug delivery that have the potential to transform healthcare.

S. Yadav · N. N. Palei
Amity Institute of Pharmacy, Amity University Lucknow Campus, Lucknow, Uttar Pradesh, India

S. C. Dinda · A. K. Dhar (✉)
School of Pharmacy, The Neotia University, Sarisa, West Bengal, India

S. Bose et al. (eds.), *Concepts in Pharmaceutical Biotechnology and Drug Development*, Interdisciplinary Biotechnological Advances,
https://doi.org/10.1007/978-981-97-1148-2_7

Keywords Nanobiotechnology · Drug delivery · Monoclonal antibody · 3D printing

7.1 Introduction

The topic of medication delivery is expanding quickly and attracting the interest of industry, scientists, and pharmaceutical companies. Pharmaceutical researchers' "holy grail" is developing efficient drug delivery systems (DDSs) that can deliver a medication precisely, safely, and to the intended location of action (Morris 2022). In fact, a plethora of new delivery systems are developed every year, and almost every bodily region has been investigated as a possible delivery system for both established and cutting-edge medications. As a result, effective methods for delivering peptides, proteins, and medicines that are poorly soluble have been developed. Furthermore, a great deal of research is now being done on innovative nasal drug delivery systems, nanodevices, bio-adhesive systems, micro-fabricated systems, implants, transdermal patches, and cell encapsulation devices. These and other innovations are causing a significant shift in the drug delivery sector. A new market for drug delivery is emerging because of increasing demand brought about by market extension, reduced drug development budgets, and novel formulations that provide a competitive edge once patents expire. In 2003, the drug delivery systems market in the US was valued at approximately $43.7 billion. It was projected to experience an average annual growth rate of 11.3% and reach $74.5 billion by 2008 (Burrows and Lambrix 2022). This information comes from a recent study from Business Communications Company Inc. Within the broad topic of biotechnology, medication delivery represents an important interface between basic science and therapeutic use (Aslam et al. 2022). The key to increasing the effectiveness of medications while reducing unwanted side effects is the capacity to precisely and efficiently distribute therapeutic chemicals to target areas within the body. Thanks to developments in biotechnological research and development, the design and implementation of DDSs have advanced significantly over the past few decades (Aslam et al. 2022). As a multidisciplinary area, biotechnology uses the ability to manipulate biological systems, cells, and live beings to create novel solutions for problems in healthcare. Biotechnology goes beyond traditional drug delivery techniques to enhance pharmaceutical substances' pharmacokinetics, bioavailability, and targeted delivery. Addressing the intricacies of many diseases, such as cancer, viral diseases, autoimmune disorders, and hereditary problems, requires a comprehensive approach. The fusion of biology, chemistry, engineering, and material science characterises the current biotechnology drug delivery landscape (Liu et al. 2022). Scholars are presently investigating novel platforms for encapsulating and delivering medicinal substances, including liposomes, micelles, nanoparticles (NPs), and biocompatible polymers. With the ability to customise these carriers, it will be possible to achieve site-specific delivery, sustained activity, and controlled release with a degree of accuracy that was previously unattainable with traditional drug administration techniques. This review addresses new technologies being researched to get around some of the present challenges with drug delivery and highlights innovative drug delivery systems that potentially enhance some traditional treatments now available on the market. We also discuss how these difficulties should be tackled to develop the profession and offer our perspective on the drug delivery research policies (Birla et al. 2022).

DDSs have come a long way, but there are still several issues that need to be resolved, which calls for continued research and development. Overcoming biological barriers to guarantee the secure and efficient delivery of therapeutic payloads to tissues or cells is one of the main obstacles. A number of complex defence systems found in the human body, such as blood-brain barrier (BBB) and other physiological barriers, can prevent medications from reaching their intended targets (Mandal et al. 2023). Nanotechnology has become an increasingly potent weapon in the medication delivery arsenal as a response to these obstacles. Nanoscale delivery agents, like NPs, have demonstrated the capacity to get past immunological reactions, cross physiological barriers, and release their payload at the desired location of action. Additionally, the surface of these nanocarriers can be altered to improve their targeting efficiency, enabling the more accurate delivery of therapeutic medicines. Finding a balance between safety and therapeutic efficacy is a significant difficulty in drug delivery. At high concentrations, many pharmaceuticals show strong effects, but this is frequently at the expense of increased toxicity and unfavourable responses. By offering regulated release and long-term maintenance of therapeutic levels in the body, drug delivery systems seek to address these problems. This improves patient adherence and quality of life while minimising side effects and increasing treatment efficacy overall (Awad et al. 2022). Drug delivery has advanced significantly in the last several years because of researchers' exploration of novel materials and methods to address long-standing issues. Because of their special physicochemical characteristics, NPs have gained a lot of interest in drug delivery studies (Park et al. 2022). These submicron-sized carriers provide protection and regulated release for a range of therapeutic substances, like proteins, nucleic acids, and tiny compounds. Another class of nanocarriers are liposomes, which are made of phospholipid bilayers and can hold both hydrophilic and hydrophobic medications. Antibiotics, vaccinations, and anticancer medications have all been delivered effectively using liposomal formulations. The fact that they can increase drug solubility, increase the duration of circulation, and improve drug accumulation at the target location has led to their extensive application in preclinical and clinical settings (Hersh et al. 2022). Drug delivery by the formation of micelles, which are created when amphiphilic molecules self-assemble in aqueous solutions, is another exciting prospect. These nanostructures can increase the bioavailability of hydrophobic medications by solubilising them. To further increase their versatility in drug delivery applications, micelles' amphiphilic nature permits the integration of both hydrophobic and hydrophilic components. Poly (lactic-co-glycolic acid) (PLGA) and polyethylene glycol (PEG), two biocompatible polymers, are now essential parts of DDS designs. Because of the adjustable qualities of these polymers, carriers with release patterns and degradation kinetics can be created. Furthermore, these materials' biocompatibility lowers the possibility of toxicity and immunological reactions, which qualifies them for a range of therapeutic uses (Sabbagh and Kim 2022).

The creation of tailored delivery systems that can precisely reach and interact with particular cells or tissues is one of the main goals of drug delivery research. Conventional systemic medication delivery frequently causes therapeutic substances to diffuse throughout the body, increasing toxicity and producing off-target effects. By delivering therapeutic molecules specifically to the desired location of action, targeted medication delivery aims to overcome these problems. Improvements in imaging technology, genetics, and molecular biology have made targeted drug delivery possible (Mohammed et al. 2022). It is now possible to create carriers that can recognise and bind to specific biomarkers linked to diseases according to research findings. This focused strategy reduces exposure to healthy tissues, which lessens side effects, while simultaneously improving drug accumulation at the illness site. Targeted drug delivery has demonstrated special promise in cancer therapy. Ligands that preferentially bind to receptors overexpressed on cancer cells can be used to functionalize NPs. By enhancing the specificity of medication delivery, this active targeting approach raises the concentration of therapeutic medicines in tumour tissues. Furthermore, the passive accumulation of NPs inside the tumour microenvironment is facilitated by the enhanced permeability and retention (EPR) effect, which is frequently observed in tumours because of leaky vasculature (Abdella et al. 2023). A new age in healthcare has begun with the development of personalised medicine, wherein patient-specific medicines are designed to meet their individual needs. Personalised medicine makes it possible to tailor therapeutic approaches for medication delivery in accordance with a patient's unique characteristics, disease profile, and genetic composition (Chilkoti et al. 2002). This strategy could reduce unfavourable reactions and maximise therapeutic results. Since genomics makes it possible to identify genetic differences that may affect drug response, personalised medicine is essential. Pharmacogenomics is the study regarding how a person's genetic makeup influences how they react to drugs, helping determine the right dosages and treatment plans. By incorporating pharmacogenomic data into drug delivery plans, customised methods that consider a patient's genetic susceptibility to particular diseases and how they react to particular medications can be developed (Wening and Breitkreutz 2011).

Personalised medicine has contributed to the development of targeted cancer medicines that utilize the unique molecular features of individual cancers. Anticancer drugs can be delivered to cancer cells directly by targeted drug delivery systems, protecting healthy tissues from the harmful effects of chemotherapy. This raises the therapeutic index and lowers the chance of developing resistance, which is a typical problem in the treatment of cancer (Mura and Couvreur 2012). With the introduction of gene therapy—a groundbreaking method with the potential to treat and perhaps cure genetic disorders—the field of drug delivery is quickly changing. To address or modify the underlying cause of an illness, gene therapy entails the insertion, modification, or deletion of genetic information within a patient's cells. One of the most crucial aspects of gene therapy is getting therapeutic genes into target cells, and developing DDSs is essential to maximising the benefits of this novel strategy.

Looking ahead to biotechnology, the use of cutting-edge technologies has the potential to change the landscape of drug delivery yet more. With its ability to analyse data, identify patterns, and make decisions, artificial intelligence (AI) is finding more and more uses in the optimisation of medication delivery systems (Mirza and Karim 2021). Therapeutic outcomes may be improved by intelligent drug delivery systems that may modify their release profiles in response to physiological inputs occurring in real time. Large-scale datasets can be analysed by machine learning algorithms to forecast patient reactions, optimise dosage schedules, and customise treatment plans for optimal effectiveness. The idea of "smart" medication delivery systems—in which nanocarriers with sensors and actuators may react dynamically to the body's microenvironment—is becoming more and more popular. Therapeutic compounds can be supplied precisely when and where they are needed thanks to these systems, which can modify their release kinetics in response to changes in temperature, pH, or the presence of particular biomarkers (Orive et al. 2003). A synergistic approach to personalised and adaptive medicine, where treatment plans are continuously improved depending on specific patient reactions, is represented by the marriage of AI and drug delivery technology. The development of precision medicine, driven by advancements in genomes and molecular diagnostics, will be crucial in determining how drugs are delivered in the future. Targeted therapies can be developed through the identification of certain biomarkers linked to illness susceptibility, progression, and response to treatment. Treatment plans can be tailored using biomarker-driven medication delivery strategies to each patient's specific genetic composition and disease profile. This strategy may improve treatment efficacy while minimising side effects, ushering in a time when there is no longer a one-size-fits-all approach to medicine. Beyond conventional small-molecule medications, biologics and nucleic acid-based medicines will continue to grow in the field of biotechnology drug delivery (Allami and Yousif 2023). Advanced DDSs will progressively incorporate gene editing technologies, RNA-based therapies, and monoclonal antibodies. The difficulties in delivering big, complicated biomolecules will be overcome by creative methods, such as specially designed NPs that can shield and carry these delicate payloads to their destinations. To sum up, medication delivery in biotechnology is a vibrant, diverse discipline that has made significant strides recently. The current environment is defined by the convergence of targeted therapies, personalised medicine, and nanotechnology, which presents new opportunities for the treatment of various diseases. Though there are still obstacles, research and technology advancements keep expanding the realm of possibility (Hamamoto et al. 2020). Future developments in artificial intelligence, precision medicine, and sophisticated medication delivery technologies have the potential to completely transform the healthcare industry. There is promise for safer, more effective, and patient-centred therapeutic treatments soon with the development of intelligent, adaptable DDSs and the achievement of completely individualised medicine. Drug delivery's path from bench to bedside is evidence of science's unwavering quest of greatness, with each discovery bringing us one step closer to a time when illnesses will be treated with never-before-seen accuracy and effectiveness.

The purpose of medications and pharmacies is to safely and consistently deliver any drug towards a specific target in the appropriate dosage at the proper time (Kaida-Yip et al. 2018). These ideal conditions, however, are more hype than reality for many medications. For instance, adequate peptide or protein medication delivery has not yet been achieved via the oral route, even though it is one of the recommended drug delivery modalities because of its non-invasive character. This is caused in part by the stomach's acidity, the liver's first-pass effect (drug loss due to metabolism processes prior to entering the systemic circulation), and the intestinal wall's resistance, which either alters, destroys, or reduces absorption of almost all macromolecules, lowering their bioavailability (Kos et al. 2018). Because of this, millions of diabetics globally are forced to administer insulin shots to themselves on a regular basis, which leads to a high rate of treatment non-compliance. Nasal medication delivery and injection are two more often used drug administration techniques. Polar compounds are poorly absorbed when administered via nasal administration, whereas injection causes discomfort and makes patients reluctant to utilise it. The application of medication through the skin is one substitute for these methods. Drugs need to get through the outer membrane, known as stratum corneum, which provides a barrier against material diffusion. Additional significant obstacles are the intracellular distribution of macromolecular medications, the long-term injection of medications into tumours, and the specific delivery of glycoproteins. In addition to affecting traditional medication administration methods and dosage forms, the absence of appropriate DDSs is impeding the development of cutting-edge therapeutic approaches including gene therapy, RNA interference (RNAi), and cell therapy (Burrows and Lambrix 2022). A virus is employed as a vector in gene therapy to introduce active and corrected genes into a patient's cells. However, the vector can create uncontrollable mutations and cancer if it inserts itself into a cell's DNA. The tragic death of a volunteer in a gene therapy experiment in 1999 (Liu et al. 2022) and the recent development of leukaemia in two children opting gene therapy for the severe combined immune deficiency illness (Birla et al. 2022) demonstrate how serious this therapy is. Likewise, for RNA interference (RNAi) therapies to fulfil their potential, researchers must devise clever mechanisms to safeguard small interfering RNAs (siRNAs) in bloodstream and direct them towards the appropriate cells. The disadvantages have incentivised numerous pharmaceutical and biotech start-ups to create DDSs that can operate both safely and effectively. However, there are still no scientific guidelines that offer a clear path forward for the development of drug delivery technology (Tsiftsoglou et al. 2013). Drug delivery in biotechnology is a dynamic and transformational frontier in the field of healthcare, revolutionising the way therapeutic agents are supplied, targeted, and controlled within the human body. The advancement of drug delivery techniques within the area of biotechnology has proven critical in improving treatment efficacy and safety while minimising side effects.

7.2 Current Drug Delivery Systems in Biotechnology

DDSs in biotechnology have made considerable advances in recent years, especially in the development of innovative nanocarriers for the effective transportation and controlled release of the drug molecules. These nanostructured systems, which include polymeric micelles, hydrogels, dendrimers, and liposomes, were developed to minimise the drug dose required to produce a given therapeutic effect, reducing costs and related adverse effects (Lombardo et al. 2019).

The biotechnology landscape of DDSs (Fig. 7.1) is characterised by a variety of sophisticated techniques aiming to improve the therapeutic effects of drugs. These systems make use of cutting-edge technology to improve the distribution of drugs, release kinetics, or target selectivity (Orive et al. 2003).

7.2.1 Nanotechnology

Nanotechnology is currently at the cutting edge of research. Polymeric research is currently playing a prominent role in the evolution of nanotechnology by providing regulated release of therapeutic drugs in constant doses over long periods of time, cyclic dosage, and adjustable release of both hydrophobic and hydrophilic drugs. Various biomaterials can be employed for this purpose, having a wide range of chemical properties and potential for further modification with NPs (Bhatia 2016).

NPs have been discovered as the most effective careers in the delivery of conventional medicines, vaccines, nucleotides, and recombinant proteins among all current advanced delivery systems. Nanoparticulate DDSs can increase drug characteristics such as pharmacokinetic parameters (such as bioavailability, biodistribution, and

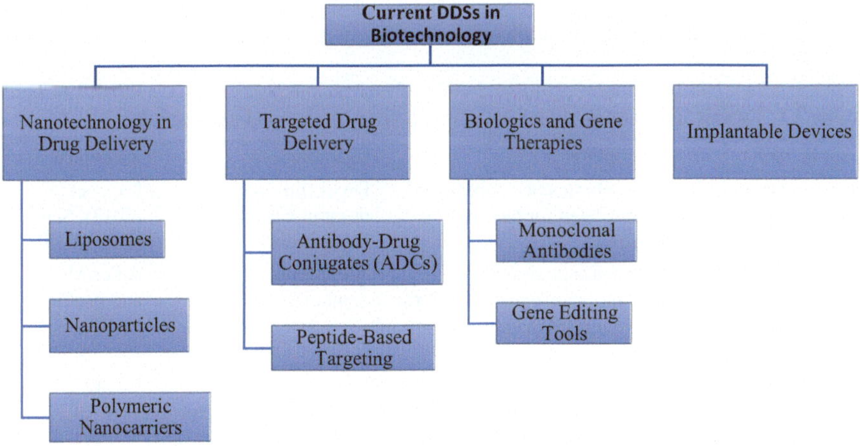

Fig. 7.1 Numerous drug delivery systems in biotechnology

drug release characteristics). Nanoparticulate DDSs enable the effective and controlled release of drugs with site-specific drug delivery, targeting cells or tissues with minimal adverse effects. As a result, the application of nanotechnology in the field of pharmaceutical biotechnology helps in the development of drug delivery strategies to solve the distribution issues of protein-based pharmaceuticals such as oligonucleotides and recombinant proteins, among others. Paradise et al. (2008) discuss some nanotechnologies used in biotechnology.

7.2.2 Liposomes

Liposomes are commonly utilised in biotechnology to support drug delivery. To initiate drug release processes, they have been functionalized using different stimuli-responsive mechanisms such as heat, ultrasound, light, magnetic fields, pH, enzymes, and redox. Furthermore, liposomes can combine cancer-targeting compounds for both therapy and diagnostic purposes, making them attractive for theranostic applications (Lombardo et al. 2019).

Furthermore, PEGylated liposomes were also exhibited to enhance stability, blood circulation time, and plasma clearance, making them useful at directing active therapeutic molecules to specific areas in biological systems mainly in the distressed tissues or tumours. The addition of ligands like peptides, monoclonal antibodies (MAbs), growth factors, and aptamers improves the specificity of liposome interactions during drug release. Overall, liposomes serve an important role in biotechnology by acting as adaptable nanocarriers for drug delivery, allowing for the targeted and regulated release of therapeutic substances with minimal adverse effects (Muthu and Feng 2013).

Liposomal nanotechnology is widely employed in biotechnology for a variety of applications (Table 7.1), utilising the unique features of liposomes to improve medication delivery, diagnostics, and therapies. Here are some significant uses of liposomal nanotechnology in biotechnology.

Table 7.1 Various applications of liposomal drug delivery system in biotechnology

Application	Description	Example	Reference
Cancer therapy	Chemotherapeutic substances encapsulated for targeted delivery	Doxil (liposomal doxorubicin) and Myocet	Cagel et al. (2017), Schütz et al. (2013)
Infectious diseases	Antibiotics or antiviral substances are encapsulated to allow targeted delivery to infected areas	AmBisome (liposomal amphotericin B)	Adler-Moore and Proffitt (2002)
Vaccine delivery	Encapsulation of the vaccine antigens for controlled release and increased stability	Epaxal (liposomal virosomal vaccine for hepatitis A)	Bovier (2008)

Table 7.1 (continued)

Application	Description	Example	Reference
Diagnostics (imaging agents)	Liposomes encapsulate imaging chemicals to improve tissue or cell visualisation	Liposomes containing Gd-DTPA for MRI (magnetic resonance imaging)	Guenoun et al. (2012)
Biological research	Liposomes can be used to deliver certain molecules into cells for study purposes	Liposomes loaded with fluorescent dye can be used for cellular research	Torchilin (2005)
Theranostic	Diagnostic and therapeutic properties combined in the same liposomal system	Doxil with imaging capabilities	Al-Jamal and Kostarelos (2011)
Biotechnological tools	Liposomes can be employed as biotechnological tools in various applications like gene transfection	Gene transfection of endothelial nitric oxide (NO) synthase using liposomes	Iwata et al. (2001)
Cosmeceuticals	Liposomes are utilised for topical administration of cosmetic and medicinal ingredients	Antioxidants incorporated in liposomes in skincare products	Van Tran et al. (2019)

7.3 Drug Delivery

Cancer Therapy—Chemotherapeutic chemicals are encapsulated in liposomes. Liposomes can aggregate selectively in tumour tissues thanks to the enhanced permeability and retention (EPR) effect, causing minimal damages to healthy cells. Myocet and Doxil (liposomal doxorubicin) are two examples (Cagel et al. 2017). Long-circulating liposomes can accumulate in infected locations with damaged vasculature via the EPR effect and have been employed for drug administration into tumours by passive accumulation on numerous occasions. Long-circulating liposomes exhibit non-saturable, dose-independent, log-linear kinetics as well as enhanced bioavailability (Torchilin 2011, Allen and Hansen 1991).

7.3.1 Vaccine Delivery

Liposomes are employed to encapsulate vaccine antigens to improve their long-term stability and immunogenicity. Controlled antigen release improves vaccine effectiveness (Schwendener 2014).

7.3.2 Gene Delivery

Liposomes are used in gene therapy applications to transport genetic material (DNA or RNA). They safeguard the genetic material while also facilitating cellular intake (Balazs and Godbey 2011).

7.4 Diagnostic Applications

7.4.1 Imaging Agents

Liposomes can incorporate imaging agents (e.g. contrast agents for MRI and fluorescent dyes to obtain optical imaging). Used to improve visibility of specific cells or tissues.

7.4.2 Theranostic

Diagnostic and therapeutic characteristics combined in the same liposomal system. It allows for simultaneous imaging and delivery of drugs, allowing for more personalised therapy (Shvets et al. 2013).

7.4.3 Biological Research

Cellular Studies—Liposomes are used as carriers in research to introduce molecules into cells. Used in the investigation of biological processes and interactions with drugs.

Intracellular Delivery—Liposomes with suitable modifications can carry drugs or biological substances within the cytoplasm of cells. Aids in intracellular targeting and cellular function regulation (Philippot and Schuber 1994).

7.4.4 Cosmeceuticals

Topical Delivery—Liposomes are employed in the topical administration of cosmetic and pharmaceuticals. Promotes active ingredient penetration into the skin (Rahimpour and Hamishehkar 2012).

7.4.5 *Nutraceuticals*

Nutrient Delivery—Liposomes may incorporate and transport nutrients or bioactive chemicals. Increases bioavailability and allows for targeted distribution to certain tissues (Khorasani et al. 2018).

7.4.6 *Research and Development*

Drug Formulation Studies—Liposomes are used at early stages of the drug development process to evaluate both the feasibility and efficacy of therapeutic formulations.

Biotechnological Tools—Liposomes are used in various biotechnological applications, including gene transfection using liposomes in research laboratories (Jesorka and Orwar 2008).

7.5 Challenges and Achievements

Circulation time and Stability—Long-circulating liposome (PEGylated liposome) formulation advances address challenges of stability and quick clearance.

Targeting Strategies—Continuous development of innovative targeting ligands and techniques for improved specificity in drug delivery.

Combination Therapies—Liposomal systems are being investigated for use in combination medicines to produce synergistic effects.

Liposomal nanotechnology in biotechnology is evolving as an outcome of ongoing research aimed at overcoming obstacles and unlocking new possibilities. Liposomes' adaptability makes them a valuable tool for adapting DDSs to meet specific biomedical needs (Shvets et al. 2013).

7.5.1 *Nanoparticles*

In the discipline of biotechnology, NPs play an important role in DDSs. They provide various advantages such as controlled release, targeted distribution, and improved therapeutic effects. NPs are particles having nm size, typically between 1 and 100 nm. Therapeutic substances can be delivered by NPs made of various types of materials, such as proteins, lipids, metals, and polymers.

Types of NPs

NPs employed in biotechnology vary in composition, size, and application. These are some of the most popular NPs used in biotechnology.

Lipid-Based NPs

Lipid-based NPs, typically made of lipids or lipid-like substances, are frequently used in the administration of drugs because of their biocompatibility. Liposomes and SLNs are few examples of lipid-based NPs. Liposomes are spherical lipid vesicles composed of lipid bilayers that are frequently utilised to encapsulate pharmaceuticals for targeted delivery like Doxil (liposomal doxorubicin) is a cancer treatment. SLNs are NPs with a solid lipid core that improves drug stability and controlled release of drugs. Hydrophobic drugs are delivered through SLN formulations (du Plessis et al. 2014).

Polymeric NPs

Polymeric NPs are biodegradable polymer-based carriers for drug delivery that provide targeted delivery and controlled release. Examples of polymeric NPs are PLGA NPs for anticancer drug delivery, and Chitosan NPs have been used to transport genes (Badwaik et al. 2019).

Metal-Based NPs

Metal-based NPs (MNPs) are made up of metallic elements and are used in imaging, diagnostics, and drug delivery. MNPs have been used in a variety of industries due to their distinct characteristics. They have many advantages because of their unique qualities, but they also have disadvantages because of their shape, size, surface area, composition, and charge. They are particularly used in medical therapy as carriers of medicines, biomolecules, and genes. MNPs have been employed as carriers and contrast agents in imaging and active or passive tumour cell targeting. These MNPs open the door for novel DDSs, site-by-site targeting, and the delivery of genes. Gold NPs are well known for their distinct optical features and are employed in photothermal therapy and imaging. Iron oxide NPs, like superparamagnetic NPs, are used in drug delivery and magnetic resonance imaging (MRI) (Sharma et al. 2022).

Dendrimers

Dendrimers are extensively branched macromolecules having a well-defined structure that, because of their controlled size and form, make them excellent for drug delivery. Examples: Polyamidoamine (PAMAM) dendrimers are widely employed in anticancer drug delivery and gene delivery applications (Boas et al. 2007).

Polymeric Micelles

Amphiphilic block copolymers self-assemble to generate polymeric micelles, which are NPs that improve the stability and solubility of pharmaceuticals. Pluronic micelles are triblock copolymers made of poly (ethylene oxide)-poly (propylene oxide)-poly (ethylene oxide) and are used to deliver poorly water-soluble drugs (Kwon 2003).

Carbon-Based NPs

Carbon-based NPs include graphene, carbon nanotubes, and fullerenes. These materials have unique characteristics that make them useful in a wide range of applications. Carbon nanotubes with functionalized surfaces enables targeted drug delivery (Majeed et al. 2020).

Silica NPs

Silica NPs are silicon dioxide particles that are well known because of their biocompatibility and flexibility of surface functionalization. Mesoporous Silica NPs (MSN) are NPs with arranged mesoporous structures that are useful for drug delivery and imaging (Rabiee et al. 2021). The versatility of NPs in biotechnology enables a wide range of applications, including drug delivery, imaging, and diagnostics. These are just a few examples of the enormous research and development that has gone into using NPs to improve medical treatments as well as diagnostics in this area of biotechnology.

Applications in Drug Delivery

NPs are playing an important role in drug delivery, with applications ranging from targeted therapy to controlled release, increased bioavailability, intracellular drug delivery, gene therapy, and imaging. NP surface functionalization enables precise targeting of certain cells or tissues, reducing side effects and enhancing therapeutic outcomes. NPs regulated and sustained release characteristics improve drug efficacy by reducing dosing frequency. NPs improve bioavailability and absorption by addressing the solubility issues of poorly water-soluble drugs (Orive et al. 2003).

Challenges and Future Prospects

The primary problem for NP-based DDSs is ensuring biocompatibility and understanding potential toxicity. Overcoming the translational gap that separates preclinical achievements and clinical applications is an ongoing problem that requires strategic advances. Future nanomedicine research will focus on personalised therapeutics, which will modify nanoparticle-based treatments to patient features. Furthermore, ongoing research efforts are aimed at developing multifunctional NPs capable of seamlessly combining diagnostic, therapeutic, and imaging characteristics into a single platform, demonstrating the expanding landscape of nanotechnology for improved theranostic. Navigating these hurdles and accepting novel approaches will be critical to understanding the full potential of NP-based drug delivery within biotechnological applications (Orive et al. 2003).

7.5.2 Polymeric Nanocarriers

Polymeric nanocarriers (PNCs) are important and versatile in biotechnology, especially in drug delivery. The short half-lives of many of these current treatments, along with the non-specific distribution and cytotoxicity of previously reported small-molecular drugs, have served as a main driving force in accelerating the development of polymeric DDSs (Kamaly et al. 2016). To date, we have observed the first bench-to-bedside transformation of targeted and adaptable controlled release polymeric NPs for small-molecular drugs, from initial proof-of-concept in vitro (Farokhzad et al. 2005) to successful in vivo investigations (Gu et al. 2008), laying the groundwork for humans being tested and ongoing clinical trials in phase II for multiple cancer types (Hrkach et al. 2012). Because PNCs are typically between 10 nm to 200 nm in size, their biodistribution throughout the human body differs significantly from that of tiny molecules. The drug's distribution throughout the human body will be determined by the PNCs, greatly increasing the therapeutic impact. PNCs have an extremely large surface area to volume ratio, which favours surface conjugation for functional moieties. In terms of stability, reducing carrier size to the nanoscale can improve water penetration and consequently breakdown in the circumstances of biodegradable carriers. NPs tend to stay in solution without considerable sedimentation due to their low density and their unique combination of the electrostatic and van der Waals forces on the surface that are relevant at this size scale, prolonging their utility compared to bigger carriers (Kowalczuk et al. 2014). These nanoscale carriers, which are primarily made of biocompatible and biodegradable polymers, provide several advantages that aid in therapeutic applications and biomedical research (Table 7.2).

Biocompatibility: PNCs are frequently made from materials that are well tolerated by the body, such as PLGA or PEG, minimising undesirable side effects.

Sustained Release: The controlled release characteristics of the PNCs allow for the administration of therapeutic drugs to be prolonged and sustained. This characteristic is especially useful in persistent medical conditions during which continuous drug administration is beneficial.

Payload Protection: PNCs can encapsulate a wide range of drugs, safeguarding them from degradation, improving stability, and allowing for targeted distribution to specific tissues or cells.

Table 7.2 Types of polymeric nanocarriers

Type of PNC	Description	Example	Reference
PLGA NPs	Biodegradable polymer NPs for controlled release of drug	PLGA NPs loaded with doxorubicin	Danhier et al. (2012)
Dendrimers	Drug delivery using highly branched macromolecules	PAMAM dendrimers for the purpose of gene delivery	Kesharwani et al. (2014)
Polymeric micelles	Self-assembled structures for the solubilisation of hydrophobic therapeutic drugs	Paclitaxel delivery using Pluronic-based micelles	Torchilin (2007)
Chitosan NPs	Chitin-derived biocompatible NPs	Chitosan NPs loaded with insulin	Wong et al. (2020)
Polymeric nanogels	3-D networks enabling stimuli-responsive drug release	pH-responsive nanogels for doxorubicin and methotrexate (anticancer) drug delivery	Salehi et al. (2015)
Polymeric nanocapsules	Hollow structures for drug encapsulation	Polymeric nanocapsules-loaded with camptothecin	Bzowska et al. (2018)
Polymeric nanospheres	Spherical solid particles for controlled release of drugs	Polymeric nanospheres-loaded with methotrexate	Dhanaraj et al. (2016)
Polymeric nanofibers	Fibrous structures for long-term delivery of drugs	Silk fibroin electrospun polymeric nanofibers for wound healing	Chouhan et al. (2018)

Tailored Release Profiles: The structure and composition of PNCs can be finely modified to achieve unique release profiles, allowing for personalisation depending on the therapeutic necessities of different drugs.

Reduced Side Effects: PNCs help to reduce side effects related to conventional DDSs by minimising systemic exposure and offering tailored delivery.

Formulation Adaptability: PNCs can be developed into a variety of shapes and sizes, such as NPs, microparticles, or hydrogels, providing adaptability for diverse routes of administration and therapeutic applications.

Biodegradability: Many PNCs are biodegradable, facilitating for the slow disintegration of the carrier material over time, in accordance with environmentally and patient-friendly principles (He et al. 2014).

Challenges and Future Prospects

The problems in polymeric nanocarriers focus around gaining precise control towards drug release kinetics and dealing with potential long-term toxicity issues, emphasising the vital necessity for biocompatible materials. Incorporating stimuli-responsive components for on-demand drug release, designing multifunctional nanocarriers for the theranostic applications, and enhancing personalised medicine through customised nanocarrier formulations based on unique patient features are the main prospects. Furthermore, improving targeted delivery systems, investigating new biocompatible materials, speeding clinical translation and scale-up operations, and investigating combination medicines are all important components of ongoing research. Addressing these issues and pursuing these future paths will be critical to achieving the full potential of PNCs, revolutionising medication delivery, and contributing to the advancement of personalised medicine. Interdisciplinary collaborations and continual developments in nanotechnology, and biomedicine are critical for determining the future landscape of PNC applications (Daglar et al. 2014).

7.6 Targeted Drug Delivery

7.6.1 Antibody-Drug Conjugates

Antibody-drug conjugates (ADCs) represent an important revolution in biotechnology's approach to targeted drug delivery (TDD). These specialised molecules integrate the precision of monoclonal antibody (mAbs) with drug cytotoxicity, enabling highly targeted therapeutic treatments (Beck et al. 2017). Here are some important instances of ADCs and their biotechnological applications.

Ado-Trastuzumab Emtansine

Kadcyla, a commercialized version of ado-trastuzumab emtansine [2], is an exceptional illustration of targeted drug delivery (TDD) in the treatment of HER2-positive metastatic breast tumors. The drug works by a precise process in which this trastuzumab antibody component binds precisely to HER2 receptors, which are found on tumour cells. This selective binding causes the drug to be internalised by the cancer cells, allowing its potent cytotoxic compound emtansine to be released. Kadcyla specifically delivers its lethal dose to HER2-positive tumour cells by utilising the HER2 receptor pathway. This focused approach reduces damage caused to healthy cells while increasing therapeutic impact. Kadcyla was approved for clinical use, confirming its efficacy, and demonstrating the promise of TDD techniques in treating certain cancer subtypes like HER2-positive metastatic breast carcinoma (Lambert and Chari 2014).

Brentuximab Vedotin

Brentuximab Vedotin, branded as Adcetris, is an effective lymphoma TDD approach. This novel treatment aims to kill CD30-positive cells, which are frequent in anaplastic large cell lymphoma and Hodgkin's lymphoma. The association of an anti-CD30 antibody along with monomethyl auristatin-E (MMAE), a strong antimitotic drug, is the mechanism of action. Brentuximab Vedotin increases drug internalisation by interacting with CD30 receptors on cancer cells, resulting in a release of the cytotoxic MMAE payload. This targeted strategy assures that this potent cytotoxic drug reaches only CD30-positive cancer cells. Adcetris has been approved by the FDA for the treatment of relapsing or refractory Hodgkin's lymphoma and certain kinds of non-Hodgkin's lymphoma, demonstrating the effectiveness of TDD for the treatment of lymphoma subtypes (Connors et al. 2018).

Trastuzumab Deruxtecan

Trastuzumab Deruxtecan, marketed as Enhertu, is a novel TDD approach for treatment in HER2-positive metastatic breast tumours. The drug is intended to specifically target tumour cells that express HER2. The combined effect of the HER2-targeting antibodies trastuzumab together with a topoisomerase-I inhibitor is responsible for the mechanism of action. Trastuzumab Deruxtecan enhances drug internalisation by connecting to HER2 receptors on cancer cells, resulting in the delivery of a cytotoxic payload to the tumour cells. This focused approach reduces damage done to healthy cells while increasing therapeutic impact. Enhertu has been granted licences for clinical use, especially for the therapeutic treatment of HER2-positive breast cancer with metastatic tumours, demonstrating the efficacy and precision of TDD techniques in treating specific disease subtypes (Shitara et al. 2020).

Polatuzumab Vedotin

Polivy, or polatuzumab vedotin, is an effective TDD approach for non-Hodgkin's lymphoma. This drug is especially developed for targeting CD79b-positive B cells, which are characteristic of this kind of lymphoma. The combined effect of an anti-CD79b antibody together with the microtubule-disrupting chemical vedotin is the mechanism. Polatuzumab Vedotin effectively delivers the cytotoxic dose to tumour cells expressing CD79b via this novel combination. This targeted approach reduces collateral damage to normal cells, hence increasing therapeutic impact. Polivy has received clinical approval for use, specifically in combination therapies for certain kinds of non-Hodgkin's lymphoma. This achievement demonstrates the efficacy of TDD systems in targeting certain cancer subtypes with more precision and enhanced therapeutic effects (Palanca-Wessels et al. 2015).

7.6.2 *Peptide-Based Targeting*

Peptide-based targeting approaches in biotechnology employ the application of peptides, short sequences of amino acids, to enable targeted and accurate drug delivery. These tactics make use of the unique attributes of peptides, like their short size, high degree of specificity, and capacity to be quickly manipulated (Chow et al. 2008). Here are some instances of how peptide-based strategies for targeting are used in biotechnology.

7.6.3 RGD Peptides

RGD peptides, particularly binding with integrins that are overexpressed on the outer surfaces of tumour cells, constitute a TDD system in cancer therapies. The approach involves RGD peptides specifically attaching to integrins, a group of cell-adhering receptors, allowing for TDD to cancer cells. Integrins have an important role in tumour angiogenesis as well as metastasis, making them potential therapeutic targets. RGD peptides are used in the enhancement of the specificity of antitumour drugs such as paclitaxel. This tailored method attempts to reduce undesirable effects and improve anticancer drug therapeutic efficacy. RGD peptide-based techniques demonstrate the possibility of precision therapy in cancer treatment by utilising the specific properties of tumour cell surfaces to obtain enhanced and targeted cancer treatment (Yin et al. 2017).

7.6.4 Cell-Penetrating Peptides

Cell-penetrating peptides (CPPs) are an important means of intracellular drug administration because they target cell membranes. CPPs facilitate the internalisation of drugs into cells, boosting cellular absorption of different payloads such as nucleic acids and smaller compounds. CPPs are important in gene delivery because they help transport genetic material throughout cell membranes during gene therapy applications. This adaptable strategy enables the efficient transport of therapeutic payloads straight within the cytoplasm or nucleus of cells, eliminating obstacles that may restrict the efficacy of traditional drug delivery systems. The widespread usage of CPPs in gene delivery emphasises their importance in promoting focused and precision therapeutic interventions for a variety of diseases, presenting them as valuable tools in biotechnology and medicine (Lehto et al. 2012).

7.6.5 Tumour-Homing Peptides

Tumour-homing peptides (THP) are a promising method for TDD in cancer. These peptides are intended to target specific components of the tumour microenvironment, such as tumour vasculature and extracellular matrix components. The peptides homing into these different traits facilitates TDD to the tumour region. This targeted approach tries to improve drug delivery precision while minimising off-target effects. THP are being studied for potential uses in tumour imaging and the delivery of drugs to enhance therapeutic outcomes. These peptides provide a personalised and focused strategy to address the intricate nature of cancer by utilising the properties of the tumour microenvironment, demonstrating their promise as significant strategies in advancing cancer detection and treatment in the area of biotechnology (Kondo et al. 2021).

7.6.6 Antimicrobial Peptides

Antimicrobial peptides (AMPs) are a powerful technique for developing innovative antibiotics to combat drug-resistant microbial infections. These peptides work specifically on microbial membranes. The method involves AMPs breaking the membranes of fungi and bacteria to demonstrate broad-spectrum antibacterial action. This disruptive activity not just provides a diverse approach that is effective towards a wide spectrum of infections, but it also possesses potential to fight against drug-resistant bacteria. The use of AMPs in biotechnology is being intensively researched as a valuable route for facilitating the development of novel antibiotics. AMPs' distinct mode of action presents them as possible alternatives for tackling the global issue of antibiotic resistance, highlighting their importance in infectious disease therapies (Luo et al. 2018).

7.6.7 Somatostatin Analogues

Somatostatin analogues are emerging to be a targeted therapeutic approach to neuroendocrine tumour treatment. Certain analogues are intended to target overexpressed somatostatin receptors in certain tumours. Somatostatin peptide analogues specifically attach to receptors on tumour cells, depending on the mechanism. This binding limits hormone secretion and triggers apoptosis, which helps to regulate symptoms and slow tumour growth. Somatostatin analogues are used to treat neuroendocrine tumours, providing a targeted strategy for managing both physiological signs and symptoms linked to hormone secretion and tumour growth. This comprehensive and targeted technique emphasises the great potential of somatostatin analogues in offering effective and personalised treatment interventions for neuroendocrine tumours in biotechnology and medicine (Wolin 2012).

7.6.8 Biologics and Gene Therapies

Biologics and gene treatments play critical roles in biotechnology, transforming disease treatment and personalised medicine. Biologics, which include MAbs and cytokines, use living organisms to provide targeted therapy in a variety of medical sectors. Advances in biotechnology in delivery systems, such as antibody-drug conjugates, improve the accuracy of biologic treatments. Gene treatments, on the other hand, include the introduction or alteration of genetic material to treat diseases at their origin. Technologies such as CRISPR facilitate precise gene editing, perhaps leading to the treatment of hereditary illnesses. The combination of biologics and gene treatments with nanotechnology allows for targeted delivery, which improves bioavailability and therapeutic effectiveness (Crommelin et al. 2020).

7.6.9 Monoclonal Antibodies

In targeted therapy, MAbs are frequently utilised, and attempts are ongoing to enhance their delivery systems. mAbs are a biotechnology cornerstone, providing precise and focused therapy by utilising antibody specificity. These antibodies, which have been engineered to preferentially attach to proteins on cell surfaces, influence immune responses or simply directly disrupt biological activities. They are widely used in biotechnology throughout therapeutic fields, most notably oncology, infectious diseases, and autoimmune disorders. It involves the development of ADCs, which allow therapeutic payloads to be delivered directly to the site of action, as well as antibody engineering for enhanced tissue penetration. These developments indicate the critical role of MAbs in biotechnology, paving the way for more effective and customised treatment techniques (Reichert 2008). Here are a few examples of commonly used MAbs (Table 7.3).

Table 7.3 Some examples of monoclonal antibodies and their applications

Monoclonal antibody	Target	Mechanism	Application	Reference
Ado-trastuzumab emtansine	HER2/neu receptor	Directly delivers a cytotoxic chemical to HER2-positive cancer cells	Breast cancer	Corrigan et al. (2014)
Rituximab	CD20 protein present on B cells	Causes B cells to undergo apoptosis; useful in treating autoimmune diseases and B-cell lymphomas	Oncology	Zhou et al. (2008)
Bevacizumab	Vascular endothelial growth factor (VEGF)	Angiogenesis is inhibited, causing tumour blood supply to be disrupted	Cancer therapy	Giantonio (2009)
Nivolumab	Programmed cell death-protein 1 (PD-1)	Like pembrolizumab, it increases the immune response towards cancer cells	Oncology	Brahmer et al. (2015)
Trastuzumab	HER2/neu receptor	Inhibits the proliferation of cancer cells that express the HER2 protein	Breast cancer	Sun et al. (2013)
Infliximab	Tumour necrosis factor alpha (TNF-α)	TNF-α is neutralised, which reduces inflammation and immune reactions	Autoimmune	Billmeier et al. (2016)
Pembrolizumab	PD-1	Inhibits the PD-1 pathway, allowing the immune system to attack cancer cells more effectively	Oncology	Billmeier et al. (2016)
Adalimumab	TNF-α	TNF-α inhibition reduces inflammation and tissue damage	Autoimmune	Cessak et al. (2014)

7.7 Gene Editing Tools

Gene editing techniques are game changers in biotechnology, providing unprecedented precision in modifying genetic material. Clustered Regularly Interspaced Short Palindromic Repeats (CRISPR) has emerged as a transformational force among these. CRISPR technology enables researchers to precisely alter DNA sequences, allowing for targeted gene changes. CRISPR has a wide range of uses in biotechnology, from generating genetically engineered creatures for research to potentially treating hereditary problems in humans. CRISPR's simplicity and versatility have accelerated advancements in functional genomics, discovering new drugs, and the generation of genetically engineered crops. The capacity to modify genes with remarkable precision expands our understanding of biological processes and offers tremendous scope for treating previously incurable hereditary disorders. As gene editing technologies advance, their incorporation into biotechnological applications reshapes the scientific landscape.

7.7.1 CRISPR-Cas9

CRISPR and CRISPR-associated Protein 9 (Cas9) is an innovative gene editing technique that allows for the exact altering of DNA sequences. This game-changing technology has transformational applications in a wide range of fields. CRISPR-Cas9 provides therapeutic promise for addressing genetic diseases by targeted genome editing. In research, it plays a crucial role in functional genomics, uncovering gene functions. Agriculture benefits as well, because CRISPR-Cas9 allows to produce genetically modified organisms (GMOs) with improved features. Furthermore, the tool advances the development and discovery of drugs by allowing researchers to comprehend more about gene functions and speed pharmaceutical research. CRISPR-Cas9 is a precision lighthouse in the wide terrain of genetic alteration, establishing the future of biotechnology.

7.7.2 CRISPR-Cas12 and CRISPR-Cas13

CRISPR-Cas12 and CRISPR-Cas13 are evolutionary versions of the CRISPR system that offer characteristics beyond the well-known Cas9. CRISPR-Cas12 is skilled in cleavage of DNA and editing, offering a unique toolkit for precise genomic alterations. CRISPR-Cas13, on the other hand, specialises in RNA targeting, allowing for the editing and control of RNA molecules. These variants improve the platform's adaptability by providing researchers with personalised solutions that address DNA and RNA alterations. CRISPR-Cas12 and CRISPR-Cas13, with applications spanning from genome engineering to RNA-level interventions, contribute greatly to the developing toolkit of molecular biology, significantly advancing our ability to accurately alter genetic material at various levels.

7.7.3 Prime Editing

Prime Editing, a recent achievement, is a versatile and extremely precise gene editing technology that inserts new genetic material directly into a specific DNA target. This approach stands out for its ability to reduce off-target effects, which represents a substantial improvement in genome editing. Prime Editing shows potential for precise genome alterations to address genetic problems with increased accuracy. The potential of technology to precisely modify individual DNA sequences brings up new possibilities for modifying genetic information with unparalleled control, demonstrating its potential to reshape the landscape of genetic modification for therapeutic and scientific purposes (van der Oost and Patinios 2023).

7.7.4 TALENs

Transcription Activator-Like Effector Nucleases (TALENs) are synthetic proteins that are designed to attach to specific DNA regions, allowing for targeted and precise gene editing. This approach has applications in a variety of fields, especially therapeutic genome editing, that has the potential to address genetic problems by customised alterations. TALENs serve a critical role in functional genomics research, unravelling the complexities of gene functions with high sensitivity. Furthermore, TALENs assist in the production of genetically edited cell lines, providing researchers with a vital tool for developing personalised cellular models. TALENs, in essence, emerge as adaptable tools for the strategic alteration of genetic material, allowing for advances concerning both research and clinical use.

7.7.5 ZFNs

Zinc Finger Nucleases (ZFNs) are designed proteins that have been precisely developed to bind to specific DNA sequences with accuracy and are paired with a nuclease allowing targeted DNA cleavage. This advanced technology has numerous uses, especially for therapeutic genome editing, thereby offering a possible pathway for precision therapies in hereditary diseases. ZFNs are also important in the production of genetically engineered organisms, allowing for specific genetic upgrades in agriculture and other industries. Furthermore, ZFNs greatly contribute to genomic research and engineering endeavours, allowing researchers to examine and change the complexities of the genetic code using a high degree of specificity. Essentially,

ZFNs emerge as potent instruments for the purposeful and precise alteration of genetic material in a variety of applications (Adli 2018).

7.8 Implantable Devices

7.8.1 Drug-Eluting Stents

Drug-eluting stents are novel medical devices that treat coronary artery disease. These stents, which are commonly comprised of metal mesh, not only give mechanical support to coronary arteries but also gradually deliver medicinal drugs. The drugs, which are frequently anti-proliferative or anti-inflammatory in nature, help to avoid restenosis by limiting excess tissue growth at the stent site. This dual functioning improves the intervention's efficacy by promoting rapid healing and minimising the possibility of problems. Drug-eluting stents became an interventional cardiology standard, considerably improving outcomes in individuals with coronary artery disease.

7.8.2 Implantable Pumps for Chronic Conditions

Implantable pumps are innovative medical devices that provide drugs in a continuous and controlled manner, particularly for chronic illnesses. These medical devices are placed by surgery just below the layer of skin and are linked to specific locations inside the body wherever drug is needed. They provide a focused and programmed strategy for drug delivery, assuring consistent and accurate dosing throughout time. Implantable pumps are employed to treat diseases such as chronic pain, spasms, and some neurological abnormalities. Their capacity to give long-term therapeutic effects while minimising adverse effects makes them useful in enhancing the quality of life of those living with chronic illnesses (Alcaraz et al. 2018).

7.9 Innovations in Smart DDSs

7.9.1 Use of Responsive NPs

The use of responsive NPs is a game changer in drug delivery. These materials, which are frequently triggered by environmental circumstances or stimuli, allow for exact control of drug release. Nanocarriers that are pH-responsive, sensitive to temperature, or enzyme-triggered provide targeted drug delivery to specified areas, improving therapeutic efficacy while minimising negative effects.

7.9.2 Artificial Intelligence in Drug Delivery Optimisation

The combination of AI and drug delivery has transformed this field. AI algorithms analyse large amounts of data, considering patient-specific variables, drug kinetics, and dynamics. This data-driven method supports in the design of personalised drug delivery regimens for specific patients, optimising dosage, timing, and formulation. AI-powered platforms increase treatment outcomes and patient experiences, signalling a paradigm shift in the personalisation of DDSs.

7.9.3 Stimulus-Triggered Drug Release Systems

Stimulus-triggered drug release systems are an advanced technique to targeted drug delivery. These systems are responsive to external stimuli like light, magnetic fields, even ultrasound, allowing medicinal drugs to be released on demand. This level of control improves drug delivery precision, allowing clinicians to strategically schedule and localise treatment, optimising therapeutic effects (Park 2016).

7.10 Future Developments and Emerging Technologies

The future of biotechnology is marked by transformative trends and emerging technologies. Advances in CRISPR technology promise precise genetic engineering, while synthetic biology opens avenues for creating synthetic organisms. Artificial intelligence is revolutionising drug discovery, enabling personalised medicine based on individual genetic profiles. Advanced bioprocessing techniques and nanomedicine applications are enhancing bioproduction and targeted medical interventions. Microbiome manipulation, 3D bioprinting, and environmental biotechnology solutions contribute to health and sustainability. Neurotechnology advancements hold promise for addressing neurological disorders. Interdisciplinary collaborations drive these innovations, shaping a dynamic landscape for the future of biotechnology.

7.10.1 Biodegradable Implants

Biodegradable implants are an advanced technique in medical technologies that are designed to degrade naturally within the body over time. Biocompatible materials are frequently used in these implants, which gradually degrade into harmless byproducts by enzymatic or hydrolysis processes. This controlled breakdown works in parallel with the recovery process, helping throughout critical stages of tissue regeneration.

Traditional drug delivery routes such as oral, transdermal, rectal, parenteral, and nasal techniques provide benefits such as convenience of administration and systemic effects. On the other hand, each route has its own set of challenges. Oral administration encounters problems such as acidity in the gastric tract, hepatic metabolism, and the growth of microbes that can cause absorption concerns, especially in chronic conditions that require regular dosage (Liu et al. 2017). Although parenteral methods provide excellent absorption, they are invasive and can result in poor patient compliance (Vhora et al. 2021). Transdermal patches can address some concerns; however, they encounter barriers such as the stratum corneum. The requirement for several daily doses causes difficulties for chronic conditions (Jeong et al. 2021). Implants with regulated drug release avoid these challenges by assuring sustained drug concentrations, providing a promising alternative for improved treatment outcomes in a patient-friendly way (García-Estrada et al. 2021).

Biodegradable implants have numerous applications in biotechnology. They are employed in orthopaedics for bone fixation, ultimately paving the way to regenerated tissues. Biodegradable polymers act as carriers in drug delivery, releasing therapeutic substances in a regulated manner. Biodegradable cardiovascular stents assist blood arteries during healing and gradually disintegrate, reducing long-term consequences. These implants reduce the number of procedures required to eliminate non-biodegradable devices, improving patient compliance. Implants provide superior safety, improved bioavailability, a controlled delivery rate, improved permeability, and minimal toxicity (Sharma et al. 2021). Examples of these biodegradable implants include drug-eluting stents, tissue engineering scaffolds, pacemakers, cardiac valves, or ocular implants (Johnson et al. 2021).

Ritasert®, an FDA-approved ocular insert intended to treat chronic uveitis, incorporated polyvinyl alcohol as a polymer for sustaining drug delivery. This device can provide drug delivery for up to 3 years. Similarly, PMMA-based implants are frequently used in bone-related scaffolds for joint replacement procedures to enable sustained drug delivery. Bioinspired cardiac valves were 3D printed after being created with silicone to serve as a polymeric agent. Under physiological heart conditions, the in vitro data revealed good hemodynamic function of the heart valves. These implants can be made of biodegradable or non-biodegradable polymers. Implantable drug delivery devices have numerous applications, including disease detection, treatment, regenerative therapies, and preventive therapies (Chavda et al. 2022).

7.10.2 Artificial Intelligence in Drug Delivery

Artificial intelligence (AI) is transforming drug delivery by improving precision, efficiency, and personalised treatment techniques. A major application is predictive modelling, which employs machine learning algorithms (MLA) to analyse large datasets to forecast drug behaviour, formulation consequences, and patient responses. This helps to improve the design of focused drug delivery systems. AI is especially important in personalised medicine, where it is used to customise drug delivery techniques to specific patient profiles by utilising genetic, genomic, and clinical data. With AI-driven platforms discovering prospective candidates and optimising formulations, the future of AI in the administration of drugs holds promise for faster drug development. This game-changing technology is expected to improve the precision and efficacy of DDSs, introducing a new era of personalised and effective therapies. Examples of AI used in drug delivery are:

IBM Watson to support Drug Discovery: Uses AI to analyse various biomedical data sources to identify drug candidates more efficiently. When domain annotators are applied to massive volumes of unstructured data using IBM Watson technology, the processing speed extracts molecules, genes, and drugs from hundreds of thousands of scientific papers and patents in hours. Human collection of this information would most likely take much longer. Annotators' information gathered from hundreds of millions of pages of text can subsequently be assembled into a wide range of runtime analytics including visualisations (Chen et al. 2016).

Atomwise: Uses AI to predict their affinity for binding of tiny compounds to specific therapeutic targets, hence speeding up drug discovery. Atomwise serves as an AI drug discovery startup that has created the AtomNet platform, which uses deep learning algorithms for identifying potential drug candidates. They announced the identification of many molecules with potential anti-disease activity in 2019, such as Ebola, cystic fibrosis, and multiple sclerosis. Certain chemicals are now being tested to see if they have the potential to be novel therapies for certain disorders (Anusha et al. 2023).

Benevolent AI: Uses AI to optimise therapeutic efficacy through drug discovery and repurposing. Analyse scientific research sets of data, then generate unique insights by forming and qualifying hypotheses (Narayanan et al. 2022).

Tempus (A Vast Data Library for Personalised Healthcare): Uses AI to analyse clinical and genomic data to provide personalised cancer treatment plans, including drug delivery optimisation. To personalise healthcare treatments, Tempus uses AI to filter through the world's largest database of clinical and biological data. The company is developing AI technologies that collect and analyse data in areas that range from the sequencing of genes to image recognition, allowing physicians to have a better understanding of treatments and cures. Tempus is presently applying AI-powered data to cancer investigation and treatment (Mohd 2022).

BERG Health: (Using AI to treat rare diseases): BERG acts as a clinical-stage, AI-powered biotech platform that diagnoses diseases to rapidly accelerate up the research and development of new treatments. BERG can develop stronger product candidates for rare diseases by integrating its "Interrogative Biology" method with standard R&D. BERG recently announced their research on the treatment of Parkinson's disease. They employed AI to discover previously discovered interactions among chemicals in the human body (Mohd 2022).

7.10.3 Exosome-Based Therapies

Exosome-based therapies are a novel approach to drug delivery in the field of biotechnology. Exosomes, composed of microscopic extracellular vesicles released by cells, are essential for intercellular communication. Their specific characteristics, including stability, biocompatibility, and the capacity to cross biological barriers (BBB), make them ideal drug delivery candidates. Exosomes are expected to revolutionise drug delivery in following ways in the future of biotechnology: They are natural drug carriers that encapsulate therapeutic payloads such as proteins and nucleic acids, providing a biocompatible framework for drug administration (El Andaloussi et al. 2013). Engineered exosomes improve specificity and reduce off-target effects, allowing for precise drug delivery. Their use extends to facilitating the transport of nucleic acid medication to target cells, including RNA interference or even gene editing tools, demonstrating versatility in biotechnological treatments (Alvarez-Erviti et al. 2011). Exosomes, specifically, can cross biological barriers, such as the formidable BBB, thereby improving drug delivery to targeted organs. Furthermore, by controlling the immune response, their immunomodulatory actions lead to improved treatment results, increasing their prospective applications. As the field of drug delivery continues to develop, continuing research efforts are centred on realising the maximum potential of exosomes over precise and personalised drug administration, providing new solutions to conventional difficulties (Pitt et al. 2017).

Exosomes high in MHC-II were produced by transfecting murine melanoma cells with the CIITA gene, which resulted in an overexpression of MHC-II at their surfaces. This is an example of how designer exosomes are used immunotherapeutically. The modified exosomes were then directed at T cells, where they improved the type 1 T-helper cells response against cancer cells. As a result, the modified exosomes' enhanced MHC-II molecule served as both a targeting peptide and a therapeutic drug (Lee et al. 2011). Alternatively, to drive exosomes towards T cells, an antibody with a light chain can be added to the exosomal surface (Raposo and Stoorvogel 2013). Curcumin-loaded exosomes have been investigated as potential means of drug delivery for Parkinson's disease therapy. Haney et al. (2015) have demonstrated the usage of naturally occurring anti-inflammatory chemicals encapsulated within exosomes (Kim et al. 2016; Haney et al. 2015). Researchers aimed to overcome multidrug resistance (MDR) in cancer cells by designing exosomes that contained doxorubicin, demonstrating their utility as anticancer drug carriers (Kim et al. 2016).

7.10.4 Synthetic Biology Approaches

The use of modified microbes in drug delivery is a cutting-edge strategy with significant future applications in the discipline of biotechnology. Synthetic biology approaches allow microorganisms to be modified to serve as sophisticated carriers for medicinal payloads. These genetically modified bacteria can be designed to navigate complicated biological surroundings, target specific tissues, and even release therapeutic chemicals in a controlled way. This novel approach has tremendous potential for facilitating the development of very precise and effective medication delivery systems. Biotechnology applications can include the targeted administration of drugs, biological products, or gene treatments for a variety of medical problems. Current advances in synthetic biology are projected to refine and increase the potential of engineered microbes, changing the future panorama of drug delivery across biotechnology (Callura et al. 2012).

The natural evolution of living organisms has resulted in the formation of increasingly sophisticated natural genetic networks for sensing the environment and responding advantageously. As a result, many synthetic biologists anticipate using this strategy to utilise the complexity of cells to create programmed therapeutic cells that are capable of independently controlling and regulating both intrinsic (that are located within cells) and extrinsic (located outside cells) signals to fight disease. To achieve this goal, there are three significant initiatives underway in the field of synthetic biology, which involve engineering cells to:

Function as sensors, sensing and reporting the availability of specific molecules that can differentiate among states of cells (i.e. healthy vs. diseased/damaged).

Perceive a specific molecule or cell state and responding by generating a controlled dosage of therapeutic biological molecules.

Function in different delivery platforms (Muldoon et al. 2021).

The key difference between synthetic biology and molecular biology is the fact that synthetic biology assembles components from molecular biology (for example, repressor proteins) to create innovative genetic frameworks that execute complex activities. Dynamic gene expression, rewiring endogenous pathways to generate biomolecules, controlling the number of cells, reporting on the condition of various surroundings, and executing Boolean logic operations are among these functions (MacDonald and Deans 2016).

Controlling gene expression reliably is critical in synthetic biology for developing functioning genetic circuits inside cells. These circuits can work independently or in conjunction with existing biological networks. For independent functioning, genetic circuits are frequently engineered to exist orthogonal to their host, to ensure the regulatory control components are distinct from the host signalling pathways. The integration of Boolean logic gates within these circuits enables exact control over time, duration, and functionality, which is made possible by specific triggered inputs. Various research techniques in molecular biology or genetic engineering act as modular elements for synthetic biologists. These methods are critical for building genomic circuits that allow sophisticated regulation of cell function, thereby advancing synthetic biology (MacDonald and Deans 2016).

7.10.5 3D Printing of DDSs

3D printing advances are revolutionising DDSs by enabling the fabrication of personalised and extremely precise dose forms. This method enables healthcare formulations to be tailored to individual patient needs, considering parameters like age, weight, and unique medical problems. Layer-by-layer application of pharmaceutical ingredients is used to construct complicated structures, allowing for unparalleled management of the design and content of drug delivery systems. By optimising drug release profiles and boosting bioavailability, this customisation improves patient compliance and overall therapeutic results. The adaptability of 3D printing for the administration of medications demonstrates an evolutionary change towards personalised treatment, providing a promising path for medicines in the future (Chia and Wu 2015).

3D printing enables the production of DD devices with varying geometries and/or customised drug release characteristics, which are difficult to create using traditional drug manufacturing procedures. The DDS design, as well as the choice of polymers and their combinations, can influence drug release kinetics. Dosing in the printed DDS can be changed to meet the demands of the patient by adjusting the size of the printed formulations and changing the composition of the solid material(s). The ability and versatility of the 3D printed DDS to independently tune the drug delivery of numerous pharmaceuticals make it suited for patient groups that require multiple drugs at the same time, such as individuals with metabolic illnesses. Designing formulations containing complicated inner structures and compartmentalised matrix systems could yield these DDS (Khaled et al. 2015).

Aprecia® Pharmaceuticals, which was created in 2003, was the first pharmaceutical business to introduce a 3D printed healthcare product. ZipDose® Technology was developed in 2008 as a patented 3D printing process primarily for producing high-dose and easy-to-swallow product for central nervous system therapies. GlaxoSmithKline (GSK) and other pharmaceutical companies have showed interest in this subject as well. Printing techniques have already been investigated for the generation of high-throughput screening matrices for drug screening and testing. Pharmaceutical and printing firms are collaborating to create more affordable and personalised healthcare solutions in the future. Furthermore, 3D printed human skin and organs provide a viable alternative to animal testing for cosmetic and biological research. Apart from 3D printing technologies, 2D printing technologies such as inkjet as well as flexographic printing have also been used in the pharmaceutical industry. Flexography makes use of rotary-type contact printing technology, whereas inkjet printers are non-contact systems based on ongoing or drop-on-demand jetting technologies (Palo et al. 2017).

7.11 Challenges and Limitations Associated with Current Drug Delivery

7.11.1 Bioavailability

Bioavailability challenges involve addressing concerns such as low drug solubility as well as degradation. Innovative formulations, such as nanoparticle-based systems and prodrugs, are critical in enhancing drug absorption and systemic distribution, ultimately increasing bioavailability.

7.11.2 Immunogenicity

Immunogenicity is a problem in biologic therapies because the immune response of the body might affect the effectiveness of therapeutic agents. A significant focus is on strategic drug design, which includes adjustments to minimise immunogenicity. This comprises structural changes as well as the application of immunomodulatory drugs to extend biologics' therapeutic lifetime.

7.11.3 Non-specific Distribution

Non-specific drug distribution can result in undesirable adverse effects. Off-target effects are minimised using advanced drug delivery technologies like liposomes and targeted formulations. However, improving these systems to facilitate accurate targeting along with controlled release is still a difficulty in drug delivery studies. Addressing these issues collectively is critical for improving DDSs and patient outcomes (Vizirianakis 2004).

7.11.4 Safety Concerns

The progress of DDSs in biotechnology creates new obstacles and ethical concerns. As novel techniques, such as nanotechnology and precision delivery methods, become fundamental to biopharmaceutical research, safety issues take centre stage. These technologies' complicated interactions with biological systems raise concerns about potential toxicity, its immunogenicity, and long-term impacts. As biotechnological DDSs affect cells and molecules, it is essential that their safety and biocompatibility be ensured. Ethical considerations include issues such as informed consent, which is especially important in the context of developing medicines such as gene editing and personalised medicine. Establishing a balance between expanding boundaries of biotechnology while protecting patient well-being necessitates thorough risk assessments, close monitoring, and constant ethical debate among the biotechnological and healthcare communities (Orive et al. 2003).

7.12 Conclusion

Biotechnology's dynamic field of drug delivery has made great progress, bringing cutting-edge approaches to healthcare concerns. Nanotechnology, liposomal nanotechnology, and TDD emerge out as powerful techniques that have the potential to revolutionise therapeutic precision and efficacy. The adaptability of liposomal nanotechnology extends to therapies for cancer, diagnostics, and cosmeceuticals, whereas incorporated NPs improve absorption and permit controlled release. With the emergence of AI, exosome-based medicines, and synthetic biology, the future holds enormous promise. These breakthroughs herald a new era of personalised and targeted interventions, with applications ranging from cancer therapy to genetics circuit regulation. However, this changing landscape raises important problems and ethical concerns, emphasising the importance of continual safety assessments and shared ethical debate. In the future, continued research and development will be essential for achieving all that is possible with these technologies. Addressing problems, improving safety protocols, as well as exploring new applications will drive the next wave of biotechnological drug delivery achievements. Researchers, physicians, ethicists, and regulators shall collaborate to design a healthcare future in which these technologies play a critical role in providing individualised and effective therapies for various healthcare-related diseases.

References

Abdella S, Abid F, Youssef SH, Kim S, Afinjuomo F, Malinga C, Song Y, Garg S (2023) pH and its applications in targeted drug delivery. Drug Discov Today 28:103414

Adler-Moore J, Proffitt RT (2002) AmBisome: liposomal formulation, structure, mechanism of action and pre-clinical experience. J Antimicrob Chemother 49:21–30. https://doi.org/10.1093/jac/49.suppl_1.21

Adli M (2018) The CRISPR tool kit for genome editing and beyond. Nat Commun 9:1911

Alcaraz J, Cinquin P, Martin DK (2018) Tackling the concept of symbiotic implantable medical devices with nanobiotechnologies. Biotechnol J 13:1800102

Al-Jamal W, Kostarelos K (2011) Liposomes: from a clinically established drug delivery system to a nanoparticle platform for theranostic nanomedicine. Acc Chem Res 44:1094–1104

Allami RH, Yousif MG (2023) Integrative AI-driven strategies for advancing precision medicine in infectious diseases and beyond: a novel multidisciplinary approach. arXiv Prepr arXiv 2307:15228

Allen TM, Hansen C (1991) Pharmacokinetics of stealth versus conventional liposomes: effect of dose. Biochim Biophys Acta 1068:133–141

Alvarez-Erviti L, Seow Y, Yin H, Betts C, Lakhal S, Wood MJA (2011) Delivery of siRNA to the mouse brain by systemic injection of targeted exosomes. Nat Biotechnol 29:341–345

Anusha K, Jasmitha KSM, Sattibabu K, Reddy G (2023) Integrating of artificial intelligence in drug discovery and development: a comparative study. Pharmacophore 14:35–40. https://doi.org/10.51847/anvmzrz4x4

Aslam H, Shukrullah S, Naz MY, Fatima H, Hussain H, Ullah S, Assiri MA (2022) Current and future perspectives of multifunctional magnetic nanoparticles based controlled drug delivery systems. J Drug Deliv Sci Technol 67:102946

Awad A, Madla CM, McCoubrey LE, Ferraro F, Gavins FKH, Buanz A, Gaisford S, Orlu M, Siepmann F, Siepmann J (2022) Clinical translation of advanced colonic drug delivery technologies. Adv Drug Deliv Rev 181:114076

Badwaik HR, Kumari L, Nakhate K, Verma VS, Sakure K (2019) Chapter 13—phytoconstituent plumbagin: chemical, biotechnological and pharmaceutical aspects. In: Atta-ur-Rahman (ed) Bioactive natural products. Elsevier, pp 415–460

Balazs DA, Godbey WT (2011) Liposomes for use in gene delivery. J Drug Deliv 2011:1

Beck A, Goetsch L, Dumontet C, Corvaïa N (2017) Strategies and challenges for the next generation of antibody–drug conjugates. Nat Rev Drug Discov 16:315–337
Bhatia S (2016) Natural polymer drug delivery systems: nanoparticles, plants, and algae
Billmeier U, Dieterich W, Neurath MF, Atreya R (2016) Molecular mechanism of action of anti-tumor necrosis factor antibodies in inflammatory bowel diseases. World J Gastroenterol 22:9300
Birla S, Singh N, Shukla NK (2022) Nanotechnology: device design and applications. CRC Press
Boas U, Christensen JB, Heegaard PMH (2007) Dendrimers in medicine and biotechnology: new molecular tools. Royal Society of Chemistry
Bovier PA (2008) Epaxal®: a virosomal vaccine to prevent hepatitis a infection. Expert Rev Vaccines 7:1141–1150
Brahmer JR, Hammers H, Lipson EJ (2015) Nivolumab: targeting PD-1 to bolster antitumor immunity. Future Oncol 11:1307–1326
Burrows R, Lambrix E (2022) mRNA vaccines: a growing and complex IP landscape. Vaccine Insights 1:191–199
Bzowska M, Karabasz A, Szczepanowicz K (2018) Encapsulation of camptothecin into pegylated polyelectrolyte nanocarriers. Colloids Surf A Physicochem Eng Asp 557:36–42
Cagel M, Grotz E, Bernabeu E, Moretton MA, Chiappetta DA (2017) Doxorubicin: nanotechnological overviews from bench to bedside. Drug Discov Today 22:270–281. https://doi.org/10.1016/j.drudis.2016.11.005
Callura JM, Cantor CR, Collins JJ (2012) Genetic switchboard for synthetic biology applications. Proc Natl Acad Sci 109:5850–5855
Cessak G, Kuzawińska O, Burda A, Lis K, Wojnar M, Mirowska-Guzel D, Bałkowiec-Iskra E (2014) TNF inhibitors–mechanisms of action, approved and off-label indications. Pharmacol Rep 66:836–844
Chavda VP, Jogi G, Paiva-Santos AC, Kaushik A (2022) Biodegradable and removable implants for controlled drug delivery and release application. Expert Opin Drug Deliv 19:1177–1181. https://doi.org/10.1080/17425247.2022.2110065
Chen Y, Elenee Argentinis JD, Weber G (2016) IBM Watson: how cognitive computing can be applied to big data challenges in life sciences research. Clin Ther 38:688–701. https://doi.org/10.1016/j.clinthera.2015.12.001
Chia HN, Wu BM (2015) Recent advances in 3D printing of biomaterials. J Biol Eng 9:1–14
Chilkoti A, Dreher MR, Meyer DE, Raucher D (2002) Targeted drug delivery by thermally responsive polymers. Adv Drug Deliv Rev 54:613–630
Chouhan D, Janani G, Chakraborty B, Nandi SK, Mandal BB (2018) Functionalized PVA–silk blended nanofibrous mats promote diabetic wound healing via regulation of extracellular matrix and tissue remodelling. J Tissue Eng Regen Med 12:e1559–e1570
Chow D, Nunalee ML, Lim DW, Simnick AJ, Chilkoti A (2008) Peptide-based biopolymers in biomedicine and biotechnology. Mater Sci Eng R Rep 62:125–155
Connors JM, Jurczak W, Straus DJ, Ansell SM, Kim WS, Gallamini A, Younes A, Alekseev S, Illés Á, Picardi M (2018) Brentuximab vedotin with chemotherapy for stage III or IV Hodgkin's lymphoma. N Engl J Med 378:331–344
Corrigan PA, Cicci TA, Auten JJ, Lowe DK (2014) Ado-trastuzumab emtansine: a HER2-positive targeted antibody-drug conjugate. Ann Pharmacother 48:1484–1493
Crommelin DJA, Mastrobattista E, Hawe A, Hoogendoorn KH, Jiskoot W (2020) Shifting paradigms revisited: biotechnology and the pharmaceutical sciences. J Pharm Sci 109:30–43
Daglar B, Ozgur E, Corman ME, Uzun L, Demirel GB (2014) Polymeric nanocarriers for expected nanomedicine: current challenges and future prospects. RSC Adv 4:48639–48659
Danhier F, Ansorena E, Silva JM, Coco R, Le Breton A, Préat V (2012) PLGA-based nanoparticles: an overview of biomedical applications. J Control Release 161:505–522
Dhanaraj SA, Muralidharan S, Kanniappan P, Hui WTS, Qi LL (2016) Formulation and evaluation of chitosan nanospheres containing methotrexate targeted drug delivery system. J Young Pharm 8:330

du Plessis HL, Marais EB, Mohammed F, Kotze AF (2014) Applications of lipid based formulation technologies in the delivery of biotechnology-based therapeutics. Curr Pharm Biotechnol 15:659–672

El Andaloussi S, Mäger I, Breakefield XO, Wood MJA (2013) Extracellular vesicles: biology and emerging therapeutic opportunities. Nat Rev Drug Discov 12:347–357

Farokhzad OC, Khademhosseini A, Jon S, Hermmann A, Cheng J, Chin C, Kiselyuk A, Teply B, Eng G, Langer R (2005) Microfluidic system for studying the interaction of nanoparticles and microparticles with cells. Anal Chem 77:5453–5459

García-Estrada P, García-Bon MA, López-Naranjo EJ, Basaldúa-Pérez DN, Santos A, Navarro-Partida J (2021) Polymeric implants for the treatment of intraocular eye diseases: trends in biodegradable and non-biodegradable materials. Pharmaceutics 13:701

Giantonio BJ (2009) Goldie–Coldman and bevacizumab beyond disease progression. Nat Rev Clin Oncol 6:311–312

Gu F, Zhang L, Teply BA, Mann N, Wang A, Radovic-Moreno AF, Langer R, Farokhzad OC (2008) Precise engineering of targeted nanoparticles by using self-assembled biointegrated block copolymers. Proc Natl Acad Sci 105:2586–2591

Guenoun J, Koning GA, Doeswijk G, Bosman L, Wielopolski PA, Krestin GP, Bernsen MR (2012) Cationic Gd-DTPA liposomes for highly efficient labeling of mesenchymal stem cells and cell tracking with MRI. Cell Transplant 21:191–205

Hamamoto R, Suvarna K, Yamada M, Kobayashi K, Shinkai N, Miyake M, Takahashi M, Jinnai S, Shimoyama R, Sakai A (2020) Application of artificial intelligence technology in oncology: towards the establishment of precision medicine. Cancers (Basel) 12:3532

Haney MJ, Klyachko NL, Zhao Y, Gupta R, Plotnikova EG, He Z, Patel T, Piroyan A, Sokolsky M, Kabanov AV (2015) Exosomes as drug delivery vehicles for Parkinson's disease therapy. J Control Release 207:18–30

He W, Hosseinkhani H, Mohammadinejad R, Roveimiab Z, Hueng D, Ou K, Domb AJ (2014) Polymeric nanoparticles for therapy and imaging. Polym Adv Technol 25:1216–1225

Hersh AM, Alomari S, Tyler BM (2022) Crossing the blood-brain barrier: advances in nanoparticle technology for drug delivery in neuro-oncology. Int J Mol Sci 23:4153

Hrkach J, Von Hoff D, Ali MM, Andrianova E, Auer J, Campbell T, De Witt D, Figa M, Figueiredo M, Horhota A (2012) Preclinical development and clinical translation of a PSMA-targeted docetaxel nanoparticle with a differentiated pharmacological profile. Sci Transl Med 4:128ra39

Iwata A, Sai S, Nitta Y, Chen M, de Fries-Hallstrand R, Dalesandro J, Thomas R, Allen MD (2001) Liposome-mediated gene transfection of endothelial nitric oxide synthase reduces endothelial activation and leukocyte infiltration in transplanted hearts. Circulation 103:2753–2759

Jeong WY, Kwon M, Choi HE, Kim KS (2021) Recent advances in transdermal drug delivery systems: a review. Biomater Res 25:1–15

Jesorka A, Orwar O (2008) Liposomes: technologies and analytical applications. Annu Rev Anal Chem 1:801–832

Johnson AR, Forster SP, White D, Terife G, Lowinger M, Teller RS, Barrett SE (2021) Drug eluting implants in pharmaceutical development and clinical practice. Expert Opin Drug Deliv 18:577–593

Kaida-Yip F, Deshpande K, Saran T, Vyas D (2018) Biosimilars: review of current applications, obstacles, and their future in medicine. World J Clin Cases 6:161

Kamaly N, Yameen B, Wu J, Farokhzad OC (2016) Degradable controlled-release polymers and polymeric nanoparticles: mechanisms of controlling drug release. Chem Rev 116:2602–2663. https://doi.org/10.1021/acs.chemrev.5b00346

Kesharwani P, Jain K, Jain NK (2014) Dendrimer as nanocarrier for drug delivery. Prog Polym Sci 39:268–307

Khaled SA, Burley JC, Alexander MR, Yang J, Roberts CJ (2015) 3D printing of five-in-one dose combination polypill with defined immediate and sustained release profiles. J Control Release 217:308–314

Khorasani S, Danaei M, Mozafari MR (2018) Nanoliposome technology for the food and nutraceutical industries. Trends Food Sci Technol 79:106–115

Kim MS, Haney MJ, Zhao Y, Mahajan V, Deygen I, Klyachko NL, Inskoe E, Piroyan A, Sokolsky M, Okolie O (2016) Development of exosome-encapsulated paclitaxel to overcome MDR in cancer cells. Nanomedicine 12:655–664

Kondo E, Iioka H, Saito K (2021) Tumor-homing peptide and its utility for advanced cancer medicine. Cancer Sci 112:2118–2125

Kos IA, Azevedo VF, Neto DE, Kowalski SC (2018) The biosimilars journey: current status and ongoing challenges. Drugs Context 7:1

Kowalczuk A, Trzcinska R, Trzebicka B, Müller AHE, Dworak A, Tsvetanov CB (2014) Loading of polymer nanocarriers: factors, mechanisms and applications. Prog Polym Sci 39:43–86. https://doi.org/10.1016/j.progpolymsci.2013.10.004

Kwon GS (2003) Polymeric micelles for delivery of poorly water-soluble compounds. Crit Rev Ther Drug Carrier Syst 20:357

Lambert JM, Chari RVJ (2014) Ado-trastuzumab emtansine (T-DM1): an antibody–drug conjugate (ADC) for HER2-positive breast cancer. J Med Chem 57(16):6949

Lee YS, Kim SH, Cho JA, Kim CW (2011) Introduction of the CIITA gene into tumor cells produces exosomes with enhanced anti-tumor effects. Exp Mol Med 43:281–290

Lehto T, Kurrikoff K, Langel Ü (2012) Cell-penetrating peptides for the delivery of nucleic acids. Expert Opin Drug Deliv 9:823–836

Liu L, Yao W, Rao Y, Lu X, Gao J (2017) pH-responsive carriers for oral drug delivery: challenges and opportunities of current platforms. Drug Deliv 24:569–581

Liu C, Chen B, Shi W, Huang W, Qian H (2022) Ionic liquids for enhanced drug delivery: recent progress and prevailing challenges. Mol Pharm 19:1033–1046

Lombardo D, Kiselev MA, Caccamo MT (2019) Smart nanoparticles for drug delivery application: development of versatile nanocarrier platforms in biotechnology and nanomedicine. J Nanomater 2019:1

Luo X, Liu Y, Qin Z, Jin Z, Xu L, Liu Y, Cao L, He F, Gu X, Ouyang X (2018) Studies on the antibacterial activity and mechanism of antimicrobial peptides against drug-resistant bacteria. J Biomed Nanotechnol 14:601–608

MacDonald IC, Deans TL (2016) Tools and applications in synthetic biology. Adv Drug Deliv Rev 105:20–34. https://doi.org/10.1016/j.addr.2016.08.008

Majeed N, Panigrahi KCS, Sukla LB, John R, Panigrahy M (2020) Application of carbon nanomaterials in plant biotechnology. Mater Today Proc 30:340–345

Mandal S, Anees M, Singh H, Unnisa A (2023) Role of nanoparticles in chemotherapy in cancer and drug delivery: current scenario and future challenges. In: Sustainable nanomaterials for biomedical engineering: impacts, challenges, and future prospects, p 227

Mirza Z, Karim S (2021) Nanoparticles-based drug delivery and gene therapy for breast cancer: recent advancements and future challenges. In: Seminars in cancer biology, vol 69. Elsevier, pp 226–237

Mohammed Y, Holmes A, Kwok PCL, Kumeria T, Namjoshi S, Imran M, Matteucci L, Ali M, Tai W, Benson HAE (2022) Advances and future perspectives in epithelial drug delivery. Adv Drug Deliv Rev 186:114293

Mohd AB (2022) Ai based advanced health care systems

Morris BA (2022) The science and technology of flexible packaging: multilayer films from resin and process to end use. William Andrew

Muldoon JJ, Kandula V, Hong M, Donahue PS, Boucher JD, Bagheri N, Leonard JN (2021) Model-guided design of mammalian genetic programs. Sci Adv 7:eabe9375

Mura S, Couvreur P (2012) Nanotheranostics for personalized medicine. Adv Drug Deliv Rev 64:1394–1416. https://doi.org/10.1016/j.addr.2012.06.006

Muthu MS, Feng S-S (2013) Theranostic liposomes for cancer diagnosis and treatment: current development and pre-clinical success. Expert Opin Drug Deliv 10:151–155

Narayanan RR, Durga N, Nagalakshmi S (2022) Impact of artificial intelligence (AI) on drug discovery and product development. Indian J Pharm Educ Res 56:s387–s397. https://doi.org/10.5530/ijper.56.3s.146

Orive G, Hernández RM, Gascón AR, Domínguez-Gil A, Pedraz JL (2003) Drug delivery in biotechnology: present and future. Curr Opin Biotechnol 14:659–664. https://doi.org/10.1016/j.copbio.2003.10.007

Palanca-Wessels MCA, Czuczman M, Salles G, Assouline S, Sehn LH, Flinn I, Patel MR, Sangha R, Hagenbeek A, Advani R (2015) Safety and activity of the anti-CD79B antibody–drug conjugate polatuzumab vedotin in relapsed or refractory B-cell non-Hodgkin lymphoma and chronic lymphocytic leukaemia: a phase 1 study. Lancet Oncol 16:704–715

Palo M, Holländer J, Suominen J, Yliruusi J, Sandler N (2017) 3D printed drug delivery devices: perspectives and technical challenges. Expert Rev Med Devices 14:685–696. https://doi.org/10.1080/17434440.2017.1363647

Paradise J, Wolf SM, Ramachandran G, Kokkoli E, Hall R, Kuzma J (2008) Developing oversight frameworks for nanobiotechnology. Minn JL Sci Tech 9:399

Park K (2016) Drug delivery research: the invention cycle. Mol Pharm 13:2143–2147

Park H, Otte A, Park K (2022) Evolution of drug delivery systems: from 1950 to 2020 and beyond. J Control Release 342:53–65

Philippot JR, Schuber F (1994) Liposomes as tools in basic research and industry. CRC Press

Pitt J, Wilcox KC, Tortelli V, Diniz LP, Oliveira MS, Dobbins C, Yu X-W, Nandamuri S, Gomes FCA, DiNunno N (2017) Neuroprotective astrocyte-derived insulin/IGF-1 stimulate endocytic processing and extracellular release of neuron-bound Aβ oligomers. Mol Biol Cell 28:2623

Rabiee N, Khatami M, Jamalipour Soufi G, Fatahi Y, Iravani S, Varma RS (2021) Diatoms with invaluable applications in nanotechnology, biotechnology, and biomedicine: recent advances. ACS Biomater Sci Eng 7:3053–3068

Rahimpour Y, Hamishehkar H (2012) Liposomes in cosmeceutics. Expert Opin Drug Deliv 9:443–455

Raposo G, Stoorvogel W (2013) Extracellular vesicles: exosomes, microvesicles, and friends. J Cell Biol 200:373–383

Reichert JM (2008) Monoclonal antibodies as innovative therapeutics. Curr Pharm Biotechnol 9:423–430

Sabbagh F, Kim BS (2022) Recent advances in polymeric transdermal drug delivery systems. J Control Release 341:132–146

Salehi R, Rasouli S, Hamishehkar H (2015) Smart thermo/pH responsive magnetic nanogels for the simultaneous delivery of doxorubicin and methotrexate. Int J Pharm 487:274–284

Schütz CA, Juillerat-Jeanneret L, Mueller H, Lynch I, Riediker M (2013) Therapeutic nanoparticles in clinics and under clinical evaluation. Nanomedicine 8:449–467

Schwendener RA (2014) Liposomes as vaccine delivery systems: a review of the recent advances. Ther Adv Vaccines 2:159–182

Sharma B, Sharma S, Jain P (2021) Leveraging advances in chemistry to design biodegradable polymeric implants using chitosan and other biomaterials. Int J Biol Macromol 169:414–427

Sharma AR, Lee YH, Bat-Ulzii A, Bhattacharya M, Chakraborty C, Lee SS (2022) Recent advances of metal-based nanoparticles in nucleic acid delivery for therapeutic applications. J Nanobiotechnology 20:1–21. https://doi.org/10.1186/s12951-022-01650-z

Shitara K, Bang Y-J, Iwasa S, Sugimoto N, Ryu M-H, Sakai D, Chung H-C, Kawakami H, Yabusaki H, Lee J (2020) Trastuzumab deruxtecan in previously treated HER2-positive gastric cancer. N Engl J Med 382:2419–2430

Shvets VI, Kubatiev AA, Shobolov DL, Balabanyan VY (2013) Biopharmaceutical technologies based on phospholipids (chemistry, biochemistry, biophysics, biotechnology, physiology, immunology, pharmacology, and production technologies of medicinal and diagnostic preparations, including nanosized preparations). Rev J Chem 3:179–206

Sun T-Y, Wang Q, Zhang J, Wu T, Zhang F (2013) Trastuzumab-peptide interactions: mechanism and application in structure-based ligand design. Int J Mol Sci 14:16836–16850

Torchilin VP (2005) Recent advances with liposomes as pharmaceutical carriers. Nat Rev Drug Discov 4:145–160

Torchilin VP (2007) Micellar nanocarriers: pharmaceutical perspectives. Pharm Res 24:1–16

Torchilin V (2011) Tumor delivery of macromolecular drugs based on the EPR effect. Adv Drug Deliv Rev 63:131–135. https://doi.org/10.1016/j.addr.2010.03.011

Tsiftsoglou AS, Ruiz S, Schneider CK (2013) Development and regulation of biosimilars: current status and future challenges. BioDrugs 27:203–211

van der Oost J, Patinios C (2023) The genome editing revolution. Trends Biotechnol 41:396–409. https://doi.org/10.1016/j.tibtech.2022.12.022

Van Tran V, Moon J-Y, Lee Y-C (2019) Liposomes for delivery of antioxidants in cosmeceuticals: challenges and development strategies. J Control Release 300:114–140

Vhora I, Khatri N, Misra A (2021) Applications of polymers in parenteral drug delivery. In: Applications of polymers in drug delivery. Elsevier, pp 221–261

Vizirianakis IS (2004) Challenges in current drug delivery from the potential application of pharmacogenomics and personalized medicine in clinical practice. Curr Drug Deliv 1:73–80

Wening K, Breitkreutz J (2011) Oral drug delivery in personalized medicine: unmet needs and novel approaches. Int J Pharm 404:1–9

Wolin EM (2012) The expanding role of somatostatin analogs in the management of neuroendocrine tumors. Gastrointest Cancer Res 5:161

Wong CY, Al-Salami H, Dass CR (2020) Formulation and characterisation of insulin-loaded chitosan nanoparticles capable of inducing glucose uptake in skeletal muscle cells in vitro. J Drug Deliv Sci Technol 57:101738

Yin H, Yang J, Zhang Q, Wang H, Xu J, Zheng J (2017) iRGD as a tumor-penetrating peptide for cancer therapy. Mol Med Rep 15:2925–2930

Zhou X, Hu W, Qin X (2008) The role of complement in the mechanism of action of rituximab for B-cell lymphoma: implications for therapy. Oncologist 13:954–966

Chapter 8
Actinomycetes as Nanofactories: Synthesis and Therapeutic Applications

Sompreeti Paul, Sindhoora Lakshmi, T. Amala, Dakshitha Akula, Mrudula Rao, Piyush Mohapatra, and Aswani Thekkangil

Abstract In recent times, the use of nanomaterials has gained prominence in the fields of agriculture, biological science, electronics, and medicine owing to their numerous applications. Their unique characteristics, such as high surface/volume ratio, specific structures, and dimensions that are similar to those of biomolecules, provide special features for biomedical and therapeutic applications. It is well known that algae, bacteria, and fungi can take metals from their surroundings and transform them into elemental nanoparticles, which can then be accumulated or released. Actinomycetes, as green nanofactories, have been regarded as an emerging field for research to produce low-cost, high-quality approaches for producing sustainable useful products with numerous applications. The distinctiveness of biologically active actinomycetes has drawn the attention of experts all around the world to better understand their possibility as effective nanofactories. As a dominant microbial phylum, actinomycetes are helpful for the breakdown and transformation of organic and inorganic substrates and can also synthesize economically *via*ble proteins, enzymes, and antibiotics. Antibiotics produced by these organisms have been utilized in almost every medical intervention and are now known to facilitate the production of nanoparticles with preferred surface and size attributes. Actinomycetes-derived

S. Paul · S. Lakshmi · D. Akula · M. Rao · A. Thekkangil (✉)
Department of Biotechnology, School of Applied Science, REVA University Bangalore, Bangalore, Karnataka, India
e-mail: aswani.t@reva.edu.in

T. Amala
Moulana College of Paramedical Sciences, Malappuram, Kerala, India

P. Mohapatra
Amity Institute of Biotechnology, Amity University, Noida, Uttar Pradesh, India

S. Bose et al. (eds.), *Concepts in Pharmaceutical Biotechnology and Drug Development*, Interdisciplinary Biotechnological Advances,
https://doi.org/10.1007/978-981-97-1148-2_8

nanoparticles have also been extensively studied for a wide range of applications related to biomedical sciences. This chapter provides an overview of the synthesis of nanomaterials from actinomycetes, characterization techniques, and various biomedical and therapeutic applications. In addition, the challenges pertaining to toxicity studies associated with the use and synthesis of biogenic metal nanoparticles are discussed.

Keywords Nanomaterials · Actinomycetes · Nano factories · Therapeutic application

8.1 Introduction

Nanoparticles (NPs) are small particles that typically range in size from one-hundred nm. These particles can exist naturally and can be synthesized using different modes. They are similar to certain things in nature, such as DNA and proteins (Khan et al. 2019). They are distinct in their sizes and surface areas.

NPs, owing to their unique characteristics, have a diverse array of applications in multiple fields. Their exceptional properties, such as distinctive optical, magnetic, and catalytic properties, along with enhanced electrical conductivity, increased reactivity, and larger surface area, make them invaluable for a wide range of applications (Khan et al. 2019). Like in the field of biomedicine, NPs are used to enhance drug delivery. Their minuscule size allows them to penetrate biological systems and reach cells or tissues effectively. Medications encapsulated within nanoparticles exhibit improved solubility, stability, and controlled release, leading to more efficient treatments with reduced side effects (Modena et al. 2019). NPs have applications in food packaging to extend the shelf life and enhance safety. They are also used in agriculture to optimize the management of fertilizers, pesticides, and growth chemicals, minimize environmental impacts, and maximize crop yield (Khan et al. 2019). NPs are indispensable for environmental purification. They can be harnessed to combat air pollution, facilitate waste management, and purify water by removing impurities. Their use provides effective and long-lasting solutions to environmental problems.

NPs can be synthesized from diverse resources like metals (gold, platinum, and silver), metal oxides (like iron oxide and titanium dioxide), semiconductors (such as tiny particles called quantum dots), plastics, carbon-based stuff (such as graphene and carbon tubes), and biological sources like plants and microbes (Jamkhande et al. 2019). There are different approaches to NP synthesis, such as chemical synthesis, physical synthesis, and green synthesis (from living things). The characteristics of NPs can also be regulated using different approaches.

8.2 Green Nanosynthesis: A Sustainable Method

NPs can be synthesized in two ways: Biological and nonbiological. Nonbiological techniques require unusual response conditions, significant energy inputs, and precarious ingredients. An eco-friendly, clean, nontoxic, and economical process is promised through the use of biological systems as nanofactories. Green synthesis has gained attention because it produces size-controlled, resilient, and dispersed nanomaterials with desired physicochemical qualities by reducing the ionic concentration through biological processes (Ying et al. 2022). Moreover, green synthesis methods are progressively replacing physical and chemical methods to forestall various setbacks such as excessive energy consumption, emissions of harmful and toxic substances, and the prerequisite of complex equipment and synthesis conditions.

Green synthesis makes use of organic resources and environmentally desirable components, such as microbial cultures and plant parts (Yosri et al. 2021). Metal nanoparticle synthesis utilizing microorganisms, specifically actinomycetes, has been attempted and succeeded in several cases. Several *Streptomycetes* species, for instance, are capable of synthesizing both AgNPs and AuNPs (Ying et al. 2022). For the conversion of nanoparticles, the proteins and polyphenols found in green sources may serve as reducing agents. The advantages of green synthesis, such as environmentally sound, economical, and stable materials, emphasize the importance of green nanoparticles (GNPs). However, there are some limitations such as increased reaction time, reduced product quality, and difficulties in raw material extraction.

8.3 Top-Down and Bottom-Up Approaches for NP Synthesis

Two basic methods have been proposed for the synthesis of metal nanoparticles: top-down and bottom-up (Fig. 8.1). Top-down methodology controls the process of manufacturing NPs by breaking down, mechanically grinding, or slicing bulk

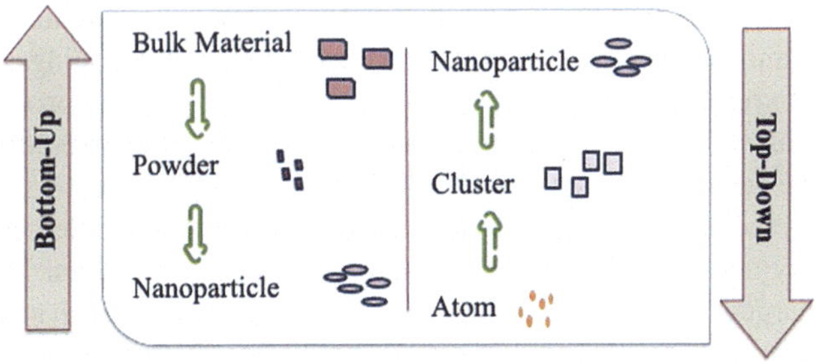

Fig. 8.1 Basic approaches for synthesis of nanoparticle

material into nanoscale particles, whereas the bottom-up approach synthesizes NPs by assembling atoms at the nanoscale (Abid et al. 2022). The physical generation of nanoparticles uses a top-down approach, whereas biological and chemical syntheses use a bottom-up approach. However, unlike physiochemical synthesis, biogenic synthesis uses a range of bioactive metabolites and enzymes from microbial and plant sources to catalyze the breakdown of metal substances to produce the correct amount of NPs under ambient conditions. This is because metal ions are reduced considerably more quickly using natural agents (Prakasham 2012).

8.4 Actinomycetes as Nanofactories

Actinomycetes, which are G^{+ve} (gram-positive) fungi-like filamentous bacteria with a high G + C (Guanine + Cytosine) content, distributed extensively in both terrestrial and marine environments (Thekkangil and Suchithra 2020; Thekkangil et al. 2021). These ubiquitous organisms are well-known for their ability to produce a wide range of distinct secondary metabolites with a wide range of applications (Aswani et al. 2019; Kumari et al. 2020). Owing to this potential, studies on the bioprospecting of actinomycetes for various purposes have garnered significant interest. The production of nanoparticles by actinomycetes is a rapidly emerging field, with various species such as *Streptomyces*, *Nocardia*, *Rhodococcus*, and *Thermomonospora* being known to produce nanometals (Sandhu and Goel 2023). These organisms are capable of generating different types of nanoparticles, including silver, gold, iron, and copper, which possess various biological activities.

8.4.1 Extracellular and Intracellular Actinomycetes-Mediated Biosynthesis of NPs

Actinomycetes are known to adapt to extreme environments and tolerate a high concentration of metals. They also have the capability to degrade inorganic elements into NPs. As part of their microbial defense strategy for cellular detoxification, these organisms are able to utilize metal ions from the surrounding environment, which further convert the ions into elemental metal form by the process of enzymatic reduction (Kumari et al. 2020). Based on the site of manifestation, NP production can be categorized as extracellular or intracellular (Table 8.1).

During extracellular synthesis, metal ions are reduced to their elemental form by microbial enzymes and proteins. During the process of intracellular NP synthesis, positively charged metal ions initially attract to the negatively charged carboxyl groups in the microbial cell wall via electrostatic force of attraction. This makes it possible for the metal ions to infiltrate through the cellular membrane and undergo further metallic reduction through intracellular proteins/enzymes within the cell

Table 8.1 Examples of extracellular and intracellular actinomycetes-mediated nanoparticles

NPs	Actinomycetes	Sources	EC/IC	References
Copper	Actinomycetes sp.	Endophytes (Palnta[a])	EC/IC	Rajivgandhi et al. (2022)
	Streptomyces sp.	Soil	EC/IC	Rajivgandhi et al. (2022)
Gold	*Rhodococcus* sp.	Water	IC	Salem et al. (2020)
	S. viridogens HM10	Soil	IC	Rasool and Hemalatha (2017)
Silver	*S. coelicolor*	Soil	–	Chackaravarthi et al. (2023)
	S. antimycoticus L-1	Endophytes (Plant[b])	EC	Salem et al. (2020)
	Nocardiopsis sp.	Water	EC	Chackaravarthi et al. (2023
	Rhodococcus sp.	Water	IC	Manimaran and Kannabiran (2017)
	S. albidoflavus CNP10	Water	EC/IC	Rasool and Hemalatha (2017)
	S. aureofaciens MTCC 365	Water	EC	Golinska et al. (2014)
	S. pseudogriseolus Acv-11	Endophytes	EC	Fouda et al. (2020)
	S. hygroscopicus BDUS49	Water	EC	Rasool and Hemalatha (2017)
	S. zaomyceticus Oc-5	Endophytes (Plant[c])	IC	Fouda et al. (2020)
	S. rochei	Water	EC	Chackaravarthi et al. (2023)
	S. capillispiralis Ca-1	Endophytes (Plant[d])	EC	Fouda et al. (2020)
	Streptomyces sp. VITBT7	Soil	EC	Manimaran and Kannabiran (2017)
	Streptomyces sp. VITPK1	Water	EC	Rajivgandhi et al. (2022)
Zinc	*Streptomyces* sp.	Soil	EC/IC	Manimaran and Kannabiran (2017)
	Streptomyces sp.	Soil/water	EC	Golinska et al. (2014)
	S. *coeruleorubidus*	Endophytes (marine)	EC	Chackaravarthi et al. (2023)

NP nanoparticle, *EC* extracellular, *IC* intracellular
[a]medicinal plant *Lonicera japonica; S* Streptomyces, *R* Rhodococcus, *N* Nocardiopsis
[b]Mentha longifolia L.
[c]*Oxalis corniculata* L.
[d]*Convolvulus arvensis* L.& *Oxalis corniculata* L.

(Venkateswaran et al. 2022). Several reports on both extracellular and intracellular synthesis have been postulated, and in every hypothesis, the major enzymes included are oxidoreductases such as sulfite reductase, nitrate reductase, and cellular transporters (Koul et al. 2021). The intracellular process entails the transfer of metal ions inside the microbial cell with the help of transporter enzymes, whereas in the extracellular method, enzymatically reduced ions accumulate on the cell surface.

Actinomycetes-mediated intracellular NP production involves the reduction of metal ions at the mycelia and cytoplasmic membrane (Kumari et al. 2020). The synthesis of AgNPs is a typical example of an intracellular method for synthesizing metal NPs. In one of the proposed mechanisms for AgNP synthesis by intracellular mechanism, first Ag^+ ions got trapped on the surface of actinomycete cells through electrostatic interactions with negatively charged carboxylate groups present in the mycelial cell wall and are further reduced to metallic AgNPs in the nuclei by the reductase enzyme. Extracellular synthesis in actinomycetes mostly occurs within the nitrogen cycle via enzymatic reduction of nitrates to nitrites. It is commonly recognized that the biogenic synthesis of AgNPs is significantly influenced by NADH-dependent nitrate reductase (Mabrouk et al. 2021). The most acceptable mechanism for AgNP synthesis is the bioreduction of Ag^+ ions to metallic Ag by nitrate reductase. This enzymatic metal-reduction reaction occurs in various electron shuttle chains (Koul et al. 2021). A related process for AuNPs is the reduction of Au ions from aqueous $AuCl_4$. It appears that electron transfer from NADH initiates the bioreduction of gold ions, which is catalyzed by NADH-dependent nitrate reductase as a carrier for electrons. Au ions are further reduced to electrons, converted to gold, and subsequently to gold NPs (Kumari et al. 2020).

Mechanisms of Nanoparticle Biosynthesis

The reduction of metal ions to their corresponding nanoparticles is facilitated by a few secondary metabolites as well as by intracellular and extracellular enzymes (Fig. 8.2). The process can be divided into three phases. In the first phase, metallic ions are either absorbed within microbial cell membranes via cationic membrane-transport systems or collected on the cell surface through electrostatic interactions with the negatively charged cell wall. In the second phase, metal ions (M^+) are reduced to zero-valent metals (M°), catalyzed by oxidoreductase enzymes, such as nitrate reductase. Which further results in the conversion to monovalent- or divalent-metal ions from zero valent states. Ultimately, the resultant NPs coalesced into a variety of morphological forms, such as cubes, ovals, triangles, hexagons, and spheres. The third step involves the stabilization of the nanoparticles via capping agents to inhibit development and agglomeration while also regulating their size and form (Messaoudi and Bendahou 2020).

8.4.2 Bioprocess of NPs Using Actinomycetes

Bioprocesses can be classified into the downstream, fermentation, and upstream subcategories. Downstream processing includes inoculum preparation and medium formulation. Actinomycetes can be isolated from various sources, such as soil, water, and

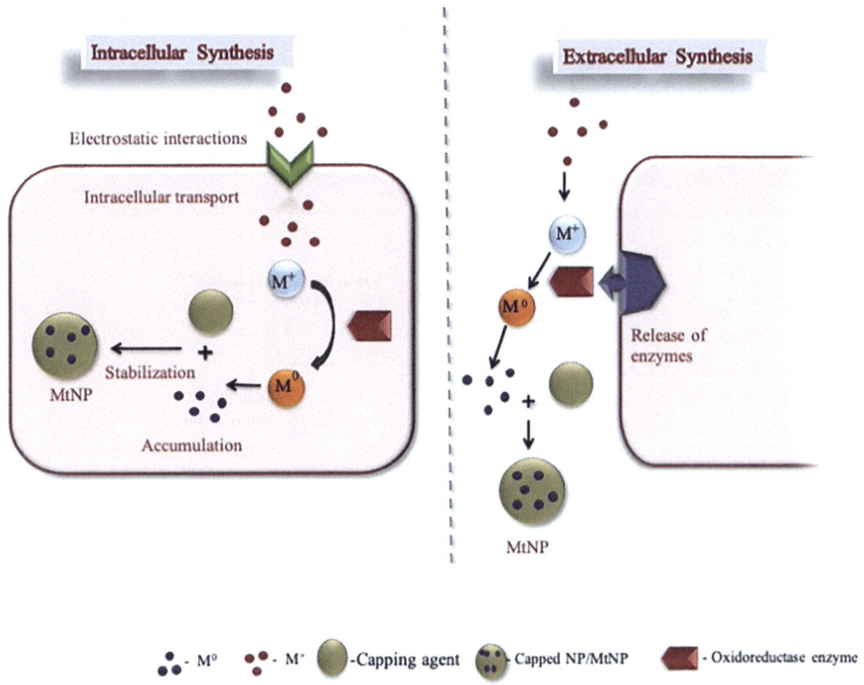

Fig. 8.2 Proposed nanoparticle synthesis mechanism: Figure illustrating the processes involved in extracellular and intracellular synthesis of MtNPs. In extracellular NP synthesis, Metal ions (M^+) are adsorbed on the cell wall and further reduced to metals (M°) by extracellular oxidoreductase enzyme. In the intracellular synthesis of MtNPs, M^+ is transferred into the cytoplasm of the cell and subsequently reduced by metabolic processes involving various oxidoreductase enzymes.

plant parts, or revived/activated from stock culture. Multiple media options exist for the growth of actinomycetes, including Starch Casein Agar, Bennett Agar, International Streptomycetes Project Mediums, and Chitin Agar (Salem et al. 2020). The cultivation conditions must be optimized for optimal NP yield, including the appropriate culture medium, pH, temperature, and agitation. Following a 3–4 day incubation period at 28 °C with shaking, mycelia were removed from the culture broth by filtration and subsequently wash with sterile distilled water (Kumari et al. 2020). Both the supernatant and biomass were tested for NP synthesis, and it was observed that microorganisms were capable of synthesizing NPs extracellularly or intracellularly.

Extracellular synthesis involves mixing the cell-free supernatant with a metal precursor salt solution, whereas intracellular synthesis involves mixing the washed biomass with a metal precursor salt solution. The color formation after incubation indicated the synthesis of nanoparticles. Centrifugation was then used to recover and collect the precipitated nanoparticles as bottom pellets. If the nanoparticles are synthesized inside the cell, an additional step of cell disruption is necessary, which can be executed through repeated cycles of ultrasonication to release the NPs from the cell for further recovery (Moholkar et al. 2021). Following recovery,

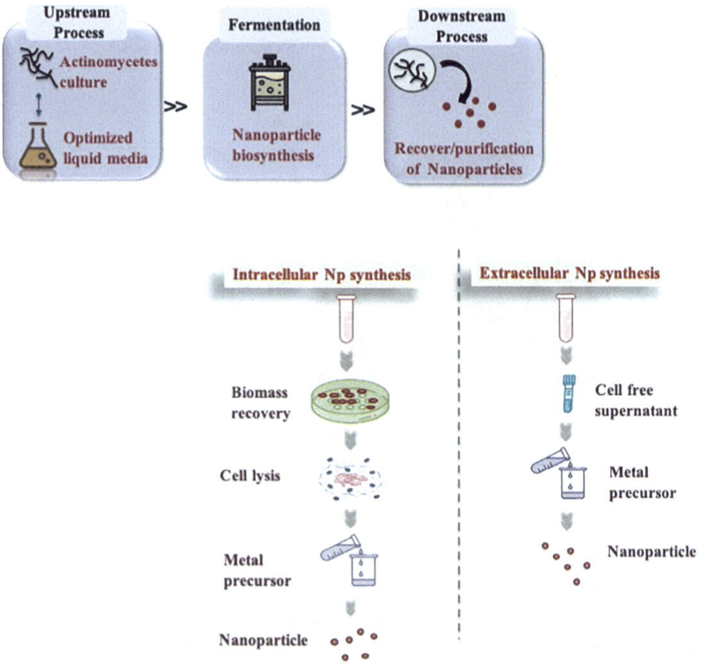

Fig. 8.3 Schematic illustration of actinomycetes-mediated green synthesis of nanoparticles

physicochemical characterization of the nanoparticles was performed to determine their structural and chemical properties. Figure 8.3 provides a visual representation of the bioprocess involved in nanoparticle synthesis mediated by actinomycetes.

8.4.3 *Factors Affecting NP Biosynthesis*

Research has indicated that the size of biosynthesized nanoparticles greatly affects their biological activity. Smaller nanoparticles have high surface/volume ratios and exhibit superior activity levels compared to larger NPs (Pandit et al. 2022). The culture techniques employed in actinomycete-assisted NP synthesis play a significant role in the synthesis of homogenous nanoparticles with improved enzymatic activity. Therefore, optimizing culture parameters, including medium composition, temperature, pH, mixing speed, and incubation time, can significantly impact the synthesis of nanoparticles with enhanced enzymatic activity (Kumari et al. 2020).

pH

pH is found to be an essential variable in the synthesis of biogenic NPs. Metal ion reduction may be impacted by a sudden change in pH because it can alter the charge of the metal ions in the reaction mixture. According to Patra and Baek 2014, the pH of the solution media can make a determent effect on the size and texture of synthesized nanoparticles (Patra and Baek 2014). An interesting finding in the biosynthesis of AgNPs was reported by Ibrahim et al. (2023), based on the observation that alterations in the pH of the solution could modify the size of the NPs (Ibrahim et al. 2023). The samples produced at pH 5 were unstable, whereas those synthesized at pH 9 were stable.

Temperature and Pressure

Temperature is another crucial factor for NP synthesis. The green synthesis of nanoparticles requires ambient temperature, and the temperature of the reaction determines the type of nanoparticle created, which has a major impact on the rate of synthesis, stability of NPs, and final size of the NPs produced (Patra and Baek 2014).

Another significant element that influences the final form of metallic NP size and shape is pressure. Research has shown that ambient pressure accelerates the reduction of metal ions compared to normal conditions (Pandit et al. 2022).

Concentration

The concentration of the cell extract is also critical for efficient NP biosynthesis, which relates the quantity of stabilizing and reducing agents (Kumari et al. 2020).

Reaction Time

Another crucial element for optimization during biosynthesis is the time necessitated to complete the bioreduction process among the metal precursor and extract. The amount of time the reaction medium is incubated offers a significant impact on the quality and structure of the NPs produced. The chemical properties of NPs are also altered according to the reaction time in a similar way. If nanoparticles are incubated for an extended period, they may aggregate, shrink, and lose some efficacy (Gupta et al. 2019).

8.4.4 Characterization Techniques for Metal NPs

Several characterization approaches have been used to study the physiochemical properties of nanoparticles. This covers a variety of analytical techniques, such as FT-IR spectroscopy, NMR spectrometry, and UV-Vis spectrometry. Optical or Zeta potential (Zp) measurements were performed using DLS, CD, and XRD. An array of microscopic analyses by AFM, SEM, TEM, and STM are also used methodically to analyze the various physical properties of NPs. Moreover, certain complementary analytical procedures, such as MS, TERS spectrometry, and SERS spectrometry,

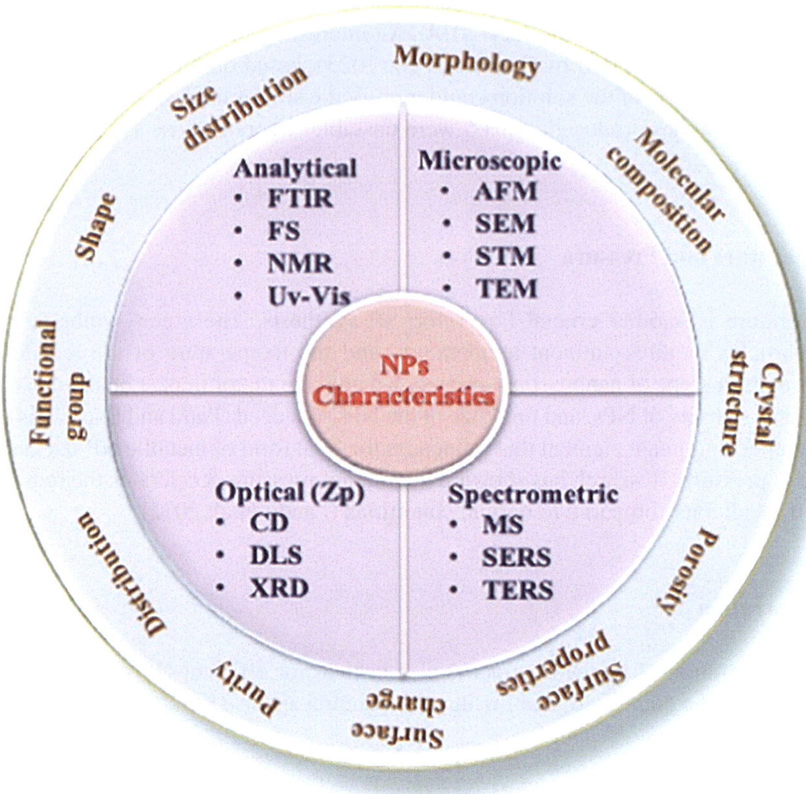

Fig. 8.4 Graphical representation of some of the physicochemical characterization and techniques of nanoparticles. *Abbreviations: NPs* nanoparticles, *ZP* zeta potential, *FTIR* Fourier transform infrared spectroscopy, *FS* fluorescence correlation spectroscopy, *NMR* nuclear magnetic resonance spectroscopy, *UV-Vis* ultraviolet-visible spectroscopy, *CD* circular dichroism, *DLS* dynamic light scattering, *XRD* X-ray crystallography, *AFM* atomic force microscopy, *SEM* scanning electron microscopy, *STM* scanning tunneling microscopy, *TEM* transmission electron microscopes, *MS* mass spectrometry, *SERS* surface-enhanced Raman spectroscopy, *TERS* tip-enhanced Raman spectroscopy

are also employed to study the details of the synthesized NPs (Fig. 8.4) (Wadhwa et al. 2022).

The process of investigation using absorption spectra in the range of 200–800 nm will help to confirm that nanoparticles were formed during synthesis. Techniques such as SEM and TEM can be employed to analyze the size, shape, distribution, and homogeneity of nanoparticles. The crystallinity and chemical composition of the synthesized NPs were determined using XRD. FTIR spectroscopy was exercised to identify the functional groups existing in the nanoparticles. DLS is an effective method for estimating the particle size, even at low concentrations. The surface area and stability of colloidal nanoparticles were examined using optical measurements (Shnoudeh et al. 2019).

8.5 Applications of NPs in Different Biotechnological Sectors

Nanoparticles have demonstrated versatility for potential applications in various biotechnological sectors, including agriculture, environment, food, and therapeutics. The utilization of metallic nanoparticles in biotechnology has increased because of their inherent biocompatibility, potent antimicrobial properties, proficiency in drug delivery, profound bioactivity, enhanced bioavailability, and effective integration into biological systems. One unique characteristic of biosynthesized metal-based nanoparticles is their lack of requirement for external stabilizing or capping agents, which has unlocked potential in previously unexplored applications in various domains (Zarina and Nanda 2014).

8.5.1 Applications of NPs in the Environmental Sector

The environmental applications of nanotechnology primarily fall into three major domains: NPs as environmentally sustainable products to combat pollution, their use in bioremediation of toxic contaminants, and their function as biosensors for environmental monitoring. Because of the detrimental effects of heavy metals on both the environment and human health, NPs have garnered a considerable amount of attention as potential biosorbent materials for the removal of heavy metals from environmental sources. Metallic nanoparticles sourced from Actinomycetes have proven to be powerful tools in the fight against biofouling, with silver nanoparticles produced by *Streptomyces rochei* HMM13 having the ability to reduce the number of bacterial cells and prevent bacterial biofilm formation (Sawada et al. 2012).

8.5.2 Applications of NPs in Food Sector

Nanotechnology has abundant applications in the food industry, including functional food development, food processing and packaging, food safety like in the detection of food pathogen, and extending the shelf life of food products. These applications can be broadly categorized into two groups: nanostructured food ingredients and nanosensing. Nanostructured food components are used in food processing, where they can be used as fillers to increase the durability of packaging materials, enhancing mechanical strength and as food additives, anticaking agents, and antimicrobial agents. Nanosensing can also be applied to enhance food safety and quality assessments (Singh et al. 2017). A potential application of green nanosynthesis in the food industry is the use of *Streptomyces*-mediated coatings to create natural antibacterial food coatings. These materials have the potential to prolong the shelf life of food products and ensure safety (Rajamanickam et al. 2012).

8.5.3 Applications of NPs in Agricultural Sector

The synthesis of green NPs has also emerged as a potential source of fertilizers, insecticides, and agents for plant growth. In recent years, there has been a trend toward using nanomaterials as substitutes for traditional pest management techniques in agriculture. Various nanomaterials and nanostructures are currently being used as biosensors to assess soil quality and fertilizer distribution. According to Karthik et al. 2014, extracellular AgNP synthesized by *Streptomyces* sp. LK3 is particularly effective in controlling pests, particularly *Haemaphysalis bispinosa* and *Rhipicephalus microplus*, in agriculture (Karthik et al. 2014).

8.5.4 Applications of NPs in Medical Sector

Nanomedicine has garnered significant attention in every branch of medicine, from diagnostics to treatment. Its applications include antimicrobial agents, biomolecule detection, biosensors, cancer treatment, diagnostic devices, drug delivery, regenerative medicine, and tissue engineering. The unique properties of nanomaterials, such as enhanced flexibility, durability, sensitivity, unparalleled performance, and distinct physicochemical properties, have been exploited in therapeutic applications for early detection of diseases, regenerative medicine, targeted clinical therapy, and tissue reconstruction. Nanomedicine operates in a manner similar to many other biological processes and has fewer side effects (Rudramurthy and Swamy 2018; Mauricio et al. 2018; Patel and Nanda 2015).

Actinomycetes can synthesize numerous metal nanoparticles, such as copper, gold, silver, and zinc, which exhibit antimicrobial effects against a diverse array of microorganisms, including multidrug-resistant strains (Table 8.2). The versatility of the antimicrobial activity against a wide range of microbes is noteworthy. Researchers have reported that AgNPs synthesized by actinomycetes demonstrate bactericidal activity against clinically significant pathogens, both inside and outside cells. For example, *Streptomyces viridochromogenes* exhibit excellent antimicrobial activity against both gram-negative and gram-positive bacteria, as well as yeasts, such as *Candida albicans* (Krishnaraj et al. 2016). Their small S/V ratio (size/volume) allows them to interact more effectively with microbial membranes, causing the cells to rupture (Chauhan et al. 2013). The increasing emergence of antibiotic resistance underscores the potential of actinobacteria-mediated NP as promising alternatives. Another study explored the synergistic potential of AgNPs derived from *Streptomyces parvulus* SSNP11. These nanoparticles can boost the effectiveness of traditional antibiotics such as Ampicillin, Cefalexin, Penicillin, and Streptomycin. Additionally, these NPs have the potential to serve as carriers for drug delivery, which is crucial in the fight against microbial infections. The anti-ESBL properties of the AgNPs isolated from *Streptomyces* sp. VITSJK10 have also been previously reported (Manikprabhu and Lingappa 2013). Biogenic ZnO oxide

Table 8.2 Examples of actinomycetes-mediated NPs with therapeutic potential

Actinomycetes sp.	NP (Size nm)	Applications	References
S. durhamensis	Ag (8–48)	Antimicrobial (*S. aureus*, *B. subtilis*, *E. coli*, *P. aeruginosa*, *K. pneumoniae*, and *P. mirabilis*)	Kumari et al. (2020)
N. sp. MBRC-1	Ag (45)	Antimicrobial (*B. subtilis*, *P aeruginosa*, and *C. albicans*)	Manivasagan et al. (2015)
S. rochei	Ag (22–85)	Antibacterial & Antibiofilm[a]	Prakasham et al. (2014)
Streptomyces sp..	Ag (21–45)	Anticandidal	Prakasham et al. (2014)
G. amicalis HS-11	Ag, au (–)	Free radical scavenging	Kumari et al. (2020)
Rhodococcus sp.	Au (10)	Antimicrobial and antioxidant	Manivasagan et al. (2015)
S. parvulus SSNP11	Ag (2–12)	Antibacterial	Prakasham et al. (2014)
Rhodococcus sp.	Au (9)	Antifungal	Kumari et al. (2020)
S. xinghaiensis OF1	Ag (5–20)	Antimicrobial	Manivasagan et al. (2015)
Streptomyces sp. VITSTK7	Ag (20–60)	Antifungal	Prakasham et al. (2014)
Streptomyces sp. LK3	Ag (–)	Acaricidal (*H. bispinosa* and *R. microplus*)	Karthik et al. (2014)
S. rochei HMM13	Ag (–)	Antibacterial and antibiofilm	Manivasagan et al. (2015)

[a] Synergistic effects with ciprofloxacin, ampicillin, streptomycin, gentamicin, tetracycline, and lincomycin; *B Bacillus*, *E E. coli*, *K Klebsiella*, *P Pseudomonas*, *C Candida*, *H Haemaphysalis*, *R Rhipicephalus*, *S Streptomycetes*, *N Nocardiopsis*, *G Gordonia*, *NP* nanoparticle, *Ag* silver, *Au* gold

nanoparticles show potential for cancer treatment and are especially effective against A549 lung cancer cells. In this biogenic synthesis, *Streptomyces* sp. acts as a reducing agent for the biosynthesis of ZnO NPs (Subbaiya et al. 2017).

Antimicrobial Activity of NP: Mechanism of Action

Actinomycetes-mediated NP has noteworthy antimicrobial properties. The mode of action of their microbicidal activity varies according to the type of metal nanoparticles and the specific pathogens they target. One critical mechanism involves the physical interaction of metal nanoparticles with the cell membranes of pathogens. This interaction impairs the integrity of the membranes, causing them to become more permeable, which allows essential components to leak out, ultimately resulting in pathogen death (Gabrielyan and Trchounian 2019). NPs also produce reactive oxygen species (ROS), which can harm cellular components, such as DNA, lipids, and proteins. Some NPs are taken up by pathogens, disrupt cellular functions, and interact with the intracellular components of the pathogen (Cui et al. 2013). Another significant mechanism is the release of toxic metal ions by nanoparticles, which bind to cellular proteins and enzymes, leading to the disruption of vital metabolic pathways of the pathogen (Nisar et al. 2019).

8.6 Toxicity of Metal Nanoparticles

Nanotoxicology is a matter of growing concern because of the potential health implications of exposure to metal nanoparticles in both humans and the environment. Exposure to nanoparticles can occur through inhalation, ingestion, or skin contact (Yao et al. 2019). The skin serves as a critical barrier against harmful substances, but nanoparticles can penetrate it depending on their size. Such penetration can lead to local inflammation, abnormal extracellular matrix production, and increased neovascularization (Sharifi et al. 2020). When ingested, nanoparticles tend to accumulate in the liver, affecting glutathione levels, mitochondrial function and increasing reactive oxygen species (ROS). Inhaled nanoparticles can adversely affect the respiratory system and induce oxidative stress (Sengul and Asmatulu 2020). Some studies have suggested that AgNPs can disrupt the integrity of the plasma membrane and disturb mitochondrial function, ultimately leading to cell death through apoptosis. Iron nanoparticle exposure can cause structural alterations in the root cell walls of plants. Several variables, including size, shape, structural characteristics, dosage, dispersion, chemical composition, and capping agents, can influence the lethal nature of nanoparticles (Yao et al. 2019). Proper dispersion and well-defined properties can render nanoparticles less toxic (Sengul and Asmatulu 2020). As the utilization of nanoparticles continues to expand, concerns about their environmental impact are also increasing, as they can disperse into water bodies, air, and soil.

8.7 Future Prospects

Although actinomycete-based metal nanoparticle production has received increasing interest of researchers worldwide in recent decades, further research is necessary to identify the putative actinomycetes. Compared to physicochemical methods, biogenic nanosynthesis is a time-consuming procedure that takes several hours. More extensive research is required to increase the synthesis rate. The size, shape, and surface area of nanoparticles play a critical role in their bioactivity; however, much research has yet to be conducted on the regulation of various biological processes that modify the size and shape of nanoparticles. Addressing concerns related to toxicity and ensuring repeatability and biosafety are essential factors to consider during the procedure.

8.8 Conclusion

Actinomycetes are well known for their capability to produce a vast array of bioactive compounds and have the potential to function as nanomachines for the synthesis of a distinct range of NPs. Biogenic production of NPs by actinomycetes offers significant possibilities for the development of nano-antimicrobials, which could prove to be effective substitutes and find widespread use in various domains. Although several studies on the biogenic synthesis of NPs from actinomycetes have been reported, there has been growing interest in actinobacterial NP synthesis over the past decade. However, further research is needed, with a greater focus on scaling up processes for actinobacterial NP synthesis to meet industrial requirements. Additionally, advancements in optimizing diverse culture conditions to enhance NP biosynthesis efficiency and provide effective control over NP size and monodispersity could pave the way for notable industrial applications.

References

Abid N, Khan AM, Shujait S, Chaudhary K, Ikram M, Imran M, Haider J, Khan M, Khan Q, Maqbool M (2022) Synthesis of nanomaterials using various top-down and bottom-up approaches, influencing factors, advantages, and disadvantages: a review. Adv Colloid Interf Sci 300:102597

Aswani T, Reshmi S, Suchithra TV (2019) Actinomycetes: its realm in nanotechnology. Microbial Nanobionics 1:127–140

Chackaravarthi G, Alharbi NS, Gnanamangai BM, Ramachandran G, Chelliah CK, Rajivgandhi G, Li WJ (2023) Anti-biofilm efficacy of marine actinomycete mediated zinc oxide nanoparticles increased the intracellular damages in biofilm forming *K. pneumoniae*. J King Saud Univ Sci 35(4):102642

Chauhan R, Kumar A, Abraham J (2013) A biological approach to the synthesis of silver nanoparticles with *Streptomyces* sp JAR1 and its antimicrobial activity. Sci Pharm 81(2):607–624

Cui L, Chen P, Chen S, Yuan Z, Yu C, Ren B, Zhang K (2013) In situ study of the antibacterial activity and mechanism of action of silver nanoparticles by surface-enhanced Raman spectroscopy. Anal Chem 85(11):5436–5443

Fouda A, Hassan SED, Abdo AM, El-Gamal MS (2020) Antimicrobial, antioxidant and larvicidal activities of spherical silver nanoparticles synthesized by endophytic Streptomyces spp. Biol Trace Elem Res 195:707–724

Gabrielyan L, Trchounian A (2019) Antibacterial activities of transient metals nanoparticles and membranous mechanisms of action. World J Microbiol Biotechnol 35(10):162

Golinska P, Wypij M, Ingle AP, Gupta I, Dahm H, Rai M (2014) Biogenic synthesis of metal nanoparticles from actinomycetes: biomedical applications and cytotoxicity. Appl Microbiol Biotechnol 98:8083–8097

Gupta A, Singh D, Singh SK, Singh VK, Singh AV, Kumar A (2019) Role of actinomycetes in bioactive and nanoparticle synthesis. In: Role of plant growth promoting microorganisms in sustainable agriculture and nanotechnology. Woodhead Publishing, London, pp 163–182

Ibrahim S, Taher H, Sayed R, Abdulla H (2023) Physical factors controlling silver nanoparticles biosynthesis using actinomycetales bacterium BC1:MZ067956. In: ISET international conference on applied science & engineering, vol 2620. AIP Publishing, Melville, NY. https://doi.org/10.1063/5.0120641

Jamkhande PG, Ghule NW, Bamer AH, Kalaskar MG (2019) Metal nanoparticles synthesis: an overview on methods of preparation, advantages and disadvantages, and applications. J Drug Deliv Sci Technol 53:101174

Karthik L, Kumar G, Kirthi AV, Rahuman AA, Bhaskara Rao KV (2014) *Streptomyces* sp. LK3 mediated synthesis of silver nanoparticles and its biomedical application. Bioprocess Biosyst Eng 37:261–267

Khan I, Saeed K, Khan I (2019) Nanoparticles: properties, applications and toxicities. Arab J Chem 12:908–931

Koul B, Poonia AK, Yadav D, Jin JO (2021) Microbe-mediated biosynthesis of nanoparticles: applications and future prospects. Biomol Ther 11(6):886

Krishnaraj C, Ji B-J, Harper SL, Yun S-I (2016) Plant extract-mediated biogenic synthesis of silver, manganese dioxide, silver-doped manganese dioxide nanoparticles and their antibacterial activity against food- and water-borne pathogens. Bioprocess Biosyst Eng 39:759–772

Kumari S, Tehri N, Gahlaut A, Hooda V (2020) Actinomycetes mediated synthesis, characterization, and applications of metallic nanoparticles. Inorg Nano-Met Chem 51:1386–1395

Mabrouk M, Elkhooly TA, Amer SK (2021) Actinomycete strain type determines the monodispersity and antibacterial properties of biogenically synthesized silver nanoparticles. J Genetic Eng Biotechnol 19:57. https://doi.org/10.1186/s43141-021-00153-y

Manikprabhu D, Lingappa K (2013) Microwave assisted rapid and green synthesis of silver nanoparticles using a pigment produced by *Streptomyces coelicolor* klmp33. Bioinorg Chem Appl 2013:341798

Manimaran M, Kannabiran K (2017) Actinomycetes-mediated biogenic synthesis of metal and metal oxide nanoparticles: Progress and challenges. Lett Appl Microbiol 64:401–408

Manivasagan P, Alam MS, Kang K-H, Kwak M, Kim S-K (2015) Extracellular synthesis of gold bionanoparticles by nocardiopsis sp. and evaluation of its antimicrobial, antioxidant and cytotoxic activities. Bioprocess Biosyst Eng 38:1167–1177

Mauricio MD, Guerra-Ojeda S, Marchio P, Valles SL, Aldasoro M, Escribano-Lopez I, Herance JR, Rocha M, Vila JM, Victor VM (2018) Nanoparticles in medicine: a focus on vascular oxidative stress. Oxidative Med Cell Longev 2018:1–20

Messaoudi O, Bendahou M (2020) Biological synthesis of nanoparticles using endophytic microorganisms: current development. In: Nanotechnology and the environment. IntechOpen, London. https://doi.org/10.5772/intechopen.93734

Modena MM, Rühle B, Burg TP, Wuttke S (2019) Nanoparticle characterization: what to measure? Adv Mater 31:1901556. https://doi.org/10.1002/adma.201901556

Moholkar ND, Havaldar VD, Potadar SR, Pawar DK (2021) Optimization of biogenic synthesis of colloidal metal nanoparticles. In: Colloids—types, preparation and applications. IntechOpen, London. https://doi.org/10.5772/intechopen.94853

Nisar P, Ali N, Rahman L, Ali M, Shinwari ZK (2019) Antimicrobial activities of biologically synthesized metal nanoparticles: an insight into the mechanism of action. JBIC J Biol Inorg Chem 24:929–941

Pandit C, Roy A, Ghotekar S, Khusro A, Islam MN, Emran TB, Lam SE, Khandaker MU, Bradley DA (2022) Biological agents for synthesis of nanoparticles and their applications. J King Saud Univ Sci 34:101869

Patel S, Nanda R (2015) Nanotechnology in healthcare: applications and challenges. Med Chem 5:12. https://doi.org/10.4172/2161-0444.1000312

Patra JK, Baek K-H (2014) Green nanobiotechnology: factors affecting synthesis and characterization techniques. J Nanomater 2014:1–12

Prakasham RS (2012) Characterization of silver nanoparticles synthesized by using marine isolate *Streptomyces albidoflavus*. J Microbiol Biotechnol 22:614–621

Prakasham RS, Kumar BS, Kumar YS, Kumar KP (2014) Production and characterization of protein encapsulated silver nanoparticles by marine isolate *Streptomyces parvulus* SSNP11. Indian J Microbiol 54:329–336

Rajamanickam U, Mylsamy P, Viswanathan S, Muthusamy P (2012) Biosynthesis of zinc nanoparticles using actinomycetes for antibacterial food packaging. In: International conference on nutrition and food sciences, vol 39. IPCBEE

Rajivgandhi G, Mythili Gnanamangai B, Heela Prabha T, Poornima S, Maruthupandy M, Alharbi NS, Kadaikunnan S, Li W-J (2022) Biosynthesized zinc oxide nanoparticles (ZnO NPS) using Actinomycetes enhance the anti-bacterial efficacy against K. pneumoniae. J King Saud Uni Sci 34:101731

Rasool U, Hemalatha S (2017) Marine endophytic Actinomycetes assisted synthesis of copper nanoparticles (cunps): characterization and antibacterial efficacy against human pathogens. Mater Lett 194:176–180

Rudramurthy GR, Swamy MK (2018) Potential applications of engineered nanoparticles in medicine and biology: an update. JBIC J Biol Inorg Chem 23:1185–1204

Salem SS, EL-Belely EF, Niedbała G, Alnoman MM, Hassan SE-D, Eid AM, Shaheen TI, Elkelish A, Fouda A (2020) Bactericidal and in-vitro cytotoxic efficacy of silver nanoparticles (AG-NPS) fabricated by endophytic Actinomycetes and their use as coating for the textile fabrics. Nano 10:2082

Sandhu A, Goel A (2023) Biosynthesis of nanoparticles by micro-organisms and its applications. J Young Pharm 15:430–440

Sawada I, Fachrul R, Ito T, Ohmukai Y, Maruyama T, Matsuyama H (2012) Development of a hydrophilic polymer membrane containing silver nanoparticles with both organic antifouling and antibacterial properties. J Membr Sci 387–388:1–6

Sengul AB, Asmatulu E (2020) Toxicity of metal and metal oxide nanoparticles: a review. Environ Chem Lett 18:1659–1683

Sharifi S, Hajipour MJ, Gould L, Mahmoudi M (2020) Nanomedicine in healing chronic wounds: opportunities and challenges. Mol Pharm 18:550–575

Shnoudeh AJ, Hamad I, Abdo RW, Qadumii L, Jaber AY, Surchi HS, Alkelany SZ (2019) Synthesis, characterization, and applications of metal nanoparticles. In: Biomaterials and bionanotechnology. Academic Press, Cambridge, MA, pp 527–612

Singh T, Shukla S, Kumar P, Wahla V, Bajpai VK, Rather IA (2017) Application of nanotechnology in food science: perception and overview. Front Microbiol 8:1501. https://doi.org/10.3389/fmicb.2017.01501

Subbaiya R, Saravanan M, Priya AR, Shankar KR, Selvam M, Ovais M, Barabadi H (2017) Biomimetic synthesis of silver nanoparticles from *Streptomyces atrovirens* and their potential anticancer activity against human breast cancer cells. IET Nanobiotechnol 11(8):965–972

Thekkangil A, Suchithra TV (2020) Antidermatophytic lead compounds from *Streptomycetes albidoflavus* STV1572a against tinea infections by *Trichophyton mentagrophytes*. Microb Pathog 142:104037

Thekkangil A, George B, Prakash SU, Suchithra TV (2021) Mechanism of *Streptomyces albidoflavus* STV1572a derived 1-heneicosanol as an inhibitor against squalene epoxidase of *Trichophyton mentagrophytes*. Microb Pathog 154:104853

Venkateswaran SP, Palaniswamy VK, Vishvanand R, Periakaruppan R (2022) Actinomycetes-assisted nanoparticles: synthesis and applications. In: Agri-waste and microbes for production of sustainable nanomaterials. Elsevier, Amsterdam, pp 375–395

Wadhwa P, Sharma S, Sahu S, Sharma A, Kumar D (2022) A review of nanoparticles characterization techniques. Curr Nanomater 7(3):202–214

Yao Y, Zang Y, Qu J, Tang M, Zhang T (2019) The toxicity of metallic nanoparticles on liver: The subcellular damages, mechanisms, and outcomes. Int J Nanomedicine 14:8787–8804

Ying S, Guan Z, Ofoegbu PC, Clubb P, Rico C, He F, Hong J (2022) Green synthesis of nanoparticles: current developments and limitations. Environ Technol Innov 26:102336

Yosri N, Khalifa SAM, Guo Z, Xu B, Zou X, El-Seedi HR (2021) Marine organisms: Pioneer natural sources of polysaccharides/proteins for green synthesis of nanoparticles and their potential applications. Int J Biol Macromol 193:1767–1798

Zarina A, Nanda A (2014) Green approach for synthesis of silver nanoparticles from marine *Streptomyces*-MS 26 and their antibiotic efficacy. J Pharm Sci Res 6(10):321

Part II
Sec B: Current Trends of Biotechnology in Drug Development

Chapter 9
Transgenesis: In the Drug Discovery Process, Including Target Identification and Target Validation

Aniruddha Mukherjee, Susovan Das, Nurul Hassan Mondal, Avijit Choudhury, and Sugato Banerjee

Abstract Transgenic targeting is the most dependable method for mouse transgenesis because it produces predictable transgene expression patterns. This method is ideal for large-scale knock-in target assessments. Furthermore, transgenic targeting is the preferred method for getting around the restrictions since it is a powerful cellular delivery strategy. The chapter introduces the concept of target validation and underscores its significance in evaluating a target's potential for therapeutic benefits. The focus is on improving the drug development process through early target validation. Target validation, a critical initial step taking few months, involves proving that altering the target in animals and cell cultures significantly improves the disease phenotype. Target validation in transgenic animals is explored in detail, emphasizing the necessity of clearly defining a model's objectives for appropriate creation, evaluation, and application. The difficulties of accurately simulating human genetic abnormalities in transgenic animals are discussed despite the possibilities offered by genome editing. The importance of

A. Mukherjee (✉)
School of Pharmacy, The Neotia University, Sarisa, West Bengal, India

S. Das
Acharya & BM Reddy College of Pharmacy, Bengaluru, India

N. H. Mondal
Dr. B. C. Roy College of Pharmacy and AHS, Dr Meghnad Saha Sarani, Durgapur, West Bengal, India

A. Choudhury
Gupta College of Technological Sciences, Asansol, West Bengal, India

S. Banerjee
Department of Pharmacology and Toxicology, National Institute of Pharmaceutical Education and Research, Kolkata, India

S. Bose et al. (eds.), *Concepts in Pharmaceutical Biotechnology and Drug Development*, Interdisciplinary Biotechnological Advances,
https://doi.org/10.1007/978-981-97-1148-2_9

validity in assessing animal models is reiterated, advocating continuous debate and improvement. The translational method is emphasized for differentiating between target validations in humans and qualifying in animals. Face and construct validity in animal models are explained, with the latter deemed more crucial. The challenges of simulating human genetic disorders in transgenic animals are acknowledged, with the recommendation to use diverse models to improve predictive validity.

Keywords Transgenic target identification · Transgenic target validation · Face validity and construct validity

9.1 Introduction

Transgenesis is a process of experimentation in which foreign genes are incorporated into the genome of an organism, resulting in the creation of pathological conditions for the observation of new drug therapeutic potential. However, transgenesis is also used to disrupt genes (knockout). The process of creating a new drug from scratch can take more than 10–15 years with uncertainty and exceed one billion dollars (based on the therapeutic application area) (Hughes et al. 2011). To increase the success rate, it is very crucial to use the whole organism early on in the process of finding new drugs. Transgenesis is the most prominent choice for the early stage of the drug discovery process because most of the drugs fail to pass phase II clinical trials because they do not meet the standards of ADME due to the multiple interactions of the in vivo study (Singh and Seed 2021).

The roots of pharmacology and medicine trace back to ancient civilizations, where the use of medicinal plants and natural remedies formed the foundation of early healthcare practices (Wadud et al. 2007). Later on, the evaluation of synthetic chemistry will strand the new drug development process. This conventional process of drug discovery has many limitations, including a higher rate of failure during testing, limiting target specificity, and, most importantly, taking many years to develop a new drug. To overcome the limitation, transgenesis is an emerging field. There are various methods available for creating transgenic animals, including microinjection, viral vectors, embryonic stem cell-mediated transgenesis, and CRISPR/Cas9 technology (Costantini 2001).

Target validation and identification are just two of the many phases of the investigation procedure that transgenic technology can help in the procedure for finding new drugs. In the case of target identification in transgenic animals carrying big DNA, fragments can be utilized to focus on genetic areas (Snaith 2002). In the case of target validation, transgenesis allowed the creation of genetically modified organisms to mimic human diseases, enabling a deeper comprehension of the underlying mechanics and potential treatments. Additionally, transgenic models have revolutionized drug discovery by serving as reliable platforms for testing new treatments before they are applied to human patients for a better understanding of disease mechanisms (Lee 2014).

9.2 History

9.2.1 Ancient Roots of Medicinal Plant Knowledge

The history of medicinal plants and their uses dates back to ancient times, and many cultures have relied on herbs and natural remedies for their medicinal properties. In Islamic tradition, there is knowledge of the properties of various herbs and their healing abilities were passed down by Allah through Adam (Alaihissalam) as the first messenger (The Qur'an 2023). Throughout history, humans have utilized plants, minerals, and animal-derived substances for medicinal purposes. Traditional herbal medicine practices laid the foundation for early drug discovery. Alternatively, during early civilizations, clay tablets from Sumeria listed hundreds of medicinal plants, while the Ebers Papyrus from Egypt describes over 850 plant-based medicines. During the eighteenth century, the establishment of the periodic table by Dmitri Mendeleev and Avogadro's atomic hypothesis laid the foundation for understanding the building blocks of molecules and their properties. Later, one isolation of morphine from opium extracts by Sertünergame changer for development of new drug compound (Drews 1979). The isolation of individual bioactive ingredients from plant materials, even when not meeting modern pharmaceutical standards, dramatically improved Byronic period medicine. In the Byronic period, medicine took a giant leap forward with the isolation of individual bioactive ingredients from plant materials despite limitations in purity (Drews 1979; Beale and Block 2011).

9.2.2 Twentieth-Century Chemical Knowledge

The twentieth century saw significant advancements in chemical knowledge, including the concept of aromaticity in benzene, which significantly influenced Paul Ehrlich's theory of "chemoreceptors." J. N. Langley presented the functional theory of receptive substances in 1905, which enables receptors to receive signals and have pharmacological or biological effects. This study spurred more investigation into medicinal chemistry, which helped establish the concept of ligand-receptor interaction (Drews 1979). Ehrlich's research opened the door for the first sensible synthetic medications, including arsphenamine, in 1908. Furthermore, Ehrlich's research group established the framework for credible biological screening and assessment methods (Beale and Block 2011).

9.2.3 Antibiotics and Drug Discoveries

Before NMR and computational chemistry, chemists had limited tools, making institutional support crucial for drug discovery. The 1928 discovery of penicillin, a blend of interconnected beta-lactams, by Alexander Fleming revealed its potent antibiotic properties. By 1942, mass production began, significantly affecting World War II sepsis management. Allied medical personnel carried penicillin during the operation, and it remains an indispensable tool in combating infections caused by Gram-positive microorganisms today (Newman et al. 2000).

A novel compound called cephalosporin was first described by Brotzu in 1948 and was used to treat infections resistant to penicillium. By that time, other businesses had expanded their microbiological capabilities, including Merck, Sandoz, and Taked, in an effort to discover alternative medications with distinct pharmacological and chemotherapeutic qualities (Drews 1979).

Starting in 1950, the landscape of medicinal chemistry underwent a pivotal transformation, delineating two significant epochs: Initially, spanning from 1950 to 1980, medicinal chemistry thrived on in vivo testing, while the subsequent era, commencing in 1980 and persisting to the present day, heralded the rise of pioneering design methodologies and advanced screening technologies (Drews 1979; Lombardino and Lowe 2004).

During the late 1960s and early 1970s, Beecham stumbled upon clavulanates, while Pfizer crafted sulbactams, molecules akin to penicillin in how they acted within the body. Beecham's discovery mirrored penicillin's traits, while Pfizer synthesized sulbactam by altering the sulfur in thiazoles to sulfone, creating a semi-synthetic compound (Newman et al. 2000). During this phase, a key focus involved assessing how drugs interacted within the human body. Despite discovering promising compounds that exhibited activity within animal models, challenges persisted in understanding how these substances were absorbed, distributed, metabolized, excreted, and interacted within the clinical setting (Lombardino and Lowe 2004).

In 1974, Annie Chang and Stanley Cohen achieved a groundbreaking feat by introducing Staphylococcus aureus genes into Escherichia coli, marking the birth of the inaugural transgenic organism (Chang and Cohen 1974). Gene transfer occurred in yeast cells for the first time among eukaryotic organisms back in 1978 (Hinnen et al. 1978). In 1979, the transformation of mouse cells marked a significant breakthrough, swiftly followed by the transformation of mouse embryos in 1980. Initially, scientists primarily employed microinjection, directly infusing DNA into cells, for these groundbreaking transmutations. However, as research progressed, diverse techniques emerged: integrating transgenes into retroviruses for cell infection, leveraging electro-infusion by utilizing electric currents to introduce foreign DNA across cell walls, employing biolistic—a technique that involves injecting DNA "bullets" into cells, or even directly putting DNA into freshly fertilized eggs. These varied methods expanded the repertoire of transformation techniques, enriching the field of genetic manipulation (Evans 2011).

Transgenes consist of three key components: a polyadenylation signal for mRNA stability and protein expression, a promoter/enhancer, and the sequence encoding the intended expression product (such as a reporter protein or regulatory RNA). While detailing the complete history of recombinant DNA technology is impractical, significant milestones include the development of enzymes enabling DNA fragment manipulation and recombination, DNA fragment cloning in bacteria leading to protein expression, chemical synthesis of DNA such as oligonucleotides, and the polymerase chain reaction's (PCR) invention (Xu et al. 2019).

Once scientists gained access to recombinant DNA technologies, the natural progression was to explore introducing this modified incorporating DNA into model systems such as organisms or cells in culture like mice. Early experiments demonstrated the feasibility technique of transfecting grown cells with externally cloned DNA to produce proteins. These studies utilized drug selection methods to pinpoint transformed cell clones containing the genetic material and expressing thymidine kinase of the herpes simplex virus (Durmaz et al. 2015).

The technique of transforming cultured cells with DNA using the process of precipitating DNA using calcium phosphate, detailed in these studies, proved effective for cultured cells but was not applicable to zygotes, the early stage of a fertilized egg (Kwon and Firestein 2013).

The creation of transgenic mice via DNA microinjection into mouse zygote pronuclei necessitated the convergence of multiple research disciplines. It involved aspects of mouse reproductive physiology (such as superovulation, embryo culture, and surgical transfer to pseudopregnant females), specialized micromanipulation equipment, and molecular genetics. Initially, microinjection techniques were employed to evaluate the developmental potential of blastomeres from two-cell and four-cell embryos by lysing some blastomeres and observing the remaining ones after implantation into pseudopregnant recipients (DeMayo et al. 2012).

The first microinjection experiment on mouse zygotes assessed the survival and development of eggs given a gamma globulin injection post-transfer. Microinjection found various applications, including studies on oocyte biology, translation of foreign mRNA species, and preimplantation developmental biology. As DNA cloning became a standardized method in molecular biology, all the necessary components for producing transgenic mice were established. Subsequently, following the initial successful creation of transgenic mice, the field rapidly expanded (Saunders 2020).

9.2.4 Advancements in Genetic Engineering

The advancement in genetic engineering technology has seen a transition from ZFNs and TALENs to the now-prevalent CRISPR/Cas9 technology because of its exceptional adaptability and user-friendliness. CRISPR/Cas9 technology has revolutionized genome targeting, particularly in the genetic modification of mice and rats. Its delivery via various forms, such as Cas9 ribonuclease protein complexes, has shown advantages, minimizing unwanted genome integration and reducing mosaicism in animal founders (Gupta and Musunuru 2014).

CRISPR/Cas9's efficiency in inducing gene knockouts or precise modifications such as point mutations, epitope tags, and conditional genes has significantly streamlined the generation of genetically modified animals. Techniques utilizing linear double-stranded DNA templates or lengthy single-stranded DNA molecules have notably improved efficiency, especially in producing floxed alleles compared to earlier methods (Arora and Narula 2017).

Controlling off-target mutations in the genome is crucial when employing CRISPR/Cas9. Careful guide RNA sequence selection and utilizing specialized Cas9 enzymes, such as enhanced specificity Cas9 or high-fidelity Cas9, have been pivotal in minimizing off-target effects. Analysis methods beyond the founder generation require meticulous attention to avoid misinterpretations due to natural genetic drift during development (Guo et al. 2023).

Despite its profound impact, CRISPR/Cas9 technology is not immune to potential replacements in the future. Alternative tools such as Cpf1 enzyme have shown promise but are less efficient than Cas9. Researchers continuously explore and optimize CRISPR/Cas9 while anticipating the possibility of even more precise and active genome editing tools down the line (Wang et al. 2022).

The overarching impact of CRISPR/Cas9 on creating genetically modified models for research is undeniable, yet the field remains open to future innovations that may further enhance precision and efficiency in genome editing (Mengstie and Wondimu 2021). Some approved and viably produced biopharmaceuticals obtained from genetically modified animals are included in Table 9.1.

Table 9.1 Biopharmaceuticals obtained from genetically modified animals

Item	Therapeutic or other use	Stage of development	Establishment	Reference
Anti-C5 monoclonal antibody	Kidney inflammation, rheumatoid arthritis	Prior to clinical	Alexion-GTC Biotherapeutics	Wang et al. (1995)
Alpha-1 antitrypsin	A genetic lack of a protein that shields the lungs' elastic components results in emphysema, a hereditary illness that seriously affects the body's digestive system, lungs, and other organs	Inactive or terminated	Bayer-PPL	Brantly et al. (1988)
Alpha-fetoprotein (AFP) tumor marker	As a result of antibodies blocking nerve-muscle transmission, the skeletal muscles of people with multiple sclerosis, a chronic autoimmune illness, get weaker	Phase II	Merrimack-GTC	Debruyne and Delanghe (2008)
Anti-EGF receptor monoclonal antibody	Cancer	Phase II	Abgenix-Amgen	Yang et al. (2001)
ABX-IL8 recombinant human monoclonal antibody	Rheumatoid arthritis	Inactive or terminated	Abgenix-GTC	Huang et al. (2002)
Antithrombin III	There are several plaque fragments or clots	Phase III	GTC	Bick (1982)
Butyrylcholinesterase	Those measures intended to mitigate risks, combat biological threats, and prepare for, respond to, and recover from bioincidents	Prior to clinical	Nexia	Darvesh et al. (2003)
Prolactin	Bolstering of immunity	I-T	GTC	Freeman et al. (2000)
Rotavirus virus-like particles	Vaccine development	Preclinical	Bioprotein	Changotra and Vij (2017)
Activator of tissue plasminogen	Breakdown of blood clots	Inactive or terminated	Genzyme	Loscalzo and Braunwald (1988)
Tissues or organs	Designed with xenotransplantation in mind	Preclinical	Alexion BresaGen Novartis Mayo Clinic	Murphy and Atala (2014)
Lactoferrin	Anti-inflammatory	Phase I	Pharming	Levay and Viljoen (1995)

(continued)

Table 9.1 (continued)

Item	Therapeutic or other use	Stage of development	Establishment	Reference
1 on the surface of merozoites	Malarial vaccine	Preclinical	GTC	Lin et al. (2014)
Polyclonal antibodies that are humanized	Various indications	Preclinical	Hematech Avigenics	Vaswani and Hamilton (1998)
huN901-DM1, ImmunoGen	Lung cancer with little cells	Preclinical	ImmunoGen-GTC	Wang et al. (2005)
Glucagon-like peptide-1	Type 2 diabetes	Inactive or terminated	PPL	Holst (2007)
Glutamic acid decarboxylase	Type 1 diabetes	Inactive or terminated	GTC-Biosyn	Erlander and Tobin (1991)
Human growth hormone	Turner's syndrome	Inactive or terminated	GTC	Strobl and Thomas (1994)
The superoxide dismutase extracellular	Ischemic reperfusion injury	Inactive or terminated	Pharming	Nozik-Grayck et al. (2005)
G-CSF	Leukopenia	Preclinical	Avigenics	Mehta et al. (2015)
Glucosidase	Glycogen storage disease	Inactive or terminated	Pharming	Lebovitz (1997)
Natalizumab	Multiple sclerosis and Crohn's disease	Preclinical	Elan-GTC	Brooks and Walker (1984)

9.3 Transgenesis: Techniques

In the intricate tapestry of genetic exploration, transgenic animals stand as invaluable allies, unraveling the secrets encoded within our DNA. By introducing exogenous genetic material into fertilized zygotes and guiding their journey through gestation, scientists have forged a powerful tool in the pursuit of understanding gene function and mimicking human diseases (Houdebine 2003). Transgenic models serve as dynamic canvases painted with specific genetic mutations that echo the intricacies of human diseases. They not only offer profound insights into molecular and physiological mechanisms but also validate potential drug targets, probe drug pharmacokinetics, and scrutinize the safety and immunogenicity of promising candidates, expediting the labyrinthine journey of drug development (Mayran and Bolt 2022).

These genetically crafted cohorts are not mere witnesses; they are active participants in the relentless pursuit of knowledge. Beyond facilitating the understanding of specific genes and paving the way for personalized medicine, transgenic models play a pivotal role in scrutinizing the labyrinthine landscape of cancer development, progression, and responses to treatment (Wijmenga and Zhernakova 2018).

As the shadows of neurodegenerative diseases loom large, transgenic models of ailments such as Alzheimer's and Parkinson's emerge as beacons of hope. They not only decipher the intricate dance of disease mechanisms but also serve as testing grounds for potential therapeutic interventions (Rani et al. 2023).

In the ever-evolving field of gene therapies, transgenesis takes center stage, providing a meticulous testing ground for the efficacy and safety of these groundbreaking approaches before they venture into the uncharted territory of clinical trials (Dunn et al. 2005).

9.4 Types of Transgenic Techniques

From the pioneering work of Herbert Boyer and Stanley Cohen in the 1970s, since the development of recombinant DNA technology, and during the past few decades, numerous techniques for creating transgenic animals have been created (Wright 1986). In the realm of drug discovery, the convergence of genomics, proteomics, and advanced reproductive biotechnologies has paved the way for groundbreaking transgenic applications in domestic animals. The methodologies employed in creating transgenic animals are meticulously tailored based on the specific intended purpose of the animal in the drug development process. There are three types of transgenic techniques for the transfer of foreign gene: gene transfer-based transgenesis, vector-based transgenesis, and targeted integration of foreign genes (Spreafico et al. 2020).

9.4.1 Gene Transfer-Based Transgenesis

The process of drug discovery often begins with the identification and isolation of specific genes that encode proteins with potential therapeutic applications. Once a gene is isolated, researchers can perform in vitro transcription to generate mRNA and, in turn, translate this mRNA into a cell-free system to produce a small amount of the corresponding protein (Spreafico et al. 2020). While this approach is useful for certain biochemical studies, it has limitations when it comes to examining the protein's biological activity in vivo or figuring out its three-dimensional structure through techniques such as crystallization (Agu et al. 2023).

To effectively decode the gene and translate it into a functional protein, the gene needs to be inserted into the cells. Cells naturally possess the necessary machinery for transcription and translation. However, the plasma membranes of several cell kinds serve as a selective barrier, controlling the uptake of compounds. Some molecules can enter cells through pores, which may be controlled in an open or closed manner (Zhang et al. 2019). Alternatively, specific carriers facilitate the transport of molecules into the cell. In some cases, molecules recognize external receptors of the plasma membrane, forming complexes that trigger endocytosis (Fig. 9.1). This process involves the absorption of the intricate and surrounding membrane (Houdebine 2003).

DNA, being a large, negatively charged molecule, is not able to pass the plasma membrane on its own. This serves as a protective mechanism for cells against foreign DNA (Houdebine 2003). However, to introduce DNA into cells for various studies, techniques collectively known as transfection are employed. Transfection

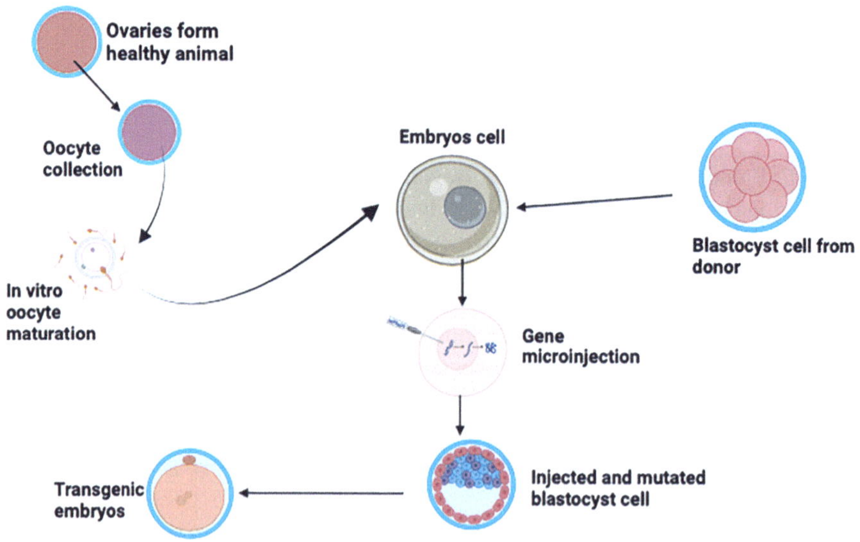

Fig. 9.1 Gene transfer to embryonic stem cells by microinjection technique

involves different methods to force DNA entry into cells and reach their nucleus. It is important to note that transfection differs from cell infection (Houdebine 2003), which is a process involving various mechanisms employed by viruses to deliver their genomes into host cells. These insights into cellular processes and gene transfer techniques are crucial in the drug discovery process, allowing researchers to study and manipulate genes for therapeutic purposes.

Several physiochemical phenomena facilitate gene transfer into cells, including cell fusion, transfection, DNA microinjection, viral vectors, electroporation, and vectors with particular ligands. Each method has its unique advantages and applications in the context of drug discovery.

9.4.2 *Vector-Based Transgenesis*

In vivo gene transfer inserts a gene via a vector into an organism, allowing researchers to investigate biological functions. However, the ultimate goal is gene therapy, which involves delivering genetic information to target cells to replace defective functions or introduce additional functions. Ex vivo gene therapy involves extracting target cells, inserting desired genes, and returning them to the patient, but has limited results.

In vivo gene therapy faces challenges due to the lack of an efficient, non-toxic gene delivery system (Fig. 9.2). Viruses can be used to transduce cells by replacing non-essential viral genes with therapeutic genes. Despite triggering a host immune response, viruses have evolved countermeasures to enter and multiply within the intended cells. In 1973, the first attempt was made to employ viruses as medicinal agents, but replication-competent viruses remain a research field (Bouard et al. 2009).

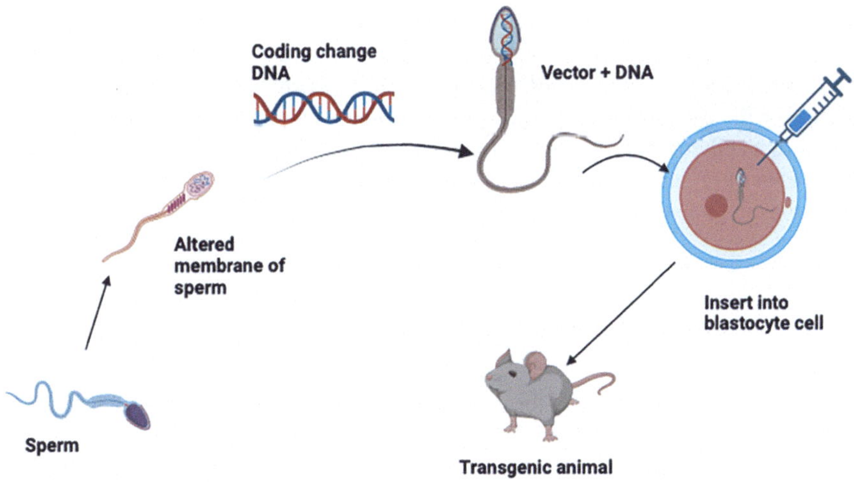

Fig. 9.2 Vector-based transgenesis

9.5 Transgenesis: Target Identification

A transgene is a gene that has been spontaneously transmitted from one creature to another or has been created by one of several genetic engineering methods. A procedure called transgenesis, which involves the insertion of a transgene, can alter an organism's phenotype. A transgene is a section of DNA that has been extracted from one organism and inserted into another, carrying a gene sequence. This non-native DNA segment may change the transgenic organism's genetic code from functioning normally, or it may preserve the transgenic organism's capacity to create RNA or protein. Usually, the DNA is integrated into the germ line of the organism (Gordon and Ruddle 1981). Gene therapy and transgenesis share the similarity of transforming cells for a certain objective, but their goals are very different: Gene therapy aims to fix a disease in cells, while transgenesis seeks to generate a genetically transformed organism by inserting a specific transgene into each cell and changing its genome (Moody et al. 1981).

Mouse molecular genetics underwent a revolution in the 1980s. The first transgenic mouse was produced by injecting cloned DNA into fertilized mouse eggs (Gordon et al. 1980). Investigators quickly reported germline transfer of transgenes via nuclear injection of DNA (Costantini and Lacy 1981). Transgenic mice have now been employed in a frenzy of research that has not stopped examining a variety of topics, such as tissue-specific gene expression, oncogenesis, and developmental alterations (Neidhardt et al. 2000).

9.5.1 Transgenic Target Identification

There are several methods for finding novel possible pharmacological targets (Debouck and Goodfellow 1999). The genetic approach and the genomic approach are two conceptually distinct methods (Schmid et al. 2021). The genetic method examines the differences in mRNA expression between, say, healthy and diseased tissues or the alterations in gene expression that occur when a substance is introduced to a cell culture medium or administered to an animal (Alberts 2017). These days, a lot of organizations employ proteomics and microarray technologies to find variations in gene expression (Lockhart and Winzeler 2000).

9.5.2 Meaning of the Target Identification

The process of gathering data for and against a target is known as target identification and validation. To gather enough information, medication developers usually spend 12–18 months (about 1 and a half years) either performing original research or replicating existing studies in the literature (Békés et al. 2022). In the absence of a small-molecule tool, genetically modified animals, neutralizing antibodies, and small-interfering ribonucleic acid (siRNA) can be used as research tools for validation; nevertheless, their capacity to understand the pharmacodynamics of the drug-target interaction is often restricted. (MacDonald et al. 1993).

9.5.3 Difference and Importance Between Target and Indicators

There are two primary reasons why drugs fail in clinical settings: they are unsafe or they do not function. Therefore, target selection and validation are among the most crucial phases in the development of a novel medication. A target is a general phrase that may be used to describe a variety of biological things, such as proteins, genes, and RNA. A good target must be "druggable," safe, suit clinical and commercial goals, and be effective. Whether a tiny molecule or a bigger biological one, a "druggable" target is reachable by the possible therapeutic chemical and, upon binding, causes a biological response that can be evaluated both in vitro and in vivo. It is now established that while antibodies are effective in obstructing protein/protein interactions, other target classes—such as GPCRs, which are connected to proteins—are more amenable to the development of small-molecule drugs. A well-defined and validated objective allows for more trust in the connection between the target and the illness and enables us to investigate whether changing the target will have side effects based on a mechanism (Henning and Beste 2002).

9.5.4 Aim of Transgenic Target Identification

While homologous integrations do happen and transfer DNA to a specific site in the genome, random integrations of imported DNA predominate. The early cases of gene targeting required stringent screening for the rare homologous occurrences, which occurred at around 10^{-3}–10^{-5} of total integrations, regardless of whether the locus was "natural" or artificially produced (Smithies et al. 1985; Lin et al. 1985). Since then, several ES cell examples have typically produced targeted frequencies of 10^{-2} or greater for the overall number of integrations. The percentage of targeted clones among selected clones can be increased by including enrichment techniques in the selection process (Mansour et al. 1988). Drug selection is crucial for isolating targeted clones since a targeting frequency of 10^{-2} of total integers corresponds to a 10^{-5}–10^{-6} absolute targeting frequency of the transfected cell population.

9.5.5 Properties of Good Gene Target

Still, it can be difficult to predict how transgenes, especially targeted ones, would express themselves. Targeting the lacZ reporter gene to the third exon of hprt was noteworthy because the targeted clones exhibited varied expressions that depended on both orientation and cell type (Shaw-white et al. 1993; Maltsev et al. 1994). Like lacZ targeted to the hprt locus, in this instance, ES clones also showed variable β-galactosidase staining patterns. In both scenarios for the variegated staining patterns, promoter occlusion effects may be accountable because it has been shown that nearby promoters might affect one another (Alam and Cook 1990; Barnes et al. 2000). It might be possible to prevent promoter occlusion effects with such a design. It is challenging to come to a definitive judgment in this regard, nevertheless, because Northern analysis was used to monitor the expression of bcl-2 rather than expression within individual cells.

9.5.6 Method and Types of Target Identification

Target validation is an ongoing, crucial procedure that begins at the beginning of the discovery phase. Various steps of this procedure need the application of different validation techniques. Cultured cells and bioinformatics technologies can be utilized in the very early stages when choices must be made among several targets. Before significant resources are committed to the possible target, there is an urgent need for in vivo TV data. Although a lot of work is being done to find quick, dependable, and universal methods for in vivo validation, transgenic animals are the only ones that have proven effective so far on a variety of targets. A target and possible medication combination cannot be considered fully "validated" until the novel medicine has been evaluated in clinical trials. The creation of genetically altered animals that either overexpress (gene addition) or lack the target (knockout animals) is a typical necessity in the TV phase. These animals offer functional in vivo data—which are frequently lacking—on a possible target. These days, varying degrees of complexity can be used to modify the target gene's expression in genetically engineered animals. One that overexpresses the transgene either broadly or selectively in a particular tissue is the fastest-growing transgenic animal. Nowadays, a wide variety of tissue-specific promoters have been used to restrict transgene expression in different cell types. (Morimoto et al. 2001; Burton et al. 1991; Field 1988). Several mechanisms exist for controlling transgenic expression using non-physiological inducers. The phosphoenolpyruvate carboxykinase promoter and the metallothionein promoter were the first inducible systems employed in transgenic animals. There were issues with background expression levels, and these systems were hard to control. With several successful trials under its belt, the tetracycline system is currently the most popular method for regulating transgenes. Although technologies for artificial dimerization of molecules and the ecdysone system have also been created, it is yet unclear how broadly these systems may be used to regulate transgenes. Furthermore, gene activity has been regulated via fusion proteins that link an effector protein (such as Cre) to a steroid receptor's ligand-binding domain (Lim et al. 1990; Ewald et al. 1996; Furth et al. 1994; Perea et al. 2001; No et al. 1996; Amara et al. 1997).

9.5.7 Steps of Transgenic Target Identification

DNA Preparation and Purification

DNA Construct/Fragment Structure

DNA microinjection appeared to be one of the only gene transfer techniques where fragment size was unaffected when compared to other techniques (especially when it came to retro-viral packaging); however, the initial limitations on fragment size were imposed by the cloning vectors cosmid (45 kb) and plasmid (12 kb). Large chromosomal segments and vectors such as the yeast artificial chromosome (YAC), P1, and bacterial artificial chromosome (BAC) can now be cloned in addition to bigger stable constructions.

Preparation of DNA for Microinjection

(a) **Purification of DNA**
    - After verifying correct cloning and confirming the nucleotide sequence, the DNA fragment undergoes purification steps.
    - The goal is to ensure that the DNA fragment is free of nicks or strand breaks and is as pure as possible.

(b) **Use of High-Quality Reagents**
    - High-quality reagents are emphasized for the final purification steps.
    - Solutions are made from tissue culture-grade, 18 MW water.

(c) **Water Purification Systems**
    - Commercial water systems are utilized, combining reverse osmosis and distillation combined with ultrafiltration or deionization.
    - Water systems from various manufacturers, such as Millipore's Milli-Q, Barnstead's NanoPure, and Corning's Mega-Pure, are mentioned.

(d) **Filtration of Solutions**
    - Solutions are processed using a 0.45- or 0.2-mm filter after preparation.

- **Host Strain for Plasmid Replication**

- It is possible to reproduce plasmid DNA in any common Escherichia coli host.
- CPLK-17, the DH5 methylase-defective strain, is recommended for maintaining the methylation state of the gene construct during cloning.

(e) **Transgene Rescue and Analysis**

- The CPLK-17 strain allows transgene rescue for toxicology applications and the analysis of flanking regions of the chromosomal insertion.
- This facilitates the localization of insertional mutants.

(f) **Cloning and Gene Construct Assembly**

- Techniques for basic cloning and gene construct assembly are referenced.
- Large-scale DNA preparation is mentioned, with references to well-documented methods.

(g) **Linearization and Microinjection Preparation**

- Before microinjection, the plasmid DNA is purified by isolating supercoiled DNA from a cesium chloride (CsCl) gradient.
- Using restriction enzyme digestion, the gene fragment is then separated from the vector sequences.

(h) **Minimization of Flanking Vector Sequences**

- The passage highlights the routine minimization of the amount of flanking vector sequences for injection fragments.
- Vector DNA sequences can influence the function of foreign genes after chromosomal integration.

(i) **Construct Design**

- The construct should be designed to facilitate easy separation of DNA bands migrated on agarose gels.

Superovulation, Egg Culture, and Harvest

(a) **Popular Mouse Hybrid for DNA Microinjection**

- Several hybrids are commonly used for DNA microinjection.
- The C57BL/6 SJL (B6SJL) F1 hybrid is specifically mentioned as being efficient in generating transgenic mice.
- References are made to studies supporting the efficiency of this hybrid.

(a) **Preference for C57BL/6 Inbred Strain in Parental Stocks**

- The C57BL/6 inbred strain is one of the parental stocks in most hybrids used for microinjection.
- This choice is based on favorable genetic and embryological characteristics associated with the C57BL/6 strain.

(b) **Advantages of Hybrid Mice**

- One of the main benefits of employing hybrid mice for microinjection is their accentuated hybrid vigor.
- Hybrid vigor not only imparts desirable reproductive characteristics but also enhances egg quality, leading to favorable microinjection characteristics.

(c) **Considerations for Genetic Background Uniformity**

- There is a cautionary note about the need to be concerned with the uniformity of the genetic background in which the transgene will be functioning.
- This consideration is particularly important for experimental designs requiring large populations or many generations of transgenic mice.

Microinjection Needles and Slides

(a) **Slide Option**

- Cut-out or depression slides can hold eggs and DNA. For cut-out slides, an equivalent area is removed from a flat glass slide instead of creating a depression in the center.

(b) **Cut-Out Slides and Disposable Coverslips**

- Cut-out slides use disposable glass coverslips for holding microdrops overlaid with oil.
- The coverslips are attached to the bottom of the cut-out slide using a ring of paraffin, positioned outside the periphery of the hole.

(c) **Advantages of Cut-Out Slides**

- Cut-out slides offer advantages over depression slides.
- Coverslips are disposable.
- There is less light refraction.

(d) **Configuration Details**

- To prevent the cover slip from contacting surfaces, on the bottom of the slide at both ends, 1–2 mm blocks or risers are employed.

(e) **Needle Manipulation Considerations**

- Due to the configuration of cut-out slides and the need for a greater needle angle when working with dishes on a microscope, needle tips must be bent at approximately a 30° angle prior to microinjection.

A HEPES-buffered medium is utilized for microinjection; examples include modified BMOC-3 plus HEPES (recipes in Table 9.1) supplemented with cytochalasin B (5 mg/mL). The egg will probably lyse quickly if the injection causes the egg membrane to be pulled into the cytoplasm. During microinjection, cytochalasin B stiffens the membranes and aids in preventing egg lysis. Conversely, 7% (v/v) ethanol produces a comparable result (P. Hoppe, private correspondence 1991). Another method for stiffening egg membranes is to utilize a stage cooler (e.g., KT Model 5000, Technology, Inc., Whitehouse Station, NJ) to reduce the slide temperature to 10 °C. As long as the eggs are returned to 37.5 °C within 45 min, neither the proportion of transgenic mice produced nor egg survival will be affected by this.

Microinjection Equipment

It is frequently employed, and a choice must be considered before making one. The brand or design of the microscope may have an impact on the final differential interference contrast (DIC) magnification, and prior training or personal experiences often determine the ideal field diameter. DIC is often only needed at 200¥ microinjection magnification. Although it offers a sizable working field for sorting eggs and priming pipettes, low-power magnification does not need DIC optics. It is thus recommended to test the final setup and any required changes (such as light filament, lens, diaphragms, and focusing lenses) prior to purchase, ideally when the manufacturer's representative and a small number of eggs are accessible at the same time. The representative may configure the DIC in a way that is deemed "optimal," but this does not guarantee the finest pronuclear pictures (Rusconi 1991, 1996).

9.5.8 Problem in Transgenic Target Identification

Since cloning has made it easier to add and replace genes, animal transgenesis has a significant influence on human health. Agricultural animal species' whole-genome sequencing provides more genes for study and helps create advantageous animal lines. Although transgenic animals are useful for researching human illnesses and creating therapeutic proteins, moral dilemmas still exist. The use of transgenesis to improve animal production presents difficulties in assessing the potential medical effects. Although gene transfer is progressing, barriers such as high costs for transgenic farm founders and delayed trait transmission remain, meaning transgenes must provide significant benefits before they can be used in practice (Houdebine 2005).

9.6 Transgenesis: Target Validation

9.6.1 Validation Method

That is, by comparing computations with a collection of experimental data, we may determine how confident we are in a code's capacity to anticipate results for a particular application (Trucano et al. 2006). Within the context of studies on the validity of aptitude tests for people evaluation, the notion of situational specificity has taken on rather varied roles. The validity coefficients were shown to differ greatly when standardized aptitude tests were first associated with performance on seemingly identical occupations in industrial validation studies (Anastasi 1986).

9.6.2 Target Validation Method

Like target identification, target validation verifies that the interaction of the target has the potential to be therapeutically beneficial. A target will not be developed further in the drug development process if it cannot be verified. Samuel Gandy, professor in the departments of neurology and psychiatry and associate director of the Alzheimer's Disease Research Center at Mount Sinai, and Reisa Sperling, professor of neurology at Harvard Medical School and director of the Center for Alzheimer's Research and Treatment, talked about two opportunities: early validation of targets and improved biomarkers. Eli Lilly and Company's Chief Scientific Officer for Tailored Therapeutics-Neuroscience, Kalpana Merchant, furthermore, provided a portfolio evaluation tool that outlined certain parameters for target validation and qualification. The crucial first phase in the development of a new medication is target validation, which normally takes two to six months to finish. In order to prove that the drug's actions on the target can provide therapeutic advantages within an acceptable safety window, a variety of procedures must be applied. The likelihood of success in clinical trials is greatly increased when early and thorough target validation is implemented, as it helps to establish a clear correlation between target alteration and disease effectiveness. A target moves on to the hit identification phase of the research after it reaches a satisfactory degree of validation and illness correlation.

9.6.3 Animal Model Validity

A scientific technique to increase a model's credibility is to validate it or assess its consistency and plausibility. Validity refers to the extent to which a test's score or measure aligns with the attribute it is intended to evaluate. It is the interpretation of the data derived from this model that is validated, not the animal model. In that regard, validity is a crucial factor to consider when assessing animal models. There is not one animal model that works for everything or in every circumstance. Because validity is limited to a particular application of the model, it must constantly be available for debate and revision (Kaplan and Saccuzzo 2001; Silva 1993). Regarding the relative importance of the various validity criteria in the model review process, there is no universal agreement. We maintain that a model's construct validity, external validity (i.e., gen-realizability), predictive validity, and reliability and replicability (internal validity) should all be taken into account throughout the validation process. Numerous papers have clarified the meaning of the term's internal validity, face, predictive, and construct validity. We will give a brief explanation of these ideas here (Wróbel et al. 2004, Van der Staay 2006).

9.6.4 *Target Validation in Animal Models*

The presence of intriguing scientific theories is inadequate to justify the levels of dedication necessary to bring new treatments to market due to the rapidly rising expenses of drug development. Beyond conclusions like those based on various messenger ribonucleic acid (mRNA) expression patterns, potential targets must be verified. It must be demonstrated that altering the target level and/or activity in animals and cell culture significantly improves the illness phenotype. Therefore, providing models that are really predictive of the condition under study is one of the biggest hurdles (especially in terms of target validation).

One of the most important questions in light of the ongoing identification of genes, gene products, and biochemical pathways that may contribute to the development or susceptibility to human illnesses is how to choose the right places of intervention (Denayer et al. 2014; Haupt et al. 2020).

9.6.5 *Target Validation in Transgenic Animal*

Animal models are created and applied in a variety of ways. To establish standards for model construction, model assessment, and model utilization, explicit declarations of the (anticipated) goals of a model are required. It is fundamental to clearly describe and identify the precise functions that an animal model must perform since this enables the establishment of a set of weighted evaluation criteria. These standards are applied in the model evaluation phase together with validity, replicability, and reliability standards. Naturally, it would not always be feasible to predict whether the model would serve the desired goal. As a result, one begins with presumptions that have to be continually verified. Evidence suggesting the intended objective or purpose cannot be achieved should be taken into consideration if it becomes apparent (Holmes 2003; Festing 2004; Massoud et al. 1998).

9.6.6 *Difference Between Face Validity and Construct Validity*

When assessing animal models for neurobehavioral diseases, there are two different notions to consider: face validity and construct validity. The degree of descriptive resemblance between the symptoms exhibited by people with a certain neurobehavioral disease and the observed behaviors in an animal model is known as face validity. It functions as a preliminary qualitative evaluation of the model's similarity to the human state. Nonetheless, face validity has come under fire for running the danger of imposing human viewpoints on animal behavior and for encouraging anthropomorphic thinking. Demanding too much face validity may sometimes be difficult, particularly when dealing with different species. Contrarily, construct validity is concerned with the theoretical underpinnings of the behavior and the degree of correspondence between the underlying processes in the model and those in the real situation that is being mimicked. It entails an experimental validation of the behavioral, pathophysiological, and/or neural aspects of the model based on theory. The authors contend that construct validity is the most important criterion for evaluating animal models since it addresses the viability of the theory underlying the model and provides a framework for comprehending generated data.

Construct validity, which is judged more important for the assessment of animal models for neurobehavioral diseases, examines the theoretical foundation and underlying mechanisms of the model, whereas face validity measures the descriptive similarity of behaviors (Brunner et al. 2002; Lewejohann et al. 2006; Matthews et al. 2005; Lubow 2005).

9.6.7 Tools for Target Identification and Validation in Transgenic Model

A number of aspects need to be carefully considered when investing in neuroscience portfolios, and Dr. Merchant offers insightful advice on how to make these decisions. The high phase II study attrition rate (about 66%) is a major obstacle to medication development, especially in the neuroscience field. This high attrition rate highlights the need for a deeper comprehension of the elements that lead to failure and the application of risk mitigation techniques.

Dr. Merchant claims that deficiencies in safety, effectiveness, the overall strategy plan, bioavailability, and pharmacokinetic qualities are the main reasons why phase II studies fail. The success of neuroscience portfolios depends on addressing these problems. Dr. Merchant supports enhanced target validation and the application of biomarkers to assist in making decisions. Dr. Merchant contends that human subjects are the most suitable for target validation, which is an essential step in the medication development process. This highlights how crucial it is to carry out thorough research and gather reliable data to verify that a particular target is pertinent to the illness being addressed. Studies on humans provide a more realistic portrayal of the intricacies of the human system and offer insights that animal models might not be able to fully convey.

Animal models are important in the larger drug development process, but Dr. Merchant notes that human trials are necessary for target validation. Target qualification, a process that establishes a target's safety and scientific validity, mostly relies on animal models. Target qualification entails determining the possible effectiveness. It is important to distinguish between target qualification in animal models and target validation in human models. It places a strong emphasis on a translational approach, in which knowledge of human biology and illness is informed and supported by discoveries from animal investigations. This strategy can assist in avoiding frequent mistakes that lead to the high attrition rate seen in phase II trials, such as choosing the incorrect target, patient group, or dosage (Texidó 2013; Lee 2014).

9.6.8 Pillars of Target Validation in Transgenic Target Validation

The four pillars are as follows: (1) confirming that the probe is sufficiently exposed within the cell; (2) proving that the chemical probe interacts with the target (i.e., target engagement); (3) demonstrating that the probe alters the target activity; and (4) according to these authors, proving that the probe modulates pertinent phenotypic changes.

As it links exposure (pillar 1) to functional pharmacology and phenotypic alterations (pillars 3 and 4, respectively), evaluating target engagement (pillar 2) is essential to target validation (Arrowsmith et al. 2015; Garbaccio and Parmee 2016).

9.6.9 Problems of Transgenic Target Validation

A multitude of human genetic disorders may now be more accurately modeled thanks to the previously unheard-of capacity to modify eukaryotic genomes. It is still difficult to represent every variation linked to a disease, and most preclinical applications probably do not need it. The model may be deemed to have excellent construct validity as long as the pathophysiology can be replicated at the molecular level and the genetic process can be duplicated through random, spontaneous, or forced mutations, as well as genome editing and engineering. This may have greater significance for several preclinical investigations than exact face validity, which in certain situations—like macular degeneration—may be practically unachievable. Nonetheless, identifying the range of genetic pathways may be possible even for monogenic diseases, which need several models that replicate various harmful alleles. Examining therapies across a range of genetic backgrounds that more closely resemble the outbred human population may improve predictive validity, the ultimate objective of preclinical work, and better represent the range of pathophysiology observed in patients. This may be especially true for more complicated disorders, when the sickness or the response to therapy may be influenced by additional risk loci in the genetic background. The context of these models' potential uses will determine how precisely these models need to be developed; nonetheless, even models that do not perfectly replicate the human situation could still be valuable (Achilli et al. 2009; Spaulding et al. 2021).

9.7 Discussion

In order to better understand gene function, disease processes, and therapeutic development, transgenic animals have been used extensively. Predicting transgenic expression presents challenges that underscore the intricacy of genetic alterations and the necessity of comprehensive validation. To manage the results and prevent unwanted consequences, precise regulatory methods for transgenic expression must be developed. Target validation is still a crucial stage in the drug development process, and using genetically modified mice offers useful in vivo data to support other validation methods.

The notion of target validation is introduced throughout the book, with a focus on its significance in assessing a target's potential for therapeutic benefit. Opportunities to improve the drug development process are highlighted, including early target validation and enhanced biomarkers. Target validation, which takes two to six months, is emphasized as an important initial step. It entails proving that changing the target in animals and cell culture considerably improves the disease phenotype. Validity is a critical component in evaluating animal models, underscoring the need to verify these models. Numerous criteria are mentioned in the book, including reliability and replicability, construct validity, external validity, and predictive validity. It is observed that no one animal model is universally applicable and that its validity is application-specific, requiring ongoing discussion and modification. The sentence emphasizes how difficult it is to provide animal models that accurately represent the disease being studied, particularly when it comes to target validation. The translational strategy that enhances the study of human biology through animal research is highlighted, along with the difference between target validation in human models and target qualifying in animal models. Target validation in transgenic animals is discussed in detail, with a focus on the necessity of clearly stating a model's objectives in order to ensure appropriate model creation, evaluation, and application. It is understood that it is crucial to regularly confirm assumptions and take into account data indicating that the planned aim cannot be accomplished. Face validity and construct validity—two essential concepts for evaluating animal models—are presented. Construct validity is concerned with the theoretical foundations and underlying mechanisms, whereas face validity is concerned with the descriptive similarity between symptoms in people and observable actions in animal models. Construct validity, according to the authors, is the most important factor to consider when evaluating animal models. The difficulties of correctly simulating human genetic abnormalities in transgenic animals are mentioned in the paragraph. Although genome editing makes molecular replication possible, it is challenging to reflect every variant associated with a disease. To increase predictive validity, it is argued that diverse models are needed to reproduce detrimental alleles and that evaluating therapies in different genetic backgrounds is important.

9.8 Conclusion

Transgenic targeting is the most dependable method for mouse transgenesis because it produces predictable transgene expression patterns and can be repeated. This method is ideal for large-scale knock-in target assessments. Furthermore, transgenic targeting is the preferred method for getting around the restrictions since it is a powerful cellular delivery strategy.

The book chapter emphasizes how important target validation is when determining treatment promises. The article highlights potential areas for drug development enhancement, including improved biomarkers and early validation. The two- to six-month process of target validation entails demonstrating the effectiveness of changing targets in cell culture and animals. Validity is considered critical to the assessment of animal models, highlighting continuous debate and improvement. It is essential to use the translational method to differentiate between target validation in humans and qualifying in animals. We talk about face and concept validity in animal models, with the latter being more important. It is accepted that there are difficulties in precisely simulating human genetic disorders in transgenic animals, highlighting the necessity of using a variety of models to improve predictive validity.

References

Achilli F, Bros-Facer V, Williams HP, Banks GT, AlQatari M, Chia R, Tucci V, Groves M, Nickols CD, Seburn KL, Kendall R (2009) An ENU-induced mutation in mouse glycyl-tRNA synthetase (GARS) causes peripheral sensory and motor phenotypes creating a model of Charcot-Marie-tooth type 2D peripheral neuropathy. Dis Model Mech 2(7–8):359–373. https://doi.org/10.1242/dmm.002527

Agu PC, Afiukwa CA, Orji OU, Ezeh EM, Ofoke IH, Ogbu CO et al (2023) Molecular docking as a tool for the discovery of molecular targets of nutraceuticals in diseases management. Sci Rep 13(1):13398. https://doi.org/10.1038/s41598-023-40160-233

Alam J, Cook JL (1990) Reporter genes: application to the study of mammalian gene transcription. Anal Biochem 188(2):245–254

Alberts B (2017) Molecular biology of the cell. Garland Science, New York

Amara JF, Clackson T, Rivera VM et al (1997) A versatile synthetic dimerizer for the regulation of protein-protein interactions. Proc Natl Acad Sci USA 94:10618–10623. https://doi.org/10.1073/pnas.94.20.10618

Anastasi A (1986) Evolving concepts of test validation. Annu Rev Psychol 37(1):1–6

Arora L, Narula A (2017) Gene editing and crop improvement using CRISPR-Cas9 system. Front Plant Sci 8:1932. https://doi.org/10.3389/fpls.2017.01932

Arrowsmith CH, Audia JE, Austin C, Baell J, Bennett J, Blagg J, Bountra C, Brennan PE, Brown PJ, Bunnage ME, Buser-Doepner C (2015) The promise and peril of chemical probes. Nat Chem Biol 11(8):536–541. https://doi.org/10.1038/nchembio.1867

Barnes LM, Bentley CM, Dickson AJ (2000) Advances in animal cell recombinant protein production: GS-NS0 expression system. Cytotechnology 32:109–123. https://doi.org/10.1023/A:1008170710003

Beale JM, Block JH (2011) Organic medicinal and pharmaceutical chemistry

Békés M, Langley DR, Crews CM (2022) PROTAC targeted protein degraders: the past is prologue. Nat Rev Drug Discov 21(3):181–200. https://doi.org/10.1038/s41573-021-00371-6

Bick RL (1982) Clinical relevance of antithrombin III. Semin Thromb Hemost 8(4):276–287. https://doi.org/10.1055/s-2007-1005058

Bouard D, Alazard-Dany N, Cosset F (2009) Viral vectors: from virology to transgene expression. Br J Pharmacol 157(2):153–165. https://doi.org/10.1038/bjp.2008.349

Brantly M, Nukiwa T, Crystal RG (1988) Molecular basis of alpha-1-antitrypsin deficiency. Am J Med 84:13–31. https://doi.org/10.1016/S0002-9343(88)80066-4

Brooks BR, Walker DL (1984) Progressive multifocal leukoencephalopathy. Neurol Clin 2(2):299–313. https://doi.org/10.1016/S0733-8619(18)31105-8

Brunner D, Nestler E, Leahy E (2002) In need of high-throughput behavioral systems. Drug Discov Today 7(18):S107–S112. https://doi.org/10.1016/S1359-6446(02)02423-6

Burton FH, Hasel KW, Bloom FE, Sutcliffe JG (1991) Pituitary hyperplasia and gigantism in mice caused by a cholera toxin transgene. Nature 350:74–77. https://doi.org/10.1038/350074a0

Chang ACY, Cohen SN (1974) Genome construction between bacterial species in vitro: replication and expression of staphylococcus plasmid genes in Escherichia coli. Proc Natl Acad Sci 71(4):1030–1034. https://doi.org/10.1073/pnas.71.4.1030

Changotra H, Vij A (2017) Rotavirus virus-like particles (RV-VLPs) vaccines: an update. Rev Med Virol 27(6):e1954. https://doi.org/10.1002/rmv.1954

Costantini F (2001) Transgenic animals. In: Brenner's encyclopedia of genetics. Elsevier, Amsterdam, pp 117–123. https://doi.org/10.1016/B978-0-12-374984-0.01560-6

Costantini F, Lacy E (1981) Introduction of a rabbit β-globin gene into the mouse germ line. Nature 294(5836):92–94. https://doi.org/10.1038/294092a0

Darvesh S, Hopkins DA, Geula C (2003) Neurobiology of butyrylcholinesterase. Nat Rev Neurosci 4(2):131–138. https://doi.org/10.1038/nrn1035

Debouck C, Goodfellow PN (1999) DNA microarrays in drug discovery and development. Nat Genet 21(1):48–50. https://doi.org/10.1038/4475

Debruyne EN, Delanghe JR (2008) Diagnosing and monitoring hepatocellular carcinoma with alpha-fetoprotein: new aspects and applications. Clin Chim Acta 395(1–2):19–26. https://doi.org/10.1016/j.cca.2008.05.010

DeMayo JL, Wang J, Liang D, Zhang R, DeMayo FJ (2012) Genetically engineered mice by pronuclear DNA microinjection. Curr Protoc Mouse Biol 2(3):245–262. https://doi.org/10.1002/9780470942390.mo110168

Denayer T, Stöhr T, Van Roy M (2014) Animal models in translational medicine: validation and prediction. New Horiz Transl Med 2(1):5–11. https://doi.org/10.1016/j.nhtm.2014.08.001

Drews J (1979) Drug discovery: a historical perspective. Science 287(5460):1960–1964. https://doi.org/10.1126/science.287.5460.1960

Dunn DA, Pinkert CA, Kooyman DL (2005) Foundation review: transgenic animals and their impact on the drug discovery industry. Drug Discov Today 10(11):757–767. https://doi.org/10.1016/S1359-6446(05)03452-5

Durmaz AA, Karaca E, Demkow U, Toruner G, Schoumans J, Cogulu O (2015) Evolution of genetic techniques: past, present, and beyond. Biomed Res Int 2015:1–7. https://doi.org/10.1155/2015/461524

Erlander MG, Tobin AJ (1991) The structural and functional heterogeneity of glutamic acid decarboxylase: a review. Neurochem Res 16:215–226. https://doi.org/10.1007/BF00966084

Evans M (2011) Discovering pluripotency: 30 years of mouse embryonic stem cells. Nat Rev Mol Cell Biol 12(10):680–686. https://doi.org/10.1038/nrm3190

Ewald D, Li M, Efrat S et al (1996) Time-sensitive reversal of hyperplasia in transgenic mice expressing SV40 T antigen. Science 273:1384–138628

Festing MF (2004) Is the use of animals in biomedical research still necessary in 2002? Unfortunately, "yes". Altern Lab Anim 32(1_suppl):733–739. https://doi.org/10.1615/CritRevNeurobiol.v15.i2.30
Field LJ (1988) Atrial natriuretic factor- SV40 T antigen transgenes produce tumors and cardiac arrhythmias in mice. Science 239:1029–1033. https://doi.org/10.1126/science.2964082
Freeman ME, Kanyicska B, Lerant A, Nagy G (2000) Prolactin: structure, function, and regulation of secretion. Physiol Rev 80:1523. https://doi.org/10.1152/physrev.2000.80.4.1523
Furth PA, St Onge L, Boger H et al (1994) Temporal control of gene expression in transgenic mice by a tetracycline-responsive promoter. Proc Natl Acad Sci USA 91:9302–9306. https://doi.org/10.1126/science.273.5280.1384
Garbaccio RM, Parmee ER (2016) The impact of chemical probes in drug discovery: a pharmaceutical industry perspective. Cell Chem Biol 23(1):10–17. https://doi.org/10.1016/j.chembiol.2015.11.011
Gordon JW, Ruddle FH (1981) Integration and stable germ line transmission of genes injected into mouse pronuclei. Science 214(4526):1244–1246. https://doi.org/10.1126/science.6272397
Gordon JW, Scangos GA, Plotkin DJ, Barbosa JA, Ruddle FH (1980) Genetic transformation of mouse embryos by microinjection of purified DNA. Proc Natl Acad Sci 77(12):7380–7384. https://doi.org/10.1073/pnas.77.12.7380
Guo C, Ma X, Gao F, Guo Y (2023) Off-target effects in CRISPR/Cas9 gene editing. Front Bioeng Biotechnol 11:1143157. https://doi.org/10.3389/fbioe.2023.1143157
Gupta RM, Musunuru K (2014) Expanding the genetic editing tool kit: ZFNs, TALENs, and CRISPR-Cas9. J Clin Invest 124(10):4154–4161. https://doi.org/10.1172/JCI72992
Haupt K, Medina Rangel PX, Bui BT (2020) Molecularly imprinted polymers: antibody mimics for bioimaging and therapy. Chem Rev 120(17):9554–9582. https://doi.org/10.1021/acs.chemrev.0c00428
Henning SW, Beste G (2002) Loss-of-function strategies in drug target validation. Curr Drug Discov 2:17–21
Hinnen A, Hicks JB, Fink GR (1978) Transformation of yeast. Proc Natl Acad Sci 75(4):1929–1933. https://doi.org/10.1073/pnas.75.4.1929
Holmes PV (2003) Rodent models of depression: reexamining validity without anthropomorphic inference. Crit Rev Neurobiol 15(2):10
Holst JJ (2007) The physiology of glucagon-like peptide 1. Physiol Rev 87(4):1409–1439. https://doi.org/10.1152/physrev.00034.2006
Houdebine L (2003) Animal Transgenesis and cloning. Wiley, New York. https://doi.org/10.1002/0470867280
Houdebine LM (2005) Use of transgenic animals to improve human health and animal production. Reprod Domest Anim 40(4):269–281. https://doi.org/10.1111/j.1439-0531.2005.00596.x
Huang S, Mills L, Mian B, Tellez C, McCarty M, Yang XD, Gudas JM, Bar-Eli M (2002) Fully humanized neutralizing antibodies to interleukin-8 (ABX-IL8) inhibit angiogenesis, tumor growth, and metastasis of human melanoma. Am J Pathol 161(1):125–134. https://doi.org/10.1016/S0002-9440(10)64164-8
Hughes J, Rees S, Kalindjian S, Philpott K (2011) Principles of early drug discovery. Br J Pharmacol 162(6):1239–1249. https://doi.org/10.1111/j.1476-5381.2010.01127.x
Kaplan RM, Saccuzzo DP (2001) Psychological testing: principles, applications, and issues. Wadsworth/Thomson Learning, Boston, MA
Kwon M, Firestein BL (2013) DNA transfection: calcium phosphate method. Methods Mol Biol 1018:107–110. https://doi.org/10.1007/978-1-62703-444-9_1018
Lebovitz HE (1997) Alpha-glucosidase inhibitors. Endocrinol Metab Clin N Am 26(3):539–551. https://doi.org/10.1016/S0889-8529(05)70266-8
Lee H (2014) Genetically engineered mouse models for drug development and preclinical trials. Biomol Ther (Seoul) 22(4):267–274. https://doi.org/10.4062/biomolther.2014.074
Levay PF, Viljoen M (1995) Lactoferrin: a general review. Haematologica 80(3):252–267

Lewejohann L, Reinhard C, Schrewe A, Brandewiede J, Haemisch A, Görtz N, Schachner M, Sachser N (2006) Environmental bias? Effects of housing conditions, laboratory environment and experimenter on behavioral tests. Genes Brain Behav 5(1):64–72. https://doi.org/10.1111/j.1601-183X.2005.00140.x

Lim IK, Dumenco LL, Yun J et al (1990) High level, regulated expression of the chimeric P-enolpyruvate carboxykinase (GTP)-bacterial O6-alkylguanine-DNA alkyltransferase (ada) gene in transgenic mice. Cancer Res 50:1701–1708

Lin FL, Sperle K, Sternberg N (1985 Mar) Recombination in mouse L cells between DNA introduced into cells and homologous chromosomal sequences. Proc Natl Acad Sci 82(5):1391–1395. https://doi.org/10.1073/pnas.82.5.1391

Lin CS, Uboldi AD, Marapana D, Czabotar PE, Epp C, Bujard H, Taylor NL, Perugini MA, Hodder AN, Cowman AF (2014) The merozoite surface protein 1 complex is a platform for binding to human erythrocytes by plasmodium falciparum. J Biol Chem 289(37):25655–25669. https://doi.org/10.1074/jbc.M114.586495

Lockhart DJ, Winzeler EA (2000) Genomics, gene expression and DNA arrays. Nature 405(6788):827–836. https://doi.org/10.1038/35015701

Lombardino JG, Lowe JA (2004) The role of the medicinal chemist in drug discovery—then and now. Nat Rev Drug Discov 3(10):853–862. https://doi.org/10.1038/nrd1523

Loscalzo J, Braunwald E (1988) Tissue plasminogen activator. N Engl J Med 319(14):925–931. https://doi.org/10.1056/NEJM198810063191407

Lubow RE (2005) Construct validity of the animal latent inhibition model of selective attention deficits in schizophrenia. Schizophr Bull 31(1):139–153. https://doi.org/10.1093/schbul/sbi005

MacDonald ME, Ambrose CM, Duyao MP, Myers RH, Lin C, Srinidhi L, Barnes G, Taylor SA, James M, Groot N, MacFarlane H (1993) A novel gene containing a trinucleotide repeat that is expanded and nun stable on Huntington's disease chromosomes. Cell 72(6):971–983. https://doi.org/10.1016/0092-8674(93)90585-E

Maltsev VA, Wobus AM, Rohwedel J, Bader M, Hescheler J (1994) Cardiomyocytes differentiated in vitro from embryonic stem cells developmentally express cardiac-specific genes and ionic currents. Circ Res 75(2):233–244. https://doi.org/10.1161/01.RES.75.2.233

Mansour SL, Thomas KR, Capecchi MR (1988) Disruption of the proto-oncogene int-2 in mouse embryo-derived stem cells: a general strategy for targeting mutations to non-selectable genes. Nature 336(6197):348–352. https://doi.org/10.1038/336348a0

Massoud TF, Hademenos GJ, Young WL, Gao E, Pile-Spellman J, Viñuela F (1998) Principles and philosophy of modeling in biomedical research. FASEB J 12(3):275–285. https://doi.org/10.1096/fasebj.12.03.275

Matthews K, Christmas D, Swan J, Sorrell E (2005) Animal models of depression: navigating through the clinical fog. Neurosci Biobehav Rev 29(4–5):503–513. https://doi.org/10.1016/j.neubiorev.2005.03.005

Mayran A, Bolt CC (2022) Transgenic model systems have revolutionized the study of disease. DNA Cell Biol 41(1):49–52. https://doi.org/10.1089/dna.2021.0514

Mehta HM, Malandra M, Corey SJ (2015) G-csf and gm-csf in neutropenia. J Immunol 195(4):1341–1349. https://doi.org/10.4049/jimmunol.1500861

Mengstie MA, Wondimu BZ (2021) Mechanism and applications of CRISPR/Cas-9-mediated genome editing. Biologics 15:353–361. https://doi.org/10.2147/BTT.S326422

Moody TW, Pert CB, Gazdar AF, Carney DN, Minna JD (1981) High levels of intracellular bombesin characterize human small-cell lung carcinoma. Science 214(4526):1246–1248. https://doi.org/10.1126/science.6272398

Morimoto S, Cassell MD, Beltz TG, Johnson AK, Davisson RL, Sigmund CD (2001) Elevated blood pressure in transgenic mice with brain-specific expression of human angiotensinogen driven by the glial fibrillary acidic protein promoter. Circ Res 89:365–372. https://doi.org/10.1161/hh1601.094988

Murphy SV, Atala A (2014) 3D bioprinting of tissues and organs. Nat Biotechnol 32(8):773–785. https://doi.org/10.1038/nbt.2958

Neidhardt L, Gasca S, Wertz K, Obermayr F, Worpenberg S, Lehrach H, Herrmann BG (2000) Large-scale screen for genes controlling mammalian embryogenesis, using high-throughput gene expression analysis in mouse embryos. Mech Dev 98(1–2):77–93. https://doi.org/10.1016/S0925-4773(00)00453-6

Newman DJ, Cragg GM, Snader KM (2000) The influence of natural products upon drug discovery (antiquity to late 1999). Nat Prod Rep 17(3):215–234. https://doi.org/10.1039/a902202c

No D, Yao TP, Evans RM (1996) Ecdysone-inducible gene expression in mammalian cells and transgenic mice. Proc Natl Acad Sci USA 93:3346–3351. https://doi.org/10.1073/pnas.93.8.3346

Nozik-Grayck E, Suliman HB, Piantadosi CA (2005) Extracellular superoxide dismutase. Int J Biochem Cell Biol 37(12):2466–2471. https://doi.org/10.1152/ajplung.00263.2015

Perea J, Robertson A, Tolmachova T et al (2001) Induced myelination and demyelination in a conditional mouse model of Charcot-Marie-tooth disease type 1A. Hum Mol Genet 10:1007–1018. https://doi.org/10.1093/hmg/10.10.1007

Rani N, Alam MM, Jamal A, Bin Ghaffar U, Parvez S (2023) Caenorhabditis elegans: a transgenic model for studying age-associated neurodegenerative diseases. Ageing Res Rev 91:102036. https://doi.org/10.1016/j.arr.2023.102036

Rusconi S (1991) Transgenic regulation in laboratory animals. Experientia 47:866–877. https://doi.org/10.1007/BF01929876

Rusconi S (1996) Transgenic regulation in laboratory animals. In: Principles of medical biology, vol 5. Elsevier, Amsterdam, pp 377–401. https://doi.org/10.1007/BF01929876

Saunders TL (2020) The history of Transgenesis. Methods Mol Biol 2066:1–26. https://doi.org/10.1007/978-1-4939-9837-1_1

Schmid KT, Höllbacher B, Cruceanu C, Böttcher A, Lickert H, Binder EB, Theis FJ, Heinig M (2021) scPower accelerates and optimizes the design of multi-sample single cell transcriptomic studies. Nat Commun 12(1):6625. https://doi.org/10.1038/s41467-021-26779-7

Shaw-white JR, Denko N, Albers L, Doetschman TC, Stringer JR (1993) Expression of the lacZ gene targeted to the HPRT locus in embryonic stem cells and their derivatives. Transgenic Res 2:1–3. https://doi.org/10.1007/BF01977675

Silva F (1993) Psychometric foundations and behavioral assessment. Sage Publications, Inc, Thousand Oaks, CA

Singh VK, Seed TM (2021) How necessary are animal models for modern drug discovery? Expert Opin Drug Discov 16(12):1391–1397. https://doi.org/10.1080/17460441.2021.1972255

Smithies O, Gregg RG, Boggs SS, Koralewski MA, Kucherlapati RS (1985) Insertion of DNA sequences into the human chromosomal β-globin locus by homologous recombination. Nature 317(6034):230–234. https://doi.org/10.1038/317230a0

Snaith MR (2002) The use of transgenic systems in pharmaceutical research. Brief Funct Genomic Proteomic 1(2):119–130. https://doi.org/10.1093/bfgp/1.2.119

Spaulding EL, Hines TJ, Bais P, Tadenev AL, Schneider R, Jewett D, Pattavina B, Pratt SL, Morelli KH, Stum MG, Hill DP (2021) The integrated stress response contributes to tRNA synthetase–associated peripheral neuropathy. Science 373(6559):1156–1161. https://doi.org/10.1126/science.abb3414

Spreafico R, Soriaga LB, Grosse J, Virgin HW, Telenti A (2020) Advances in genomics for drug development. Genes (Basel) 11(8):942. https://doi.org/10.3390/genes11080942

Strobl JS, Thomas MJ (1994) Human growth hormone. Pharmacol Rev 46(1):1–34

Texidó G (2013) Genetically engineered animal models for in vivo target identification and validation in oncology. Methods Mol Biol 986:281–305

The Qur'an (2023) Translated by Dr. Mustafa Khattab. https://quran.com/al-baqarah/2-31. Accessed 12 Dec 2023

Trucano TG, Swiler LP, Igusa T, Oberkampf WL, Pilch M (2006) Calibration, validation, and sensitivity analysis: what's what. Reliab Eng Syst Saf 91(10–11):1331–1357. https://doi.org/10.1016/j.ress.2005.11.031

Van der Staay FJ (2006) Animal models of behavioral dysfunctions: basic concepts and classifications, and an evaluation strategy. Brain Res Rev 52(1):131–159. https://doi.org/10.1016/j.brainresrev.2006.01.006

Vaswani SK, Hamilton RG (1998) Humanized antibodies as potential therapeutic drugs. Ann Allergy Asthma Immunol 81(2):105–119. https://doi.org/10.1016/S1081-1206(10)62794-9

Wadud A, Prasad PV, Rao MM, Narayana A (2007) Evolution of drug: a historical perspective. Bull Indian Inst Hist Med Hyderabad 37(1):69–80

Wang Y, Rollins SA, Madri JA, Matis LA (1995) Anti-C5 monoclonal antibody therapy prevents collagen-induced arthritis and ameliorates established disease. Proc Natl Acad Sci 92(19):8955–8959. https://doi.org/10.1073/pnas.92.19.8955

Wang L, Amphlett G, Blättler WA, Lambert JM, Zhang WE (2005) Structural characterization of the maytansinoid–monoclonal antibody immunoconjugate, huN901–DM1, by mass spectrometry. Protein Sci 14(9):2436–2446. https://doi.org/10.1110/ps.051478705

Wang SW, Gao C, Zheng YM, Yi L, Lu JC, Huang XY et al (2022) Current applications and future perspective of CRISPR/Cas9 gene editing in cancer. Mol Cancer 21(1):57. https://doi.org/10.1186/s12943-022-01518-8

Wijmenga C, Zhernakova A (2018) The importance of cohort studies in the post-GWAS era. Nat Genet 50(3):322–328. https://doi.org/10.1038/s41588-018-0066-3

Wright S (1986) Recombinant DNA technology and its social transformation, 1972-1982. Osiris 2:303–360. https://doi.org/10.1086/368659

Wróbel A, Nowak G, Ossowska G, Danilczuk Z, Zebrowska-Lupina I, Wielosz M (2004) Effect of chronic treatment with dexamethasone on brain dopamine receptors in mice. Pol J Pharmacol 1984(56):399–405. https://doi.org/10.1007/BF00427414

Xu ZJ, Jia YL, Wang M, Yi DD, Zhang WL, Wang XY et al (2019) Effect of promoter, promoter mutation and enhancer on transgene expression mediated by episomal vectors in transfected HEK293, Chang liver and primary cells. Bioengineered 10(1):548–560. https://doi.org/10.1080/21655979.2019.1684863

Yang XD, Jia XC, Corvalan JR, Wang P, Davis CG (2001) Development of ABX-EGF, a fully human anti-EGF receptor monoclonal antibody, for cancer therapy. Crit Rev Oncol Hematol 38(1):17–23. https://doi.org/10.1016/S1040-8428(00)00134-7

Chapter 10
Drug Design Models for Human Diseases by Transgenic Animals

Sabyasachi Banerjee, Subhasis Banerjee, Sankhadip Bose, Avik Das, Santanu Banerjee, and Rajendra Gyawali

Abstract The procedures of drug research and preclinical trials are difficult, and a significant majority of medication candidates are unable to receive authorization from the United States Food and Drug Administration. To enhance the likelihood of success in the new drug development process, it is necessary to use efficient and predictive techniques that can identify high-quality targets. An effective approach to address the difficulties encountered in the advancement of novel medications and combination treatments is using cost-effective and easily controllable animal models for in vivo experimentation. The use of transgenic animals in the drug discovery laboratory as a model is considered an important tool for researching human diseases and determining responses to therapeutic intervention. These animals are genetically modified organisms that exhibit characteristics resembling symptoms of certain human illnesses. It offers genetic models of many human illnesses that are crucial for comprehending diseases and establishing novel targets. Since the 1980s, scientists have achieved the ability to manipulate the mouse genome via genetic modification, namely by eliminating or replacing a particular gene. ATryn, the first medication derived from transgenic

S. Banerjee (✉) · S. Banerjee
Department of Pharmaceutical Chemistry, Gupta College of Technological Sciences, Asansol, West Bengal, India

S. Bose
School of Pharmacy, The Neotia University, Sarisa, West Bengal, India

A. Das · S. Banerjee
Department of Pharmacology, Gupta College of Technological Sciences, Asansol, West Bengal, India

R. Gyawali
Department of Pharmacy, Kathmandu University, Dhulikhel, Nepal

S. Bose et al. (eds.), *Concepts in Pharmaceutical Biotechnology and Drug Development*, Interdisciplinary Biotechnological Advances,
https://doi.org/10.1007/978-981-97-1148-2_10

animals, was produced and authorized by the United States Food and Drug Administration within 20 years of its creation. This breakthrough has paved the way for the development of pharmaceuticals derived from transgenic animals. One such application is the creation of biologically safe medications that are based on human regulatory proteins. Many issues, such as ethical concerns, regulatory requirements, and patents pertaining to the usage of transgenic animals, will get additional attention in the coming years. Transgenic animals play a critical role in drug discovery and development with potential applications in xenotransplantation, clinical trials, and other areas.

Keywords Transgenic animals · Genetically engineered mice · Models · Drug development · Drug design

10.1 Introduction

The development of a new marketable drug product is becoming costly because of the biological complexities of human diseases, research and development process, preclinical and clinical studies, compound manufacturing process, regulatory policy, and prolonged time for market entry. The attrition rate of these procedures is substantial, with fewer than 10% of the substances examined in clinical trials receiving clearance from the United States Food and Drug Administration (US FDA) (Zambrowicz and Sands 2003; Sharpless and DePinho 2006). The expenditure associated with introducing a novel pharmaceutical surpasses $1 billion. Hence, it is crucial to have technical advancements that enable convenient and effective identification and validation of targets in the first stages of drug development. This is necessary in order to decrease the risk of failure and the expenses associated with introducing a new medication (Begley and Ellis 2012). Utilizing appropriate methodologies that simulate the behavior and forecast the effectiveness of medication would enhance the overall rate of success and provide a consistent approach to the advancement of drugs.

An effective approach to address the significant obstacles encountered in the development of novel medications and combination treatments is using inexpensive and easily controllable animal models in vivo (Hansen and Khanna 2004; Suggitt and Bibby 2005). For drug testing, wild-type or spontaneously altered mice were the only options until 1980. The animal model systems were crucial in the development of several chemotherapeutic drugs, such as alkylating and other DNA-damaging compounds, that are now in use (Esteller et al. 2000). Immunocompromised mice were recently used to evaluate potential anticancer drugs (Sharpless and DePinho 2006; Suggitt and Bibby 2005). Nevertheless, there was significant variation in the findings of these screens when comparing the reactions in mice and humans. These models demonstrated limited capability in forecasting the outcomes of clinical trial Phase I or II (Richmond and Su 2008).

Genetically engineered mouse models (GEMMs) of human ailments have the potential to enhance medication development (Politi and Pao 2011; Kucherlapati 2012). Various kinds of GEMMs, including transgenic, knock-in, and knockout mouse models, have been used in drug research and preclinical studies since the 1980s (Politi and Pao 2011; Frese and Tuveson 2007). Transgenic innovation enables the integration of additional DNA that contains the desired gene into the mice genome. Knock-in or knockout technology allows for the targeted removal or modification of specific sections of the mice genome.

The advancement of new transgenic techniques provides better and more genetically altered animals and is employed in a broad range of biological, pharmaceutical, and safety testing scenarios. The genomes of three widely studied organisms—the human, rat, and mouse have been deciphered, speeding up this development (Houdebine 2009). Pharmaceutical businesses encounter the obstacle that around 10% of substances examined in clinical trials successfully reach the marketplace, and among those, only a small portion will provide substantial profits. The expense associated with discovering novel drugs is substantial, amounting to around $800 million in the United States. Of this amount, 80% is allocated toward clinical trials and improvement. There is hope that transgenic technology might reduce drug development failures by improving the quality of drug targets and compounds (Snaith and Törnell 2002).

In the last three decades, the methods for modifying the genetic code of mice and their embryos have advanced significantly. For instance, researchers have constructed inducible and conditional models that allow them to temporarily activate or deactivate genes, either in specific locations or for a certain period of time (Hansen and Khanna 2004; Frese and Tuveson 2007; Sun et al. 2007). Knock-in models include the insertion of transgenes at a specific location or the introduction of specific mutations into target genes. The advanced mouse models created via these advancements in mice engineering approaches are valuable for drug development and preclinical studies.

In recent years, GEMMs have become an indispensable tool for determining target function, verifying selectivity, and making predictions about toxicology and drug development. Over the last three decades, researchers have created several mouse models (Kucherlapati 2012; Frese and Tuveson 2007). The primary roles of GEMMs in drug development are target confirmation, pharmacodynamic indications of drug action identification, toxicity identification, and safety assessment (Politi and Pao 2011) (Table 10.1). GEMMs have also been used to verify assumptions about the plausibility of a novel gene as a possible therapeutic target.

Table 10.1 Applications of GEMMs in drug development (Lee 2014)

Drugs	GEM phenotype	Target gene
Evista, Premarin	Knockout mice exhibited impaired fertility and decreased bone mineral densities	Estrogen Esr1, Esr2 receptors
Epogen, Procrit	Individuals with a targeted null mutation in both sets of the gene show a decrease in the production of early red blood cells and do not survive beyond embryonic day 13 owing to a disruption in the formation of red blood cells in the fetal liver	Erythropoietin
Zyrtec, Allegra, Claritin	Knockout mice exhibited reduced B- and T-cell reactivity, decreased vigilance, and modified activity levels	Histamine H1 receptor
Celebrex	Knockout mice exhibited diminished inflammation, a significant decrease in collagen-induced arthritis, reduced fever response, and reduced production of polyps	Cox2
Celexa, Effexor, Zoloft, Paxil, Prozac	Individuals with a targeted null mutation in a homozygous state have significantly reduced amounts of serotonin in the brain and demonstrate increased levels of anxiety	Serotonin transporter (5-HTT)

10.2 Preclinical Trials and Drug Development Using New Genetically Modified Mouse Models

Although GEMMs have significant benefits in drug design and preclinical examinations, there are certain technological constraints that hinder the creation of new GEMMs (Beard et al. 2006). Transgenic mice exhibit significant variability among different individuals, whereas knockout mice may not always display the anticipated phenotypes and may sometimes exhibit entirely new or unexpected traits. Another significant issue is the duration of generation for GEMMs. The timescales necessary for the production of transgenic and knockout mice are very lengthy, spanning from 12 to 24 months from the original design phase to the formation of the first experimental group. Several researchers interested in genetically engineering mice have explored novel approaches to altering the mouse genome and embryos in order to create and utilize GEMMs. Targeted transgenesis is a method that decreases the expense and speeds up the process of generating GEMMs (Beard et al. 2006; Seibler et al. 2005). This method further enables the transgene's long-term incorporation and expression in the mouse genome. RNA interference (RNAi) and inducible transgenic mouse models are two examples of the new types of mice made possible by targeted transgenesis. These models allow for the regulation of specific genes, either by knocking them down or inducing their expression, in a temporal or spatial manner (Lewandoski 2001; Kleinhammer et al. 2011). These models are expected to boost the usefulness of GEMMs in drug discovery and open up new avenues for their use. The generation of GEMMs may be streamlined and made more efficient by targeted transgenesis. These approaches will be used to create many transgenic and RNAi mouse models.

10.2.1 Advantages and Disadvantages of Generating GEMMs Using Engineered Nucleases

Engineered nucleases provide new approaches for creating GEMMs. Currently, there are three types of engineered nucleases that have been created: transcription activator-like effector (TALEN) nuclease, zinc-finger nuclease (ZFNs), and clustered regularly interspaced short palindromic repeats (CRISPR)/CRISPR-associated (Cas) system (Gaj et al. 2013). These techniques enable the production of live knock-in or knockout mice without the need for chimeras, resulting in time and cost savings compared to traditional procedures that use Embryonic stem (ES) cells and homologous recombination.

There are other concerns that need to be thoroughly evaluated before producing and using these mice with the use of modified nuclease. First, a significant problem arises from the large occurrence of off-target effects, which result from the inherent characteristics of these approaches. Various techniques have been attempted and refined to minimize the unintended impacts. Furthermore, because of the heterogeneous nature of the mutations produced by these approaches, it is necessary to separate and identify these changes. Furthermore, these techniques induce bi-allelic mutation at the desired location. It would be impossible to generate mutant mice using these methods if the mutations had a detrimental impact on embryonic development or neonatal survival. Despite significant technological advancements and several benefits in generating GEMMs using these approaches, the applicability of GEMMs created by these methods is limited to specific circumstances due to the various difficulties mentioned earlier.

10.3 Production of Transgenic Animals

In the scientific literature, three distinct categories of experimental animal models are discussed. These may be categorized as induced, spontaneous, and transgenic models. Induced models are created by laboratory techniques such as administering medicine or chemicals, providing specialized diets, or performing surgical operations. Spontaneous models arise as a result of spontaneous mutations, whereas transgenic models belong to the third category (Bagle et al. 2012). Transgenic refers to the procedure of introducing complementary deoxyribonucleic acid (cDNA) obtained from specific messenger ribonucleic acid (mRNA) into cells (Welfare NIoHOoLA, Association AREN 2002). Animals that have had their genes altered by the insertion of recombinant DNA are called transgenic (Colman 1996).

The development of transgenic animals typically follows a similar sequence, regardless of the species (Houdebine 2009; Mohan et al. 2017):

1. Determination and integration of the exogenous gene and associated promoter sequences.

2. Introducing DNA straight within the pronucleus of a single fertilized egg using several techniques.
3. Transplantation of these genetically modified cells within surrogate mothers.
4. Demonstrating the successful development of the embryo till birth, confirming the stable and inheritable integration of foreign DNA into the DNA of at least a portion of the offspring at birth.
5. Establishing the gene's efficient regulation in order to effectively operate in its new surroundings.

Microinjections or transposon techniques may be used to introduce the foreign DNA into the embryo's cytoplasm or pronucleus. Sperm cells, cloning, lentivirus, and pluripotent cells are other ways in which DNA may be transferred. The latter three methods allow the incorporation of genes either randomly or by targeted integration using homologous recombination or gene substitution, leading to genetic alteration (Houdebine 2009; Ghanghas et al. 2021). Targeted mutation refers to the intentional modification of a specific gene, achieved by either eliminating or deactivating it (knocked out) or sometimes by rendering it functional (knocked in) (Welfare NIoHOoLA, Association AREN 2002; Niemann and Kues 2007). A multiple-transgenic organism is an organism that has been genetically transformed with more than one transgene (Niemann and Kues 2007). These approaches do not lead to the establishment of new species, but instead provide methods for producing novel varieties of animals that contain distinct genetic information (Jube and Borthakur 2006).

10.4 Disease-Specific Transgenic Animal Models

Animal models are live, nonhuman organisms utilized for scientific study and exploration of human ailments, with the aim of enhancing our comprehension of sickness without subjecting a human being to potential injury throughout the entirety of developing and discovering drugs. Fetal eggs are fertilized with human DNA and then inserted into the oviducts of faux-pregnant females to develop into transgenic animal models (Sulabh and Kumar 2018). Various illnesses have diverse models of transgenic animals.

10.4.1 Angiogenesis

The use of mouse models allows for the study of vascular development in relation to arterial stenosis, angiogenesis, thrombosis, atherosclerosis, thrombolysis, and bleeding. This involves the assessment of various methodologies (Snaith and Törnell 2002). Currently, the suppression of the formation of new blood vessels represents a significant potential for the development of novel cancer treatments. Angiogenesis transgenic animal models are used to identify inhibitors that target particular pathways of angiogenesis (Snaith and Törnell 2002).

10.4.2 Human Immunodeficiency Virus (HIV)/Acquired Immunodeficiency Syndrome (AIDS)

The transgenic mouse model for testing some pathogenic manifestations is highly useful in drug discovery research. The first transgenic model was developed with the express purpose of investigating HIV, where the Tg26 HIVAN Mouse Model was released to the scientific community in 1991. Newer variants have been released since then. Rosenstiel et al. provide a comprehensive overview of 32 transgenic murine HIVAN models that have been established (Rosenstiel et al. 2009). These transgenic animals have the ability to produce HIV-1 proteins and exhibit symptoms and immunological deficits that are comparable to the signs of AIDS in people. There are two further versions available: the AIDS Mouse and the Smart Mouse (Bagle et al. 2012).

10.4.3 Cardiovascular Disease

Given the wide range of models available, nowadays, the best strategy is to utilize the transgenic models to investigate cardiovascular diseases by manipulating the expression of genes involved in cardiovascular regulation, such as endothelin, angiotensin, calcium binding-signaling, natriuretic peptides, catecholamines, nitric oxide synthesis, and sodium channel transporters. These models allow for the study of both gain and loss of function in these genes (Bader et al. 2000). Transgenic models of heart failure and hypertrophy have been established, including gene over expression of Calmodulin, gene mutation of alpha cardiac myosin heavy chain, and knockout gene model of transforming growth factor (Hasenfuss 1998). A mutation in the ApoE gene, which plays a crucial role in the absorption of chylomicrons and very low-density lipoprotein particles, leads to the development of atherosclerotic lesions that are histologically identical to those seen in humans (Snaith and Törnell 2002).

Cancer

The oncomouse was the first transgenic mammal to get a patent. For any cancer drug, although they are active during preclinical studies, the nuances of testing these drugs in the whole animal setting are equally important. The germ cells and somatic cells of the animal include a human oncogene sequence that was introduced at the early embryonic stage. This assures the presence of the oncogene in every cell of the organism (Bagle et al. 2012). Tumor development and metastasis may be facilitated by E-cadherin and other adhesion molecules by using different transgenic deletion mice (Welfare NIoHOoLA, Association AREN 2002).

Alzheimer's Disease

Prior to the use of transgenic technology, there were no animal models available for the condition. Alzheimer's disease is also regarded as the ideal disease for modeling in transgenic animals. Vaccinating transgenic mice with Amyloid precursor protein A42 demonstrated the promise of immunization as a treatment strategy for Alzheimer's disease. In order to better understand Alzheimer's disease, several animal models have been developed. These include Alzheimer's mice; TAU transgenic mice such as 7TauTg and ALZ7 mice; amyloid pathology animals such as Tg2576 and PDAPP mice; and presenilin transgenic mice such as ApoE knockout (Spires and Hyman 2005; Götz et al. 2004; Schenk 2002).

Diabetes Mellitus

In the transgenic mice, during the study, a variety of mechanisms ranging from biochemistry to histopathology and different models are used to represent the diversity seen in human diabetic patients. It is necessary to construct transgenic models in order to examine the genes involved in peripheral insulin action. Various models have been created to study insulin secretion in type 2 diabetes, including glucokinase, islet amyloid polypeptide, and hepatic glucose synthesis (Srinivasan and Ramarao 2007). A transgenic mouse model has been designed to exhibit Insulin-Dependent Diabetes Mellitus by introducing a viral gene during the animal egg stage (Etuk 2010). Additional models, such as the beta receptor knockout mice and the uncoupling protein (UCP1) knockout mouse, exist for evaluating antidiabetic drugs. Furthermore, there are acute and chronic models available for this purpose (Srinivasan and Ramarao 2007; Henson and O'Brien 2006; Kumar et al. 2012; Eddouks et al. 2012).

Transgenic animal models are employed to evaluate mutagenicity and carcinogenicity, as well as to investigate metabolic enzymes and receptors (Boverhof et al. 2011). The World Health Organization has authorized transgenic animal models for mutagenicity experiments, such as the LACI transgenic model (using the Big Blue® construct) and the LACZ transgenic model (using the Muta™ Mouse construct). A diverse array of transgenic animals has been created using several approaches in the field of experimental immunotherapy for cancer. Each of these tactics is designed to stimulate distinct components of the immune system. Several of these models consist of transgenic rodents that express tumor-associated antigens, such as MUC1 transgenic mice, as well as transgenic mice with oncogenes to investigate immunotherapeutic approaches. Additionally, there are transgenic mice that express immune effector cell molecules, such as Fc-receptor transgenic mice (McLaughlin et al. 2001). The preclinical transgenic model investigates the bioactive products of Matrix Metalloproteinase (MMP) inhibitors and their potential impact on cellular function (Pavlaki and Zucker 2003). Animal models are also created for skeletal muscle disease, Huntington's disease, and other disorders (Ramaswamy et al. 2007; Horton 2003). Medical research requires disease models to identify specific targets for medication development. Manipulating the genetic composition of these animals by adding or deleting genes grants them novel characteristics that become valuable in gaining a deeper comprehension of diseases or developing curative solutions. Conducting the first testing in people is both unethical and unsafe; hence, transgenic animals are used instead. Animal research is essential for the first evaluation of novel vaccinations and treatments, making disease models invaluable (Tiwari et al. 2012).

10.5 Transgenic Animal Products

The primary focus of studying transgenic organisms is to explore their scientific uses and evaluate their potential for commercial pharmaceutical production. Below are a few examples of transgenic animals now undergoing research and the goods they are expected to produce.

Cow: Factors-VIII and -IX, recombinant antithrombin III (rATIII), protein C, human milk protein, and recombinant human serum albumin (HSA) (Lillico et al. 2005).

Chicken and Egg: Interferons, cytokines, vaccines, insulin, HSA, and monoclonal antibodies (MAbs) (Lillico et al. 2005).

Goat: MAbs, tPA (tissue Plasminogen Activator), Ig fusion proteins, and ATryn (recombinant human antithrombin III) are examples of transgenic recombinant proteins. ATryn, in particular, has the distinction of being the first transgenic recombinant protein from a transgenic animal to get approval from the US Food and Drug Administration (FDA) in January 2009 (Lillico et al. 2005; Ormandy et al. 2011).

Sheep: Sheep milk contains fibrinogen, which is a key component along with thrombin and Factor XIII. It also contains human Factor-VII, -IX, α1-antitrypsin, and activated protein C (Lillico et al. 2005; Hunter et al. 2005).

Rabbit: Recombinant forms of human C1 inhibitor, human α-antitrypsin, human erythropoietin, tPA, α-glucosidase, human interleukin 2, and human growth hormone (Lillico et al. 2005).

Mice: Production of many proteins for the creation of a malaria vaccine. These proteins include ATIII, MAbs, β-interferon, Factor X, myelin basic protein, HSA, cystic fibrosis transmembrane regulator, fibrinogen, tPA, prolactin, and antineoplastic urine protein (Lillico et al. 2005).

Different Other Species: Some of the most often used model species for studying promoter elements and gene transfer strategies include frogs, nematodes, and marine invertebrates (Lillico et al. 2005). At present, the majority of pharmaceutical firms verify a significant number of targets to different degrees and advance them to the high-throughput screening stage. The medications that target these specific sites are then used in clinical trials, which often have a high incidence of attrition, resulting in significant costs. Increasing the use of transgenic models is thought to decrease the volume of work needed to achieve success, resulting in a substantial reduction in expenses (Jube and Borthakur 2006).

10.6 Applications of Drugs Derived from Transgenic Animals and Other Sources

The use of proteins as medications started in the 1920s when insulin derived from pig pancreas was first employed. During the early 1980s, scientists produced human insulin using recombinant bacteria, which is today the primary form of insulin used by the majority of diabetes patients. The success was restricted because bacteria lack the ability to produce intricate proteins like monoclonal antibodies or coagulation blood factors, which need posttranslational changes to become functional or stable inside a living organism. These objectives can only be completely accomplished in mammalian cells that can be cultivated in fermenters or used in live animals. Currently, two methods are being deployed for the production of recombinant proteins in transgenic animals. The first option is the milk produced by transgenic animals, while the second option is the chicken egg white method. Transgenic animals are being used to generate a diverse range of recombinant proteins, including vaccines, cytokines, blood factors, antibodies, hormones, fibrinogen, milk proteins, growth factors, enzymes, collagen, and more (Houdebine 2009; Hunter et al. 2005).

The mammary gland is the optimal location for production due to its ability to create large amounts of protein and the availability of well-established techniques for extracting and purifying these proteins. In order to study the genes involved in peripheral insulin action, transgenic models are developed. The α-glucosidase enzyme derived from transgenic rabbits' milk has effectively treated Pompe's illness (Niemann and Kues 2007; Hunter et al. 2005). Other potential systems for analysis include the silk gland, seminal plasma, blood, urine, and insect larvae hemolymph. Blood often has limited capacity to retain elevated amounts of recombinant proteins, which are inherently unstable. Additionally, the presence of biologically active proteins in the blood might potentially impact the health of the animals. Milk is now the most advanced technique for producing recombinant proteins from transgenic species, thereby circumventing these issues. Current investigations confirm that egg whites may serve as a viable source of foreign proteins, including recombinant vaccines (Houdebine 2009; Hunter et al. 2005).

10.7 Blood Replacement

The existing blood product-producing process relies on donated human blood, which is constrained by concerns related to diseases, a shortage of eligible donors, and regulatory challenges. Animals that have been genetically modified, such as cattle that possess human antibody genes and can produce human polyclonal antibodies that offer the possibility of a consistent source of polyclonal antibodies for treating different medical and infectious conditions, including cancer, autoimmune diseases, organ transplant rejection, and other ailments (Van Eenennaam 2008). Currently, there are a minimum of 33 medications undergoing clinical testing. Some of these treatments are in pivotal trials and incorporate changeable sections from transgenic mice that are encoded by human sequences. Furthermore, the USFDA has authorized a total of 17 therapeutic MAbs that are now undergoing various stages of pharmacological development (Lonberg 2005). Transgenic pigs have successfully developed functional human hemoglobin. The transgenic protein isolated from pig blood had oxygen-binding properties comparable to those of native human hemoglobin. However, only a minuscule fraction of swine red blood cells carried the human variant of hemoglobin (Niemann and Kues 2007).

10.7.1 Transplanting Pig Organs into Humans (Xenotransplantation)

Presently, almost 250,000 individuals owe their survival only to the successful allotransplantation of a suitable human organ. Nevertheless, advancements in organ transplanting technology have resulted in a severe scarcity of suitable organs, and both deceased and living organ donations are insufficient to satisfy the need. In order to address the increasing disparity between the need for suitable human organs and their limited availability, porcine xenografts derived from farmed pigs are regarded as the most optimal option (Niemann and Kues 2007).

The necessary requirements for a successful xenotransplantation are (Niemann and Kues 2007; Niemann and Kues 2003):

1. Zoonotic transmission prevention
2. Compatibility of the donated organs in terms of their anatomical and physiological characteristics
3. Resolving the immunologic rejections of the transplanted organs

Xenotransplantation is regarded as the preferred approach due to the high demand and limited supply of suitable organs. The pig is considered an ideal candidate for organ donation due to the similarity in size between pig and human organs, as well as the comparable architecture and physiology. Additionally, pigs can be maintained at high sanitary standards at very low costs (Niemann and Kues 2003; Fung et al. 1997). The long-term suppression of hyperacute rejection (HAR) has been effectively researched using two basic ways. Antigenic features on swine cells called gal epitopes are removed in order to prevent HAR, and transgenic pigs that generate human complement regulating proteins are created (Niemann and Kues 2003; Cooper 2003).

Issues pertaining to medications derived from transgenic animals include the inability to produce erythropoietin in transgenic animal's mammary glands. The retrieval rates of Factor-VIII protein were suboptimal (Niemann and Kues 2003). Another worry arises from the potential leaking of a specific protein into the bloodstream via the mammary epithelial cells. This may be assessed by seeing elevated levels of the protein in the animal's plasma, which is intended to be exclusively expressed in its milk. There is also a potential for infection to be transmitted from animals to humans (Hunter et al. 2005). Some animals that produce recombinant proteins in their mammary gland may have a premature lactational shutdown, which is a distinct problem (Van Eenennaam 2008). Although transgenic animals present some challenges, the advantages they provide far outweigh these issues. Furthermore, with advancements in technology, these challenges may be effectively addressed.

10.8 Clinical Trials of Medications Derived from Transgenic Animals

The USFDA's clearance of ATryn has paved the way for the introduction of additional pharmaceuticals derived from transgenic animals (Ormandy et al. 2011). The transgenic rabbits have effectively produced the recombinant C1 inhibitor in their milk. This inhibitor has passed phase III trials and is expected to get registration (Hunter et al. 2005; Niemann and Kues 2003). A successful phase III clinical trial has been conducted for a topical antibiotic that specifically targets Streptococcus mutans, a bacterium responsible for causing dental caries. This antibiotic is intended for both preventing and treating dental caries (Niemann and Kues 2003). The vaccine administered in mice with Alzheimer's disease has successfully improved their neurological function. Currently, the vaccine is undergoing phase II clinical trials in humans. The clinical study phase II/III for Pompe's illness is now evaluating the effectiveness of α-Glucosidase derived from rabbits (Hunter et al. 2005; Niemann and Kues 2003). α-antitrypsin, which is used in the treatment of cystic fibrosis and α-AT deficiency, together with tPA, which is utilized for coronary clotting, are now completing clinical trial phase II or III. These goods are expected to become commercially accessible in the next years (Niemann and Kues 2003).

10.9 Transgenic Animals and Ethical Considerations

The ethical concerns surrounding genetic alteration are often less prominent when it comes to microorganisms and plants. Nevertheless, when it pertains to the genetic alteration of animals and especially humans, a greater number of concerns are raised. There is a concern that advancements in transgenic animal technology may actually result in an increase, rather than a reduction, in the number of animals utilized for research. This is due to the fact that medical conditions and diseases may be studied with the help of new technology (Einsiedel 2005). The oncomouse presents a moral conundrum since it is routinely subjected to discomfort for the sake of collecting useful data, which goes against the ideals of animal rights (Gupta and Maurya 2018). The Adenomatous polyposis coli knockout mutant mice display conventional clinical features until the development of intestinal polyps, at which juncture they manifest anemia and endure a decline in body weight. Every new transgenic strain has the ability to provoke detrimental health issues and discomfort in the animals. Hence, it is essential to use strategies focused on mitigating animal distress (Mertens and Rulicke 2007). Additional ethical considerations are the violation of species boundaries and the recognition that animal life should not be seen as a chemical commodity that can be genetically modified and patented for commercial gain. Genetic engineering of animals also disrupts the animal's integrity or telos. Telos refers to the collection of genetically determined wants and interests that are manifested in the environment and together shape the style of life shown by an animal. The fulfillment or obstruction of these needs and interests is significant to the animal (Van Eenennaam 2006).

In India, the treatment of animals is influenced by religious and ethical beliefs, making it important to address religious sensitivities and public awareness (Giridharan et al. 2000). The 3R principles (Reduction, Refinement, Replacement) are designed to decrease pain and suffering endured by animals involved in experimental procedures (Ormandy et al. 2011). Despite the aforementioned issues, transgenic animals may be considered an "improvement" compared to many conventional experimental illness models in which animals endure significant pain. A genotype serves as a very effective representation of disease, specifically focusing on certain bodily processes at the cellular or molecular level. In contrast, the corresponding phenotype exhibits total health. Therefore, it is imperative to contemplate the ethical ramifications of creating such a species, as well as implementing strategies to mitigate animal distress (Mertens and Rulicke 2007).

10.10 Transgenic Animal Patents

The patenting of animal models is crucial at present as it serves as an essential tool for evaluating the effectiveness of new molecules in treating different illnesses. The assessment of the therapeutic effectiveness of a new molecule is mostly reliant on identifying a significant pathological state in animals. Eliminating inert ingredients makes it simpler for pharmacologists to evaluate the therapeutic potential and examine the toxicological aspects of new chemical or biological molecules (Gupta and Maurya 2018). The two primary considerations for awarding patents to animal models are ethicality and replicability. Additional issues such as the imposition of restricted licensing on patents might impede the dissemination of information (Hunter et al. 2005).

Preclinical animal models play an important role in drug discoveries since they provide the foundation for human trials. Patents continue to be a crucial method for pharmaceutical corporations to recoup their research expenses. According to Sects. 3i and 3j of the Indian Patent Law, it is specified that surgical methods and animals are not eligible for patents. Therefore, animal models cannot be patented in India. If appropriate modifications are implemented, animal models might become eligible for patent protection in India, hence creating new opportunities for research in the country (Gupta and Maurya 2018).

10.11 Regulating Transgenic Animals

Animals are used in pharmaceutical research in India, with support from the National Institute of Immunology, the Department of Biotechnology, and the Environmental Protection Act. The Animal Welfare Board is an administrative board with a Committee for the Protection and Care of Experimental Animals (CPCSEA). The CPCSEA is in charge of ensuring that all aspects of animal research adhere to the law and are conducted ethically. Research animal care standards are in accordance with those established by the International Committee for Laboratory Animal Science (ICLAS). However, cloned and transgenic animals are not governed by any hard and fast laws or principles. Providing it is focused on important scientific goals, the Indian Council of Medical Research's 2000 study supports transgenic animal research (Fenwick et al. 2009; Tait et al. 2007).

The modifications in the drug discovery process have led to the creation of animal models that closely resemble human diseases. As a result, US patents on these animal models serve as an incentive for scientists in America and Europe to develop such models, which greatly aid in the drug development process (Gupta and Maurya 2018). The Indian regulatory authorities must anticipate and be ready to address the ethical, regulatory, and patent-related difficulties associated with transgenic animals.

10.12 Future Prospects

The potential exists for successful cell-based therapy in human patients using cells generated from xenogeneic sources, namely pigs. There has been some long-term success in transplanting pig islet cells into diabetic patients. Patients with neurological disorders including Huntington's disease, Parkinson's disease, focal epilepsy, and stroke have had neural cells from pig embryos transplanted into their brains.

One significant benefit of pig brain cells over human brain cells is their widespread availability (Niemann and Kues 2003). Diseases including hypochondroplasia, Turner syndrome, and intrauterine growth retardation have been linked to deficits in growth hormone-releasing hormone, and the pig provides a useful model for studying these conditions (Niemann and Kues 2007). Eggs provide a noninvasive alternative to traditional harvesting methods. Interferon β-1a and -a humanized monoclonal antibody (miR24) were generated in significant concentrations in the egg whites of genetically engineered chickens. miR24 is now being developed as a treatment for malignant melanoma (Begley and Ellis 2012).

Transgenic animals have been created to produce specialized biological materials, such as polymers derived from spider silks. These materials have potential use in face and orthopedic reconstructive surgery, serving as sutures or plastic materials (Van Eenennaam 2006; Goldman et al. 2004). It is very improbable that enough plasma butyrylcholinesterase (BChE) will be available to cure those who have been confronted with organophosphorus substances utilized in agricultural and chemical warfare operations. Milk from transgenic animals is being studied for its potential to produce recombinant BChE (Huang et al. 2007). If 90,000 human blood samples were used instead of transgenic goats, the yearly supply of human antithrombin III would be equivalent to 1. Several human outputs may be replaced by the products of transgenic animals due to their high productivity and usefulness. Advancements in transgenic technology include inducible gene expression, artificial chromosomes, and progress in nuclear transfer (Niemann and Kues 2003; Cooper 2003; Park 2007).

10.12.1 Emerging Transgenic Technologies (Niemann and Kues 2007; Hunter et al. 2005; Park 2007)

- Lentiviral transfection is used to introduce genetic material into oocytes and zygotes.
- Ribonucleic acid interference.
- Chimera formation is achieved by injecting pluripotent cells into blastocysts.
- Culturing spermatogonia and transferring them to male recipients.

Scientists are utilizing transgenic animals to create treatments for the ailments mentioned before, as well as additional conditions such as emphysema, anemia, malaria, hemophilia, and rheumatoid arthritis (Bagle et al. 2012).

10.13 Conclusion

The ongoing improvement of reproductive biotechnologies provides better predictive models for the success and safety of transgenic animals. Genetically engineered animals will have far-reaching consequences for the future of biomedicine, notably in the fields of animal disease modeling, drug development, xenotransplantation, gene pharming, antibody production, and blood replacements (Niemann and Kues 2007). It is important to carefully analyze the regulatory concerns and ethical implications while using transgenic animals. Research suggests that goods obtained from pigs and transgenic animals, like as milk and eggs, have promise in the development of medicinal techniques. Drugs derived from transgenic mice may improve the efficiency of target and chemical combinations, reducing the dropout rate in clinical trials and facilitating a more rapid move from the discovery to the development stages. The usage of transgenic technology has the capability to impact several phases of the discovery process, including identifying and validating targets. Additionally, it offers models specifically created to promptly notify researchers about possible issues related to medication metabolism and toxicity. This will aid in the development of improved models for human disorders (Snaith and Törnell 2002).

The area of transgenics is fast progressing to generate better disease models and has successfully developed the first medicine for transgenic animals along with the recognition of the US FDA. Hence, the utilization of transgenic animals has the capability to surmount the present and forthcoming requirements in the field of medicine and is now an imperative rather than a discretionary concern.

References

Bader M, Bohnemeier H, Zollmann FS, Lockley-Jones OE, Ganten D (2000) Transgenic animals in cardiovascular disease research. Exp Physiol 85(6):713–731

Bagle T, Kunkulol R, Baig M, More S (2012) Transgenic animals and their application in medicine. Int J Med Res Health Sci 2(1):107–116

Beard C, Hochedlinger K, Plath K, Wutz A, Jaenisch R (2006) Efficient method to generate single-copy transgenic mice by site-specific integration in embryonic stem cells. Genesis 44(1):23–28

Begley CG, Ellis LM (2012) Raise standards for preclinical cancer research. Nature 483(7391):531–533

Boverhof DR, Chamberlain MP, Elcombe CR, Gonzalez FJ, Heflich RH, Hernández LG et al (2011) Transgenic animal models in toxicology: historical perspectives and future outlook. Toxicol Sci 121(2):207–233

Colman A (1996) Production of proteins in the milk of transgenic livestock: problems, solutions, and successes. Am J Clin Nutr 63(4):S639–SS45

Cooper DK (2003) Clinical xenotransplantation—how close are we? Lancet 362(9383):557–559

Eddouks M, Chattopadhyay D, Zeggwagh NA (2012) Animal models as tools to investigate anti-diabetic and anti-inflammatory plants. Evid Based Complement Alternat Med 2012:1

Einsiedel E (2005) Public perceptions of transgenic animals. Rev Sci Tech Office Int Des Epizoot 24(1):149

Esteller M, Garcia-Foncillas J, Andion E, Goodman SN, Hidalgo OF, Vanaclocha V et al (2000) Inactivation of the DNA-repair gene MGMT and the clinical response of gliomas to alkylating agents. N Engl J Med 343(19):1350–1354

Etuk E (2010) Animals models for studying diabetes mellitus. Agric Biol JN Am 1(2):130–134

Fenwick N, Griffin G, Gauthier C (2009) The welfare of animals used in science: how the "three Rs" ethic guides improvements. Can Vet J 50(5):523

Frese KK, Tuveson DA (2007) Maximizing mouse cancer models. Nat Rev Cancer 7(9):654–658

Fung J, Rao A, Starzl T (1997) Clinical trials and projected future of liver xenotransplantation. World J Surg 21:956–961

Gaj T, Gersbach CA, Barbas CF (2013) ZFN, TALEN, and CRISPR/Cas-based methods for genome engineering. Trends Biotechnol 31(7):397–405

Ghanghas A, Gahlawat S, Gahlawat N, Poonia A (2021) Global rules, regulations and intellectual property rights on diagnostic methods. In: Advances in animal disease diagnosis. CRC Press, Boca Raton, pp 299–315

Giridharan N, Kumar V, Muthuswamy V (2000) Use of animals in scientific research. Indian Counc Med Res:1–27

Goldman IL, Kadulin SG, Razin SV (2004) Transgenic animals in medicine: integration and expression of foreign genes, theoretical and applied aspects. Med Sci Monit 10(11):RA274–RA285

Götz J, Streffer J, David D, Schild A, Hoerndli F, Pennanen L et al (2004) Transgenic animal models of Alzheimer's disease and related disorders: histopathology, behavior and therapy. Mol Psychiatry 9(7):664–683

Gupta A, Maurya H (2018) Fabrication techniques and utilization of transgenic animal. Cell (Fig 2) 16:17

Hansen K, Khanna C (2004) Spontaneous and genetically engineered animal models: use in preclinical cancer drug development. Eur J Cancer 40(6):858–880

Hasenfuss G (1998) Animal models of human cardiovascular disease, heart failure and hypertrophy. Cardiovasc Res 39(1):60–76

Henson MS, O'Brien TD (2006) Feline models of type 2 diabetes mellitus. ILAR J 47(3):234–242

Horton WA (2003) Skeletal development: insights from targeting the mouse genome. Lancet 362(9383):560–569

Houdebine L-M (2009) Production of pharmaceutical proteins by transgenic animals. Comp Immunol Microbiol Infect Dis 32(2):107–121

Huang Y-J, Huang Y, Baldassarre H, Wang B, Lazaris A, Leduc M et al (2007) Recombinant human butyrylcholinesterase from milk of transgenic animals to protect against organophosphate poisoning. Proc Natl Acad Sci 104(34):13603–13608

Hunter CV, Tiley LS, Sang HM (2005) Developments in transgenic technology: applications for medicine. Trends Mol Med 11(6):293–298

Jube S, Borthakur D (2006) Recent advances in food biotechnology research. In: Food biochemistry food process. Wiley, New York, pp 35–70

Kleinhammer A, Wurst W, Kühn R (2011) Constitutive and conditional RNAi transgenesis in mice. Methods 53(4):430–436

Kucherlapati R (2012) Genetically modified mouse models for biomarker discovery and preclinical drug testing. Clin Cancer Res 18(3):625–630

Kumar S, Singh R, Vasudeva N, Sharma S (2012) Acute and chronic animal models for the evaluation of anti-diabetic agents. Cardiovasc Diabetol 11(1):1–13

Lee H (2014) Genetically engineered mouse models for drug development and preclinical trials. Biomol Ther 22(4):267

Lewandoski M (2001) Conditional control of gene expression in the mouse. Nat Rev Genet 2(10):743–755

Lillico SG, McGrew MJ, Sherman A, Sang HM (2005) Transgenic chickens as bioreactors for protein-based drugs. Drug Discov Today 10(3):191–196

Lonberg N (2005) Human antibodies from transgenic animals. Nat Biotechnol 23(9):1117–1125

McLaughlin PM, Kroesen B-J, Harmsen MC, de Leij LF (2001) Cancer immunotherapy: insights from transgenic animal models. Crit Rev Oncol Hematol 40(1):53–76

Mertens C, Rulicke T (2007) Welfare assessment and phenotype characterization of transgenic mice. ALTEX 24:46–48

Mohan G, Kumar S, Deginal R, Prasad K, Kumar A (2017) Advances in transgenic animal production and applications. Int J Sci Environ Technol 6(4):2567–2575

Niemann H, Kues WA (2003) Application of transgenesis in livestock for agriculture and biomedicine. Anim Reprod Sci 79(3–4):291–317

Niemann H, Kues WA (2007) Transgenic farm animals: an update. Reprod Fertil Dev 19(6):762–770

Ormandy EH, Dale J, Griffin G (2011) Genetic engineering of animals: ethical issues, including welfare concerns. Can Vet J 52(5):544
Park F (2007) Lentiviral vectors: are they the future of animal transgenesis? Physiol Genomics 31(2):159–173
Pavlaki M, Zucker S (2003) Matrix metalloproteinase inhibitors (MMPIs): the beginning of phase I or the termination of phase III clinical trials. Cancer Metastasis Rev 22:177–203
Politi K, Pao W (2011) How genetically engineered mouse tumor models provide insights into human cancers. J Clin Oncol 29(16):2273
Ramaswamy S, McBride JL, Kordower JH (2007) Animal models of Huntington's disease. ILAR J 48(4):356–373
Richmond A, Su Y (2008) Mouse xenograft models vs GEM models for human cancer therapeutics, vol 1. The Company of Biologists Limited, Cambridge, pp 78–82
Rosenstiel P, Gharavi A, D'Agati V, Klotman P (2009) Transgenic and infectious animal models of HIV-associated nephropathy. J Am Soc Nephrol 20(11):2296–2304
Schenk D (2002) Amyloid-β immunotherapy for Alzheimer's disease: the end of the beginning. Nat Rev Neurosci 3(10):824–828
Seibler J, Küter-Luks B, Kern H, Streu S, Plum L, Mauer J et al (2005) Single copy shRNA configuration for ubiquitous gene knockdown in mice. Nucleic Acids Res 33(7):e67
Sharpless NE, DePinho RA (2006) The mighty mouse: genetically engineered mouse models in cancer drug development. Nat Rev Drug Discov 5(9):741–754
Snaith MR, Törnell J (2002) The use of transgenic systems in pharmaceutical research. Brief Funct Genomics 1(2):119–130
Spires TL, Hyman BT (2005) Transgenic models of Alzheimer's disease: learning from animals. NeuroRx 2(3):423–437
Srinivasan K, Ramarao P (2007) Animal models in type 2 diabetes research: an overview. Indian J Med Res 125(3):451–472
Suggitt M, Bibby MC (2005) 50 years of preclinical anticancer drug screening: empirical to target-driven approaches. Clin Cancer Res 11(3):971–981
Sulabh S, Kumar A (2018) Application of transgenic animals in animal production and health. Agric Rev 39(2):169
Sun Y, Chen X, Xiao D (2007) Tetracycline-inducible expression systems: new strategies and practices in the transgenic mouse modeling. Acta Biochim Biophys Sin 39(4):235–246
Tait J, Wield D, Bruce A, Chataway J (2007) OECD international futures project on "the bio-economy to 2030: designing a policy agenda". Health Biotechnology to 2030
Tiwari G, Tiwari R, Sriwastawa B, Bhati L, Pandey S, Pandey P et al (2012) Drug delivery systems: an updated review. Int J Pharm Invest 2(1):2
Van Eenennaam A (2006) What is the future of animal biotechnology? Calif Agric 60(3):132–139
Van Eenennaam A (2008) Genetically engineered animals: an overview. Department of Animal Science, University of California, Davis, CA
Welfare NIoHOoLA, Association AREN (2002) Institutional animal care and use committee guidebook. Office of Laboratory Animal Welfare, Bethesda
Zambrowicz BP, Sands AT (2003) Knockouts model the 100 best-selling drugs—will they model the next 100? Nat Rev Drug Discov 2(1):38–51

Chapter 11
Role of Genomics and Proteomics in Drug Discovery

Jeevandran Sundarasekar and Geethaa Sahgal

Abstract Genomics and proteomics are two powerful technologies that are revolutionizing the drug discovery process. Genomics is the study of the entire genome of an organism, including the DNA sequence and gene expression patterns. Proteomics is the study of all the proteins in an organism, including their structure, function, and interactions. Both genomics and proteomics can be used to identify new drug targets. A drug target is a molecule that is involved in the disease process. By understanding the molecular basis of disease, scientists can identify potential drug targets that can be modulated to treat the disease. This chapter provides an overview of the role of genomics and proteomics in drug discovery. The chapter begins by discussing the basics of genomics and proteomics. It then goes on to discuss how these technologies can be used to identify new drug targets, develop new drugs, and improve the safety and efficacy of existing drugs. The chapter also discusses the challenges and future directions of genomics and proteomics in drug discovery.

Keywords Genomics · Proteomics · Drug discovery · Biotechnology

J. Sundarasekar
Faculty of Applied Science, AIMST University, Bedong, Kedah, Malaysia
e-mail: jeevan@aimst.edu.my

G. Sahgal (✉)
Faculty of Pharmacy, AIMST University, Bedong, Kedah, Malaysia

S. Bose et al. (eds.), *Concepts in Pharmaceutical Biotechnology and Drug Development*, Interdisciplinary Biotechnological Advances,
https://doi.org/10.1007/978-981-97-1148-2_11

11.1 Introduction

Drug discovery is a multi-stage, complex process with the goal of finding, creating, and commercializing new therapeutic medications. It entails the methodical identification and improvement of substances with the potential to treat particular illnesses or medical situations. There are usually numerous important steps to the drug development process, which will be covered in more detail below. Genomics and proteomics are two potent technologies that have transformed biology and are now having a significant influence on drug discovery. The study of an organism's entire genome, including its DNA sequence and patterns of gene expression, is known as genomics. The study of all the proteins in an organism, together with their structure, use, and connections, is known as proteomics.

Proteomics and genomics are complementary fields of study. Proteomics gives information on the proteins that are really made and expressed by the organism, whereas genomics provides information about the genetic code of an organism. These data can be utilized to find novel pharmacological targets, create new medications, and enhance the potency and security of currently available medications (Baloch et al. 2023).

11.1.1 Importance of Genomics and Proteomics in Drug Discovery

Scientists screen millions of molecules to locate a select handful that have the required therapeutic effect in the traditional drug discovery process. This method is time-consuming and expensive, and it frequently results in the creation of medications with unanticipated side effects. By giving researchers a greater understanding of the molecular causes of disease, genomics and proteomics can aid in overcoming these difficulties. With the help of these data, new medication targets that are more likely to be both successful and safe can be found.

There are many approaches to using genomics and proteomics to find new therapeutic targets. Finding genes that have been altered or are overexpressed in sick tissues is one approach. More likely to be implicated in the illness process are genes that have undergone mutations or are overexpressed in sick tissues, which makes them attractive therapeutic targets.

Studying protein-protein interactions is another method for leveraging genetics and proteomics to find new therapeutic targets. For proteins to function, they must interact with one another. Scientists can pinpoint proteins that are crucial to the development of disease by researching protein-protein interactions. The potential as therapeutic targets for these proteins can subsequently be explored.

The discovery of novel medication targets for a number of diseases, such as cancer, Alzheimer's disease, and HIV/AIDS, has been facilitated by the use of genomics and proteomics (Wang et al. 2019). For instance, by using genomics, researchers have been able to pinpoint the genes that are altered in cancer cells. These genes are currently being researched as potential cancer therapeutic targets. For example, the medication Herceptin targets the overexpressed HER2 protein found in many breast cancer cells (Dean-Colomb and Esteva 2008).

Proteomics has also been utilized by researchers to pinpoint protein-protein interactions that are crucial for the replication of HIV. New HIV/AIDS medications are currently focusing on these interactions. For instance, HIV needs the protease protein to replicate; hence, the medication Prezista targets it (Clark et al. 2017).

Genomics can be utilized to identify biomarkers. Biomarkers are molecules that can be used to identify disease, forecast how the disease will proceed, or track how well a treatment is working (Vasan 2006). Genes linked to disease can be found using genomics, and biomarkers based on these genes can be created.

Proteomics can be used to identify drug resistance mechanisms as well. A significant problem in the treatment of cancer is drug resistance. Proteins associated with drug resistance can be found via proteomics, and this knowledge can then be applied to the development of countermeasures (Li et al. 2011).

On top of this, knowledge in genomics and proteomics can also be used in creating a personalized medicine approach. Personalized medicine is a medical strategy that considers a patient's unique genetic and molecular make-up. Based on each patient's unique genetic and molecular profile, individualized treatment strategies can be created using genomics and proteomics.

11.2 Applications of Genomics in Drug Discovery

11.2.1 Genomic Technologies and Techniques

Our understanding of genetics, gene expression, and the function of genes in health and disease has been completely transformed by genomic technologies and methodologies. With their invaluable insights into target identification, biomarker discovery, and customized medicine, these cutting-edge technologies and techniques have become indispensable in the drug discovery process.

Sanger sequencing was the first technique for DNA sequencing that was widely used. It still has a place in some applications and involves chain termination. DNA can be sequenced in massively parallel using next-generation sequencing (NGS) technologies such as Illumina and Ion Torrent, which accelerate and lower the cost of the process. Research in genomics has tremendously advanced thanks to NGS (Liu et al. 2012).

RNA sequencing (RNA-Seq) enables analysis of the dynamics of gene expression under varied situations by profiling gene expression at the transcriptome level. Technology using CRISPR-Cas9 method is used due to its precise genome-editing capabilities. By using CRISPR-Cas9, it is now possible to selectively alter genes and investigate their functions (Corsi et al. 2022).

Single-nucleotide polymorphism (SNP) analysis is another method used in genomic variation analysis. Understanding genetic differences linked to illnesses and medication responses requires SNP genotyping. Copy number variation (CNV) analysis can also be used to analyse genomic variation. The effects of differences in the number of copies of particular genes or genomic regions on health and disease are revealed by CNV analysis (Wheeler et al. 2013).

In order to find conserved areas and evolutionary changes, comparative genomics compares the genomes of several species. It aids in comprehending the importance of genes in terms of function (Alföldi and Lindblad-Toh 2013). The genetic make-up of entire microbial populations is studied using metagenomics. It can be used in the drug discovery process, especially in determining how the human microbiome affects health (Chiu and Miller 2019). DNA methylation analysis is a method used in the field of epigenomics. Gene regulation is greatly influenced by DNA methylation patterns. To analyse DNA methylation, methods such as bisulfite sequencing are employed. Another technique in epigenomics is chromatin immunoprecipitation sequencing (ChIP-Seq). Mapping of protein-DNA interactions and locating epigenetic changes are both facilitated by ChIP-Seq (Lhoumaud et al. 2019).

Functional annotation and pathway analysis can be carried out in drug discovery by using methods such as gene ontology and pathway analysis. Gene Ontology (GO) analysis classifies genes according to their functions, supporting the interpretation of massive genomic data. In pathway analysis, drug target development depends on the identification of biological networks and pathways connected to disease processes (Zhou et al. 2017).

Single-cell RNA sequencing (scRNA-Seq) and long-read sequencing can be used in the single-cell genomic drug discovery process. Researchers may examine gene expression at the single-cell level using scRNA-Seq, which sheds light on cellular heterogeneity and development (Kharchenko 2021). Long-read sequencing is another technology made available by PacBio and Oxford Nanopore that increases the precision of genome assembly and corrects structural variations (Amarasinghe et al. 2020). In genomic research, efficient data integration and analysis are essential. The interpretation of genomic data requires the use of bioinformatics tools and databases.

Personalized medicine methods and the discovery of prospective drug targets have both been made possible by the advancement of genomic technology and methodologies, which have increased our understanding of the genetic basis of disease. These technologies have a lot of potential to advance medication discovery and enhance patient outcomes as they develop further.

11.2.2 Human Genome Project and Its Impact

One of the most important scientific endeavours of the twentieth century is the Human Genome Project (HGP). It was a cooperative worldwide project comprising scientists from all around the world that began in 1990 and ended in 2003. The main objective of the HGP was to fully decode the human genome, revealing the DNA sequence that forms the genetic code of Homo sapiens. Science, medicine, and society have all been profoundly impacted by this enormous project (Wilson and Nicholls 2015).

In order to determine the precise sequence of the 3 billion base pairs of DNA that make up our genetic code, the HGP sought to identify and map every gene in the human genome. In order to provide a complete picture of our genetic make-up, this effort required the sequencing of both the coding (exonic) and non-coding (intronic and intergenic) sections of the genome (Wilson and Nicholls 2015). Advanced genomic technologies and procedures, such as high-throughput DNA sequencing techniques such as next-generation sequencing (NGS), were made possible by the HGP. These technologies, which enable quicker, more economical sequencing and a variety of genomic studies, have emerged as crucial instruments in genomic research.

The project shed insights into genetic variation by demonstrating the extent of copy number variations (CNVs) and single-nucleotide polymorphisms (SNPs) in the human population. Understanding the genetic basis of many diseases, pharmacological reactions, and human variety has benefited greatly from this information (Vemula et al. 2023). In order to better understand how genes are expressed and how proteins function in health and disease, functional genomic and proteomic studies were made possible thanks to the HGP. It has been possible to find prospective therapeutic targets and biomarkers due to this knowledge.

Personalized medicine has greatly benefited from the availability of a reference human genome. Healthcare professionals can customize therapies to a patient's particular genetic profile by comparing an individual's genome to the reference genome, optimizing therapeutic outcomes, and reducing side effects. Research into the genetic causes of illnesses such as cancer, cardiovascular ailments, and unusual genetic conditions has also advanced because of the HGP. It has aided in the discovery of fresh medication targets and the advancement of specialized treatments.

In addition to worries about genetic privacy, discrimination, and the appropriate use of genetic information, the HGP presented significant ethical and legal issues. To shield people from genetic prejudice, laws such as the Genetic Information Nondiscrimination Act (GINA) were passed in the USA (Johnson and Shaw 2023). The HGP has thus significantly influenced public understanding of genomics and science education. Future generations of scientists have been motivated by it, and the public is now more interested in genomics and genetics.

Overall, the Human Genome Project has had a profound impact on society, science, and medicine. It has improved our knowledge of human genetics, sped up advancements in medical science, and ushered in the era of customized medicine. The HGP's legacy persists as proof of the value of interdisciplinary scientific investigation and discovery as genomic research develops.

11.2.3 Functional Genomics in Drug Discovery

In the context of a complete organism, functional genomics is a subfield of genomics that focuses on understanding how genes, their products (proteins and RNA), and their interactions operate. By shedding light on how genes and gene products affect health, disease, and treatment responses, it plays a crucial role in drug discovery. Through the identification of prospective therapeutic targets, the elucidation of disease mechanisms, and the development of more efficient treatments, functional genomic approaches have completely changed the way that drugs are developed.

Identification and verification of putative therapeutic targets are made possible by functional genomics. Researchers are able to identify the genes that are directly responsible for an illness by methodically examining the way those genes act. In order to evaluate the effect of particular genes on processes relevant to disease, methods such as RNA interference (RNAi) and CRISPR-Cas9 gene editing allow for the selective silencing or change in those genes (Knudsen et al. 2023).

Drug development requires a thorough understanding of the mechanisms by which medications work. The particular pathways and biological processes impacted by potential therapeutic candidates are clarified by functional genomics (Pearson et al. 2016). Mechanism of action (MoA) studies help to refine medication candidates and lower the risk of side effects by revealing unwanted off-target effects.

Finding genes, proteins, or genetic variants that act as markers of disease, the course of disease, or the response to treatment is crucial to the development of biomarkers. Functional genomics performs this role to perfection (Choy et al. 2019). By directing treatment choices in accordance with a patient's particular genetic profile, biomarkers enable customized medicine. High-throughput screening (HTS) tests incorporate functional genomic approaches to quickly screen hundreds of chemicals for their impact on cellular functioning. Lead compounds are identified, and their potential as therapeutic candidates is evaluated utilizing HTS employing functional genomic data (Wildey et al. 2017).

By identifying novel indications for current medications, functional genomics aids efforts to repurpose existing medications. Researchers can find potential candidates for drug repurposing in various disorders by examining how approved medications affect diverse biological pathways.

Pharmacogenomics, which studies how genetic variants affect a person's response to medications, relies heavily on functional genomics. It aids in identifying patients who might have negative side effects or ineffective treatments, and it directs the choice of the most suitable treatments. Systems biology, a comprehensive method that models the intricate relationships between genes, proteins, and other molecules, incorporates functional genomics. The thorough study of disease networks and medication responses is made possible by systems biology.

For analysis and interpretation, functional genomic experiments create a tremendous amount of data that calls for sophisticated bioinformatics tools. Our understanding of pharmacological effects is improved by the integration of genomic data with other "omics" data (such as proteomics and metabolomics). Lastly, functional genomics improves the selection of substances that move forward to clinical trials by validating the efficacy and safety of therapeutic candidates in preclinical studies.

Functional genomics has developed into a crucial tool in the drug discovery process, making it easier to find new drug targets, create focused medicines, and improve treatment plans. Functional genomics will become more and more important as technology develops, helping to bring safer and more effective medications to market and ultimately enhancing patient outcomes and customized treatment.

11.2.4 Pharmacogenomics and Personalized Medicine

Pharmacogenomics is a subfield of genomics that examines how a person's genetic make-up affects how they react to medications. It aims to comprehend how genetic variants impact the effectiveness, safety, and propensity for unfavourable medication reactions. In order to optimize therapeutic advantages while limiting potential negative effects, medical therapies can be customized using pharmacogenomics, ushering in the era of personalized medicine.

Every person has a different genetic make-up, and this genetic variability greatly affects how medications are digested, absorbed, and used by the body. Drug pharmacokinetics (how the body processes the drug) and pharmacodynamics (how the drug interacts with its target) can be affected by genetic differences such as single-nucleotide polymorphisms (SNPs) and copy number variations (CNVs) (Raj 2019). Therefore, pharmacogenomics enables medical professionals to customize medicine recommendations based on a person's genetic profile. Genetic testing can be used to determine the best medication, dosage, and potential hazards for a certain patient, resulting in more targeted and efficient therapies.

In the field of medicine, adverse drug reactions (ADRs) are a major concern. Pharmacogenomics can aid in identifying patients who are more susceptible to ADRs (Daly 2013). Healthcare professionals can improve patient safety by steering clear of medications that potentially have serious adverse responses in some patients. Pharmacogenomics has also completely changed the way cancer is treated in oncology by detecting specific genetic alterations in tumours (Zaimy et al. 2017). In order to better effectively and safely treat cancer, targeted medicines and immunotherapies are created to specifically target the genetic defects that fuel cancer growth.

Cardiovascular medicine can benefit from pharmacogenomics as well, which uses genetic information to direct the choice of anticoagulants, antiplatelet medications, and lipid-lowering medications. Reduced risk of cardiovascular events is the goal of personalized cardiovascular therapy (Weeke and Roden 2014). Psychiatric drugs can have unpredictable results and negative effects. The best choice of antidepressants, antipsychotics, and mood stabilizers can also be made with the aid of pharmacogenomics (Alchakee et al. 2022). It increases the likelihood that people with mental health disorders may find a medicine that works well and has fewer negative effects.

It is necessary to address issues including cost-effectiveness, accessibility to genetic testing, and the interpretation of intricate genetic data before pharmacogenomics may be used in clinical practice. Education and standards for medical practitioners are necessary for widespread adoption. Besides, pharmacogenomic information is delicate and gives rise to privacy issues. To safeguard patients' rights and maintain confidence in customized therapy, genetic information must be kept secure.

In the near future, pharmacogenomic developments will enhance our knowledge of drug–gene interactions and improve therapeutic strategies. Pharmacogenomics combined with artificial intelligence and data from other "omics" systems may improve personalized treatment even more. A paradigm revolution in health care is being ushered in by pharmacogenomics, which replaces the generalized model of medicine with one that is more customized and precise. Pharmacogenomics promises to optimize treatment approaches, improve patient outcomes, and lessen the burden of adverse drug reactions, heralding a paradigm-shifting chapter in medical practice as our understanding of genetics and pharmacology advances.

11.3 Proteomics in Drug Discovery

11.3.1 Proteomic Technologies and Techniques

Understanding the functions, relationships, and dynamics of proteins in health and disease depends heavily on the field of proteomics, which is the study of all the proteins present in a biological system. Because of the substantial advancements in proteomic technology and methods, scientists are now able to investigate the enormous complexity of the proteome. These techniques are crucial in a variety of disciplines, including systems biology, biomarker identification, and drug discovery.

Liquid chromatography-mass spectrometry (LC-MS) is an effective method for identifying and measuring proteins. It assesses the mass-to-charge ratios of proteins and differentiates them according to their chemical characteristics. Proteins and peptides in their intact state are analysed using matrix-assisted laser desorption/ionization (MALDI). It includes using a laser to ionize samples and weigh the results (McEwen et al. 2014). 2D-PAGE, however, produces a 2D map of the proteome by separating proteins according to their isoelectric point and molecular weight. This method is helpful for comparing the levels of protein expression in various samples (Issaq and Veenstra 2008). Protein-protein interactions, protein binding, and antibody-antigen interactions can all be thoroughly examined using protein microarrays. They help researchers understand signalling pathways and find possible therapeutic targets (Stoevesandt et al. 2009).

Several mass spectrometry-based proteomic workflow techniques can be used in proteomic drug discovery such as shotgun proteomics, label-free quantification, and isobaric tagging. In shotgun proteomics, proteins are broken down into peptides and then analysed by MS to find and count the proteins in a complicated mixture. Using their MS signal intensity in label-free quantification, label-free approaches compare the quantity of peptides in various samples. In isobaric tagging (e.g. TMT and iTRAQ), labelling peptides with isobaric tags that can be differentiated during MS analysis enables the relative quantification of proteins (Li et al. 2012; Rauniyar and Yates III 2014).

Protein sequence databases are also used for protein identification. The NCBI's RefSeq database and UniProt are well-known databases. Experimental MS data are matched to sequences in these databases by search techniques such as Mascot, SEQUEST, and MaxQuant (Gevaert and Vandekerckhove 2000). Techniques such as yeast two-hybrid (Y2H) assays and co-immunoprecipitation (co-IP) are used to identify and validate protein-protein interactions in protein-protein interaction (PPI) studies. PPI studies help elucidate cellular pathways and networks (Struk et al. 2019).

Proteins between various circumstances or samples can also be accurately quantified using stable isotope labelling techniques such as SILAC and iTRAQ. They are crucial for examining dynamic proteome alterations (Chahrour et al. 2015). Quantifying particular targeted proteins or peptides of interest is the main goal of targeted proteomics. Common techniques include multiple reaction monitoring (MRM) and selected reaction monitoring (SRM). It is helpful for clinical tests and biomarker validation (Shi et al. 2016).

Three-dimensional structures of protein can be ascertained using methods such as nuclear magnetic resonance (NMR) spectroscopy and X-ray crystallography. Understanding protein function and drug design requires this structural information (Harris et al. 2009). Proteomic data also must be processed, analysed, and interpreted using cutting-edge bioinformatics methods. They support the investigation of pathways, functional annotations, and important protein alterations.

Our understanding of biology and illness has been completely transformed by proteomic technology and methodologies. They are crucial to the development of new drugs, the identification of biomarkers, and customized treatment. Proteomics will become more and more important as technology develops because it can help us understand the complexity of the proteome, which will lead to better treatments and outcomes for patients.

11.3.2 Protein Expression Profiling

A crucial method in proteomics is protein expression profiling, which includes systematically examining the kinds and amounts of proteins expressed in a given biological sample, cell, tissue, or organism at a certain time or under particular circumstances. This method aids researchers in understanding how changes in protein expression relate to numerous physiological processes, illnesses, and therapeutic responses by offering insightful information about the functional state of a biological system.

Techniques for Protein Expression Profiling

In 2D gel electrophoresis method, a 2D map of the proteome can be created by dividing proteins according to their isoelectric point and molecular weight. It enables researchers to contrast the patterns of protein expression in various samples. In mass spectrometry (MS) method, protein identification and quantification are performed using LC-MS and other MS-based techniques. Comprehensive data regarding the proteome, such as the relative abundance of proteins in a sample, can be obtained using MS. Using protein microarrays methods, immobilized proteins are included in protein microarrays, which enable researchers to probe them with labelled samples to analyse protein-protein interactions or quantify protein levels.

In western blotting technique, antibodies are used to locate particular proteins in a sample. It can be applied to analyse protein expression in both qualitative and quantitative ways. Label-free quantification method focuses on assessing the ion intensities of proteins or peptides in MS spectra to enable relative protein quantification across samples without the use of chemical labelling. Stable isotope labelling methods by isotope labelling amino acids in cell culture (SILAC) and isobaric tags for relative and absolute quantification (iTRAQ) are used to assess protein expression levels quantitatively between various samples (Sydor and Nock 2003).

Applications of Protein Expression Profiling

In disease research, the identification of biomarkers linked to conditions including cancer, Alzheimer's disease, and diabetes depends on protein expression profiling. Prognostic and diagnostic information can be obtained from changes in protein expression. For the purpose of drug discovery, knowing how proteins react to medication aids in identifying drug targets and creating novel therapeutic approaches is important. Additionally, it may show possible medication resistance pathways.

In stem cell research, researchers can describe stem cells and their differentiated offspring by using protein expression profiling to track changes in protein levels during cellular differentiation. For functional genomics, the addition of protein expression data to genomic and transcriptome information offers a more complete picture of cellular processes and regulatory systems. Lastly, in clinical proteomics, protein expression profiling can be utilized in the clinical setting to stratify patients, choose treatments, and keep track of therapy outcomes (Sydor and Nock 2003).

Challenges and Future Directions

Understanding the functional importance of changes in protein expression requires integrating protein expression data with other "omics" data, such as genomes and transcriptomics, which is a difficult task. Advanced bioinformatics tools and algorithms are needed for data processing, statistical analysis, and interpretation for analysing large-scale protein expression data.

Single-cell proteomic techniques have improved, allowing for the investigation of individual cells and the discovery of cell-specific protein expression and cellular heterogeneity (Theilgaard-Moench et al. 2011). In biological research and clinical applications, protein expression profiling is a potent method that offers a deeper comprehension of the molecular pathways underlying health and disease. Protein expression profiling will become increasingly important in enhancing our understanding of proteomics and its implications for personalized medicine and drug development as technologies grow and become more sophisticated.

11.3.3 Structural Proteomics

The three-dimensional (3D) structures of proteins and their complexes are the main focus of the branch of proteomics known as structural proteomics. It entails figuring out how atoms are arranged spatially within proteins and comprehending how their structures connect to their functions. The study of biological pathways, the creation of targeted medicines, and many other scientific fields depend heavily on structural proteomics.

A well-known method that has proved essential in figuring out the atomic structures of proteins is X-ray crystallography. The 3D structure is reconstructed using diffraction patterns created by generating protein crystals and exposing them to X-ray photons. Protein structures in solution are determined using nuclear magnetic resonance (NMR) spectroscopy. NMR is a powerful tool for determining the structural details of proteins by observing the interactions between their atomic nuclei.

In recent years, cryo-electron microscopy (cryo-EM) has become a popular structural biology approach. In order to get high-resolution structures of proteins and complexes, even those that cannot crystallize, it entails imaging protein samples at cryogenic temperatures. Homology modelling (comparative modelling) is used to estimate a protein's three-dimensional structure based on the structures of related proteins when experimental methods are not practical (Manjasetty et al. 2012).

Designing medications that specifically target particular binding sites requires an understanding of the 3D structure of proteins, especially enzymes and receptors. Rational drug design and virtual screening are made easier by using this structural proteomics. Knowing the relevant proteins' 3D structures is essential for locating illness biomarkers. This may result in the creation of focused medicines and diagnostic tests. Furthermore, understanding the structure of a protein can reveal information about its catalytic mechanisms, ligand binding, and interactions with other molecules. Models of biological processes and protein networks are created using protein structures. Clarifying biological processes is made easier by comprehending the architecture of important signalling proteins (Manjasetty et al. 2012).

There are many challenges ahead before proteomics can fully be exploited. The future direction of proteomics lies in certain niche areas. Research in structural proteomics is currently focused on figuring out the structures of membrane proteins, which are frequently difficult to crystallize. For full systems biology and knowledge of the molecular causes of diseases, structural proteomic data must be integrated with genomes, transcriptomics, and metabolomics data.

Researchers can now examine protein structures at the single-cell level thanks to the development of single-cell proteomics, which sheds light on cellular heterogeneity and diversity. As technologies such as cryo-EM and NMR spectroscopy continue to advance, we are better able to understand complicated protein structures and protein-protein interactions (Sali et al. 2003). Between genomes, transcriptomics, and functional proteomics, structural proteomics fills a critical gap. It aids in the discovery of biomarkers, therapeutic development, and our comprehension of biological systems by providing precise information about protein structures. As technology develops, structural proteomics will become increasingly important in deciphering the proteome's complexity and enabling customized therapy.

11.3.4 Proteomics in Target Identification and Validation

Target identification and validation, which are essential phases in the drug discovery process, are made possible by proteomics. These processes entail locating prospective molecular targets for therapeutic development and confirming their applicability to a particular illness or ailment. Here, we examine the application of proteomic methods to this key stage of drug discovery.

Protein profiles in biological materials can be thoroughly analysed using proteomic techniques. Researchers can find proteins that are differentially expressed in disease states by comparing the proteomes of healthy and diseased tissues or cells, providing prospective therapeutic targets (Kumar et al. 2016). This technique is known as proteome profiling. Biomarkers are proteins that can signal the presence, progression, or severity of a disease. Proteomics aids in the discovery of these markers. These biomarkers could potentially be used as drug development targets (Cominetti et al. 2016). Proteomics can also be used to identify important nodes (proteins) in protein-protein interaction networks that are linked to particular diseases by analysing these networks. If these proteins are targeted, disease-related processes may be disrupted (Yugandhar et al. 2019).

Investigating the functionalities of possible pharmacological targets involves the use of proteomic methods. This entails evaluating a protein's function in disease mechanisms and determining whether it is a suitable therapeutic target. However, to determine how a particular protein fits into a larger biological system, proteomic data can be combined with data from other omics disciplines (such as genomics and transcriptomics). This supports why it is a worthwhile aim. Proteomic analysis of phosphorylation events can reveal signalling networks and protein regulatory mechanisms. Identification of essential proteins for drug development can be aided by phosphoproteomics (Xu et al. 2019).

High-throughput screening (HTS) can be used after prospective targets have been found and validated. In order to test hundreds of chemicals for their capacity to influence the activity of the target proteins, HTS assays can be designed using proteomic data. Finding lead compounds that can be developed further into medication candidates requires the use of HTS (Wildey et al. 2017). The creation of small compounds or biologics that specifically interact with the target and modulate its activity can be guided by the 3D structures of verified therapeutic targets, which were discovered by structural proteomic approaches. The development of targeted therapeutics, in which medications are created to specifically hit a particular target with few off-target effects, depends heavily on proteomic data (Wang et al. 2005).

Drug candidate efficacy in disease contexts is validated using animal models and cell-based assays. To evaluate changes in protein expression and activity in response to the treatment, proteomics is used. These studies support the validity of the target's and medication candidates' potential therapeutic usefulness (Filiou et al. 2011). Proteomic approaches can be used to validate the clinical relevance of pharmacological targets that advance to the clinical development stage. Here, target expression in patient samples is evaluated, and its relationship to therapeutic outcomes is examined (Ni et al. 2015).

By offering a systematic and data-driven method for choosing the most promising targets, proteomics in target identification and validation speeds up the drug discovery process. It assists in ensuring that possible therapeutic targets are both clinically and physiologically meaningful. Proteomic technologies will become more and more important in the creation of innovative treatments and the advancement of precision medicine as they continue to improve.

11.4 Biomarker Discovery

11.4.1 Importance of Biomarkers in Drug Development

Biomarkers are quantifiable indications that reveal details about different biological processes, ailments, or disorders. Biomarkers are essential tools in the drug development process that provide information on the security, effectiveness, and mode of action of proposed medications (Kraus 2018). They are crucial because they greatly speed up and enhance the drug development process in the following ways.

Biomarkers can help in early disease identification and diagnosis, enabling prompt intervention and therapy. For diseases such as cancer, Alzheimer's disease, and cardiovascular illnesses, for instance, certain proteins or genetic markers can be used as diagnostic biomarkers. Biomarkers are essential for identifying and verifying pharmacological targets. Researchers can identify possible biological targets for drug development by examining biomarkers linked to a specific disease. Using illness features, genetics, or therapy responsiveness as criteria, biomarkers can assist in dividing patients into subgroups. This makes it possible to create personalized, targeted therapy for each patient.

Biomarkers offer a quantitative and unbiased way to evaluate a drug's effectiveness. Researchers can identify whether a medicine is having the desired therapeutic effect by observing changes in particular biomarkers. Biomarkers are used to track the safety of medications and evaluate any potential side effects. Particular biomarkers may show organ damage, metabolic abnormalities, or other drug-related toxicities. By assisting researchers in locating lead compounds and evaluating their efficacy and safety earlier in the pipeline, biomarkers hasten the medication development process. As a result, developing drugs takes less time and money.

The idea of personalized medicine is based on biomarkers. Biomarkers can improve treatment effectiveness and reduce negative effects by customizing therapies to each patient's particular genetic and molecular profile. By easing patient selection, stratification, and endpoint assessment, biomarkers support the design of clinical trials. Clinical studies as a result are more effective and instructive. When reviewing new drug applications, regulatory organizations such as the Food and Drug Administration (FDA) and European Medicines Agency (EMA) are increasingly taking biomarker data into account. Evidence based on biomarkers can speed up the regulatory approval procedure.

Biomarkers enable clinicians to keep track of patients' responses both during and after pharmacological therapy. This aids in modifying treatment plans and evaluating long-term results. It can cut the price of drug development and prevent late-stage failures by enabling the early detection of dangerous or inefficient therapeutic candidates. Research on biomarkers helps us better understand how diseases work, and that knowledge can help us create new therapeutic strategies.

In conclusion, biomarkers are crucial to medication development at every stage, from target identification to clinical trials and post-market surveillance. They improve the speed, accuracy, and safety of drug development procedures, which eventually results in the identification and approval of safer and more potent medicines. The landscape of drug development and personalized medicine is still being shaped by the ongoing developments in biomarker research and technology.

11.4.2 Genomic and Proteomic Approaches to Biomarker Discovery

The early detection, diagnosis, prognosis, and treatment of diseases all depend on biomarkers. The identification of particular molecular markers connected to a variety of medical disorders has been made possible by genomic and proteomic methods, which have revolutionized the area of biomarker research (Huizar et al. 2020). In order to find certain genetic differences linked to diseases, genomic biomarkers analyse a person's DNA. This can include copy number variations (CNVs), insertions/deletions, and single-nucleotide polymorphisms (SNPs). Genome-wide association studies (GWAS) can be used to find genetic variations linked to certain diseases by analysing the entire genome. Researchers can identify potential genetic biomarkers by contrasting the genomes of people with a disease and those without one.

Next-generation sequencing (NGS) enables a thorough examination of a person's DNA. Using whole-genome and whole-exome sequencing, it is possible to find genetic variations that can be used as biomarkers for disease susceptibility and hereditary disorders. The study of gene expression patterns in various disease states is the main objective of transcriptomics. Changes in RNA levels can serve as biomarkers for disease diagnosis and prognosis, and these changes can be detected using microarrays and RNA sequencing (RNA-Seq). Additionally, epigenomics investigates how variations in histone modifications, non-coding RNA, and DNA methylation patterns can affect gene expression. Biomarkers for cancer and other disorders can be derived from altered epigenetic marks (Huizar et al. 2020; He 2006).

The study of proteins, which are more specific indicators of biological processes and disorders, is referred to as proteomic biomarkers. By monitoring protein levels, post-translational changes, and interactions, proteomics can detect biomarkers. Mass spectrometry (MS) is a potent method for finding proteome biomarkers. It can recognize and measure proteins, their isoforms, and post-translational changes, making it possible to find biomarkers linked to disease. To pinpoint individual antibody-antigen interactions, protein microarrays can be probed with patient samples that contain immobilized proteins. Finding diagnostic and prognostic protein biomarkers may result from this.

Using targeted proteomics, specific proteins of interest are quantified. High sensitivity and accuracy are achieved by using the techniques of selected reaction monitoring (SRM) and multiple reaction monitoring (MRM). Functional proteomic studies the role of proteins in various illness conditions. It can uncover possible biomarkers and reveal the mechanisms driving illness progression. Advanced bioinformatics tools are crucial for analysing the massive amounts of data produced during the proteomic biomarker development process. Our understanding of disease mechanisms is improved by the integration of genomes and proteomic data (Huizar et al. 2020; Ion et al. 2016).

However, there are many challenges that need to be addressed and focused on push this technology in the right direction. Researchers are looking into some areas that have the potential to drive genomics and proteomics in biomarker discovery. For example, managing and analysing enormous amounts of genomes and proteomic data are difficult tasks that call for sophisticated computational resources and tools. To guarantee their dependability and relevance, biomarker candidates need thorough validation.

Integrating genomes, proteomics, and other "omics" data is essential for gaining a thorough understanding of disease mechanisms and for the identification of trustworthy biomarkers. New methods in single-cell genomics and proteomics also allow for the analysis of single cells, revealing details on cellular heterogeneity and cell-specific biomarkers. The development of targeted medicines, personalized medicine, and our understanding of the molecular basis of diseases are all being driven by genomic and proteomic approaches to biomarker identification. With the potential for better diagnostics, prognostics, and therapies, these technologies have changed the face of health care. Genomic and proteomic research is likely to elucidate a variety of new biomarkers that will benefit patients and the medical community as a whole as time goes on (Mäbert et al. 2014; Matthews et al. 2016).

11.4.3 Case Studies of Successful Biomarker Utilization

In several medical specialties, biomarkers have been useful in enhancing diagnosis, choice of therapy, and patient outcomes. Here are a few noteworthy case studies that demonstrate how biomarkers can be used effectively.

Biomarker: Human epidermal growth factor receptor 2 (HER2) gene amplification and protein overexpression. Case Study: Herceptin, a monoclonal antibody, has been used to treat HER2-positive breast cancer patients with great success. This treatment primarily targets cancer cells that are HER2-positive. A one-size-fits-all strategy for treating breast cancer existed prior to the use of this biomarker. Patients with HER2-positive breast cancer now have much better survival rates and therapeutic results because of Herceptin (Wilson et al. 2018).

Biomarker: Epidermal growth factor receptor (EGFR) mutations. Case Study: Patients with particular EGFR mutations in non-small cell lung cancer (NSCLC) benefit from targeted treatments such as gefitinib and erlotinib. These mutations are easier to identify, which aids oncologists in choosing the best course of action and increases response rates and progression-free survival (Codony-Servat et al. 2019).

Biomarker: BCR-ABL fusion gene. Case Study: The introduction of imatinib (Gleevec) therapy for CML transformed the way that cancer is treated. Imatinib targets this genetic anomaly's BCR-ABL fusion protein in particular. Many CML patients may now properly manage their illness and enjoy regular lives as a result (Abdulmawjood et al. 2021).

Biomarker: Mutations in the BRCA1 and BRCA2 genes. Case Study: Breast and ovarian cancer risk are enhanced in the presence of BRCA1 and BRCA2 mutations. These mutations can be identified, allowing for early discovery and customized risk-reduction measures. Prophylactic operations and surveillance programmes can considerably lower the incidence of cancer in high-risk persons (Lee et al. 2014).

Biomarker: PSA levels in blood. Case Study: A common biomarker for prostate cancer is PSA. It assists in early disease detection and therapy response monitoring. By detecting prostate cancer at an earlier, more treatable stage, PSA testing has increased survival rates (Saini 2016).

Biomarker: Haemoglobin A1c (HbA1c) levels. Case Study: A biomarker called HbA1c is used to track the long-term management of blood sugar in diabetics. Maintaining ideal HbA1c levels aids in avoiding problems and customizing treatment plans, improving disease management and patient outcomes (Kaiafa et al. 2021).

Biomarker: Human papillomavirus (HPV) DNA. Case Study: Cervical cancer screening has been improved by HPV testing. It pinpoints the disease's high-risk HPV subtypes, enabling early detection and treatment. The prevalence of cervical cancer has dramatically decreased as a result of this strategy (Shah et al. 2020).

These case studies highlight the significant influence that biomarkers have on diagnosing diseases, choosing treatments, and providing for patients. By enabling personalized medicine, early diagnosis, and focused therapies, the use of biomarkers has changed the face of health care, ultimately enhancing patient outcomes and quality of life. The pursuit of new biomarkers promises to provide more indicators that will continue to influence medical advancements.

11.5 Target Identification and Validation

11.5.1 Identifying Potential Drug Targets Through Genomics and Proteomics

A crucial phase in the drug discovery process is the identification of prospective therapeutic targets, and genomics and proteomics have completely changed this area of pharmaceutical research. These cutting-edge technologies have sped up the identification of new therapeutic targets and aided in the creation of more accurate and potent medications.

Genomic and proteomic approaches in identifying drug targets are the same as the approaches to biomarker discovery as discussed in Sect. 11.4.2. The following is a figure showing the list of approaches in genomics and proteomic drug targeting (Kramer and Cohen 2004; Sleno and Emili 2008) (Fig. 11.1 and 11.2).

For a thorough understanding of prospective therapeutic targets, the combination of genomes and proteomic data is crucial. The identification of genes and proteins that are consistently linked to disease and are functionally relevant is made possible by the combination of genetic information (genomics) with data on protein expression (proteomics). Finding high-confidence pharmacological targets is where this integrated method excels.

In conclusion, genomics and proteomics have changed the way prospective therapeutic targets are found. By enabling the identification of therapeutic targets at the genetic and protein levels, these technologies improve the accuracy and efficiency of drug development. The convergence of these two fields, together with developments in computational biology and bioinformatics, continues to spur drug discovery innovation and raises the prospect of more successful and focused treatments in the future.

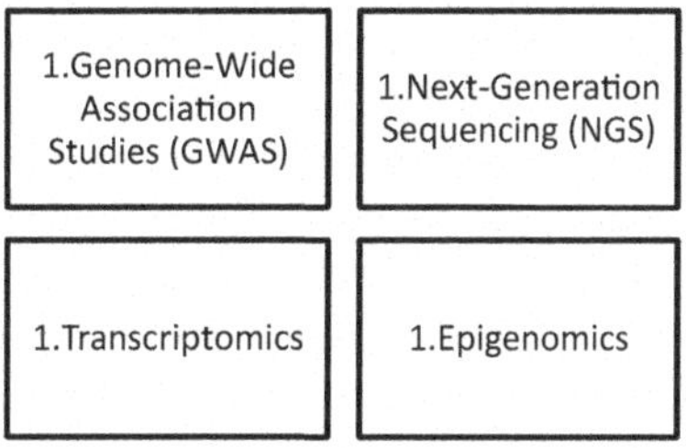

Fig. 11.1 Approaches in genomic drug targeting

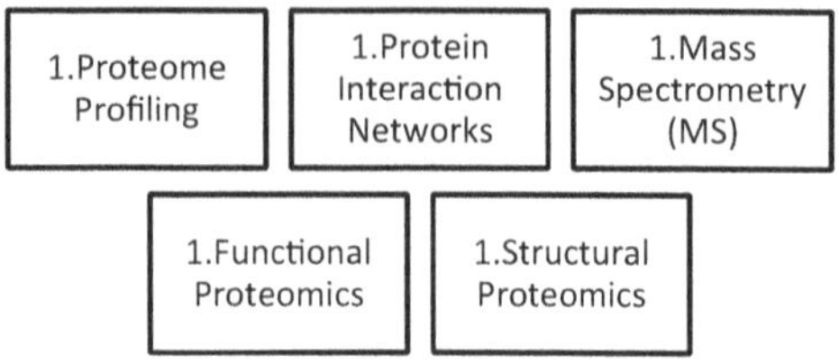

Fig. 11.2 Approaches in proteomic drug targeting

11.5.2 Validation of Drug Targets

A crucial phase in the drug development process is validating drug targets. It entails proving that a particular biological molecule, frequently a protein, is not only connected to a disease but also offers a promising target for therapeutic action. Target validation makes sure that resources are used wisely and that there is a good chance that the drug development process will succeed. The main elements and techniques involved in validating pharmacological targets are listed below (Uitdehaag et al. 2012).

Genetic and genomic evidence such as genetic associations, loss-of-function studies, and gain-of-function studies can be used in the validation of drug targets. Genome-wide association studies (GWAS), for example, can find genetic variants linked to a disease. These studies give solid evidence for validation if a target gene or protein is repeatedly connected to a certain ailment.

Loss-of-function studies such as genetic knockout experiments in cellular systems or model organisms can show that a target is required for the development of a disease. The disease phenotype should be prevented or improved by the absence of the target. In gain-of-function studies, by overexpressing a gene or protein as a therapeutic target, researchers can track the emergence or aggravation of disease symptoms (Minikel et al. 2020).

Biochemical and cellular evidence can also be used in the validation of drug targets. Proteomic methods can show variations in the levels of protein expression, post-translational alterations, and connections linked to the disease. Its validation is aided by repeated changes in the target protein. By creating functional assays to gauge the target protein's activity in disease-related pathways, it is possible to confirm the protein's role in pathogenesis (Waduge et al. 2022).

Drug validation can also be studied by using preclinical models. Using animal models, researchers can examine the effects of target modulation (such as inhibition or activation) on the development of disease. Positive outcomes in animals provide evidence that therapeutic intervention in humans may be effective. Alternatively, in a controlled setting, the effects of targeting the protein can be evaluated using cultured cells or cell lines obtained from patients (Lacombe et al. 2011).

Pharmacological evidence can be used to validate drug targets. In preclinical models, the potential of small molecules to modify the disease can be explored by using chemical probes. These compounds must specifically interact with the target protein. These chemical probes serve as instruments for validating targets. The creation of drug candidates that target the validated molecule can add to the evidence supporting the validity of the molecule as a therapeutic target. The drugs' effectiveness and safety in both preclinical models and clinical trials add to the target's validity (Roy 2018).

Examining patient tissues and samples for the presence and activity of the target protein, as well as its relationship to the severity or prognosis of the condition, is a straightforward means to confirm the target's applicability to human biology. The strongest support for this theory comes from medications that successfully treat the molecule in clinical trials. These studies evaluate the safety, effectiveness, and dosage in patients, further demonstrating the target's function in the treatment of disease (Lee et al. 2005).

Finding biomarkers linked to the target can act as an indirect form of validation. These biomarkers may alter in response to target modulation, which can be a sign of a successful treatment (Kraus 2018). To verify a pharmacological target, several lines of evidence from several methodologies (genetic, biochemical, preclinical models, and clinical trials) should ideally converge. Cross-validation supports the argument that the target is appropriate.

The pharmaceutical industry has difficulties guaranteeing that the chosen targets are both physiologically and therapeutically relevant because the validation of pharmacological targets is a continuous process. When a target is not validated, expensive drug research efforts may be abandoned. However, the establishment of innovative medicines that can lessen disease burden and enhance patient outcomes is made possible by the successful validation of a target. The precision and effectiveness of target validation continue to be improved by developments in genetics, proteomics, and drug discovery technologies.

11.5.3 Role of Bioinformatics in Target Validation

Target validation is a vital phase in the drug discovery process, and bioinformatics is essential to this process. Target validation seeks to establish the viability and relevance of a particular biological molecule—often a protein—as a target for therapeutic intervention. This procedure is streamlined and improved by bioinformatics, which makes use of computer tools, data analysis, and integration of multiple "omics" data. Here is how bioinformatics helps validate targets (Jiang and Zhou 2005).

Bioinformatics tools enable researchers to merge data from the domains of genomics, transcriptomics, proteomics, and other "omics". A more complete picture of a target's participation in a disease can be revealed by integrating these various multi-omics datasets. Bioinformatics software may associate genes and proteins with certain networks and pathways, assisting researchers in comprehending how the target functions within the broader scheme of disease-related activities (Wang et al. 2021).

Bioinformatics also makes it easier to find genes and proteins that express themselves differently in healthy and diseased conditions. Prioritizing prospective therapeutic targets is made easier by analysing extensive differential expression data.

Bioinformatics can as well identify potential interaction partners and downstream effectors by analysing the patterns of co-expression of a target gene or protein with others (Jiang and Zhou 2005).

Bioinformatics tools offer functional annotation and enrichment analysis, helping to evaluate the functions of target genes or proteins in cellular processes and pathways associated with disease (Lin et al. 2017). Using bioinformatics approach, it is possible to build protein-protein interaction networks, which can show how connected a target is to other proteins in a larger landscape of protein-protein interactions. Understanding the regulatory components that regulate a target gene or protein, such as transcription factors or microRNAs, can shed light on the molecular processes behind disease (Chen et al. 2016). In text mining, the software uses natural language processing techniques to glean pertinent data from the scientific literature. This aids researchers in collecting data from published studies that show a target is related to disease (Zheng et al. 2019).

In comparative genomics, genomes from different species may be compared thanks to bioinformatics. Genes or proteins that have undergone evolutionary conservation are frequently more likely to be crucial for biological processes and make for desirable targets (Wright et al. 2013). By combining genetic and molecular data with clinical data, it is possible to find associations between target expression or variation and patient outcomes, assisting in target validation for personalized therapy (Yang et al. 2012). Using a variety of factors, including biological relevance, druggability, and specificity, bioinformatics tools may rank and score candidate drug targets (Mathai et al. 2020). Bioinformatics in silico drug screening method also enables virtual testing of small compounds against target protein 3D structures. The validated target's prospective medication candidates are identified through this method (Rao and Srinivas 2011).

Bioinformatics technologies produce visual representations of data, facilitating the analysis of large datasets and the production of insightful findings by researchers (Koscielny et al. 2017). Target validation is accelerated by bioinformatics thanks to its organized and data-driven methodology. It aids decision-making by researchers regarding the applicability and significance of particular targets for drug development. Bioinformatics expedites the discovery of interesting drug targets by combining enormous volumes of biological data with powerful computer approaches and ultimately aids in the creation of safer and more effective medicines.

11.6 High-Throughput Screening

11.6.1 Automation and Robotics in High-Throughput Screening

High-throughput screening (HTS): A crucial stage in the drug development process is HTS, which entails evaluating a large number of chemical compounds or biological agents for possible therapeutic candidates. This procedure has been transformed by the use of automation and robotics in HTS, making it quicker, more economical, and more efficient. This is how robotics and automation are changing HTS.

Automation via increased throughput makes it possible to handle multiple samples at once, which greatly expands the number of chemicals that can be screened in a shorter amount of time. This speed is necessary to find promising medication candidates quickly (Wildey et al. 2017). Dispensing reagents and samples using automated equipment is extremely accurate and consistent, which minimizes human mistakes. Ensuring the quality of data and generating dependable outcomes depend heavily on this reproducibility. Assays can be made smaller by using fewer amounts of chemicals and reagents thanks to automation. This allows for the screening of rare or costly molecules while also conserving resources (Mayr and Bojanic 2009).

A vast array of HTS assays can be performed by robotic systems, which are flexible and easy to program. They are appropriate for different phases of drug development since they can handle a variety of plate layouts and test types. Automation makes it feasible to include imaging systems in high-content screening (HCS), which in turn permits high-content screening. This makes it possible to analyse several characteristics at the single-cell level and so learn more about the compound effects (Fox et al. 2006). A lot of data is gathered and stored by automated systems. In order to handle, analyse, and interpret these data and assist researchers in making defensible conclusions, bioinformatics tools can be included.

For big compound libraries, automation helps with the tracking, retrieval, and storage of chemical compounds. It guarantees that samples are ready for screening at any time. Automated methods can be utilized for follow-up assays, such as dose-response studies, selectivity testing, and lead optimization, once hits have been identified in primary screens (Wildey et al. 2017). 3D drug screening is also made possible by robotic systems' ability to handle 3D spheroids and cell culture models, which enables more physiologically accurate drug testing. Automation is helpful for combination screens, which examine different combinations of chemicals to find synergistic effects, and for repurposing of already-approved medications for new applications.

By minimizing human exposure to potentially dangerous compounds, automated systems improve safety. They are capable of handling substances that need certain environmental conditions or containment (Du et al. 2020). High-throughput phenotypic screening, in which chemicals are evaluated according to their impact on the phenotype of cells or organisms that are being examined, is making a comeback thanks to automation and robots. This method frequently finds targets and mechanisms that are hidden from view in conventional target-based displays.

Automated HTS can be used to realign current medications for novel conditions, hence reducing development time and costs. The ability of robotic systems to run constantly makes screening possible around the clock, which is necessary for quick, extensive HTS campaigns (Datta et al. 2018). The drug discovery process has been greatly expedited by the incorporation of automation and robotics in HTS, enabling researchers to efficiently and precisely screen millions of molecules. This technical breakthrough leads to the development of novel and life-saving drugs by speeding up the identification of possible drug candidates and lowering the cost of drug research. Automation and robots in HTS will continue to be at the forefront of pharmaceutical research as long as technology keeps developing.

11.6.2 Combining Genomics and Proteomics with High-Throughput Screening

High-throughput screening (HTS) has been transformed by the merging of genomes and proteomics, which offers a comprehensive strategy to find and evaluate prospective drug targets, explore chemical interactions, and expedite the development of novel treatments. This combination improves drug discovery's effectiveness and success rates in a number of ways (Bleicher et al. 2003).

In genomics, genes and genetic variants linked to diseases can be identified using genomic data. These observations provide a basis for determining possible therapeutic targets. By verifying proteomic targets' existence and importance at the protein level, proteomic data can aid in their validation. It enables scientists to determine if the expression of a gene results in a functioning protein and whether that protein is a promising target for therapy.

Under the biomarker discovery mechanism, genetic markers found by genomics may be used as prospective illness biomarkers. These markers can be used to track the course of an illness or evaluate how well a treatment is working. Proteomic profiling is able to recognize particular proteins or protein patterns as biomarkers that signify the occurrence, development, or response to therapy of a given disease.

For drug target identification, combining the strengths of proteomics and genomics allows for a thorough understanding of a disease's molecular landscape. Finding the most potential pharmacological targets may result from combining genetic and protein expression data. Using mechanism of action (MoA) studies, proteomic methods can show how medications or other substances interact with particular proteins in a cell. Understanding a compound's molecular weight at equilibrium is essential for medication development. Researchers can evaluate the effects of substances on several biological parameters at the cellular or subcellular level by combining high-content screening with proteomics and genomics. This method offers a more thorough understanding of chemical action.

Genetic differences unique to each patient that affect how they respond to treatment can be found using genomic data. By identifying patient-specific patterns of protein expression, proteomic data might help doctors customize a patient's course of care. Additionally, compounds that particularly target genes or genetic variations linked to disease can be found using genomic data. Information pertaining to the expression and function of particular proteins within cells can then be inferred from proteomic data.

Proteomic and genomic methods help with lead optimization by determining and defining the most promising compounds for additional research and development. Combining HTS with other analytical tools enables the evaluation of a compound's possible toxicity at the protein and genetic levels. By guaranteeing that the most pertinent targets and molecules are given priority, the coupling of these omics technologies with HTS speeds up the drug development process.

Through the integration of high-throughput screening with genomics and proteomics, researchers can leverage the copious amounts of data produced by these technologies to make better-informed decisions at every stage of the drug discovery process. By improving target selection, validation, and lead optimization, this integrative method eventually contributes to the development of safer and more potent medicines. By customizing medicines to each patient's own genetic and proteomic profile, it also advances the development of personalized medicine. This integrated approach will further transform the drug discovery area as long as technology and data analysis methodologies keep up their current pace of advancement (Bleicher et al. 2003; Blay et al. 2020).

11.6.3 Lead Compound Identification

Lead compound identification is a crucial stage in the drug development process that entails choosing a few promising compounds from a wider group of possible therapeutic candidates. The basis for additional research, development, optimization, and testing is provided by these lead molecules. Identification of lead compounds is a complicated procedure that uses a variety of scientific methods and tools. An outline of the main components is provided here (Noah 2010).

The process of identifying lead compounds frequently starts with a well-defined molecular target, such as a particular protein or genetic sequence linked to a disease. The choice of target is essential since it establishes the direction of the lead compound search. Compound libraries with a variety of chemical entities are maintained by biotechnology corporations, research institutes, and pharmaceutical companies. Natural substances, artificial molecules, and proprietary compounds can all be found in these libraries. These libraries are screened by researchers to find lead compounds with the right characteristics (Bogatcheva et al. 2011).

The high-throughput screening (HTS) technique enables the quick assessment of thousands or even millions of compounds for their efficacy against a particular target. These tests are carried out using automated robotic equipment, which makes HTS an extremely effective method. Based on a compound's structure, computer algorithms are used in silico, or virtual screening, to forecast a compound's possible binding affinity to a target. This method reduces the number of chemicals that need to be tested experimentally. Compounds that precisely fit into the target's active site can be designed thanks to knowledge of the target's three-dimensional structure, which is often gained by methods such as X-ray crystallography. This methodical approach to design produces lead compounds that have a better chance of succeeding.

In the fragment-based drug design approach, tiny, low molecular weight molecules are screened in order to find fragments that bind to the target. Lead compounds can be synthesized from these pieces (Doak et al. 2016). After possible lead compounds or "hits" are found, medicinal chemists carry out a process known as "hit-to-lead optimization". This entails modifying the substance chemically and conducting studies on the structure-activity relationship (SAR) to enhance its pharmacokinetics, selectivity, and binding affinity. In order to ascertain how lead compounds are absorbed, transported, metabolized, and removed by the body, their absorption, distribution, metabolism, and excretion (ADME) properties are evaluated. It is more probable for compounds with favourable ADME profiles to advance to the following phases of development (Edlin et al. 2012).

To make sure lead compounds are safe for human usage, they go through extensive toxicity testing. This entails examining the compound's safety margin and any negative effects. Pharmacokinetic (PK) and pharmacodynamic (PD) profiling investigations shed light on the lead compound's metabolism and in vivo interactions with the target. Predicting its treatment success requires an understanding of these factors (Eddershaw et al. 2000). To assess a compound's safety and effectiveness in a living creature, lead compounds are tested in animal models. The dose, toxicity, and efficacy for possible human application are all improved by these investigations.

A single lead molecule is selected to proceed to the next stage of drug development based on the results of all the aforementioned evaluations. We now refer to this chemical as the "development candidate". Businesses may submit patent applications at this stage in order to safeguard their lead compounds and other intellectual property (Raj et al. 2015). Lead compound identification is a highly dynamic and iterative process, and not all substances that enter the pipeline will eventually turn into effective medications. Finding the lead molecule with the most promising therapeutic potential while adhering to safety and regulatory requirements is the aim. After selecting a lead chemical, it moves on to the preclinical and clinical phases of development, ultimately resulting in the production of a novel pharmaceutical product that will benefit patients.

11.7 Drug Development and Clinical Trials

11.7.1 Translating Genomic and Proteomic Findings into Drug Candidates

One of the most important phases in the drug development process is the translation of genomic and proteomic data into drug candidates, where knowledge gained from the analysis of genes and proteins is applied to create possible medicinal agents. This multi-phase translation procedure is essential to the release of novel medications on the market. This is a summary of the process by which medication candidates are developed from genomes and proteomic research.

Finding targets in drug discovery is an essential first step in creating novel medications. Through validation, genes or genetic variants linked to certain diseases are found through genomic research and may be developed into therapeutic targets. Proteins implicated in disease pathways are identified by proteomic research; proteins with changed expression, activity, or post-translational changes may also be suitable targets for therapeutic intervention. To guarantee that a gene or protein is relevant to a disease, target validation is necessary. Numerous techniques, including functional genomic assays and knockout experiments in model species, are used to confirm genetic connections. Experiments showing the functional relevance of proteins in disease mechanisms justify the involvement of proteomics in disease (Minikel et al. 2020).

Compounds from various libraries are tested using high-throughput screening (HTS) to see if they have the ability to modify the verified target (Bleicher et al. 2003). Compounds with promising action are called hits. In hit-to-lead optimization, hits are optimized by medicinal chemistry to enhance their drug-like qualities. This procedure entails pharmacological profiling, structure-activity relationship (SAR) research, and chemical changes. To make sure lead compounds are safe for use in toxicology and safety evaluation, they go through a rigorous toxicological study. These investigations determine a safety margin and evaluate possible adverse effects. Studies on pharmacokinetics and pharmacodynamics (PK/PD) shed light on how a substance is digested and interacts with its target in living things (Eddershaw et al. 2000). For additional development, the top-performing lead compound—often referred to as the "development candidate"—is chosen.

Companies frequently submit patent applications during the lead compound identification and development process to safeguard their therapeutic prospects and intellectual property (Raj et al. 2015). Preclinical research is conducted on the development of candidates in animal models to assess safety and effectiveness. The efficacy, toxicity, and proper dosage are all determined in part by these investigations. In regulatory considerations, businesses must take regulatory requirements into account and make plans for the data and documents required for regulatory approvals (Siegel and Lakings 2008).

When preclinical research is successful, clinical trials start when the medication candidate is examined on living people. Multiple steps are included in clinical studies in order to assess safety and efficacy. The medication candidate may be granted regulatory permission for commercialization and distribution if clinical trials provide positive results. Following approval, the medication is subject to ongoing post-market surveillance to ensure its efficacy and safety (Lemmens and Gibson 2014).

It usually takes many years to convert the results of genomes and proteomic research into viable treatment possibilities. This process is difficult and resource-intensive. However, the field of drug discovery has undergone a revolution because of the application of genomes and proteomics, which has allowed for the development of more specialized and potent medicines for a range of illnesses. The ultimate objective is to introduce novel medications that enhance patients' health and quality of life.

11.7.2 Clinical Trial Design and Patient Selection

Patient selection and clinical trial design are essential steps in the medication development process. In order to produce relevant data, guarantee patient safety, and prove the safety and effectiveness of a novel medical intervention, proper trial design and patient selection are crucial. Here is a summary of these elements.

Clinical Trial Design

Research studies called clinical trials are used to evaluate the efficacy and safety of novel therapies or interventions for a range of illnesses or ailments. Typically, clinical trials are divided into four phases, each of which has a distinct function. Phase I trials feature a limited number of healthy volunteers or patients and concentrate on safety and dose. Phase II trials assess safety and efficacy in a broader patient population. Phase III trials are large-scale studies that evaluate treatment effectiveness and track side effects in a variety of patient populations. Phase IV trials, sometimes referred to as post-marketing trials, monitor a medication's effectiveness and safety even after it has been approved and is being used widely (Streiner and Norman 2009).

Several techniques are frequently used in clinical studies to guarantee the reliability and validity of the findings. By randomly allocating patients to treatment or control groups, randomization serves to reduce bias and ensures group comparability. Blinding is a strategy that reduces bias and produces more dependable results by keeping either the patient or the researcher (double-blind) or both (single-blind) uninformed of the treatment being provided. In a form of trial known as a placebo control, the treatment group is compared to a placebo group that is given an inactive drug. Determining the intervention's actual efficacy depends on its design. In a crossover design trial, patients get several treatments one after the other. This can lower the number of patients needed and allow for comparisons between subjects. A trial with an adaptive design enables the research design to be changed as data are gathered and analysed. This adaptability may increase the trial's effectiveness (Enck and Klosterhalfen 2019).

The selection of endpoints or the outcomes that are measured to evaluate treatment efficacy and safety is another crucial component of clinical trials. Clinical (such as survival or symptom relief), surrogate (such as biomarkers), or patient-reported (such as quality of life) endpoints are all possible. Selecting the right primary and secondary endpoints is essential for evaluating the intervention's risks and benefits. To guarantee that the study has the statistical power to identify significant effects, choosing the right sample size is also crucial. A number of variables, including the estimated impact size, the significance threshold, and the outcome's unpredictability, affect how big of a sample to use (Chow et al. 2017).

Patient Selection

A key component of clinical trial design is patient selection, which establishes the validity and generalizability of the findings. Patient selection is influenced by a number of factors, including informed consent, patient safety, patient preferences, demographics, disease severity, comorbidities, genetic and biomarker testing, performance status, and ethical issues (Ivy et al. 2010).

The features that patients must fulfil in order to be included in the trial and those that rule them out are outlined in the inclusion and exclusion criteria (exclusion). Age, gender, disease stage, medical history, and prior therapy are a few examples of criteria. The features of the patient population, such as age, gender, race, and ethnicity, are referred to as demographics. To understand how the treatment might influence different groups, it is important to ensure that the patient population is diverse. The degree of disease severity reflects the stage of the patient's disease or progression. To evaluate how the treatment affects the disease at different phases, patients might be categorized based on the severity of their conditions (Kirsten et al. 2016).

Comorbidities are additional illnesses that people may experience in addition to their main ailment. Comorbidities may have an impact on a patient's study eligibility and how the results are interpreted. Testing for genetic traits and biomarkers is one way to determine a patient's or a disease's molecular characteristics. Genetic and biomarker testing may be utilized in precision medicine studies to identify individuals who are more likely to respond well to the prescribed course of action. A patient's capacity to do everyday tasks and general state of well-being are reflected in their performance status (Duma et al. 2019).

Patient performance status might affect their eligibility and prognosis in a trial. It is commonly quantified using scales such as the Eastern Cooperative Oncology Group (ECOG) performance status. Patients must give their informed permission before they may partake in a clinical trial. This involves explaining the trial's goals, methods, possible dangers, and rewards. For patients to take part in a clinical trial, informed permission is required. Ensuring patient safety in clinical trials involves safeguarding them against danger or injury. It is crucial to make sure patients fulfil stringent safety requirements. For instance, women who are capable of bearing children may need to take pregnancy tests or use contraception (Jin et al. 2017).

A patient's values and expectations surrounding a clinical trial are reflected in their preferences and quality of life considerations. Patient selection and adherence to the study procedure may also be influenced by patient preferences and quality of life factors. Clinical trial procedures and patient selection are governed by ethical considerations. Protecting the rights and welfare of patients in clinical trials is the goal of ethical concepts such as informed consent, respect for autonomy, justice, and minimizing harm and maximizing benefit (Samuel et al. 2022).

Reliable and significant results can only be achieved by carefully planning clinical trials and selecting volunteers, all while reducing participant risks. These procedures are essential for expanding our understanding of medicine, enhancing patient care, and developing and introducing novel, secure, and efficient medicines. Furthermore, continual assessment and improvement of patient selection standards and trial designs optimize the medication development process and support the ongoing advancement of health care interventions.

11.7.3 Monitoring and Assessing Drug Efficacy and Safety

As a crucial and ongoing process, monitoring and evaluating drug efficacy and safety occur across a medication's whole life cycle, from preclinical development to clinical trials and post-marketing surveillance. This all-encompassing strategy guarantees that medications are safe for patients to take and are effective in addressing medical ailments. This is a summary of the main elements involved in tracking and evaluating the safety and efficacy of drugs (Burnier and Wuerzner 2015).

Drug development is a multi-phase, multi-stage procedure that starts with preclinical evaluation and ends with post-marketing surveillance. A medication candidate must go through rigorous preclinical research, including in vitro and in vivo studies, to evaluate its efficacy, mechanism of action, and potential toxicities before it can move on to human trials. Clinical trials are carried out in several stages, each with a distinct population and goal. Phase II evaluates efficacy and adverse effects in a wider patient population, while Phase I concentrates on safety and dosage in a small sample of healthy volunteers. Large-scale, randomized, controlled studies are conducted in Phase III to assess efficacy, track side effects, and verify safety across a range of patient populations. Finding out if a treatment has the desired therapeutic impact is the main objective of clinical trials. This is usually done by utilizing clinical endpoints, which include survival rates, symptom relief, or other disease-specific markers (Streiner and Norman 2009).

The medication's safety is closely monitored both during and after the clinical trial phase, as well as following regulatory approval. This includes keeping an eye on and disclosing any unanticipated reactions, side effects, and adverse events. Studies on pharmacokinetics and pharmacodynamics look at the drug's interactions with the body and target and how it is absorbed, distributed, metabolized, and removed from the body (Eddershaw et al. 2000).

During the post-marketing surveillance phase, the medicine is continuously observed in real-world settings to ensure safety and efficacy. These data support label changes and aid in the detection of uncommon or chronic adverse effects. Pharmacovigilance systems help with the continuous evaluation of drug safety by gathering, analysing, and reporting adverse events and side effects. A comparative effectiveness study is sometimes carried out to assess the safety and efficacy of new drugs in relation to current treatments (Moore et al. 2019).

For the purpose of making regulatory decisions, a drug's benefit-to-risk ratio must be continuously assessed. For a medication to be sold, it must offer more advantages than disadvantages. Regulatory control is provided by health organizations such as the FDA in the USA and the EMA in Europe, who make sure that medications fulfil strict safety and efficacy requirements. Adverse occurrences and side effects can be reported by patients and medical experts, which helps with the continuous monitoring of drug safety. The drug's label may be amended to reflect any new information regarding safety or efficacy, so patients and healthcare providers are kept informed (Lemmens and Gibson 2014).

The dynamic processes of monitoring and evaluating drug safety and efficacy seek to strike a balance between the possible hazards and the advantages of treatment. For patient safety and the ongoing advancement of health care, this continuous review is essential. It guarantees that medications in circulation continue to be safe, effective, and compliant with the most up-to-date scientific research.

11.8 Challenges and Future Directions

11.8.1 Ethical and Regulatory Considerations

The advancement of medical knowledge and the creation of novel medications depend on clinical trials. To preserve the rights and welfare of participants and guarantee the validity and reliability of the results, they also present ethical and legal issues that need to be resolved. Among the most important moral and legal concerns in clinical trials are discussed herein.

The ethical precept of informed consent states that patients and clinical trial participants must provide their assent after being fully informed about the research's goals, methods, possible dangers, and potential rewards. Additionally, regulatory bodies around the world, including the FDA in the United States and the EMA in Europe, have mandated it as a regulatory requirement (Lema et al. 2009). Patient safety is the most important ethical precept, according to which any possible dangers connected to the experimental medication must be thoroughly evaluated, reduced, and made public. Regulatory agencies are also required by law to closely monitor and assess the safety of pharmaceuticals during their development, and safety data must undergo a thorough evaluation before being approved for sale (Siegel and Lakings 2008; Burnier and Wuerzner 2015).

Clinical trial design ethical guideline states that clinical trials should be planned to minimize participant risk while providing an effective and efficient response to scientific issues. Regulatory bodies are also required by law to offer recommendations and evaluate trial procedures in order to guarantee that the trials are ethically sound and well-designed (Siegel and Lakings 2008; Ivy et al. 2010). Transparency and disclosure are the ethical precepts that complete disclosure of trial results—both favourable and unfavourable—is necessary to further medical understanding. Transparency in the pharmaceutical sector is becoming increasingly important, and regulatory bodies frequently demand the disclosure of clinical trial results (Siegel and Lakings 2008). Independent ethical review boards (IRBs) or ethics committees are responsible for supervising clinical trials and ensuring that the study design and participant protections are suitable. Regulatory authorities must also obtain IRB permission in order for clinical trials to move forward (Kim 2012).

Placebo use refers to the ethical precept that, in order to prevent withholding potentially helpful therapy, the use of placebos in clinical trials should be minimized, particularly when effective therapies are available. Regulations that prioritize patient welfare and outline the circumstances in which placebos may be used are also mandatory (Enck and Klosterhalfen 2019). The data integrity ethical concept states that academics and the pharmaceutical business have basic ethical duties to maintain data integrity and to present results honestly. Additionally, regulatory bodies are required to carefully examine the accuracy of data provided for drug approval; failure to do so may result in harsh penalties (Kim 2012). On top of that, regulatory bodies ought to have the power and means to thoroughly assess the quality, safety, and efficacy of drugs. Additionally, it is mandated by law that regulatory bodies, such as the FDA in the United States and the EMA in Europe, evaluate and approve new pharmaceuticals in accordance with scientific and ethical standards (Siegel and Lakings 2008).

Data sharing ethical tenet of data sharing holds that patients can gain from scientific advancements and clinical trial data can be shared while maintaining patient privacy and confidentiality. Additionally, several regulatory bodies increasingly mandate data sharing as a condition of their post-market transparency and surveillance initiatives. Trials involving vulnerable populations, such as children, pregnant women, or the elderly, must take extra precautions and provide special concern to these groups. It is also required by law that regulatory rules give disadvantaged populations certain protections and requirements (Jin et al. 2017).

Regulations are based on ethical concepts, and ethical values themselves are entwined with regulatory considerations. Ensuring that new medication development upholds the greatest standards of ethics, openness, and patient safety while benefiting patients and society is the aim. These factors are dynamic and ever-evolving, mirroring the rapidly changing fields of science, technology, and medicine.

11.8.2 Emerging Technologies in Genomics and Proteomics

At the front edge of biomedical research, the dynamic fields of genomics and proteomics provide profound insights into the molecular underpinnings of health and illness. Scientists may now study biological systems with never-before-seen accuracy and depth because of emerging tools in proteomics and genomics (Chung et al. 2007).

Genomics

The study of an organism's whole genetic make-up, or genomics, is one of the most fascinating areas of modern biology. The development of sequencing technology has transformed genomics by enabling unprecedentedly fast, accurate, and high-resolution analysis of DNA and RNA. Among the most innovative technologies for sequencing are discussed here.

Single-cell sequencing technique enables the examination of individual cells, exposing heterogeneity inside the cell and illuminating the roles played by many cell types in intricate biological processes, including development, cancer, and immunological responses. Companies such as Oxford Nanopore Technologies provide nanopore sequencing, which allows real-time study of DNA or RNA as it flows through a nanopore. Because of its great portability, this technology is useful for quick diagnostics and field applications.

Beyond DNA sequencing, the field of epigenomics is studying changes in epigenetic marks, such as DNA methylation and histone modifications, to gain a better understanding of how genes are regulated and how diseases are caused by these changes. Metagenomics technology makes it possible to sequence whole microbial populations in intricate settings. Grasping the significance of the microbiome in human health and disease and environmental ecosystems requires a grasp of metagenomics.

In the long-read sequencing method, longer DNA reads can be performed by technologies such as Oxford Nanopore and PacBio, which lessen the need for intricate genome assembly and make it easier to identify repetitive sequences and structural changes. These technological advancements are revolutionizing our understanding of biology and creating new opportunities for study and creativity (Berriman et al. 2007; Huang 2007).

Proteomics

A potent technology known as mass spectrometry imaging (MSI) blends the analytical powers of mass spectrometry with the spatial resolution of imaging methods. Through the simultaneous detection and localization of hundreds of proteins in tissue slices, MSI makes it possible to determine the distribution and quantity of these proteins across various cell types and regions. Understanding the molecular causes of diseases at the tissue level, such as cancer, neurodegeneration, and infection, is made possible thanks in large part to MSI.

The process of designing and producing new proteins and antibodies with desired features and functions is known as protein engineering. Protein engineering modifies the genetic coding and structure of proteins by using cutting-edge methods such as phage display and CRISPR-Cas9. In biotechnology, medicine, and agriculture, protein engineering has several uses, including the creation of novel medications, diagnostic tools, vaccines, enzymes, and biosensors.

A developing method called "top-down proteomics" examines intact proteins without first breaking them down or digesting them. Compared to the conventional bottom-up method, top-down proteomics has a number of benefits, including the preservation of data on protein isoforms and post-translational modifications (PTMs). PTMs and isoforms play a crucial role in controlling the function and activity of proteins and are frequently linked to a number of illnesses. A more thorough and precise description of the proteome can be obtained using top-down proteomics.

The goal of the cutting-edge field of single-molecule proteomics is to measure and describe individual proteins in intricate biological systems. The sensitivity, specificity, and dynamic range constraints of traditional bulk technologies can be circumvented by single-molecule approaches. Rare protein variations and heterogeneity that are otherwise hidden by averaging effects can be found by single-molecule proteomics. The temporal and spatial dynamics of protein interactions and conformational changes can also be captured by single-molecule proteomics.

One of the most important methods for understanding the relationships and functional roles of proteins in cells is protein-protein interaction mapping. Complex networks of protein-protein interactions control a range of biological functions, including transcription, translation, metabolism, and signalling. Protein-protein interaction mapping has the ability to identify biomarkers and new therapeutic targets in addition to revealing previously unidentified pathways and mechanisms underlying health and illness. Proximity labelling and cross-linking mass spectrometry are two emerging technologies that make it possible to explore protein-protein interactions comprehensively and efficiently.

The area of proteomics known as structural proteomics is devoted to figuring out the three-dimensional configurations of proteins and their complexes. Understanding the architecture of proteins can help us understand their relationships, roles, and modes of action. Structure-based medication design, which looks for or optimizes molecules that bind to particular protein targets, is also made easier by structural proteomics. The area of structural proteomics is progressing because of methods such as X-ray crystallography and cryo-electron microscopy (cryo-EM), which allow for the high-resolution imaging of huge and complicated protein complexes (Jeffery and Bogyo 2003; Thomford et al. 2018).

Integration

The integration of several forms of omics data, including transcriptomics, proteomics, metabolomics, and genomes, is one of the newer developments in omics research. The goal of these multi-omics techniques is to offer a more thorough and integrated understanding of intricate biological processes, including the onset of disease, the response to drugs, and cellular signalling. However, because of the high dimensionality, variability, and noise of the datasets, analysing and interpreting multi-omics data presents a number of difficulties (Waller et al. 2007).

Artificial intelligence (AI) and machine learning approaches are being used more and more in omics data analysis to address these issues, opening the door to the identification of unexpected patterns, fresh perspectives, and prediction models. By fusing omics data with clinical and phenotypic data, AI and machine learning can also help with customized medicine, drug discovery, and biomarker identification (Iskar et al. 2012).

Creating artificial genes, proteins, and cellular pathways is known as synthetic biology, and it is another exciting technique that is revolutionizing omics research. Customized medicinal medicines and diagnostic instruments that target particular biological pathways or mechanisms can be made via synthetic biology. The building of novel biological systems and functionalities made possible by synthetic biology has the potential to broaden the focus and use of omics research.

Scientific research is being accelerated, and health care is being revolutionized by emerging technologies in genomics and proteomics. These developments could lead to a better understanding of diseases, more effective medication development, and the emergence of customized medicine, in which a patient's therapy is based on their unique proteome and genetic profile. These technologies will surely change the course of clinical treatment and biomedical research as they develop further (Duarte et al. 2019).

11.8.3 Future Trends in Drug Discovery

Drug research is a dynamic subject that is always changing due to scientific discoveries, technological advancements, and the growing need for novel and efficient treatments. These are a few of the major developments in drug discovery that are worth watching.

Artificial intelligence and machine learning are two fields that are revolutionizing drug development through their ability to analyse large amounts of data, predict drug-target interactions, find possible therapeutic candidates, and optimize clinical trial designs. AI-driven medication discovery lowers costs and speeds up the process (Gupta et al. 2021). Integration of multi-omics information from many omics fields, including proteomics, metabolomics, transcriptomics, and genomes, allows for a more thorough knowledge of disease mechanisms. Using a multi-omics approach will help identify novel pharmacological targets and individualized treatment plans.

By allowing for the precise change of genes linked to specific diseases, CRISPR-Cas9 gene editing technology holds the potential to completely transform the medication discovery process. It opens up new possibilities for researching the genetic components of disease and creating tailored treatments (Zhang et al. 2021). There is also a growing body of research on the human microbiome. The development of microbiome-based therapies will result from the understanding that altering the microbiota can affect a broad spectrum of medical disorders. Thanks to developments in RNA interference and RNA editing technology, RNA-based therapeutics show great promise. These consist of RNA editing methods, RNAi-based therapies, and mRNA vaccinations (Qadir et al. 2020).

The 3D bioprinting and organoid technologies have made it possible to create intricate, physiologically accurate in vitro models for drug testing. Preclinical testing can be enhanced by this technology, which can lessen the need for animal models (Rae et al. 2021). Post-marketing surveillance will be improved, and treatment decisions will be informed by the use of real-world data evidence, such as wearables, electronic health records, and patient-reported data. As immunotherapy treatments progress, they provide new avenues for the treatment of autoimmune disorders and cancer. Examples of these medicines include immune checkpoint inhibitors and CAR-T cell therapy (Di Trani et al. 2022).

By drug repurposing, finding new uses for currently available pharmaceuticals is becoming more efficient, affordable, and quick with the use of computer methods and big data analysis. Nanotechnology is enabling tailored therapy, enhancing drug solubility, and creating new paths for drug administration. Additionally, it makes the advancement of customized medicine made possible by nanomedicine techniques easier. By taking a step into patient-centric drug discovery, the process of developing new drugs is increasingly including patients. Drug development and access are increasingly influenced by patient advocacy groups, real-world evidence, and patient-reported results.

By using algorithms in AI-driven drug synthesis to forecast effective synthetic routes for drug molecules, drug synthesis can be accelerated and the time it takes to introduce new pharmaceuticals to the market can be decreased (Iskar et al. 2012). In response to the quickly changing environment, regulatory bodies are putting more flexible strategies into place to expedite the licensing of drugs and simplify the research and development of ground-breaking treatments by regulatory adaptation methods. All of these upcoming developments in drug discovery are pointing towards a more patient-centred, focused, and effective method of creating new treatments. A new age in health care is expected to develop as a result of the convergence of technology, data, and scientific knowledge, which promises to uncover creative solutions to some of the most difficult health issues. These solutions will enhance patient outcomes.

11.9 Conclusion

In conclusion, the symbiotic integration of genomics and proteomics has unequivocally revolutionized the landscape of drug discovery. The wealth of information derived from genomic and proteomic analyses has unveiled intricate details of cellular functions, disease mechanisms, and potential therapeutic targets. Genomics, by decoding the blueprint of an individual's genetic make-up, has facilitated the identification of genetic variations associated with diseases, enabling the development of targeted therapies. Concurrently, proteomics has elucidated the dynamic and complex protein networks governing cellular processes, offering insights into protein structures and functions crucial for drug design.

The collaborative efforts of genomics and proteomics have expedited the identification and validation of biomarkers, paving the way for personalized medicine. The ability to tailor treatments based on individual genetic and proteomic profiles holds immense promise for enhancing therapeutic efficacy while minimizing adverse effects. However, amidst these remarkable advancements, challenges such as data integration, standardization, and ethical considerations persist. The evolving field of multi-omics approaches demands continued interdisciplinary collaboration and technological innovation to harness its full potential. As we navigate the intricate web of genomics and proteomics in drug discovery, it is evident that this holistic approach marks a paradigm shift, offering unparalleled opportunities to uncover novel drug targets and revolutionize the development of innovative and more effective therapeutic interventions. The future of drug discovery is undoubtedly intertwined with the continued exploration and refinement of genomics and proteomics, propelling the pharmaceutical industry into an era of precision medicine and personalized treatment strategies.

References

Abdulmawjood B, Costa B, Roma-Rodrigues C, Baptista PV, Fernandes AR (2021) Genetic biomarkers in chronic myeloid leukemia: what have we learned so far? Int J Mol Sci 22(22):12516
Alchakee A, Ahmed M, Eldohaji L, Alhaj H, Saber-Ayad M (2022) Pharmacogenomics in psychiatry practice: the value and the challenges. Int J Mol Sci 23(21):13485
Alföldi J, Lindblad-Toh K (2013) Comparative genomics as a tool to understand evolution and disease. Genome Res 23(7):1063–1068
Amarasinghe SL, Su S, Dong X, Zappia L, Ritchie ME, Gouil Q (2020) Opportunities and challenges in long-read sequencing data analysis. Genome Biol 21(1):1–6
Baloch AR, Feugang JM, Rodríguez-Osorio N (2023) Genomic and epigenomic applications in animal and veterinary sciences. Front Vet Sci 10:1167079
Berriman M, Lustigman S, McCarter JP (2007) Genomics and emerging drug discovery technologies. Expert Opin Drug Discov 2(sup1):S83–S89
Blay V, Tolani B, Ho SP, Arkin MR (2020) High-throughput screening: today's biochemical and cell-based approaches. Drug Discov Today 25(10):1807–1821
Bleicher KH, Böhm HJ, Müller K, Alanine AI (2003) Hit and lead generation: beyond high-throughput screening. Nat Rev Drug Discov 2(5):369–378
Bogatcheva E, Hanrahan C, Nikonenko B, De Los Santos G, Reddy V, Chen P, Barbosa F, Einck L, Nacy C, Protopopova M (2011) Identification of SQ609 as a lead compound from a library of dipiperidines. Bioorg Med Chem Lett 21(18):5353–5357
Burnier M, Wuerzner G (2015) Drug adherence monitoring in clinical trials: a necessity for a correct assessment of the efficacy and safety of antihypertensive therapies. J Hypertens 33(12):2395–2398
Chahrour O, Cobice D, Malone J (2015) Stable isotope labelling methods in mass spectrometry-based quantitative proteomics. J Pharm Biomed Anal 113:2
Chen S, Jiang H, Cao Y, Wang Y, Hu Z, Zhu Z, Chai Y (2016) Drug target identification using network analysis: taking active components in Sini decoction as an example. Sci Rep 6(1):24245
Chiu CY, Miller SA (2019) Clinical metagenomics. Nat Rev Genet 20(6):341–355
Chow SC, Shao J, Wang H, Lokhnygina Y (2017) Sample size calculations in clinical research. CRC, Boca Raton, FL
Choy CT, Wong CH, Chan SL (2019) Embedding of genes using cancer gene expression data: biological relevance and potential application on biomarker discovery. Front Genet 9:682
Chung CH, Levy S, Chaurand P, Carbone DP (2007) Genomics and proteomics: emerging technologies in clinical cancer research. Crit Rev Oncol Hematol 61(1):1–25
Clark KM, Jenkins JL, Fedoriw N, Dumont ME (2017) Human CaaX protease ZMPSTE 24 expressed in yeast: structure and inhibition by HIV protease inhibitors. Protein Sci 26(2):242–257
Codony-Servat J, Codony-Servat C, Cardona AF, Giménez-Capitán A, Drozdowskyj A, Berenguer J, Bracht JW, Ito M, Karachaliou N, Rosell R (2019) Cancer stem cell biomarkers in egfr-mutation–positive non–small-cell lung cancer. Clin Lung Cancer 20(3):167–177

Cominetti O, Núñez Galindo A, Corthésy J, Oller Moreno S, Irincheeva I, Valsesia A, Astrup A, Saris WH, Hager J, Kussmann M, Dayon L (2016) Proteomic biomarker discovery in 1000 human plasma samples with mass spectrometry. J Proteome Res 15(2):389–399
Corsi GI, Gadekar VP, Gorodkin J, Seemann SE (2022) CRISPRroots: on-and off-target assessment of RNA-seq data in CRISPR–Cas9 edited cells. Nucleic Acids Res 50(4):e20
Daly AK (2013) Pharmacogenomics of adverse drug reactions. Genome Med 5:1–2
Datta A, Kim H, McGee L, Johnson AE, Talwar S, Marugan J, Southall N, Hu X, Lal M, Mondal D, Ferrer M (2018) High-throughput screening identified selective inhibitors of exosome biogenesis and secretion: a drug repurposing strategy for advanced cancer. Sci Rep 8(1):8161
Dean-Colomb W, Esteva FJ (2008) Her2-positive breast cancer: herceptin and beyond. Eur J Cancer 44(18):2806–2812
Di Trani CA, Fernandez-Sendin M, Cirella A, Segués A, Olivera I, Bolaños E, Melero I, Berraondo P (2022) Advances in mRNA-based drug discovery in cancer immunotherapy. Expert Opin Drug Discov 17(1):41–53
Doak BC, Norton RS, Scanlon MJ (2016) The ways and means of fragment-based drug design. Pharmacol Ther 167:28–37
Du Y, Li X, Niu Q, Mo X, Qui M, Ma T, Kuo CJ, Fu H (2020) Development of a miniaturized 3D organoid culture platform for ultra-high-throughput screening. J Mol Cell Biol 12(8):630–643
Duarte Y, Márquez-Miranda V, Miossec MJ, González-Nilo F (2019) Integration of target discovery, drug discovery and drug delivery: a review on computational strategies. Wiley Interdiscip Rev Nanomed Nanobiotechnol 11(4):e1554
Duma N, Kothadia SM, Azam TU, Yadav S, Paludo J, Vera Aguilera J, Gonzalez Velez M, Halfdanarson TR, Molina JR, Hubbard JM, Go RS (2019) Characterization of comorbidities limiting the recruitment of patients in early phase clinical trials. Oncologist 24(1):96–102
Eddershaw PJ, Beresford AP, Bayliss MK (2000) ADME/PK as part of a rational approach to drug discovery. Drug Discov Today 5(9):409–414
Edlin CD, Morgans G, Winks S, Duffy S, Avery VM, Wittlin S, Waterson D, Burrows J, Bryans J (2012) Identification and in-vitro ADME assessment of a series of novel anti-malarial agents suitable for hit-to-lead chemistry. ACS Med Chem Lett 3(7):570–573
Enck P, Klosterhalfen S (2019) Placebos and the placebo effect in drug trials. Handb Exp Pharmacol 260:399–431
Filiou MD, Turck CW, Martins-de-Souza D (2011) Quantitative proteomics for investigating psychiatric disorders. Proteomics Clin Appl 5(1–2):38–49
Fox S, Farr-Jones S, Sopchak L, Boggs A, Nicely HW, Khoury R, Biros M (2006) High-throughput screening: update on practices and success. J Biomol Screen 11(7):864–869
Gevaert K, Vandekerckhove J (2000) Protein identification methods in proteomics. Electrophoresis 21(6):1145–1154
Gupta R, Srivastava D, Sahu M, Tiwari S, Ambasta RK, Kumar P (2021) Artificial intelligence to deep learning: machine intelligence approach for drug discovery. Mol Divers 25:1315–1360
Harris RK, Wasylishen RE, Duer MJ (eds) (2009) NMR crystallography. Wiley, New York
He YD (2006) Genomic approach to biomarker identification and its recent applications. Cancer Biomark 2(3–4):103–133
Huang Z (2007) Drug discovery research: new frontiers in the post-genomic era. John Wiley & Sons, New York
Huizar CC, Raphael I, Forsthuber TG (2020) Genomic, proteomic, and systems biology approaches in biomarker discovery for multiple sclerosis. Cell Immunol 358:104219
Ion A, Popa IM, Papagheorghe LM, Lisievici C, Lupu M, Voiculescu V, Caruntu C, Boda D (2016) Proteomic approaches to biomarker discovery in cutaneous T-cell lymphoma. Dis Markers 2016:9602472
Iskar M, Zeller G, Zhao XM, van Noort V, Bork P (2012) Drug discovery in the age of systems biology: the rise of computational approaches for data integration. Curr Opin Biotechnol 23(4):609–616

Issaq HJ, Veenstra TD (2008) Two-dimensional polyacrylamide gel electrophoresis (2D-PAGE): advances and perspectives. BioTechniques 44(5):697–700
Ivy SP, Siu LL, Garrett-Mayer E, Rubinstein L (2010) Approaches to phase 1 clinical trial design focused on safety, efficiency, and selected patient populations: a report from the clinical trial design task force of the national cancer institute investigational drug steering committee. Clin Cancer Res 16(6):1726–1736
Jeffery DA, Bogyo M (2003) Chemical proteomics and its application to drug discovery. Curr Opin Biotechnol 14(1):87–95
Jiang Z, Zhou Y (2005) Using bioinformatics for drug target identification from the genome. Am J Pharmacogenomics 5:387–396
Jin S, Pazdur R, Sridhara R (2017) Re-evaluating eligibility criteria for oncology clinical trials: analysis of investigational new drug applications in 2015. J Clin Oncol 35(33):3745
Johnson DE, Shaw DJ (2023) The dangerous use of genetic information. J Inf Commun Ethics Soc 21:533
Kaiafa G, Veneti S, Polychronopoulos G, Pilalas D, Daios S, Kanellos I, Didangelos T, Pagoni S, Savopoulos C (2021) Is HbA1c an ideal biomarker of well-controlled diabetes? Postgrad Med J 97(1148):380–383
Kharchenko PV (2021) The triumphs and limitations of computational methods for scRNA-seq. Nat Methods 18(7):723–732
Kim WO (2012) Institutional review board (IRB) and ethical issues in clinical research. Korean J Anesthesiol 62(1):3–12
Kirsten N, Bulai Livideanu C, Richard MA, Konstantinou MP, Khemis A, Balluteaud C, Goujon C, Beylot-Barry M, Paul C, French Psoriasis Research Group (2016) Inclusion and exclusion criteria in phase III trials with systemic agents in psoriasis: the external validity of drug development. Br J Dermatol 175(3):636–638
Knudsen DR, Raman P, Ettefa F, De Ravin L, Jose AM (2023) Target-specific requirements for RNA interference can be explained by a single regulatory network. bioRxiv 2023:527351
Koscielny G, An P, Carvalho-Silva D, Cham JA, Fumis L, Gasparyan R, Hasan S, Karamanis N, Maguire M, Papa E, Pierleoni A (2017) Open targets: a platform for therapeutic target identification and validation. Nucleic Acids Res 45(D1):D985–D994
Kramer R, Cohen D (2004) Functional genomics to new drug targets. Nat Rev Drug Discov 3(11):965–972
Kraus VB (2018) Biomarkers as drug development tools: discovery, validation, qualification and use. Nat Rev Rheumatol 14(6):354–362
Kumar D, Bansal G, Narang A, Basak T, Abbas T, Dash D (2016) Integrating transcriptome and proteome profiling: strategies and applications. Proteomics 16(19):2533–2544
Lacombe O, Videau O, Chevillon D, Guyot AC, Contreras C, Blondel S, Nicolas L, Ghettas A, Bénech H, Thevenot E, Pruvost A (2011) In vitro primary human and animal cell-based blood–brain barrier models as a screening tool in drug discovery. Mol Pharm 8(3):651–663
Lee JW, Weiner RS, Sailstad JM, Bowsher RR, Knuth DW, O'Brien PJ, Fourcroy JL, Dixit R, Pandite L, Pietrusko RG, Soares HD (2005) Method validation and measurement of biomarkers in nonclinical and clinical samples in drug development: a conference report. Pharm Res 22:499–511
Lee JM, Hays JL, Annunziata CM, Noonan AM, Minasian L, Zujewski JA, Yu M, Gordon N, Ji J, Sissung TM, Figg WD (2014) Phase I/Ib study of olaparib and carboplatin in BRCA1 or BRCA2 mutation-associated breast or ovarian cancer with biomarker analyses. J Natl Cancer Inst 106(6):dju089
Lema VM, Mbondo M, Kamau EN (2009) Informed consent for clinical trials: a review. East Afr Med J 86(3):133–142
Lemmens T, Gibson S (2014) Decreasing the data deficit: improving post-market surveillance in pharmaceutical regulation. McGill Law J 59(4):943–988

Lhoumaud P, Sethia G, Izzo F, Sakellaropoulos T, Snetkova V, Vidal S, Badri S, Cornwell M, Di Giammartino DC, Kim KT, Apostolou E (2019) EpiMethylTag: simultaneous detection of ATAC-seq or ChIP-seq signals with DNA methylation. Genome Biol 20(1):1–2
Li XH, Li C, Xiao ZQ (2011) Proteomics for identifying mechanisms and biomarkers of drug resistance in cancer. J Proteome 74(12):2642–2649
Li Z, Adams RM, Chourey K, Hurst GB, Hettich RL, Pan C (2012) Systematic comparison of label-free, metabolic labeling, and isobaric chemical labeling for quantitative proteomics on LTQ Orbitrap Velos. J Proteome Res 11(3):1582–1590
Lin Y, Mehta S, Küçük-McGinty H, Turner JP, Vidovic D, Forlin M, Koleti A, Nguyen DT, Jensen LJ, Guha R, Mathias SL (2017) Drug target ontology to classify and integrate drug discovery data. J Biomed Semantics 8:1–6
Liu L, Li Y, Li S, Hu N, He Y, Pong R, Lin D, Lu L, Law M (2012) Comparison of next-generation sequencing systems. J Biomed Biotechnol 2012:1
Mäbert K, Cojoc M, Peitzsch C, Kurth I, Souchelnytskyi S, Dubrovska A (2014) Cancer biomarker discovery: current status and future perspectives. Int J Radiat Biol 90(8):659–677
Manjasetty BA, Büssow K, Panjikar S, Turnbull AP (2012) Current methods in structural proteomics and its applications in biological sciences. 3Biotech 2:89–113
Mathai N, Chen Y, Kirchmair J (2020) Validation strategies for target prediction methods. Brief Bioinform 21(3):791–802
Matthews H, Hanison J, Nirmalan N (2016) "Omics"-informed drug and biomarker discovery: opportunities, challenges and future perspectives. Proteomes 4(3):28
Mayr LM, Bojanic D (2009) Novel trends in high-throughput screening. Curr Opin Pharmacol 9(5):580–588
McEwen AB, Henson CM, Wood SG (2014) Quantitative whole-body autoradiography, LC–MS/MS and MALDI for drug-distribution studies in biological samples: the ultimate matrix trilogy. Bioanalysis 6(3):377–391
Minikel EV, Karczewski KJ, Martin HC, Cummings BB, Whiffin N, Rhodes D, Alföldi J, Trembath RC, van Heel DA, Daly MJ (2020) Evaluating drug targets through human loss-of-function genetic variation. Nature 581(7809):459–464
Moore N, Berdaï D, Blin P, Droz C (2019) Pharmacovigilance–the next chapter. Therapies 74(6):557–567
Ni YH, Ding L, Hu QG, Hua ZC (2015) Potential biomarkers for oral squamous cell carcinoma: proteomics discovery and clinical validation. Proteomics Clin Appl 9(1–2):86–97
Noah JW (2010) New developments and emerging trends in high-throughput screening methods for lead compound identification. Int J High Throughput Screen 1:141–149
Pearson AD, Herold R, Rousseau R, Copland C, Bradley-Garelik B, Binner D, Capdeville R, Caron H, Carleer J, Chesler L, Geoerger B (2016) Implementation of mechanism of action biology-driven early drug development for children with cancer. Eur J Cancer 62:124–131
Qadir MI, Bukhat S, Rasul S, Manzoor H, Manzoor M (2020) RNA therapeutics: identification of novel targets leading to drug discovery. J Cell Biochem 121(2):898–929
Rae C, Amato F, Braconi C (2021) Patient-derived organoids as a model for cancer drug discovery. Int J Mol Sci 22(7):3483
Raj GM (2019) Pharmacogenetics, pharmacogenomics, and personalized medicine. In: Introduction to basics of pharmacology and toxicology: Volume 1: general and molecular pharmacology: principles of drug action. Springer, Cham, pp 235–259
Raj GM, Priyadarshini R, Mathaiyan J (2015) Drug patents and intellectual property rights. Eur J Clin Pharmacol 71:403–409
Rao VS, Srinivas K (2011) Modern drug discovery process: an in silico approach. J Bioinform Seq Anal 2(5):89–94
Rauniyar N, Yates JR III (2014) Isobaric labeling-based relative quantification in shotgun proteomics. J Proteome Res 13(12):5293–5309
Roy A (2018) Early probe and drug discovery in academia: a minireview. High Throughput 7(1):4
Saini S (2016) PSA and beyond: alternative prostate cancer biomarkers. Cell Oncol 39:97–106

Sali A, Glaeser R, Earnest T, Baumeister W (2003) From words to literature in structural proteomics. Nature 422(6928):216–225

Samuel JN, Booth CM, Eisenhauer E, Brundage M, Berry SR, Gyawali B (2022) Association of quality-of-life outcomes in cancer drug trials with survival outcomes and drug class. JAMA Oncol 8(6):879–886

Shah UJ, Nasiruddin M, Dar SA, Khan MK, Akhter MR, Singh N, Rabaan AA, Haque S (2020) Emerging biomarkers and clinical significance of HPV genotyping in prevention and management of cervical cancer. Microb Pathog 143:104131

Shi T, Song E, Nie S, Rodland KD, Liu T, Qian WJ, Smith RD (2016) Advances in targeted proteomics and applications to biomedical research. Proteomics 16(15–16):2160–2182

Siegel EB, Lakings DB (2008) Regulatory considerations. Preclinical development handbook: toxicology. Wiley, New York, pp 945–963

Sleno L, Emili A (2008) Proteomic methods for drug target discovery. Curr Opin Chem Biol 12(1):46–54

Stoevesandt O, Taussig MJ, He M (2009) Protein microarrays: high-throughput tools for proteomics. Expert Rev Proteomics 6(2):145–157

Streiner DL, Norman GR (2009) Drug trial phases. Commun Oncol 6(1):36–40

Struk S, Jacobs A, Sánchez Martín-Fontecha E, Gevaert K, Cubas P, Goormachtig S (2019) Exploring the protein–protein interaction landscape in plants. Plant Cell Environ 42(2):387–409

Sydor JR, Nock S (2003) Protein expression profiling arrays: tools for the multiplexed high-throughput analysis of proteins. Proteome Sci 1(1):1–7

Theilgaard-Moench K, Boultwood J, Ferrari S, Giannopoulos K, Hernandez-Rivas JM, Kohlmann A, Morgan M, Porse B, Tagliafico E, Zwaan CM, Wainscoat J (2011) Gene expression profiling in MDS and AML: potential and future avenues. Leukemia 25(6):909–920

Thomford NE, Senthebane DA, Rowe A, Munro D, Seele P, Maroyi A, Dzobo K (2018) Natural products for drug discovery in the 21st century: innovations for novel drug discovery. Int J Mol Sci 19(6):1578

Uitdehaag JC, Verkaar F, Alwan H, de Man J, Buijsman RC, Zaman GJ (2012) A guide to picking the most selective kinase inhibitor tool compounds for pharmacological validation of drug targets. Br J Pharmacol 166(3):858–876

Vasan RS (2006) Biomarkers of cardiovascular disease: molecular basis and practical considerations. Circulation 113(19):2335–2362

Vemula D, Singothu S, Bhandari V (2023) Concepts in pharmacogenomics: tools and applications. In: Recent advances in pharmaceutical innovation and research, vol 12. Springer Nature Singapore, Singapore, pp 41–76

Waduge P, Tian H, Webster KA, Li W (2022) Profiling disease-selective drug targets: from proteomics to ligandomics. Drug Discov Today 28:103430

Waller CL, Shah A, Nolte M (2007) Strategies to support drug discovery through integration of systems and data. Drug Discov Today 12(15–16):634–639

Wang Y, Chiu JF, He QY (2005) Proteomics in computer-aided drug design. Curr Comput Aided Drug Des 1(1):43–52

Wang T, Liu XH, Guan J, Ge S, Wu MB, Lin JP, Yang LR (2019) Advancement of multi-target drug discoveries and promising applications in the field of Alzheimer's disease. Eur J Med Chem 169:200–223

Wang Y, Yang Y, Chen S, Wang J (2021) Deep DRK: a deep learning framework for drug repurposing through kernel-based multi-omics integration. Brief Bioinform 22(5):bbab048

Weeke P, Roden DM (2014) Applied pharmacogenomics in cardiovascular medicine. Annu Rev Med 65:81–94

Wheeler E, Huang N, Bochukova EG, Keogh JM, Lindsay S, Garg S, Henning E, Blackburn H, Loos RJ, Wareham NJ, O'Rahilly S (2013) Genome-wide SNP and CNV analysis identifies common and low-frequency variants associated with severe early-onset obesity. Nat Genet 45(5):513–517

Wildey MJ, Haunso A, Tudor M, Webb M, Connick JH (2017) High-throughput screening. Annu Rep Med Chem 50:149–195

Wilson BJ, Nicholls SG (2015) The human genome project, and recent advances in personalized genomics. Risk Manag Healthc Policy 8:9–20

Wilson FR, Coombes ME, Brezden-Masley C, Yurchenko M, Wylie Q, Douma R, Varu A, Hutton B, Skidmore B, Cameron C (2018) Herceptin®(trastuzumab) in HER2-positive early breast cancer: a systematic review and cumulative network meta-analysis. Syst Rev 7(1):1–7

Wright PR, Richter AS, Papenfort K, Mann M, Vogel J, Hess WR, Backofen R, Georg J (2013) Comparative genomics boosts target prediction for bacterial small RNAs. Proc Natl Acad Sci 110(37):E3487–E3496

Xu W, Comhair SA, Chen R, Hu B, Hou Y, Zhou Y, Mavrakis LA, Janocha AJ, Li L, Zhang D, Willard BB (2019) Integrative proteomics and phosphoproteomics in pulmonary arterial hypertension. Sci Rep 9(1):18623

Yang Y, Adelstein SJ, Kassis AI (2012) Target discovery from data mining approaches. Drug Discov Today 17:S16–S23

Yugandhar K, Gupta S, Yu H (2019) Inferring protein-protein interaction networks from mass spectrometry-based proteomic approaches: a mini-review. Comput Struct Biotechnol J 17:805–811

Zaimy MA, Saffarzadeh N, Mohammadi A, Pourghadamyari H, Izadi P, Sarli A, Moghaddam LK, Paschepari SR, Azizi H, Torkamandi S, Tavakkoly-Bazzaz J (2017) New methods in the diagnosis of cancer and gene therapy of cancer based on nanoparticles. Cancer Gene Ther 24(6):233–243

Zhang HW, Lv C, Zhang LJ, Guo X, Shen YW, Nagle DG, Zhou YD, Liu SH, Zhang WD, Luan X (2021) Application of omics-and multi-omics-based techniques for natural product target discovery. Biomed Pharmacother 141:111833

Zheng S, Dharssi S, Wu M, Li J, Lu Z (2019) Text mining for drug discovery. Bioinformatics and drug. Discovery:231–252

Zhou T, Yao J, Liu Z (2017) Gene ontology, enrichment analysis, and pathway analysis. In: Bioinformatics in aquaculture: principles and methods. Wiley, New York, pp 150–168

Chapter 12
Immunoinformatics for the Diagnosis and Monitoring of Autoimmune Diseases

Shalesh Gangwar, Neha Sharma, and Devinder Toor

Abstract A complicated and diverse set of illnesses known as autoimmune diseases occur when the immune system attacks healthy cells and tissues, causing tissue damage and chronic inflammation. The diversity of clinical manifestations and a dearth of specific biomarkers for autoimmune disease impose difficulty in the diagnosis and surveillance of it. A fresh approach known as "immunoinformatics" has been developed to assist with these difficulties. The current scenario of diagnosis and monitoring of autoimmune diseases poses several disadvantages such as difficulty in accurate diagnosis and often relies on expensive and time-consuming laboratory techniques which makes the idea of personalized treatments a distant reality. On the other hand, immunoinformatics serves as a potential and improved alternative to traditional approaches as it leverages computational techniques to analyse large-scale biological data, helping identify disease-specific biomarkers and prediction of immune system responses. This enables more precise diagnosis as well as assists in the planning of more personalized treatment strategies for each particular case. Immunoinformatics also aids in monitoring disease progression through continuous data analysis, allowing for adjustments to treatment plans. In this chapter, we have explored the ways in which cutting-edge digital technologies might be utilized to identify disease-specific markers for the treatment and diagnosis of autoimmune diseases.

Keywords Autoimmune diseases · Immunoinformatics · Molecular mimicry · Epitope prediction · Vaccine design · Multiple sclerosis

---

S. Gangwar
Department of Computer Science, Jamia Millia Islamia, New Delhi, India

N. Sharma · D. Toor (✉)
Amity Institute of Virology and Immunology, Amity University Uttar Pradesh, Noida, India
e-mail: dtoor@amity.edu

S. Bose et al. (eds.), *Concepts in Pharmaceutical Biotechnology and Drug Development*, Interdisciplinary Biotechnological Advances,
https://doi.org/10.1007/978-981-97-1148-2_12

12.1 Introduction

Autoimmune illnesses are a major cause of concern for the global population, affecting 14% of the whole population of the United States (NIEHS n.d.), 10% of the population of the United Kingdom (Conrad et al. 2023), and 5% of the total population of New Zealand and Australia (ASCIA 2023). The lack of identifiable biomarkers and clinical heterogeneity provides considerable obstacles in effectively identifying and monitoring autoimmune diseases (Chatanaka et al. 2022).

The existing diagnostic landscape for autoimmune illnesses comes with difficulties, such as the reliance on expensive and time-consuming laboratory procedures. This condition makes personalized treatments difficult to implement, contributing to a gap in addressing the particular nature of many diseases. However, "immunoinformatics", a novel technique meant to address these diagnostic complications, represents a prospective paradigm change.

Immunoinformatics, which uses computer tools to examine large amounts of biological data, has the potential to be a game changer. This method aids in the identification of disease-specific biomarkers and the prediction of immune system responses, allowing for more exact diagnoses and the development of personalized treatment plans customized to each patient's unique characteristics. Notably, immunoinformatics' continuous data analysis capabilities facilitate ongoing disease progression monitoring, allowing for dynamic modifications to treatment programs based on current information (Brusic and Petrovsky 2005).

In this chapter, we explore the fundamental aspects of autoimmune diseases, provide insights into the core principles of immunoinformatics, investigate its application in diagnosis, discuss its role in disease monitoring, highlight case studies and applications, and finally address the future directions and challenges within this transformative field.

12.2 Understanding Autoimmune Diseases

A complicated and diverse set of illnesses known as autoimmune diseases arise when our immune system targets healthy cells and tissues, causing tissue damage and chronic inflammation. These sets of diseases can affect various organs and organ systems, leading to varying complications. The underlying causes of many autoimmune diseases are still unknown, but an association of genetic, environmental, and immunological variables is thought to have a significant role in their progression and development, making their association a key for the effective management and treatment of these diseases.

The aetiology of autoimmune diseases also involves an interaction between genetic predisposition and epigenetic factors. A substantial risk factor is genetic predisposition, as demonstrated by the familial clustering of some autoimmune diseases, including systemic lupus erythematosus (SLE) and rheumatoid arthritis (Anaya et al. 2016). Studies have identified various human leukocyte antigens (HLA) and various non-HLA genes associated with various autoimmune diseases (Vojdani 2014).

However, autoimmunity cannot be triggered solely by the presence of these genetic markers, highlighting the significance of environmental variables. The onset and development of autoimmune reactions have been linked to environmental variables, including infections, hormone fluctuations, and exposure to specific chemicals. For example, molecular mimicry (Fig. 12.1), the process by which viral antigens resemble self-antigens and trigger cross-reactive immune responses, has been used to link viral infections to the onset of type 1 diabetes (Fujinami and Oldstone 1989). Moreover, changes in the makeup of the gut microbiota have been associated with the development of various autoimmune diseases, highlighting the importance of the gut-immune axis in preserving immunological homeostasis (Wu and Wu 2012).

Regulatory T cells (Tregs), a subset of T cells, are crucial in controlling the immune system and keeping it from attacking its own tissues of the body (Sakaguchi et al. 2008). In addition to being generated from naive T cells in the periphery, the thymus, a gland in the chest, also plays a role in Treg production (Sakaguchi et al. 2008; Baecher-Allan 2004).

The ability to remain tolerant and reject self-antigens requires Tregs. Tregs are necessary to prevent the immune system from attacking the body's own tissues,

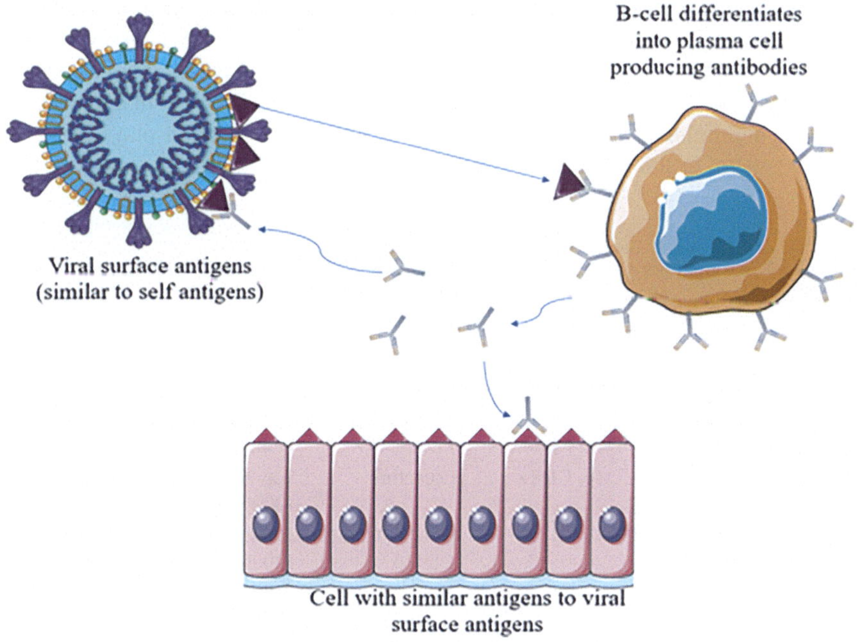

Fig. 12.1 Illustration of molecular mimicry

which can result in autoimmune illnesses like rheumatoid arthritis (Hirota et al. 2007), multiple sclerosis, and type 1 diabetes (Szanya et al. 2002). Additionally, Tregs are involved in the suppression of hyperactive immunological reactions to external antigens, including those triggered by infections or allergic reactions (Belkaid et al. 2002). Tissue damage might result from the immune system overreacting and not having Tregs. Immune tolerance maintenance and the prevention of autoimmune disorders depend on regulatory T cells or Tregs. Additionally, they aid in reducing overreactions of the immune system to foreign antigens. Numerous inflammatory and immunological disorders can result from Treg imbalance (Sakaguchi et al. 2008).

Also, in case of infections like tuberculosis and cold sores brought on by herpesviruses, the immune response does not always result in the total clearance of infections. In vulnerable people, these infections can colonize for an extended period of time. In these circumstances, studies of antigen-specific immune cells show significant distinctions between both naive and effector cells. There are two main ways in which aggressive immune responses are suppressed in these situations. First, cells that are actively combating the infection display an intrinsic shutdown mechanism, which is identified by a rise in the surface expression of co-inhibitory receptors (Table 12.1). These receptors send out signals that stop attacking processes like cell death, which stops local tissue damage and elimination of pathogen-infected cells. This balance allows for the tolerance of infected cells while minimizing further harm.

The second type of regulation occurs extrinsically through the expansion of antigen-specific anti-inflammatory cytokines released by these cells to limit immunity and promote tissue repair. Tissue antigens activate regulatory cells, which are essential for preserving the homeostasis of the immune system. Mutations causing impairments in these cells can lead to serious autoimmune disorders in humans and animals, highlighting the critical role immune response regulation plays in maintaining general health (Nicholson 2016).

Table 12.1 Co-inhibitory molecules involved in the intrinsic shutdown mechanism of immune response

Receptor family	Molecules	Receptor-expressing cell during infection	Ligands	References
CD28	CTLA-4 PD1	CD4+ CD4+, CD8+	CD80, CD86 PD-L1, PD-L2	Kaufmann et al. (2007), Kassu et al. (2010)
TIM	Tim-3	CD4+, CD8+	Galectin9	Kassu et al. (2010), Zhu et al. (2005), Jones et al. (2008)
CD226	TIGIT	CD4+, CD8+	CD155, CD112, CD113	Kassu et al. (2010), Day et al. (2006), Palmer et al. (2013), Shimauchi et al. (2007), Yasuma et al. (2016)

CTLA-4 cytotoxic T-lymphocyte-associated protein 4, *PD1* programmed cell death 1, *TIGIT* T-cell immunoglobulin and ITIM domain

12.3 Immunoinformatics Fundamentals

The multidisciplinary topic of immunoinformatics combines immunology and bioinformatics to interpret and evaluate the intricate workings of the immune system. Researchers can look at several facets of the immune response, from antigen–antibody interactions to the identification of immunogenic epitopes, by methodically applying computational and informatics methods.

12.3.1 Computational Analysis of Immune Receptors and Molecules

Immunoinformatics allows the analysis of immune receptors, including the highly diverse T-cell receptors (TCRs). TCRs are diverse due to recombination in their respective gene segments, that is, V (variable), D (diversity), and J (joining). This recombination causes diversity in hypervariable complementarity determining region (CDR), which is essential in binding to specific antigens (Davis and Bjorkman 1988). The utilization of sequence alignment algorithms, such as BLAST and FASTA, facilitates the identification of conserved and variable regions within TCR and BCR (B-cell receptor) sequences, enabling researchers to understand their structural diversity and functional implications.

The immune system's distinctive ability to identify foreign targets (nonself) without harming the host's tissues (self) is primarily mediated by T cells. T cells identify their targets through interactions between T-cell receptors (TCRs) and peptide-major histocompatibility complex (pMHC) molecules displayed by antigen-presenting cells (APCs). In the thymus, a broad range of peptide-specific, MHC-restricted TCR repertoires is formed through somatic gene rearrangements. Each of these repertoires is represented by an individual T-cell clone, which undergoes positive and negative selection processes in the thymus. During these processes, T cells are conditioned to recognize self-MHC and to tolerate any self-peptides presented. This results in the development of a naive T-cell repertoire that is self-MHC restricted and equipped to identify foreign peptides encountered in the peripheral tissues (Davis et al. 2007). The prediction of major histocompatibility complex (MHC)-binding peptides, essential for antigen presentation, is a critical component of immunoinformatics research. Tools like NetMHC (Andreatta and Nielsen 2016), SYFPEITHI (Class I) (Rammensee et al. 1999), and NetMHCIIpan (Class II) (Jensen et al. 2018) aid in the accurate prediction of peptide-MHC-binding affinities, thereby assisting in the designing of peptide-based vaccines and immunotherapies (Fig. 12.2).

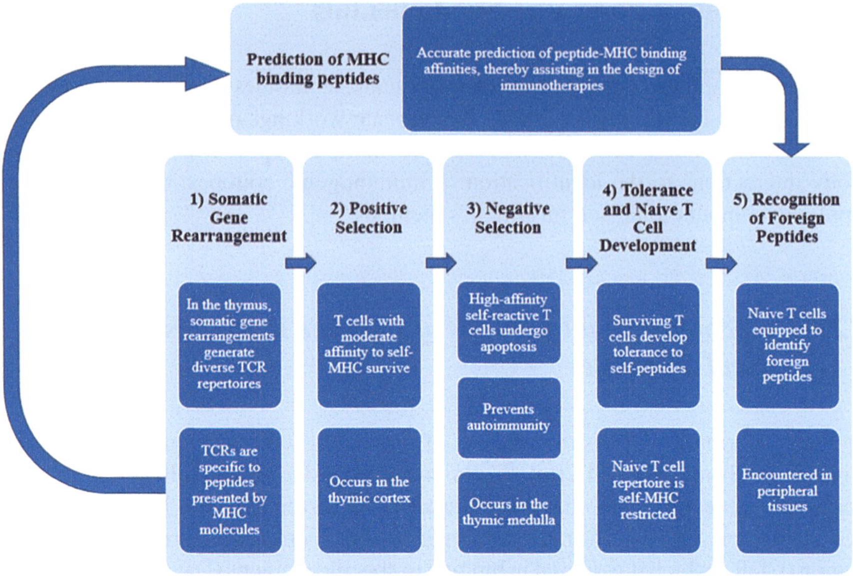

Fig. 12.2 T-cell development and selection (traditional method) and immunoinformatics-based computational analysis of immune receptor assisting in the design of immunotherapies

12.3.2 Epitope Prediction and Vaccine Design

A key focus of immunoinformatics is the prediction and characterization of epitopes. Epitope prediction algorithms, such as ElliPro, leverage sequence and structural data for the prediction of antibody epitopes in protein antigens, facilitating the rational design of vaccines (Ponomarenko et al. 2008).

Rappuoli led reverse vaccinology (RV) in the early 1990s, a genome-centred technique for identifying feasible protein vaccine candidates against Group B meningococcus (MB) (Rappuoli 2000). Finding antigens that mostly reside in the extracellular or outer membrane areas and elicit a humoral antibody response was the main goal at first. This required a thorough examination of every open reading frame obtained from the MB strain MC58 genome sequence in order to identify proteins that were anticipated to be lipoproteins, secreted, or surface exposed.

The RV technique, which incorporates computerized analysis of pathogenic protein sequences as the first phase of the process, has transformed vaccine development. This facilitates the process of choosing a subset of potential vaccine candidates (PVCs), which are promising antigens.

Several computational tools have been developed to predict potential vaccine candidates such as VaxiJen (Doytchinova and Flower 2007), Vaxign (He et al. 2010), NERVE (New Enhanced Reverse Vaccinology Environment) (Vivona et al. 2006), Bowman-Heinson (Bowman et al. 2011), and VacSol (Rizwan et al. 2017).

12.3.3 *Immune Repertoire Analysis and High-Throughput Sequencing*

The most recent developments in high-throughput sequencing technology have transformed the analysis of immune repertoires, enabling the comprehensive profiling of T- and B-cell receptor repertoires at unprecedented depth and scale. Immunoinformatics pipelines, such as TraCeR (Stubbington et al. 2016), V'DJer software (Mose et al. 2016), and MiXCR (Bolotin et al. 2015), facilitate the processing and analysis of large-scale immune repertoire sequencing data, providing insights into clonal diversity, somatic hypermutation, and immune response dynamics (Bolotin et al. 2017).

Deep learning (DL) and bioinformatics are being introduced into the field of immunology while TCR repertoire sequencing technology is progressing. This is mostly due to the fact that a single person's repertoire data typically consist of numerous sequences. Machine learning (ML) is a useful technique for deriving significant insights from large data sets. It is now essential for repertoire sequencing analysis. Additionally, ML has made it easier to create a new repertoire of sequencing-based applications, such as the production of tailored cancer vaccines (also known as neoantigen vaccines) (Ott et al. 2017) and fresh approaches to diagnosing infections such as COVID-19 (Gittelman et al. 2022). Beyond their use in medicine, machine-learning-based analysis techniques are advancing basic science in the area. The application of machine learning (ML) to repertoire sequencing is expanding quickly (Katayama et al. 2022).

12.4 Immunoinformatics in Autoimmune Disease Diagnosis

The complicated and diverse nature of autoimmune disorders makes diagnosing these conditions extremely difficult. Conventional diagnostic techniques frequently depend on the identification of autoantibodies and clinical signs, which does not offer enough specificity and sensitivity to reliably classify diseases. By enabling the identification of disease-specific biomarkers and the development of cutting-edge diagnostic tools, immunoinformatics, i.e., an emerging interdisciplinary field that integrates immunology, bioinformatics, and computational biology, has the potential to transform the diagnosis of autoimmune diseases. Researchers can look at several aspects of the immune response, from antigen–antibody interactions to the identification of immunogenic epitopes, by applying computational and informatics methods.

12.4.1 *Identification of Biomarkers*

The identification and validation of novel biomarkers for the diagnosis of autoimmune diseases are greatly aided by immunoinformatics. Researchers can uncover potential biomolecular signatures linked with particular autoimmune disorders by analysing high-dimensional omics data, such as transcriptomics, proteomics, and genomes, using computational algorithms and data mining techniques. As an example, integrated analysis of gene expression profiles and protein–protein interaction networks in the context of rheumatoid arthritis (RA) has resulted in the identification of candidate biomarkers with high diagnostic and prognostic value (Song et al. 2023).

In clinical practice, biomarkers are essential, and finding biomarkers can be sped up by applying immunoinformatics to data about proteins and the immune system. Previous investigations have identified autoantibodies as serum biomarkers for systemic lupus erythematosus (SLE) (Zhu et al. 2015; Huang et al. 2012); however, the entire pattern information required for diagnosis is absent from the preliminary evaluation tests. In order to solve this, a novel method was created that precisely distinguishes SLE patients using the k-nearest neighbour (kNN) algorithm, allowing them to obtain timely and reliable diagnoses (Binder et al. 2005).

12.4.2 Peptide Microarray-Based Diagnostics

Peptide microarray technology powered by immunoinformatics has become a viable method for multiplexed autoantibody detection against certain peptide epitopes linked to autoimmune disorders. Researchers can find autoantibody signatures that differentiate between various autoimmune disorders by synthesizing vast libraries of disease-relevant peptides and applying high-throughput screening techniques. In the case of rheumatoid arthritis (RA), researchers analysed gene expression data and identified several key genes associated with immune response and inflammatory signalling pathways using microarray data sets. Notably, they highlighted the diagnostic significance of specific genes such as CCR5, CCL5, CXCL9, CXCL10, CXCL13, PNOC, TLR8, and CD52 in RA. The study also emphasized the critical role of the chemokine system in mediating leukocyte migration and the potential implications of genes like TLR8 and CD52 in the immune response and inflammation in RA. Overall, the researchers aimed to uncover potential biomarkers and therapeutic targets to advance the treatment of RA (Ren et al. 2020).

12.4.3 Computational Model Integrating Immunoinformatics with Imaging Techniques

In the medical field, radiomics includes the extraction of valuable information from radiographic images such as CT (computed tomography), MR (magnetic resonance), and PET (positron emission tomography) scans, along with machine-learning-based quantification of lesion areas and important immunologic data (Uribe et al. 2019). This procedure aids in disease diagnosis, categorization, and grading (van Helden et al. 2018; Lambin et al. 2012). When it comes to predicting treatment response, radiomic characteristics from PET imaging work better than more conventional metrics like tumour diameter, volume, and metastases. By lowering picture noise and obtaining more precise radiation characteristics, deep learning applied to PET image reconstruction and post-processing of traditionally recovered images along with immunologic data can improve the quality of PET images. However, the low occurrence of certain diseases leads to difficulties in differentiating between affected and unaffected lesion areas in PET scans (Mayerhoefer et al. 2020).

12.5 Immunoinformatics for Disease Monitoring

In several pathological situations, such as cancer, autoimmune disorders, and infectious diseases, disease monitoring is essential for evaluating the effectiveness of treatment, the course of the disease, and the likelihood of reappearance. The multidisciplinary discipline of immunoinformatics, which combines computational biology, bioinformatics, and immunology, provides cutting-edge methods and instruments for the methodical study and tracking of immune responses in both health and illness.

Multiple sclerosis (MS) is a multifaceted condition shaped by the interactions of environmental and genetic elements, leading to disruptions in immune function. The prevailing theory regarding its pathogenesis suggests that T-lymphocytes, which are auto-reactive and activated in the periphery, especially those displaying Th1 and Th17 phenotypes, infiltrate the central nervous system (CNS) by exploiting a compromised blood–brain barrier (BBB) (Yadav et al. 2015). Once within the brain, these auto-reactive immune cells recruit macrophages, microglia, and B lymphocytes to promote inflammation, leading to the generation of reactive oxygen species (ROS), reactive nitrogen species (RNS), and inflammatory cytokines, which cause trauma to tissues (Dendrou et al. 2015).

Immunomodulating or immunosuppressive therapies are the basis of MS treatment. Over the last two decades, a number of monoclonal antibodies and oral medications have changed the therapeutic landscape significantly. Thirteen active ingredients are present in the approximately 20 medications that have been licensed for as therapeutic of multiple sclerosis (MS), as dimethyl fumarate, glatiramer acetate, and interferon 1 beta are marketed under different brand names.

The vision of early therapeutic intervention has gained popularity in the MS community in recent years since there is mounting evidence that brain damage starts as soon as the disease manifests and continues throughout the course of the illness, even in clinically silent phases (Trapp et al. 1998; Filippi et al. 2004; Knier et al. 2016). Studies conducted over an extended period of time show that patients with earlier treatment outcomes saw reduced rates of relapse, disability accumulation, and loss of brain volume (Ziemssen et al. 2015). There are first-line medications with varying modes of action, but it is still difficult to predict individual reactions, and no laboratory examination has been proven to be a good indicator of drug responsiveness or treatment failure in multiple sclerosis. Consequently, there is a continuous search for fresh and trustworthy instruments to inform therapy choices. Although tailored techniques seem promising, their current deployment in clinical practice is limited due to the large number of variables that need to be taken into account (Pennisi et al. 2013). In order to tackle this issue, a number of computer models have been put out to depict the dynamics of multiple sclerosis (MS), particularly the relapsing-remitting MS (RRMS) kind. These models are intended to provide customized simulations when particular medications are administered.

Researchers developed a simulation framework capable of predicting the dynamics of relapsing-remitting multiple sclerosis (MS) patients undergoing specific drug treatments. The goal of the study was to use a mechanistic model to predict treatment outcomes for individual MS patients by performing a thorough investigation of the immune system's behaviour. A number of first-line medications, such as teriflunomide, fingolimod, and IFN-β1a, as well as several second-line medications, such as natalizumab and ocrelizumab, were included in the drug analysis. Customized simulations were made possible by the simulation framework, which offered multiple ways to mimic an actual patient in a virtual environment. The simulation in a patient was able to replicate several immunological repertoires, which is essential for all simulations involving the dynamics of the immune system (Pappalardo et al. 2020).

12.6 Case Studies and Application

12.6.1 Immunoinformatics Association of Human Coronaviruses and Autoimmunity

Researchers explained the importance of conserved epitopes found in many coronaviruses and how they could affect immune response modulation and possibly even trigger autoimmunity (Mathew et al. 2022). After applying bioinformatics screening to 351 whole human coronavirus genome sequences, 108 conserved areas were found. Out of them, 16 were B-cell immunogenic epitopes and 19 were T-cell immunogenic epitopes that may have an association with autoimmunity and cross-reactivity.

Interestingly, these epitopes were detected in structural and non-structural proteins of SARS-CoV-2 in addition to the spike protein. The presence of B-cell epitopes across a range of viral proteins suggests a complex immune response to the virus. Furthermore, the potential impact of pre-existing immunity to other viruses on the immunological response to SARS-CoV-2 is intriguing. Some of the identified B-cell epitopes exhibit potential cross-reactivity with other viruses, such as human herpesvirus 4 and H1N1. This cross-reactivity could play a role in shaping the overall immune response and has implications for understanding the broader dynamics of immunity to SARS-CoV-2.

Additionally, the study demonstrated the similarity between human proteins and some of the discovered B-cell epitopes, raising the possibility that molecular mimicry is a crucial factor in the autoimmune diseases observed in COVID-19 patients (Marino Gammazza et al. 2020). According to a number of studies, there may be a cross-reactivity between SARS-CoV-2 and the sections of the human proteome. This suggests that SARS-CoV-2 may cause cross-reactivity with the host's self-antigens, which could result in a range of clinical symptoms and autoimmune disorders (Venkatakrishnan et al. 2020; Kanduc 2020).

The analysis of T-cell-mediated immune response revealed that some of the conserved T-cell epitopes of SARS-CoV matched T-cell epitopes in response to other viruses. This emphasizes that prior exposure to other viruses may influence the immune response to SARS-CoV-2 and that there may be cross-reactivity among coronaviruses (Grifoni et al. 2020; Braun et al. 2020). Additionally, the research demonstrated that a sizable percentage of the discovered SARS-CoV-2 peptides bound to HLA class I molecules, possibly triggering CD8+ and CD4+ T-cell responses (Anderson et al. 2021). This result suggested that COVID-19 and HLA types linked to autoimmune disorders and hyper-inflammatory states may be related. Most of the conserved B- and T-cell epitopes were found within the functional domains of the SARS-CoV-2 proteins, notably close to the spike protein's receptor-binding region, according to additional structural analysis. This emphasizes how these epitopes may cause protective antibodies to be produced, which could prevent or cause an allosteric effect on ACE2 binding, so affecting the infectiousness of the virus (Galeotti and Bayry 2020). The study's authors stressed the need for more research, especially experimental studies to evaluate functional antibodies and the immunogenic effect in vivo, even though the results they reported were based on immunoinformatic predictions. The research contributed to the refinement of knowledge about SARS-CoV-2 pathophysiology and potential consequences by illuminating the intricate relation between viral epitopes and the immune system.

12.6.2 Inferring Microorganisms Associated with Rheumatoid Arthritis (RA)

In order to understand how human-associated microorganisms may cause rheumatoid arthritis (RA) through a process known as molecular mimicry, a study has been conducted using an immunoinformatics investigation (Repac et al. 2021). The findings offer insight into a variety of species that may contribute to the initiation of the disease process by activating autoreactive T cells specific to RA. Notably, the research highlights the critical role of autoreactive T cells in the disease's inception, even while B cells contribute to RA pathogenesis in later stages (van Hamburg and Tas 2018; Chemin et al. 2019; Rashid et al. 2017).

Numerous potential RA-triggering organisms were found in the study, including commensal species, opportunistic pathogens, and well-known pathogens including *P. mirabilis* (Rashid and Ebringer 2007) and *M. tuberculosis* (Lan and Wu 1992). It implies that components of the human microbiome may play a direct role in the pathophysiology of RA, maybe going beyond the scope of their traditional roles in inflammation. Remarkably, certain components of the gut and oral microbiota, including *B. fragilis* and several strains of *Prevotella,* were found to be putative triggers (Scher and Abramson 2011). This is consistent with recent research that associates RA with disturbed gut and oral microbiota. The investigation also brought to light the significant potential of fungi, which are frequently disregarded as possible initiators of the onset of RA. A number of new fungal diseases were suggested by the study as potential RA triggers because of their closer evolutionary relationship with humans and their capacity for molecular mimicry since they share homologous peptides with human proteins. The investigation found putative T-cell epitopes across many viral groups, including the *Marseillevirus* group, indicating their potential significance in inducing autoimmune reactions, even though the results were less clear-cut for viruses (Colson and Raoult 2010).

Strict filtering guidelines were used in the study to guarantee the validity of the results, taking into account the 80% sequence identity cut-off and the lack of alignment gaps. Despite these drawbacks, the work provides insightful information about the contribution of fungi to autoimmunity as well as the possible direct involvement of bacterial commensals in the pathophysiology of RA. It also emphasizes the need to identify strong autoantigens, which are essential to comprehend a range of autoimmune diseases (Repac et al. 2021).

12.7 Future Directions and Challenges

Through computational models and informatics tools, immunoinformatics, a rapidly developing field, plays a critical role in furthering research in the field of immunology (Tong and Ren 2009). Numerous obstacles and diverse paths become apparent as we investigate its present patterns and potential future developments.

12.7.1 Future Directions

Enhanced Epitope Prediction

Immunoinformatics will likely focus on refining epitope prediction methods by utilizing advancements in artificial intelligence and machine learning. This could lead to more accurate predictions of immunogenic epitopes (Alkaff et al. 2020).

Personalized Immunotherapy

Immunoinformatics may aid in the development of individualized immunotherapies, which adjust treatments to individuals' immunological profiles. This has the potential to transform disease management and improve therapy results (Oli et al. 2020).

Systems Immunology Integration

Immunoinformatics integrates with systems like immunology, providing a thorough understanding of immune responses at the system level. This combination has the potential to identify complex regulatory networks and improve immune prediction (Tong and Ren 2009).

Multi-model Data Fusion

The future of immunoinformatics involves harnessing the power of multi-modal data fusion. By combining data from various sources such as genomics, transcriptomics, and proteomics, researchers can gain a more holistic view of immune system dynamics and responses (Chakraborty et al. 2021).

Enhanced Vaccine Design Strategies

Immunoinformatics will refine vaccine design strategies in the future, with a focus on producing vaccinations against difficult infections. This includes aspects such as antigenic variation and the evolution of infectious agents (Prawiningrum et al. 2022).

Machine Learning in Immunoinformatics

Machine learning algorithm developments will be critical in enhancing immunoinformatics predictions. Deep learning and neural networks could lead to more precise epitope predictions and comprehensive immune system behaviour studies (Alkaff et al. 2020).

12.7.2 Challenges in Immunoinformatics

Complexity of Antigen Recognition

An epitope is a complicated structure with numerous interactions going on; understanding the complicated mechanics of antigen recognition is a challenge for immunoinformatics. More advancement is needed to understand the many interactions between immune system components and antigens (Tong and Ren 2009).

Data Integration and Standardization

Immunoinformatics is based on algorithms and data, and integration of non-standardized immunological data poses a significant challenge. Standardizing data formats and terminologies is vital for effective collaboration and interpretation of results (Tong and Ren 2009).

Vaccine Safety

Designing effective vaccines demands overcoming challenges such as safety, even though ADME (absorption, distribution, metabolism, excretion) can be analysed virtually with some degree of success, it is still far behind the real-world scenario, which will eventually require in vitro and in vivo analysis.

12.8 Discussion

Autoimmune diseases pose significant challenges due to their complex aetiology, diverse clinical manifestations, and lack of specific diagnostic markers. Traditional diagnostic approaches often rely on time-consuming and expensive laboratory techniques, hindering the development of personalized treatment strategies. Immunoinformatics, an interdisciplinary field that integrates immunology, bioinformatics, and computational biology, offers a promising solution to address these challenges by providing powerful tools for analysing immune system data and identifying novel targets for diagnosis and therapy. The development of autoimmune disorder is due to the complex interplay of genetic, environmental, and immunological factors. Regulatory T cells (Tregs) play a crucial role in maintaining immune system balance and preventing autoimmune reactions.

Fundamentals of immunoinformatics focuses on computational analysis of immune receptors, epitope prediction, and vaccine design. The prediction of major histocompatibility complex (MHC)-binding peptides for antigen presentation can be done by utilizing tools like NetMHC and SYFPEITHI. The advancements in high-throughput sequencing technology and the role of deep learning in analysing immune repertoires.

Applications of immunoinformatics range from the identification of biomarkers through the analysis of omics data and offer more specific and sensitive diagnostic tools to peptide microarray-based diagnostics powered by immunoinformatics.

Immunoinformatics is emerging as a transformative approach in the diagnosis and monitoring of autoimmune diseases, offering personalized and efficient solutions through the integration of computational techniques with immunological insights. Ongoing advancements in this field hold the promise of revolutionizing our understanding and the management of autoimmune disorders in the future.

References

Alkaff AH, Saragih M, Fardiansyah MA, Tambunan USF (2020) Role of Immunoinformatics in accelerating epitope-based vaccine development against dengue virus. Open Biochem J 14:9–18

Anaya J-M, Ramirez-Santana C, Alzate MA, Molano-Gonzalez N, Rojas-Villarraga A (2016) The autoimmune ecology. Front Immunol 7:139. https://doi.org/10.3389/fimmu.2016.00139

Anderson EM, Goodwin EC, Verma A et al (2021) Seasonal human coronavirus antibodies are boosted upon SARS-CoV-2 infection but not associated with protection. Cell 184:1858–1864.e10

Andreatta M, Nielsen M (2016) Gapped sequence alignment using artificial neural networks: application to the MHC class I system. Bioinformatics 32:511–517

ASCIA (2023) ASCIA Autoimmune diseases

Baecher-Allan C (2004) Human CD4+CD25+ regulatory T cells. Semin Immunol 16:89–98

Belkaid Y, Piccirillo CA, Mendez S, Shevach EM, Sacks DL (2002) CD4+CD25+ regulatory T cells control Leishmania major persistence and immunity. Nature 420:502–507

Binder SR, Genovese MC, Merrill JT, Morris RI, Metzger AL (2005) Computer-assisted pattern recognition of autoantibody results. Clin Vaccine Immunol 12:1353–1357

Bolotin DA, Poslavsky S, Mitrophanov I, Shugay M, Mamedov IZ, Putintseva EV, Chudakov DM (2015) MiXCR: software for comprehensive adaptive immunity profiling. Nat Methods 12:380–381

Bolotin DA, Poslavsky S, Davydov AN et al (2017) Antigen receptor repertoire profiling from RNA-seq data. Nat Biotechnol 35:908–911

Bowman BN, McAdam PR, Vivona S et al (2011) Improving reverse vaccinology with a machine learning approach. Vaccine 29:8156–8164

Braun J, Loyal L, Frentsch M et al (2020) SARS-CoV-2-reactive T cells in healthy donors and patients with COVID-19. Nature 587:270–274

Brusic V, Petrovsky N (2005) Immunoinformatics and its relevance to understanding human immune disease. Expert Rev Clin Immunol 1:145–157

Chakraborty C, Sharma AR, Bhattacharya M, Sharma G, Lee S-S (2021) Immunoinformatics approach for the identification and characterization of T cell and B cell epitopes towards the peptide-based vaccine against SARS-CoV-2. Arch Med Res 52:362–370

Chatanaka MK, Ulndreaj A, Sohaei D, Prassas I (2022) Immunoinformatics: pushing the boundaries of immunology research and medicine. ImmunoInformatics 5:100007

Chemin K, Gerstner C, Malmström V (2019) Effector functions of CD4+ T cells at the site of local autoimmune inflammation—lessons from Rheumatoid Arthritis. Front Immunol 10:353. https://doi.org/10.3389/fimmu.2019.00353

Colson P, Raoult D (2010) Gene repertoire of amoeba-associated Giant viruses. Intervirology 53:330–343

Conrad N, Misra S, Verbakel JY et al (2023) Incidence, prevalence, and co-occurrence of autoimmune disorders over time and by age, sex, and socioeconomic status: a population-based cohort study of 22 million individuals in the UK. Lancet 401:1878–1890

Davis MM, Bjorkman PJ (1988) T-cell antigen receptor genes and T-cell recognition. Nature 334:395–402

Davis MM, Krogsgaard M, Huse M, Huppa J, Lillemeier BF, Li Q (2007) T cells as a self-referential, sensory organ. Annu Rev Immunol 25:681–695

Day CL, Kaufmann DE, Kiepiela P et al (2006) PD-1 expression on HIV-specific T cells is associated with T-cell exhaustion and disease progression. Nature 443:350–354

Dendrou CA, Fugger L, Friese MA (2015) Immunopathology of multiple sclerosis. Nat Rev Immunol 15:545–558

Doytchinova IA, Flower DR (2007) VaxiJen: a server for prediction of protective antigens, tumour antigens and subunit vaccines. BMC Bioinformatics 8:4

Filippi M, Rovaris M, Inglese M, Barkhof F, De Stefano N, Smith S, Comi G (2004) Interferon beta-1a for brain tissue loss in patients at presentation with syndromes suggestive of multiple sclerosis: a randomised, double-blind, placebo-controlled trial. Lancet 364:1489–1496

Fujinami RS, Oldstone MBA (1989) Molecular mimicry as a mechanism for virus-induced autoimmunity. Immunol Res 8:3

Galeotti C, Bayry J (2020) Autoimmune and inflammatory diseases following COVID-19. Nat Rev Rheumatol 16:413–414

Gittelman RM, Lavezzo E, Snyder TM et al (2022) Longitudinal analysis of T cell receptor repertoires reveals shared patterns of antigen-specific response to SARS-CoV-2 infection. JCI Insight 7:e151849. https://doi.org/10.1172/jci.insight.151849

Grifoni A, Weiskopf D, Ramirez SI et al (2020) Targets of T cell responses to SARS-CoV-2 coronavirus in humans with COVID-19 disease and unexposed individuals. Cell 181:1489–1501.e15

He Y, Xiang Z, Mobley HLT (2010) Vaxign: the first web-based vaccine design program for reverse vaccinology and applications for vaccine development. J Biomed Biotechnol 2010:1–15

Hirota K, Yoshitomi H, Hashimoto M et al (2007) Preferential recruitment of CCR6-expressing Th17 cells to inflamed joints via CCL20 in rheumatoid arthritis and its animal model. J Exp Med 204:2803–2812

Huang W, Hu C, Zeng H, Li P, Guo L, Zeng X, Liu G, Zhang F, Li Y, Wu L (2012) Novel systemic lupus erythematosus autoantigens identified by human protein microarray technology. Biochem Biophys Res Commun 418:241–246

Jensen KK, Andreatta M, Marcatili P, Buus S, Greenbaum JA, Yan Z, Sette A, Peters B, Nielsen M (2018) Improved methods for predicting peptide binding affinity to MHC class II molecules. Immunology 154:394–406

Jones RB, Ndhlovu LC, Barbour JD et al (2008) Tim-3 expression defines a novel population of dysfunctional T cells with highly elevated frequencies in progressive HIV-1 infection. J Exp Med 205:2763–2779

Kanduc D (2020) From anti-SARS-CoV-2 immune responses to COVID-19 via molecular mimicry. Antibodies 9:33

Kassu A, Marcus RA, D'Souza MB, Kelly-McKnight EA, Golden-Mason L, Akkina R, Fontenot AP, Wilson CC, Palmer BE (2010) Regulation of virus-specific CD4+ T cell function by multiple costimulatory receptors during chronic HIV infection. J Immunol 185:3007–3018

Katayama Y, Yokota R, Akiyama T, Kobayashi TJ (2022) Machine learning approaches to TCR repertoire analysis. Front Immunol 13:858057. https://doi.org/10.3389/fimmu.2022.858057

Kaufmann DE, Kavanagh DG, Pereyra F et al (2007) Upregulation of CTLA-4 by HIV-specific CD4+ T cells correlates with disease progression and defines a reversible immune dysfunction. Nat Immunol 8:1246–1254

Knier B, Berthele A, Buck D, Schmidt P, Zimmer C, Mühlau M, Hemmer B, Korn T (2016) Optical coherence tomography indicates disease activity prior to clinical onset of central nervous system demyelination. Mult Scler J 22:893–900

Lambin P, Rios-Velazquez E, Leijenaar R et al (2012) Radiomics: extracting more information from medical images using advanced feature analysis. Eur J Cancer 48:441–446

Lan JL, Wu CH (1992) Detection of Mycobacterium tuberculosis antigen in synovial fluid of patients with Rheumatoid Arthritis. Rheumatology 31:615–618

Marino Gammazza A, Légaré S, Lo Bosco G, Fucarino A, Angileri F, Conway de Macario E, Macario AJ, Cappello F (2020) Human molecular chaperones share with SARS-CoV-2 antigenic epitopes potentially capable of eliciting autoimmunity against endothelial cells: possible role of molecular mimicry in COVID-19. Cell Stress Chaperones 25:737–741

Mathew S, Fakhroo AD, Smatti M, Al Thani AA, Yassine HM (2022) Immunoinformatics prediction of potential immunodominant epitopes from human coronaviruses and association with autoimmunity. Immunogenetics 74:213–229

Mayerhoefer ME, Materka A, Langs G, Häggström I, Szczypiński P, Gibbs P, Cook G (2020) Introduction to radiomics. J Nucl Med 61:488–495

Mose LE, Selitsky SR, Bixby LM, Marron DL, Iglesia MD, Serody JS, Perou CM, Vincent BG, Parker JS (2016) Assembly-based inference of B-cell receptor repertoires from short read RNA sequencing data with V'DJer. Bioinformatics 32:3729–3734

Nicholson LB (2016) The immune system. Essays Biochem 60:275–301

NIEHS (n.d.) Autoimmune Diseases. Natl Inst Environ Heal Sci

Oli AN, Obialor WO, Ifeanyichukwu MO, Odimegwu DC, Okoyeh JN, Emechebe GO, Adejumo SA, Ibeanu GC (2020) Immunoinformatics and vaccine development: an overview. Immunotargets Ther 9:13–30

Ott PA, Hu Z, Keskin DB et al (2017) An immunogenic personal neoantigen vaccine for patients with melanoma. Nature 547:217–221

Palmer BE, Neff CP, LeCureux J, Ehler A, Dsouza M, Remling-Mulder L, Korman AJ, Fontenot AP, Akkina R (2013) In vivo blockade of the PD-1 receptor suppresses HIV-1 viral loads and improves CD4+ T cell levels in humanized mice. J Immunol 190:211–219

Pappalardo F, Russo G, Pennisi M, Parasiliti Palumbo GA, Sgroi G, Motta S, Maimone D (2020) The potential of computational modeling to predict disease course and treatment response in patients with relapsing multiple sclerosis. Cell 9:586

Pennisi M, Rajput A-M, Toldo L, Pappalardo F (2013) Agent based modeling of Treg-Teff cross regulation in relapsing-remitting multiple sclerosis. BMC Bioinformatics 14:S9

Ponomarenko J, Bui H-H, Li W, Fusseder N, Bourne PE, Sette A, Peters B (2008) ElliPro: a new structure-based tool for the prediction of antibody epitopes. BMC Bioinformatics 9:514

Prawiningrum AF, Paramita RI, Panigoro SS (2022) Immunoinformatics approach for epitope-based vaccine design: key steps for breast cancer vaccine. Diagnostics 12:2981

Rammensee H-G, Bachmann J, Emmerich NPN, Bachor OA, Stevanović S (1999) SYFPEITHI: database for MHC ligands and peptide motifs. Immunogenetics 50:213–219

Rappuoli R (2000) Reverse vaccinology. Curr Opin Microbiol 3:445–450

Rashid T, Ebringer A (2007) Rheumatoid arthritis is linked to proteus—the evidence. Clin Rheumatol 26:1036–1043

Rashid T, Ebringer A, Wilson C (2017) The link between Proteus mirabilis, environmental factors and autoantibodies in rheumatoid arthritis. Clin Exp Rheumatol 35:865–871

Ren C, Li M, Du W, Lü J, Zheng Y, Xu H, Quan R (2020) Comprehensive bioinformatics analysis reveals hub genes and inflammation state of rheumatoid arthritis. Biomed Res Int 2020:1–13

Repac J, Mandić M, Lunić T, Božić B, Božić Nedeljković B (2021) Mining the capacity of human-associated microorganisms to trigger rheumatoid arthritis—a systematic immunoinformatics analysis of T cell epitopes. PLoS One 16:e0253918

Rizwan M, Naz A, Ahmad J, Naz K, Obaid A, Parveen T, Ahsan M, Ali A (2017) VacSol: a high throughput in silico pipeline to predict potential therapeutic targets in prokaryotic pathogens using subtractive reverse vaccinology. BMC Bioinformatics 18:106

Sakaguchi S, Yamaguchi T, Nomura T, Ono M (2008) Regulatory T cells and immune tolerance. Cell 133:775–787

Scher JU, Abramson SB (2011) The microbiome and rheumatoid arthritis. Nat Rev Rheumatol 7:569–578

Shimauchi T, Kabashima K, Nakashima D, Sugita K, Yamada Y, Hino R, Tokura Y (2007) Augmented expression of programmed death-1 in both neoplastic and non-neoplastic CD4 + T-cells in adult T-cell leukemia/lymphoma. Int J Cancer 121:2585–2590

Song X, Zhang Y, Zhao L et al (2023) Analyzation of the peripheral blood mononuclear cells atlas and cell communication of rheumatoid arthritis patients based on single-cell RNA-Seq. J Immunol Res 2023:1–20

Stubbington MJT, Lönnberg T, Proserpio V, Clare S, Speak AO, Dougan G, Teichmann SA (2016) T cell fate and clonality inference from single-cell transcriptomes. Nat Methods 13:329–332

Szanya V, Ermann J, Taylor C, Holness C, Fathman CG (2002) The subpopulation of CD4+CD25+ Splenocytes that delays adoptive transfer of diabetes expresses L-selectin and high levels of CCR7. J Immunol 169:2461–2465

Tong JC, Ren EC (2009) Immunoinformatics: current trends and future directions. Drug Discov Today 14:684–689

Trapp BD, Peterson J, Ransohoff RM, Rudick R, Mörk S, Bö L (1998) Axonal transection in the lesions of multiple sclerosis. N Engl J Med 338:278–285

Uribe CF, Mathotaarachchi S, Gaudet V, Smith KC, Rosa-Neto P, Bénard F, Black SE, Zukotynski K (2019) Machine learning in nuclear medicine: part 1—introduction. J Nucl Med 60:451–458

van Hamburg JP, Tas SW (2018) Molecular mechanisms underpinning T helper 17 cell heterogeneity and functions in rheumatoid arthritis. J Autoimmun 87:69–81

van Helden EJ, Vacher YJL, van Wieringen WN, van Velden FHP, Verheul HMW, Hoekstra OS, Boellaard R, der Houven M-v, van Oordt CW (2018) Radiomics analysis of pre-treatment [18F]FDG PET/CT for patients with metastatic colorectal cancer undergoing palliative systemic treatment. Eur J Nucl Med Mol Imaging 45:2307–2317

Venkatakrishnan AJ, Kayal N, Anand P, Badley AD, Church GM, Soundararajan V (2020) Benchmarking evolutionary tinkering underlying human–viral molecular mimicry shows multiple host pulmonary–arterial peptides mimicked by SARS-CoV-2. Cell Death Discov 6:96

Vivona S, Bernante F, Filippini F (2006) NERVE: new enhanced reverse vaccinology environment. BMC Biotechnol 6:35

Vojdani A (2014) A potential link between environmental triggers and autoimmunity. Autoimmune Dis 2014:1–18

Wu H-J, Wu E (2012) The role of gut microbiota in immune homeostasis and autoimmunity. Gut Microbes 3:4–14

Yadav SK, Mindur JE, Ito K, Dhib-Jalbut S (2015) Advances in the immunopathogenesis of multiple sclerosis. Curr Opin Neurol 28:206–219

Yasuma K, Yasunaga J, Takemoto K, Sugata K, Mitobe Y, Takenouchi N, Nakagawa M, Suzuki Y, Matsuoka M (2016) HTLV-1 bZIP factor impairs anti-viral immunity by inducing co-inhibitory molecule, T cell immunoglobulin and ITIM domain (TIGIT). PLoS Pathog 12:e1005372

Zhu C, Anderson AC, Schubart A, Xiong H, Imitola J, Khoury SJ, Zheng XX, Strom TB, Kuchroo VK (2005) The Tim-3 ligand galectin-9 negatively regulates T helper type 1 immunity. Nat Immunol 6:1245–1252

Zhu H, Luo H, Yan M, Zuo X, Li Q-Z (2015) Autoantigen microarray for high-throughput autoantibody profiling in systemic lupus erythematosus. Genomics Proteomics Bioinformatics 13:210–218

Ziemssen T, De Stefano N, Sormani MP, Van Wijmeersch B, Wiendl H, Kieseier BC (2015) Optimizing therapy early in multiple sclerosis: an evidence-based view. Mult Scler Relat Disord 4:460–469

Chapter 13
Erythropoietin in the Treatment of Anaemia

Nibir Ghosh, Gourab Dey, Pallabita Rakshit, and Sourav De

Abstract Anaemia is a prevalent clinical presentation in individuals suffering from chronic kidney disease (CKD). This issue is linked to an elevated incidence of illness and death in patients. Erythropoietin is a renal hormone that is essential for the synthesis of red blood cells (RBCs). Anemia occurs due to a lack of production and premature destruction of RBC. The basic role of erythropoietin is to stimulate the bone marrow and increase the production of RBC when the body has a low oxygen level in the circulatory system. In anaemia, there is a deficiency of RBC or haemoglobin. In this condition, the body may release more erythropoietin in response to low oxygen levels. If the anaemia is related to the production of RBC or CKD, erythropoietin therapy can be used. This therapy will stimulate the bone marrow to produce more RBC. The first recombinant human erythropoietin was epoetin alpha approved by the US FDA in 1989. A substance similar to that of epoetin alpha is known as erythropoietin stimulating agent (EPA). Nowadays, treatment with ESA agents has become very popular for the treatment of anaemia that occurs in CKD patients. This chapter focuses on the role of ESA therapy in the treatment of anaemic patients with CKD.

Keywords Anaemia · Chronic kidney disease · Erythropoietin · ESA agent

N. Ghosh · G. Dey · S. De (✉)
Department of Pharmaceutical Technology, Eminent College of Pharmaceutical Technology, Kolkata, India

P. Rakshit
Department of Pharmaceutical Technology, Jadavpur University, Kolkata, India

S. Bose et al. (eds.), *Concepts in Pharmaceutical Biotechnology and Drug Development*, Interdisciplinary Biotechnological Advances,
https://doi.org/10.1007/978-981-97-1148-2_13

13.1 Introduction

A number of chronic illnesses, including chronic kidney disease (CKD), might have anaemia as a side effect. An absolute decline in the total quantity of red blood cells (RBCs) in circulation is referred to as anaemia. For clinical purposes, anaemia is diagnosed by considering factors like haemoglobin concentration, haematocrit, or RBC count. The existence of an illness or disease is indicated by this condition in the laboratory; anaemia by itself should not be regarded as a diagnosis (Barany 2007). Three aetiologic categories—increased RBC breakdown, decreased RBC generation, and blood loss—are typically utilised to classify anaemia. Anemia can also be associated with several chronic diseases, such as chronic kidney disease (CKD). Renal failure, which is caused by the loss of kidney function in whole or in part, poses a serious risk to human health (Eschbach et al. 1989). The term "renal anaemia" describes the range of conditions in which toxic uremic plasma chemicals or erythropoietin-related compounds disrupt the metabolism of RBCs, resulting in anaemia. Chronic renal deficiency, characterised by different levels of impaired kidney function, often leads to renal anaemia. The main chain of the 34,000 Da glycoprotein erythropoietin (EPO) contains 193 amino acids. About 90% of renal interstitial cells release EPO, a hormone-like material, whereas 10% of liver cells do the same. The kidney has a vital function in regulating the amount of EPO in the blood. Reduced EPO synthesis, subsequent hypoxia, and impaired blood flow might result from injury to the renal glomerular or tubular tissues. This, in turn, might trigger the creation of more EPOs, exacerbating the situation. Following its synthesis, EPO cannot be kept directly inside the body; rather, it promptly undergoes metabolic processing (Brugnara 2003). The regulation of serum EPO levels is primarily influenced by the feedback mechanism involving red blood cell pressure and haemoglobin. Disruption of this control results in fluctuating levels of EPO, causing a prolonged elevation in serum EPO levels, finally resulting in kidney injury (Collins et al. 2000). Erythropoietin therapy is also used in the therapy of various disease-induced anaemia such as CHF and diabetes mellitus.

13.2 Pathophysiology of Anaemia in CKD Patients

Anaemia in CKD is caused by various factors. It has long been believed that the steady decline in endogenous erythropoietin (EPO) levels is the primary factor. Anaemia in CKD patients can be caused by various factors, including absolute iron inadequacy because of impaired iron absorption or blood losses, systemic inflammation resulting from CKD and associated comorbidities, shortened red cell lifespan, decreased bone marrow response to EPO because of uremic toxins, inefficient utilisation of iron stores caused by elevated hepcidin levels, and deficiencies in folic acid or vitamin B_{12} (KDIGO Anemia Working Group 2012).

13.3 Hypoxia-Induced Factor

EPO assembled in the bone marrow's erythroid progenitor cells via attaching to their surface receptor is a glycoprotein with a molecular weight of 30.4 kDa—primarily works as a primary stimulant for the differentiation, proliferation, and survival of red blood cells. The kidneys' fibroblast-like interstitial peritubular cells and the liver's perisinusoidal cells are the primary sources of EPO, which is generated in response to changes in tissue oxygen concentration (Pan et al. 2011; Rankin et al. 2009). Transcription of the EPO gene regulates the amount of EPO produced. The hypoxia-inducible factor (HIF) system is an important regulator of its expression, as its activity is dependent upon the level of oxygen present in the tissue (Provenzano et al. 2016), (Fig. 13.1).

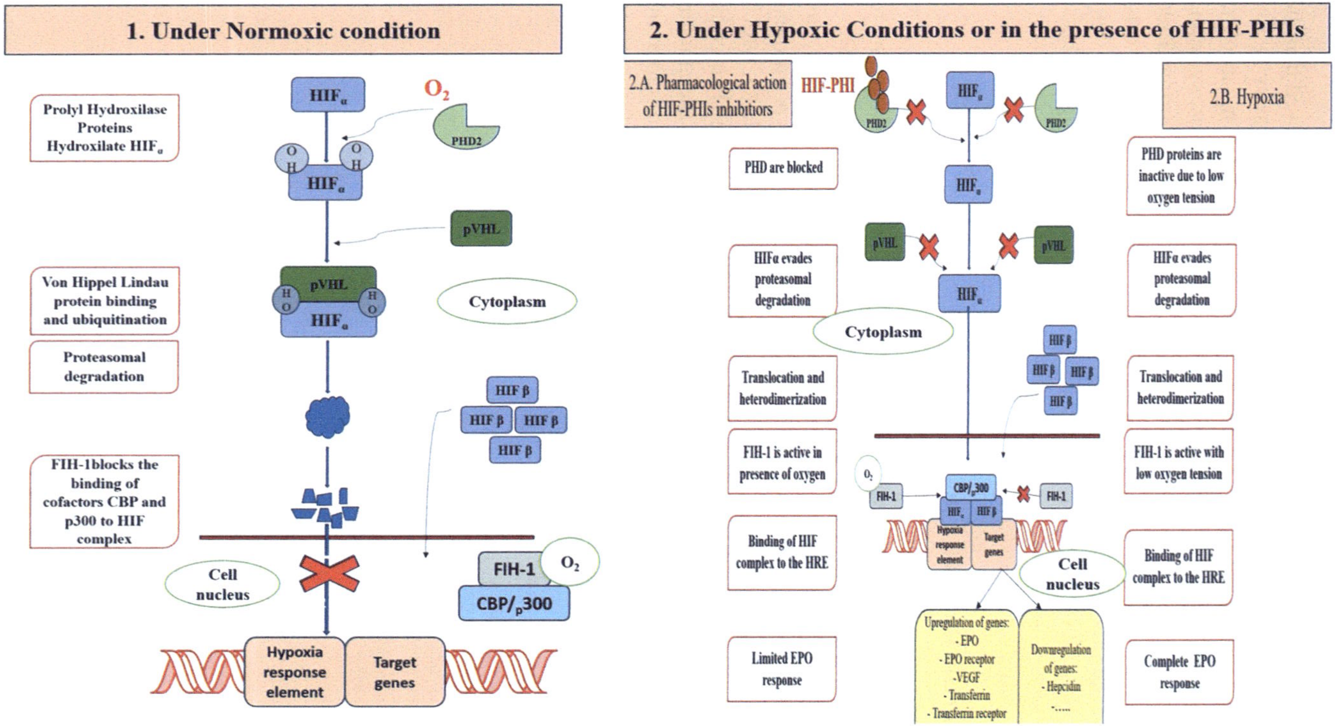

Fig. 13.1 The control of HIF at normal oxygen levels, the impact of hypoxia-inducible factor prolyl hydroxylase inhibitors (HIF-PHIs) on the body's response to low oxygen levels, and the regulation of HIF during low oxygen levels. Abbre*via*tion: *HIF* hypoxia-inducible factor, *OH* hydroxyl, *PHD* prolyl hydroxylase domain protein, O_2 oxygen, *FIH-1* factor inhibiting HIF, *CBP* CREB-binding protein, *pVHL* Von Hippel Lindau protein, *HIF-PHI* hypoxia-inducible factor prolyl hydroxylase inhibitor

13.4 Erythropoietin Production in CKD

Insufficiently low EPO levels are associated with severe anaemia in patients with chronic kidney disease (CKD). CKD begins with an EPO deficiency, but it seems to get worse when the estimated glomerular filtration rate (eGFR) is reduced below 30 mL per minute per 1.73 m^2 (Wenger and Hoogewijs 2010). A reduction in EPO production or mistakes in EPO sensing may be the cause of an absolute EPO shortage. Reduced blood flow to the kidneys causes an imbalance in oxygen delivery, which is linked to chronic kidney disease (CKD). Renal tissue adapts as a result, consuming less oxygen and allowing the tissue oxygen gradient to remain normal. PHD enzymes continue to function as a result, and neither the HIF heterodimer nor the EPO gene is activated. Empirical evidence has shown the presence of cytokines such as transforming growth factor-beta (TGF-β), interleukin-lβ (IL-lβ), IL-la, and tumour necrosis factor (TNF)-α decrease the generation of EPO caused by hypoxia (Fandrey and Jelmann 1991). The immunological activation molecule hypoxia-induced EPO is suppressed, and it is well known that CKD causes an increase in inflammation (Rao et al. 2007; Souma et al. 2016).

13.5 Role of Erythropoietin in CKD-Related Anaemia

EPO has the capacity to stimulate the differentiation and proliferation of bone marrow cells. It serves as a humoral factor that facilitates the generation of RBCs. Furthermore, EPO has the ability to enhance the production of haemoglobin, increase blood circulation in peripheral arteries, and encourage the process by which stem cells may develop into RBCs. To facilitate the transformation of the red line into fully developed RBCs, the hormone EPO activates EPO receptors, which in turn promote the differentiation and proliferation of burst-forming unit-erythroid (BFU-E) cells. This chapter basically highlights the role of erythropoietin agents in the treatment of anaemia in CKD patients along with patients with diabetic anaemia, CHF anaemia, and inflammatory bowel disease anaemia (Batchelor et al. 2020).

13.6 Types of Various Erythropoietin Stimulating Agent Use in CKD Anaemia

The clinical management of CKD-induced anaemia has been significantly improved with the introduction of the first human recombinant EPO, especially for dialysis patients. ESA therapy accounted for 11% of all Medicare ESRD costs in 2004 and was the greatest single Medicare drug expenditure in the USA, with a total of US$1.8 billion. There are three generations of EPO agents: first generation, second generation, and third generation. The first-generation EPO agents have a shorter half-life and require frequent dosing. Hence, second-generation EPOs are modified with a higher short life to reduce the frequent dose timing.

Table 13.1 Various erythropoietin agents' generation for clinical use in CKD anaemia

Manufacturing company	Agent name	Generation	Marketing agency
Ortho biotech	Epoetin alfa (Procrit)	First	US
Amgen	Epoetin alfa (Epogen)	First	US
Janssen-Ortho	Epoetin alfa (Eprex)	First	Worldwide except US
Roche	Epoetin beta (NeoRecormon)	First	Europe
Amgen	Darbepoetin (Aranesp)	Second	North America, Europe, Canada
Roche	Erythropoietin receptor activator	Third	US

Recently, third-generation EPO agents have been modified into ESA agents for a better therapeutic effect, and it is also known as EPO receptor activators (Eschbach et al. 1989; Macdougall 2008). There are several EPO agents that are available in the market (Table 13.1).

13.7 Role of Erythropoietin with Combination Drug Therapy in CKD Anaemia

13.7.1 Iron Sucrose

Clinically, there are two main categories for administering iron supplements: oral and intravenous. Oral delivery is often insufficient to fully fulfil the iron needs in the bone marrow of individuals with uraemia and diseases related to iron release and absorption in the gastrointestinal tract. On the other hand, intravenous iron supplementation has a relatively high utilisation rate and few gastrointestinal side effects (Van Wyck et al. 2005). This can boost iron reserves and enhance iron use, helping to get beyond the reticuloendothelial system's iron release barrier. In the reticuloendothelial system, active macrophages help release iron ions from intravenous iron complexes. Iron utilisation in the body increases when a portion of the iron combines with apoferritin to produce serum ferritin reserves inside cells and when another portion of the iron connects with transferrin-to-transferrin receptors located on the surface of juvenile red blood cells. Clinical observation has demonstrated that starting with EPO treatment and gradually increasing the iron sucrose treatment can both significantly improve clinical efficacy and improve symptomatic treatment by mitigating the adverse reaction to low and mild levels of the sugar iron treatment (Xue et al. 2015).

13.7.2 L-Carnitine

L-carnitine is an amino acid that is widely distributed throughout the tissues of the body. The compound has the potential to impact the acylation and phospholipid processes as well as the deformability of the erythrocyte membrane. Additionally, it can stabilise the membrane of RBCs and increase the number of red blood cells in order to improve the body's ability to withstand different types of pressures. A shortage of L-carnitine at the cell level can have harmful effects, shorten life expectancy, raise normal red cell osmotic fragility, and perhaps lead to EPO resistance. In patients who are in remission, L-carnitine has been shown to reduce micro-inflammation, tumour necrosis factor (TNF), interleukin 1 (IL-1), and C-reactive protein (CRP), effectively decreasing renal anaemia (Boran et al. 1996; Emami Naini et al. 2012).

13.7.3 Iron Dextran

An intravenous iron preparation is known as iron dextran. Patients with anaemia who get erythropoietin treatment have imbalanced iron storage and excessive iron consumption. IV iron supplementation has good efficacy, yet there is a risk of both acute adverse reactions and delayed reactions that resemble allergic reactions. These reactions include hypotension, urticaria, chest discomfort, respiratory issues, and vasogenic oedema. When iron dextran dispersion tablets are used in conjunction with erythropoietin and L-carnitine injection, the effectiveness of EPO is greatly increased. As a consequence, there is a drop in EPO levels, a notable enhancement in the condition of anaemia, a reduction in adverse effects, and a decrease in dialysis syndrome (Larramendi et al. 2006).

13.7.4 Reduced Glutathione

Persons with chronic renal failure have decreased antioxidant activity, increased oxidant activity, and a buildup of metabolic products, leading to reduced levels of glutathione and increasing amounts of oxidised glutathione. Disulfide linkages are created by molecules in the membrane of red blood cells, resulting in the aggregation of macromolecules on the membrane and causing deformation of the RBCs. The three-peptide complex glutathione is made up of glycine, cysteine, and glutamic acid. It serves as a significant metabolic regulator inside the cell. It may provide cysteine and thiol while preserving the cell's normal metabolic and physiological processes, hence preserving the integrity of the cell membrane (Zachara et al. 2006). Furthermore, it may eliminate lipid peroxide by interacting with pro-electron or reactive-oxygen species, protecting the cell. Exogenous reductive glutathione treatment for chronic renal failure has many potential benefits, including stabilising the membrane of red blood cells, increasing their longevity, decreasing levels of oxidised glutathione, and improving red blood cell reduction. Supplemental reductive glutathione, as opposed to EPO alone, has profound effects on anaemia, according to studies (Ghorbani et al. 2016).

13.7.5 Irbesartan and Enalapril

Patients with various forms of renal anaemia and hypertension suffer from chronic renal insufficiency. Angiotensin-converting enzyme can be inhibited with enalapril, which hydrolyses into enalapril and lowers the amount of angiotensin II (AT-II), alleviating hypertension by relaxing blood arteries and decreasing blood pressure. Irbesartan, an AT- II receptor blocker, inhibits the vascular angiotensin-converting enzyme (ACE) 1 receptor, thereby preventing the conversion of vascular angiotensin I to angiotensin II (Peters et al. 2015). It lowers blood pressure by directly inhibiting the binding of ACE-1 receptors and angiotensin II, resulting in reduced constriction of blood vessels and decreased release of aldosterone. Patients undergoing haemodialysis for chronic renal illness who are treated with enalapril and irbesartan have higher haemoglobin and red blood cell counts, although these increases were not as great as those shown in the group that only used EPO once. To guarantee that enalapril is successful, hypertensive individuals with chronic renal function disorders need to be administered the proper dose of erythropoietin (Ordaz-Medina et al. 2010).

13.7.6 Vitamin C

Vitamin C (ascorbic acid) is uptaken by the body via the upper section of the small intestine after dissolving in water. For reducing metabolic oxidative stress, vitamin C is absolutely essential. Vitamin C has a crucial role in the production of collagen, the prevention of scurvy, maintaining gum and dental health, preventing muscle wasting, inhibiting artery hardening, synthesising antioxidants, and treating anaemia. Iron consumption in the liver is enhanced by vitamin C, which also lowers ferric iron and increases intestinal iron absorption. Lack of vitamin C lowers dialysis clearance in certain haemodialysis patients (Eiselt et al. 2001).

13.8 Route of Administration

When providing ESA to patients with CKD, the administration route should be considered. Research indicates that the subcutaneous approach is generally more economical than the intravenous approach, requires fewer injections overall, and minimises the need for extra support after injections. Despite this, since ESA may be comfortably given during dialysis, a lot of dialysis patients and possibly nephrology professionals choose the intravenous method. For example, most dialysis patients in the USA take ESA therapy intravenously. However, the option is restricted to subcutaneous administration in peritoneal dialysis patients and in the non-dialysis CKD population since patients are frequently told to use the least intrusive, subcutaneous, mode of injection at home without supervision. If a patient still fails to respond well to intravenous ESA administration after their dose is adjusted, they may consider switching to the subcutaneous method. The subcutaneous dose should be lowered by 30% when considering this move to prevent the chance of an abrupt increase in haemoglobin levels above target (Tonelli et al. 2003).

13.9 Risk Ratio of Treatment with ESA Agent-Related Hypertension

About 5% of individuals may experience an increase in hypertension after using an ESA, necessitating a change in dosage or another ESA. Seldom does complete ESA withdrawal become essential in the age of innovative and potent antihypertensive therapy. It is believed that the mechanism of ESA-induced hypertension involves ESA stimulating the vascular endothelium, which raises endothelin levels in the blood. Furthermore, vasospasm may develop from an increase in blood viscosity brought on by the haemoglobin increase linked to ESA medication. Thus, it is essential that patients getting ESA medication have their blood pressure checked on a regular basis (Public Health Advisory; Information for Healthcare Professionals) (Krapf and Hulter 2009).

13.10 Role of Erythropoietin Agent in CHF Patient with Anaemia

With almost half a million new cases each year, congestive heart failure (CHF) is a fast-expanding public health issue that affects approximately five million people in the USA alone. Seniors have the highest prevalence of CHF; ten out of every 1000 people over 65 have the condition. One modifiable concomitant condition of heart failure has been identified as anaemia. Based on the World Health Organization's (WHO) definition of anaemia, which is defined as haemoglobin <13 g/dL in men and < 12 g/dL in women, it has been estimated that up to 55% of CHF patients had anaemia overall, and up to 79% of patients with advanced, New York Heart Association (NYHA) class IV CHF (Silverberg et al. 2008). Anaemia has been related to adverse outcomes such as left ventricular dilatation and hypertrophy, as well as an increase in heart failure. Renal insufficiency is a typical occurrence in people with CHF; up to 50% of patients have renal impairment. As renal failure progresses, circulating erythropoietin levels drop, which in turn lowers haemoglobin levels and bone marrow erythrocyte synthesis. The EPO gene has been assigned a specific location on human chromosome 7, namely at position 7q21. The therapeutic use of recombinant EPO was first introduced in 1987 for patients suffering from anaemia caused by chronic renal insufficiency (Anand 2008).

13.11 Anaemia in Patients with Congestive Heart Failure

CHF-associated anaemia, despite being quite prevalent and having negative implications, may not receive sufficient recognition. About 22% of patients in the extensive Studies of Left Ventricular Dysfunction (SOLVD) experiment were diagnosed with anaemia. The severity of anaemia was directly linked to higher mortality rates. Lower haemoglobin levels were associated with a higher risk of mortality or hospitalisation for people with chronic heart failure, according to the findings of a different study named Outcome of Prospective Trial of Intravenous Milrinone for Exacerbations of Chronic Heart Failure (OPTIME-CHF). Specifically, for every 1 g/dL decrease in haemoglobin, there was a 12% increased risk (Silverberg et al. 2006). There are numerous factors that can produce anaemia linked with CHF. CHF is a persistent inflammatory condition characterised by significant increases in cytokines, including interleukins and TNF-α. The presence of these inflammatory cytokines hinders the generation and effectiveness of EPO. Additionally, they have the potential to impede the release of iron from tissue reserves. In addition, the enlargement of the heart and the associated kidney dysfunction might lead to a dilution of the blood. Anaemia in congestive heart failure can potentially be caused by the pharmacological medications administered for the treatment of heart failure. Angiotensin receptor blockers (ARBs) and angiotensin-converting enzyme (ACE) inhibitors are the backbone of contemporary treatment. These drugs can disrupt the production of red blood cells either by directly suppressing bone marrow activity or by reducing the effectiveness of the erythropoietin receptor. The administration of ACE inhibitors or ARBs in individuals with hypertension may lead to a reduction in haemoglobin levels of up to 0.3 mg/dL. Additional possible factors that can contribute to the condition are prolonged aspirin use leading to blood loss in the gastrointestinal system and bowel oedema causing reduced absorption of iron and other nutrients (Groenveld et al. 2008a). Furthermore, the reduced synthesis of EPO that occurs in conjunction with chronic renal illness can potentially be a contributing factor to the development of anaemia. On the other hand, having a high level of creatinine in the blood is a significant and separate factor that increases the risk of death in individuals with heart failure. Uraemia is linked to a higher accumulation of atherosclerotic plaque and, as a result, a greater occurrence of coronary lesions and events. The main cause of death for individuals with end-stage renal disease (ESRD) is cardiovascular events. Anaemia is made worse by chronic kidney disease (CKD), which increases the chance of heart failure (Nemeth 2008).

13.12 Growth Factor's Role in Treating CHF-Linked Anaemia

The implementation of EPO has had a substantial influence on the treatment of anaemia. Initially, EPO was employed for the treatment of anaemia in individuals with ESRD. Afterwards, the focus has turned towards its utilisation in anaemia due to congestive heart failure (National Kidney Foundation 2006). The implementation of neuro-hormonal inhibition in current therapy for CHF has led to substantial reductions in both morbidity and death. Nevertheless, addressing anaemia linked with CHF may result in further advantages. Echocardiograms in anaemic patients with chronic renal insufficiency who received EPO treatment showed a reduction in left ventricular (LV) mass and a decrease in LV dilatation. An experiment conducted by Silverberg et al. (2002) examined anaemic patients with advanced CHF who were treated with EPO and intravenous (IV) iron (Silverberg et al. 2002). The study found that an increase in haemoglobin (Hb) levels was linked to improvements in left ventricular ejection function (LVEF), a reduction in functional class, and a decrease in the frequency of hospitalisations. In a separate trial conducted by Mancini et al. (2003), anaemia was treated by administering EPO and oral ferrous gluconate (Mancini et al. 2003). This treatment resulted in improved exercise parameters, as seen by an elevation in peak oxygen consumption levels (VO^2). In addition to its ability to improve anaemia, EPO also provides other benefits that are not related to the production of red blood cells. It reduces the occurrence of programmed cell death in cardiac muscle cells. Recruiting endothelial progenitor cells improves the formation of new blood vessels, known as angiogenesis, and consequently promotes neovascularisation (Groenveld et al. 2008b). Patients with CHF may experience positive effects from the latter option due to the amelioration of ischaemia associated with coronary and cerebrovascular atherosclerosis. Because it may lower levels of cytokines and interleukins, EPO may have an anti-inflammatory effect. Hence, EPO may prevent myocardial infarction (MI) damage from progressing any further. The retina and other brain tissues have also shown evidence of EPO receptors. Animal studies have shown that EPO may help prevent kidney and central nervous system (CNS) damage caused by acute ischaemia. Comparable results have been observed in patients treated with darbepoetin for both anaemia and CHF. The preliminary results of a randomised, double-blind experiment that was carried out across several sites suggested that darbepoetin alfa raised haemoglobin levels, which improved quality of life (Tanner et al. 2002).

13.13 Conclusions

ESA treatment is recommended for anaemia in individuals with CKD to enhance the quality of life and reduce morbidity and death. Several concerns highlighted in this article must be considered while administering ESA for the treatment of anaemia in CKD. Moreover, as growing data indicate ESA medication may not be entirely harmless, it is crucial to strictly follow instructions and maintain haemoglobin levels within the suggested target ranges. Regularly monitoring the patient is necessary to enhance overall outcomes. Despite the use of appropriate pharmaceutical management, CHF remains a significant public health issue. Anaemia associated with CHF is a significant coexisting condition that, when addressed, might potentially reduce the chances of heart failure. The administration of growth factors has been linked to increased levels of haemoglobin, which in turn has been related to better outcomes in patients with advanced heart failure. Effective identification and treatment of this disease may require the cooperation of primary care physicians, nephrologists, and cardiologists. We need further research on the impact of growth factors on mortality and morbidity, the cost-efficiency of darbepoetin and EPO, and other topics. This should be done through larger studies that are randomised, double-blind, and placebo-controlled. Furthermore, the ideal goal levels for Hb have not been determined.

References

Anand IS (2008) Anemia and chronic heart failure. J Am Coll Cardiol 52(7):501–511

Barany H-J (2007) Muller maintaining control over hemoglobin levels: optimizing the management of anemia in chronic kidney disease. Nephrol Dial Transplant 22:iv10–iv18

Batchelor EK, Kapitsinou P, Pergola PE, Kovesdy CP, Jalal DI (2020) Iron deficiency in chronic kidney disease: updates on pathophysiology, diagnosis, and treatment. J Am Soc Nephrol 31:456–468. https://doi.org/10.1681/ASN.2019020213

Boran M, Dalva I, Gonenc F, Cetin S (1996) Response to recombinant human erythropoietin (r-Hu EPO) and L-carnitine combination in patients with anemia of endstage renal disease. Nephron 73:314–315

Brugnara C (2003) Iron deficiency and erythropoiesis: new diagnostic approaches. Clin Chem 49:1573–1578

Collins AJ, Ma JZ, Ebben J (2000) Impact of hematocrit on morbidity and mortality. Semin Nephrol 20:345

Eiselt J, Racek J, Trefil L, Opatrny K Jr (2001) Effects of a vitamin E-modified dialysis membrane and vitamin C infusion on oxidative stress in hemodialysis patients. Artif Organs 25:430–436

Emami Naini A, Moradi M, Mortazavi M, Amini Harandi A, Hadizadeh M, Shirani F, Basir Ghafoori H, Emami NP (2012) Effects of oral L-carnitine supplementation on lipid profile, anemia, and quality of life in chronic renal disease patients under hemodialysis: a randomized, double-blinded, placebo-controlled trial. J Nutr Metab 2012:1

Eschbach JW, Kelly MR, Haley NR et al (1989) Treatment of the anemia of progressive renal failure with recombinant human erythropoietin. N Engl J Med 20:158–163

Fandrey J, Jelmann WE (1991) Interleukin-1 and tumor necrosis factor-alpha inhibit erythropoietin production *in vitro*. Ann N Y Acad Sci 628:250–255. https://doi.org/10.1111/j.1749-6632.1991.tb17252.x

Ghorbani M, Vatannejad A, Khodadadi I, Amiri I, Tavilani H (2016) Protective effects of glutathione supplementation against oxidative stress during cryopreservation of human spermatozoa. Cryo Letters 37:34–40

Groenveld HF, Januzzi JL, Damman K et al (2008a) Anemia and mortality in heart failure patients: a systematic review and meta-analysis. J Am Coll Cardiol 52(10):818–827

Groenveld HF, Januzzi JL, Damman K et al (2008b) Anemia and mortality in patients with heart failure a systematic review and meta-analysis. J Am Coll Radiol 52:818–827

KDIGO Anemia Working Group (2012) KDIGO clinical practice guideline for anemia in chronic kidney disease. Kidney Int 2:279–335. https://doi.org/10.1038/kisup.2012.38

Krapf R, Hulter HN (2009) Arterial hypertension induced by erythropoietin and erythropoiesis-stimulating agents (ESA). Clin J Am Soc Nephrol 4:470

Larramendi CH, Marco FM, Garcia-Abujeta JL, Mateo M, de la Vega A, Sempere JM (2006) Acute allergic reaction to an iron compound in a milk-allergic patient. Pediatr Allergy Immunol 17:230–233

Macdougall IC (2008) Novel erythropoiesis-stimulating agents: a new era in anemia management. Clin J Am Soc Nephrol 3:200–207

Mancini DM, Katz SD, Lang CC et al (2003) Effect of erythropoietin on exercise capacity in patients with moderate to severe chronic heart failure. Circulation 107:294–299

National Kidney Foundation (2006) KDOQI Clinical practice guidelines and clinical practice recommendations for anemia in chronic kidney disease. Am J Kidney Dis 5:S1–S3

Nemeth E (2008) Iron regulation and erythropoiesis. Curr Opin Hematol 15(3):169–175

Ordaz-Medina SM, Gonzalez-Plascencia J, Martin del Campo F, Rojas-Campos E, Montanez-Fernandez JL, Espinoza-Gomez F, Cueto-Manzano AM (2010) Is systemic inflammation of hemodialysis patients improved with the use of enalapril? Results of a randomized, double-blinded, placebo-controlled clinical trial. ASAIO J 56:37–41

Pan X, Suzuki N, Hirano I, Yamazaki S, Minegishi N, Yamamoto M (2011) Isolation and characterization of renal erythropoietin- producing cells from genetically produced anemia mice. PLoS One 6:e0025839. https://doi.org/10.1371/journal.pone.0025839

Peters CD, Kjaergaard KD, Jensen JD, Christensen KL, Strandhave C, Tietze IN, Novosel MK, Bibby BM, Jespersen B (2015) Short and long-term effects of the angiotensin II receptor blocker Irbesartan on intradialytic central hemodynamics: a randomized double-blind placebo-controlled one-year intervention trial (the SAFIR study). PLoS One 10:e0126882

Provenzano R, Besarab A, Sun CH, Diamond SA, Durham JH, Cangiano JL et al (2016) Oral hypoxia-inducible factor prolyl hydroxylase inhibitor roxadustat (FG-4592) for the treatment of anemia in patients with CKD. Clin J Am Soc Nephrol 11:982–991. https://doi.org/10.2215/CJN.06890615

Rankin EB, Rha J, Selak MA, Unger TL, Keith B, Liu QHV (2009) Hypoxia-inducible factor 2 regulates hepatic lipid metabolism. Mol Cell Biol 29:4527–4538. https://doi.org/10.1128/MCB.00200-09

Rao M, Wong C, Kanetsky P, Girndt M, Stenvinkel P, Reilly M et al (2007) Cytokine gene polymorphism and progression of renal and cardiovascular diseases. Kidney Int 72:549–556. https://doi.org/10.1038/sj.ki.5002391

Silverberg DS, Wexler D, Iaina A (2002) The importance of anemia and its correction in the management of severe congestive heart failure. Eur J Heart Fail 4:681–686

Silverberg DS, Wexler D, Iaina A, Schwartz D (2006) The interaction between heart failure and other heart diseases, renal failure, and anemia. Semin Nephrol 26(4):296–306

Silverberg DS, Wexler D, Iaina A, Schwartz D (2008) The role of correction of anaemia in patients with congestive heart failure: a short review. Eur J Heart Fail 10(9):819–823

Souma T, Nezu M, Nakano D, Yamazaki S, Hirano I (2016) Erythropoietin synthesis in renal myo fibroblasts is restored by activation of hypoxia signaling. JASN 27:428–438. https://doi.org/10.1681/ASN.2014121184

Tanner H, Moschovitis G, Kuster GM et al (2002) The prevalence of anemia in chronic heart failure. Int J Cardiol 86:115–121

Tonelli M, Winkelmayer WC, Jindal KK et al (2003) The cost-effectiveness of maintaining higher hemoglobin targets with erythropoietin in hemodialysis patients. Kidney Int 64:295–304

Van Wyck DB, Roppolo M, Martinez CO, Mazey RM, McMurray S, United States Iron Sucrose Clinical Trials G (2005) A randomized, controlled trial comparing IV iron sucrose to oral iron in anemic patients with nondialysis-dependent CKD. Kidney Int 68:2846–2856

Wenger RH, Hoogewijs D (2010) Regulated oxygen sensing by protein hydroxylation in renal erythropoietin-producing cells. Am J Physiol Ren Physiol 298:F1287–F1296. https://doi.org/10.1152/ajprenal.00736.2009

Xue CE, Shen QH, Wang Y, Zhang JY, Lin FR (2015) Clinical significance of the serum EPO level in patients with iron deficiency anemia. Zhongguo Shi Yan Xue Ye Xue Za Zhi 23:1410–1414

Zachara BA, Gromadzinska J, Wasowicz W, Zbrog Z (2006) Red blood cell and plasma glutathione peroxidase activities and selenium concentration in patients with chronic kidney disease: a review. Acta Biochim Pol 53:663–677

Chapter 14
Gene Therapy in Molecular Biology and Drug Delivery

Dinesh Kumar Lakshmi Narayanan

Abstract The objective of gene therapy is to manipulate/modify the gene expression or modify living cell characteristics for therapeutic utilization. It aims to achieve therapeutic effects by either inserting novel genes, rectifying or substituting the genes that are abnormal, or modifying the pre-existing gene expression. The therapy shows potential in addressing a diverse array of genetic and acquired disorders. This includes certain types of cancer, hereditary genetic diseases like muscular dystrophy and cystic fibrosis, and other circumstances where genetic factors have a significant influence. Gene therapy can be accomplished by several mechanisms, namely, gene editing, gene replacement, gene addition, and gene silencing. Gene replacement involves the replacement of malfunctioning genes with the functional copy in the patient's cells. Gene editing pertains to technologies like CRISPR-Cas9 to edit the specific gene in the patient's cells. In a genetic condition caused by a missing gene, the gene addition procedure can add a functioning copy of that gene to the patient's cells to compensate for the deficiencies. In genetic disorders resulting from the overexpression of genes, gene silencing can be applied to reduce the expression or downregulation of these genes by techniques such as RNA interference (RNAi). Gene therapy holds promising possibilities and represents an innovative approach to drug delivery. It has the capacity to eliminate the necessity for frequent injections of proteins or medications, thereby alleviating the challenges associated with adhering to external drug schedules.

Keywords Gene editing · Gene addition · Gene silencing · Drug delivery · Molecular biology

D. K. L. Narayanan (✉)
Faculty of Pharmacy, AIMST University, Bedong, Kedah, Malaysia
e-mail: dineshkumar@aimst.edu.my

S. Bose et al. (eds.), *Concepts in Pharmaceutical Biotechnology and Drug Development*, Interdisciplinary Biotechnological Advances,
https://doi.org/10.1007/978-981-97-1148-2_14

14.1 Introduction

Gene therapy refers to the control of gene expression within a cellular environment, accomplished through the introduction of new or altered genetic material or the replacement of existing genes in order to cure the disease or to mitigate the clinical condition of a patient. The field of molecular medicine, specifically gene therapy, holds great potential for making substantial advancements in human health in the future. New approaches to therapy are expected to be developed for a wide array of hereditary and acquired disorders. The mechanism by which gene therapy works are (a) Gene Replacement in which a mutated or malfunctioning gene is replaced with a healthy gene copy, (b) Gene Silencing involves the inactivating or "silencing" the hyperactive or malfunctioning gene, and (c) Introduction of new or modified gene into the cell to treat the disease (Verma and Weitzman 2005).

Gene defects can result in a variety of diseases, such as hemophilia, muscular dystrophy, cystic fibrosis, combined immunodeficiency syndromes, and various cancers. Gene therapy can be employed to rectify or substitute defective genes. Gene therapy has shown remarkable and long-lasting therapeutic improvement when employed to treat combined immunodeficiency disorders. Gene therapy has the ability to eradicate cancerous cells, prevent cardiovascular diseases, ameliorate neurological disorders, and completely eradicate pathogens (Cotrim and Baum 2008).

Various methodologies, including addition, editing (repair), deletion or knockout (suppression), and others, can be implemented, dependent upon the underlying genetic issue.

Mechanisms Involved in Gene Therapy

(i) Adding the normal form of gene to enhance gene expression. For example, introducing the normal allele for human clotting factor IX with the aim of facilitating the expression of factor IX in individuals affected by hemophilia type B disease.
(ii) Introducing a genetically altered gene into cells to elicit novel traits in them. For example, inserting CAR (chimeric antigen receptor) structures in CAR-T cells and inducing immune reaction to target specific cancer cells.
(iii) Introducing suicide genes or gene-directed enzyme prodrug therapy (GDEPT) into cells to induce cytotoxicity substance or enzyme. For example, inserting thymidine kinase to induce salvage pathway thereby targeting cancer cells.
(iv) Gene therapies also involve vaccination by the insertion of a specific antigen to induce immunity, e.g., COVID-19 vaccine (Arabi et al. 2022).

14.1.1 Brief History of Gene Therapy

Waclaw Szybalski, a Polish American scientist, made important contributions to our understanding of genetic processes. He was aware that cells can integrate foreign DNA into their own genomes. He studied lambda phages to gain insight into gene transfer. A decade following the initial observation that bacteriophages possess the ability to transfer genetic material between bacteria, Howard Temin made a significant discovery that demonstrated the inheritance of specific genetic changes through viral infection. The researcher reached the conclusion that chicken cells that have been infected with the Rous sarcoma virus (RSV) exhibited stable inheritance of viral-specific gene changes, which included the necessary information for the production of RSV progeny, based on his experimental observations. The enormous therapeutic potential of introducing DNA sequences into patients' cells to address hereditary disorders was highlighted by renowned scientists Theodore Friedmann and Richard Roblin. Nearly 18 years after considerable study on gene therapy prospects, the retrovirus was used to generate a neomycin-resistant marker on tumor-infiltrating lymphocytes so that infiltrating lymphocytes could be detected in melanoma treatment. The first clinical experiment employing gene therapy was performed to treat severe combined immunodeficiency (SCID) disorder, which is caused by a defective adenosine deaminase (ADA) gene. In 1990, to treat a 4-year-old kid, a functioning copy of the adenosine deaminase (ADA) gene using a viral vector was injected. The success of this clinical trial showcased the possibility of gene therapy. The success of the above-mentioned trial also paved the way for many trials in the gene therapy field (Wirth et al. 2013). However, severe adverse effects were also reported while employing vectors of viral origin. These vectors apart from the severe adverse effects also caused immunological reactions as well as insertional mutagenesis. These incidents did hinder the advancement of this type of therapy. In 2000, a clinical trial was conducted in Paris for X-linked severe combined immunodeficiency, which is caused by mutations in the IL2RG gene located in the X chromosome. The vector employed to carry the normal copy of the gene induced the oncogene resulting in leukemia in 5 out of 20 infants who received treatment. In a separate gene therapy trial done in London, Thrasher and colleagues reported an additional instance of leukemia following retroviral-mediated gene therapy for SCID. An unfortunate case of death of an 18-year old happened in 1999 who was diagnosed with ornithine transcarbamylase deficiency and volunteered for the ornithine transcarbamylase gene encoding adenoviral vector gene transfer trial. While these setbacks hindered the progress of gene therapy, they also revealed the promising future and potential of this therapeutic approach. Many gene therapy products were approved for the treatment of various disorders. The first gene therapy product to receive regulatory approval was Gendicine, approved in China in 2003 for the treatment of head and neck squamous cell carcinoma. Following this, Strimvelis gained conditional marketing authorization from the European Medicines Agency (EMA) in 2016 for the treatment of severe combined immunodeficiency. In 2017, the USFDA approved Kymriah, a chimeric antigen receptor T-cell (CAR-T) therapy, for pediatric and young adult patients with relapsed or refractory acute lymphoblastic leukemia (ALL). Later in 2017, the FDA also approved Luxturna for the treatment of inherited retinal dystrophy caused by mutations in the RPE65 gene. Zolgensma, a gene therapy for spinal muscular atrophy (SMA), received FDA approval in 2019. These approvals represent significant milestones in the evolving field of gene therapy, showcasing its potential for treating a range of genetic and

Table 14.1 Examples of gene therapy products that are currently approved for use

Gene therapy product	Details of gene transferred/ modification	Indication
Vitravene (Fomivirsen)	An antisense oligonucleotide designed to target the UL123 gene of cytomegalovirus (CMV)	Cytomegalovirus (CMV) retinitis in patients with AIDS (acquired immunodeficiency syndrome) (withdrawn in 2002)
Gendicine	p53 gene (tumor suppressor gene)	Head and neck cancer
Oncorine (H101)	E1B-deleted adenovirus (genetically modified oncolytic adenovirus)	Late-stage refractory nasopharyngeal cancer
Rexin-G (Mx-dnG1/ DeltaRex-G)	A genetically altered version of cyclin G1	Soft tissue sarcoma and osteosarcoma
Neovasculgen (Cambiogenplasmid/ PI-VEGF165)	Vascular endothelial growth factor (VEGF) (angiogenic factor)	Peripheral vascular disease and limb ischemia
Imlygic (Talimogene Laherparepvec), melanoma	Addition of GMCSF	Advanced melanoma
Strimvelis (GSK2696273)	Adenosine deaminase gene	Severe combined immunodeficiency disorder
Collategene (Beperminogene perplasmid)	Hepatocyte growth factor (HGF)	Critical limb ischemia

acquired disorders. Several products are also now being reviewed by regulatory agencies (Arabi et al. 2022) (Table 14.1).

14.2 Gene Delivery System

Effective gene therapy relies on the significance of gene delivery systems, as they facilitate the transportation and introduction of therapeutic genetic material into precise target cells. The selection of the delivery mechanism is based on specific cell types, the characteristics of the genetic material involved, and safety concerns. They can be categorized broadly as viral vectors and nonviral vectors. Both the delivery systems have their own merits and demerits based on which system is selected.

14.2.1 Viral Vectors

A wide variety of viral vectors are available as delivery vehicles. They are specifically engineered to facilitate either temporary or sustained long-term expression. It comprises viruses composed of RNA and DNA, either double-stranded or single-stranded genomes. Vectors from the following five types of viruses are commonly employed in gene therapy: oncoretroviral, lentivirus, adenoviruses, adeno-associated viruses (AAVs), and herpes virus. Lentiviruses and oncoretroviruses are capable of integrating into the chromatin of the host cell, while adenoviruses, adeno-associated viruses (AAVs), and herpesviruses persist within the nucleus of the cell (Ghosh et al. 2020).

Adeno Virus and Adeno-Associated Virus (AAV) Vectors

Adenoviruses serve as frequently employed viral vectors in gene delivery systems, accommodating up to 7.5 kb of foreign DNA and demonstrating the capacity to initiate expression within the host cell. Adeno-associated virus (AAV) vectors are extensively utilized in gene transfer for the treatment of human diseases. Adenoviruses have a complex icosahedral capsid structure and a double-stranded DNA genome; they are nonenveloped, larger viruses (36–40 kilobase pairs (kb)). Persistent gene expression over an extended period is accomplished by employing adeno-associated virus vectors, and this capability extends to both dividing and nondividing cells. In contrast, adenoviruses are predominantly regarded for transient gene expression. Adeno-associated Viruses (AAVs) are nonenveloped, smaller viruses (4.7 kilobase pairs (kb) in size) with a more basic capsid structure and a single-stranded DNA genome. Recent advancements in the production of therapeutically valuable AAV capsids, the optimization of genome designs, and the application of novel biotechnologies have all made substantial contributions to the growth of the field. AAV's standing as a prospective therapeutic vector has been enhanced by preclinical and clinical achievements pertaining to gene silencing, editing, and replacement facilitated by AAV (Wang et al. 2019).

The adenovirus is a type of virus that does not have a protective outer layer. It mainly affects the upper respiratory tract, but it can also move to other parts of the body like the brain and bladder. The protein capsid of this virus is shaped like an icosahedron and has double-stranded linear DNA genome. The length of the genome can vary between 26 and 45 kilobases. In the adenovirus genome, inverted terminal repeats (ITRs) encapsulate the genetic material. These ITRs are like hairpins and can range in length from 30 to 371 base pairs. ITRs help with DNA replication without needing primases by acting as self-priming structures. To package the viral genome, it is important to have a packaging signal located on the left side of the genome. The adenovirus genome can synthesize 35 various proteins at different times during the viral gene transcription process, both in the early and late phases. The adenovirus genome encompasses five genes recognized as "early-phase" genes: E1A, E1B, E2, E3, and E4. The E1A gene holds significance in activating other viral genes, including E1B, E2, E3, and E4. These genes contribute to viral DNA synthesis and influence the host's gene expression. Specifically, E1B plays a crucial role in preventing apoptosis by interacting with and inactivating p53, facilitating virus replication. These viral vectors have been employed in the treatment of hereditary conditions such as hemophilia, muscular dystrophy, and certain forms of blindness.

Advantages of Adenovirus-Associated Vectors (Bulcha et al. 2021)

1. Excellent transduction effectiveness in both quiescent and proliferating cells
2. Long-term transgenic expression by chromosomal integration
3. Adaptability to a wide range of tissue targets
4. The availability of production methods that can be scaled up
5. Low pathogenicity and toxicity

Disadvantages of Adenovirus-Associated Vectors (Lundstrom 2018)

1. Small packaging capacity
2. Immunogenicity of adenoviruses during continuous administration
3. Rare risk of insertional mutagenesis
4. Some individuals may have pre-existing immunity to AAV due to previous exposure

Lentivirus

Lentiviruses, which have a packaging capacity of 8 kilobase pairs (kb), are a subgroup of retroviruses. These vectors consist of a positive-sense RNA sequence, which is a single strand transcribed into DNA and subsequently integrated into the host genome (Lundstrom 2019). Scientists have engineered lentiviral vectors from different types of lentiviruses to make it easier for them to deliver genes. Lentivirus possesses gag, pol, and env genes as its fundamental components. These genes are tasked with encoding various elements of the virus. The gag gene is responsible for encoding structural proteins, and the pol gene encodes enzymes essential for processes like reverse transcription and integration into the host cell's genome. The env gene encodes the glycoproteins found in the viral envelope. Lentivirus vectors are composed of long terminal repeats (LTRs), serving as cis-acting components necessary for packaging and integration. These vectors carry the gene of interest, regulated by a promoter. Pseudotyping of lentiviral vector envelopes enhances their tropism or cell-targeting capabilities. Examples of frequently used envelope proteins include the lymphocytic choriomeningitis virus G protein (LCMV-GP) and the feline endogenous virus RD114 envelope. Another example is the vesicular stomatitis virus glycoprotein (VSV-G), which broadens the host range. Lentivectors also contain packaging cassettes with trans-elements transcribed from the gag and pol genes, encompassing structural and enzymatic proteins crucial for vector particle generation and effective transduction of target cells. Accessory proteins, additional trans-elements encoded by lentiviruses, are essential for successful infection and replication within the host organism. Plasmid genes are modifiable to prevent insertional mutagenesis. Due to their ability to invade nondividing cells, lentiviral vectors are applied to target a wide range of cell and tissue types. They have found extensive use in gene therapy applications, including the treatment of specific immunodeficiencies and the development of chimeric antigen receptor T-cell therapies for cancer treatment (Cockrell and Kafri 2007).

Advantages of Lentivirus Vectors (Zheng et al. 2018)

1. They exhibit proficient transduction in various tissues.
2. They also exhibit stable and long-term transgene expression.
3. A diverse array of proliferating and nondividing cells can be efficiently infected by lentiviral vectors.
4. Lentiviral vectors are capable of accommodating transgene sequences of larger size.
5. Lentiviral vectors are quite safe and have a low immunogenicity.
6. Lentiviruses are compatible with housekeeping gene promoters as well as promoters for specific cells or tissues.
7. Lentiviral vectors swiftly induce transgene expression.

Disadvantages of Lentivirus Vectors (Schlimgen et al. 2016)

1. LVV-infected cells may activate oncogenes or inactivate tumor suppressor genes.
2. Risk of developing a replication-competent virus: lentiviral particles are capable of replicating in host cells.
3. Insertional mutagenesis.
4. Risk of insertional mutagenesis and oncogenesis in professional healthcare professionals and the exposed individual.
5. Integration of lentiviral vectors poses a risk of altering cellular gene expression patterns in the host genome.

14.2.2 Nonviral Vectors

Nonviral vectors, as employed in gene therapy, are delivery techniques that do not involve viral agents. Consistently regarded as less hazardous than viral vectors, these vectors elicit fewer adverse immune reactions and possess a diminished propensity for immunogenicity. The potential risk involved with viral vectors such as transgene mutagenesis also led to seeking a nonviral delivery system. Polymers, lipids, nanoparticles, inorganic particles, or combinations of these are some of the nonviral carriers that are being researched extensively (Zu and Gao 2021).

Polymer-Based Vectors

Notable nonviral gene therapy vectors are mostly cationic polymers. Due to their potentially enormous information-holding ability and flexible chemical structure, they have long sparked the interest of scientists. A complex (polyplex) capable of conveying the payload to a specific cell can be produced by neutralizing the negatively charged genetic material (Zu and Gao 2021).

Nonbiodegradable polymers such as polyethyleneimine (PEI) and biodegradable polymers such as poly(lactic-co-glycolic acid) (PLGA) are employed as vectors. The surface of polyethyleneimine (PEI) exhibits a substantial quantity of positive charge, enabling its adsorption in conjunction with negatively charged nucleic acid medications via electrostatic interactions. The adsorption process acts to protect the nucleic acid pharmaceuticals against degradation within the lysosomes. Yet it should be noted that in vivo degradation of PEI is not feasible, and its application development is hindered by its considerable toxicity. As it is not biodegradable and builds up around the cells, it will result in cytotoxicity. Biodegradable polymeric vectors have less cytotoxicity and can be administered continuously. PLGA is a biodegradable polymer that can be used to encapsulate genetic material and release it over time. As far as their chemical structure and batch consistency are concerned, synthetic polymers display a remarkable variety; nonetheless, it is possible that they have limited contact with the constituents of cells. Natural polymers have a notable degree of biocompatibility, yet they provide challenges in terms of batch-to-batch variance owing to disparities in their origins. Therefore, to ensure the quality of the product, it is necessary to implement control measures for the major properties of natural polymers.

Lipid-Based Vectors

Liposomes, lipid-based vectors, are the most commonly employed nonviral vectors. Liposomes are synthetic vesicles that replicate the structure of cell membranes by having lipid bilayers. These platforms offer a flexible and biocompatible method for encapsulating and transporting nucleic acids. The synthesis of a wide variety of lipids led to the creation of a system that can deliver genomes. The liposomal system is very adaptive and versatile. Liposomes have proven to be the most effective nonviral vectors to date. Various liposomal formulations have been previously developed, with cationic lipid-based liposomes exhibiting the highest efficacy in delivery. The lipids of cationic nature can be used to generate cationic liposomes, either alone or with additional lipids (known as helper lipids), depending on their chemical composition and the conditions used for liposome manufacturing. The interaction between cationic liposomes and DNA results in the spontaneous formation of complexes, known as lipoplexes, exhibiting a DNA loading efficiency of 100%. The complex is thought to be a resultant of the interaction between the negative charge of DNA with the positive charge groups of the liposomes. The proportion of lipids to DNA and the overall concentrations of lipids used to create these complexes are important factors in ensuring successful gene delivery. These factors can vary depending on the specific applications.

Liposomes have several advantages in gene therapy such as (i) inexpensive production as they are synthetic; (ii) they possess the ability to safeguard DNA against degradation facilitated by nucleases; (iii) they can transport chromosomal-sized DNA fragments; and (iv) they have the capability to selectively target specific cells or tissues. Several clinical trials have used a range of liposome-based vectors for the treatment of cancer. Despite their advantages, lipid-based vectors may face challenges such as potential toxicity at high concentrations, stability issues during storage, and limitations in cargo capacity compared to viral vectors (Guo and Huang 2012).

Nanoparticles

Nanoparticles are utilized in gene delivery due to their capacity to protect and transport genetic material to target cells. Nanoparticles can be designed with different materials and structures to enhance stability, improve cellular uptake, and ensure controlled release of the genes they carry. Nanoparticles are currently being studied as potential gene delivery devices due to their advantageous qualities such as safety, flexibility, affordability, and more importantly their property of nonimmunogenicity. These qualities make nanoparticles a promising solution for overcoming challenges in gene delivery. Significant progress in the fields of nanotechnology and genomics has resulted in the emergence of nanoparticle-based gene therapies for treating diverse diseases. Nanoparticle-based gene delivery systems consist of three main components: a cationic polymer that binds to nucleic acids, a polyethylene glycol steric stabilization agent, and a targeting ligand that attaches specific target cells via their receptors. Nanoparticles play a crucial role in gene therapy, being used for various purposes such as treating genetic disorders, cancer, and infectious diseases. They are also utilized in the development of RNA-based vaccines, including the COVID-19 vaccine (Chen et al. 2016).

14.3 CRISPR/Cas9 Gene Editing

CRISPR/Cas9 is a gene-editing methodology in which accurate modifications to DNA within living organisms can be performed. The CRISPR (Clustered Regularly Interspaced Short Palindromic Repeats) system, when coupled with the "CRISPR associated protein 9" enzyme, provides scientists with improved command over genetic material through the selective modification of genes. CRISPR-associated protein 9 (Cas9) is an RNA-guided DNA endonuclease. Its primary function is to act as molecular scissors that can precisely cut DNA at specific locations. The guide RNA directs Cas9 to the targeted DNA sequence, and the enzyme then induces a cut at that precise location. Deactivated Cas9, also referred to as Cas9 nuclease, has the remarkable ability to precisely modify the epigenetic landscape. This makes it a versatile tool for RNA-guided DNA targeting, genome control, and imaging applications. This tool is innovative as it allows for the precise rewriting of the epigenetic state within specific genetic sequences. Researchers can effectively utilize the programmable nature of Cas9 to leverage its RNA-guided capabilities for the precise targeting and manipulation of specific genomic regions of interest. This not only creates opportunities for genome editing but also offers a strong foundation for visualizing and controlling gene expression at the molecular level (Wang et al. 2016).

CRISPR/Cas, an acquired defense mechanism against viruses and phages mediated by Cas nucleases and crRNA-based DNA recognition, was first identified in bacteria and archaea. The CRISPR/Cas system is present in 40% of bacterial genomes and 90% of archaeal genomes. The CRISPR locus comprises a sequence of spacers, which are conserved repeated sequences interspersed with Cas9 protein and various nonrepetitive sequences. Cas nuclease is employed to break down infiltrating foreign DNA into minute fragments. These fragments are subsequently incorporated into the host genome CRISPR loci as spacers. The spacers function as transcriptional templates to generate crRNA in reaction to phage and virus infections. This crRNA instructs Cas to cleave target DNA sequences that are present on the invading viruses and phages. The functions of Cas proteins include crRNA biogenesis, spacer incorporation, target recognition, and DNA cleavage. The CRISPR/Cas system functions in a multistep process. Initially, it captures and integrates genetic material from viruses or plasmids into its CRISPR array. The array is then transcribed and processed into guide RNA molecules (crRNA). During interference, crRNA guides Cas proteins to specific sequences in invading DNA or RNA, leading to targeted cleavage or repression. This precise action provides immunity and establishes a memory of encountered threats. The system evolves by adapting to new challenges through the acquisition of additional spacers. The system is employed in genetic engineering as a powerful tool for precise genome editing.

Applications

1. Genome editing: The system offers an effective and versatile method for editing the genome to accurately alter the genomic segments and expedite the investigation of the roles played by target genes in biological processes and diseases. The technique has been employed to introduce specific genomic alterations in human cells by co-delivering plasmids encoding crRNA and Cas9.
2. Transcription regulation: The technique is utilized for examining the operation of genes and the transcriptional network. They can alter the transcription of specific genes by disrupting functional sites associated with transcription, which can lead to permanent modifications due to their irreversible nature modification.
3. Gene therapy: The technique can be employed for incorporating new protective genes, repairing disease-causing mutations, or disrupting endogenous disease-causing genes, thereby editing the genome. Genome modification possesses the capacity to eradicate diseases permanently.

Challenges and Considerations

1. Off-target effects: Cas9 can cleave at random locations in the genome.
2. Repair efficiency: The repair process does not always result in the desired alterations.
3. Delivery methods: Efficient distribution of CRISPR components into target cells might be difficult.
4. Ethical issues: Concerns about germline editing and unexpected consequences.

CRISPR/Cas9 has transformed genetic research and has significant therapeutic potential. Researchers are still refining and expanding its capabilities while tackling obstacles and ethical concerns linked with its use. The field of research keeps evolving, with CRISPR/Cas systems other than Cas9 being investigated for increased precision and utility (Zhang et al. 2014).

14.4 RNA Interference (RNAi)

RNA interference (RNAi) is a natural cellular process that involves the silencing or regulation of gene expression. It primarily occurs at the post-transcriptional level and is mediated by small RNA molecules. The small RNAs, including small interfering RNAs (siRNAs) and microRNAs (miRNAs), play a significant role in controlling the activity of genes. RNAi involves the pairing of a short RNA sequence with endogenous mRNA. RNA interference (RNAi) molecules are intended to prevent the translation of mRNA into proteins (Bobbin and Rossi 2016). RNA interference occurs through two distinct phases: the initiation phase and the effector phase. During the initiation phase, lengthy double-stranded RNAs (dsRNAs) are cleaved into discrete small RNA fragments of approximately 21–25 nucleotides, known as siRNAs. During the effector phase, guide strands of siRNA or miRNA are utilized and integrated into RNAi effectors like the RNA-induced silencing complex (RISC). RISC functions by cleaving mRNA and inhibiting translation, while RNA-induced transcriptional silencing (RITS) is responsible for regulating heterochromatin assembly (Xu et al. 2019).

14.4.1 Therapeutic Applications of RNA Interference (RNAi) Molecules

siRNA-mediated RNAi can be employed to target organs such as the lung, eye, ear, heart, pancreas, liver, kidney, tumors, blood, central nervous system, and peritoneum.

Cancer

A substantial amount of research focuses on cancer treatment, with the primary goal of preventing tumor growth. In general, target molecules consist of growth factors, receptors, antiapoptotic proteins, and other genes that are found to be significant in the proliferation of tumors. RNAi can be employed to silence specific genes such as oncogenes or genes responsible for promoting tumor growth. Furthermore, its utility in cancer treatment extends to the inhibition of cancer-specific gene expression by RNA interference.

Antiviral Treatment

RNAi can inhibit the replication of viruses by targeting and degrading their DNA. Novel approaches using RNA interference (RNAi) to combat viral infections depend on the targeted suppression of virus-specific genes using specific small interfering RNAs (siRNAs).

Genetic Disorders

RNAi can target and silence the expression of mutated genes responsible for genetic disorders.

Other Targets

RNAi can be employed and presently investigated in the treatment of neurodegenerative diseases, macular degeneration, hypercholesterolemia, rheumatoid arthritis, inflammatory bowel diseases, atherosclerosis, etc.

RNA interference has been demonstrated to be an exceptionally effective and targeted method for silencing genes of interest that are physiologically or pathologically significant. This is particularly true regarding genes that are considered "non-druggable," thereby creating innovative therapeutic opportunities. However, the therapeutic efficacy and utility of siRNAs will be significantly contingent upon their safe and efficient in vivo distribution, with the avoidance of any adverse effects (Aigner 2007).

14.5 Gene Therapy

Approximately 20 gene therapy products have received approval, and there are currently more than 2000 clinical trials in progress. These advancements provide enormous hope for the treatment of genetic diseases and incurable disorders. Although it has faced several challenges and many obstacles in the past 30 years, it has made significant advancements in the discipline of modern medicine and is currently making progress in clinical settings.

14.5.1 Hematological Disorders Treatment by Gene Therapy

Gene therapy has been investigated widely for the treatment of genetic and acquired hematological disorders. For this therapy, ex vivo gene therapy involving hematopoietic stem cells (HSCs) has been the focus. Red blood cells (HSCs) are generated by HSCs, which are also capable of self-renewal. The stable integration of therapeutic genes in the host genome is necessary for their presence in every progeny cell. The ability of these cells to integrate is facilitated through their cyclic divisions, which serve either as a means of self-renewal or to endure differentiation into a distinct cell type (Mendell et al. 2021).

Human allogeneic stem cell therapy is dependent on the functional attributes of hematopoietic stem cells. However, achieving successful engraftment of these stem cells in the recipient and preventing graft-versus-host disease requires overcoming immunological barriers. Ongoing and forthcoming research endeavors will maintain their emphasis on the discovery of regulators of hematopoietic stem cells, as well as the development of techniques for manipulating these stem cells both in laboratory settings and within living organisms. This includes the use of genome editing approaches to broaden the range, capabilities, and safety of therapeutic applications involving hematopoietic stem cells. Clinical trials are being conducted to improve the synthesis of hemoglobin to treat beta-thalassemia. In addition, trials are being conducted to modify the genetic mutation responsible for sickle cell disorder. Gene therapy involving HSCs provides significant potential for the treatment of various hereditary blood diseases and immunological deficiencies. Ongoing research and clinical trials are focused on developing techniques and addressing issues to enhance the safety and accessibility of these medications (Ng and Alexander 2017).

14.5.2 Cancer Treatment by Gene Therapy

Gene therapy trials for cancer have employed strategies to target tumors. Gene therapy in cancer treatment involves the introduction or modification of genetic material in cancer cells to inhibit their growth, induce cell death, or enhance the body's immune response against cancer. Currently, there are various effective strategies being used in gene therapy to target cancer. The approaches comprise the following:

(a) Gene expression to augment tumor's susceptibility to conventional drug or radiation therapy or to elicit apoptosis (cell death)
(b) Introducing a wild-type tumor suppressor gene
(c) By inhibiting the expression of an oncogene via RNA or DNA antisense
(d) Immune cell reactivity stimulation by increasing the immunogenicity of the tumor

Therapeutic genes comprise a transducing gene that imparts cytotoxic activity to tumor cells via an enzyme that converts a nontoxic prodrug into one. Immunogene therapy involves enhancing systemic immunity to tumors. The abnormalities caused by cancer are polygenic. This implies that substantial genetic diversity exists not only among individuals but also among tumors discovered in various anatomical sites within the same patient. Chemotherapies and gene therapies are dissimilar in that the former exhibits a restricted rate of success when applied to cancer patients. Consequently, every gene therapy treatment is thought of as an orphan drug. Also, one more disadvantage of cancer gene therapy in cancer is their inability to target the specific cell. It has been determined that the application of gene therapy methods to target metastases is futile for a variety of reasons. The challenges encompass the identification of dispersed metastases throughout the body, the emergence of resistance, and the existence of genetic and epigenetic heterogeneity. Gene therapy ought to possess the capacity to efficaciously address both primary and disseminated tumors while concurrently mitigating any adverse effects on healthy cells. By utilizing a gene product that can specifically induce its own translation in cancer cells or by stimulating the immune system to prevent the spread of cancer to other organs and tissues, this issue can be resolved (Das et al. 2015).

14.5.3 Central Nervous System and Neuromuscular Disorders Treatment by Gene Therapy

The efficacy of gene therapy in preventing and treating neurological disorders has been the subject of several randomized controlled trials. Initial attempts faced challenges, potentially attributed to the insufficient distribution of genetic material in targeted tissues. Some initial trials did not achieve the desired therapeutic outcomes, possibly due to the inadequate distribution of genetic material within targeted tissues. However, advancements in vector delivery techniques have led to improved transgene expression and enhanced safety in gene treatments. Gene therapy experiments in animal studies involving various neurodegenerative disorders have shown significant promise recently. Transgene methods offer a versatile means of delivering a range of nucleic acids for diverse therapeutic applications in gene therapy. This includes small interfering RNA (siRNA) for targeted gene silencing, complementary DNA (cDNA) to introduce functional copies of genes, microRNA for post-transcriptional gene regulation, guide RNA facilitating precise gene editing through systems like CRISPR/Cas, RNA or DNA editing enzymes enabling targeted genetic modifications, docking sites for DNA-binding proteins to regulate gene expression, antisense oligonucleotides for modulating RNA function, and short hairpin RNA (shRNA) serving as genomic cargo to interfere with specific gene expression. These wide-range options of the transgene delivery method allow for a tailored approach to address various genetic conditions and advancing treatments for diseases, particularly in the field of neurodegenerative disorders.

Extensive research has been conducted for employing gene addition as a potential treatment option for spinal muscular atrophy, Alzheimer's disease, Parkinson's disease, and Canavan disease. Gene addition effectiveness has been assessed in the precise delivery of complementary DNA (cDNA) for key genes such as aromatic l-amino acid decarboxylase (AADC), survival motor neuron (SMN), human aspartoacylase (ASPA), and nerve growth factor (NGF). It has been shown to be successful and well tolerated, leading to sustained clinical improvement in long-term studies.

At present, researchers are investigating the potential use of altered transcriptional regulators and gene editing techniques as novel therapeutic approaches for neurodegenerative diseases. The zinc finger proteins (ZFPs) are particularly intriguing in clinical experiments because of the similarities observed between rat and human proteins, as well as the relatively compact genomes of these proteins. Although a clinical trial was carried out to introduce the iduronate 2-sulfatase (IDS) gene into albumin loci as a potential therapy for mucopolysaccharidosis II, successful gene editing is likely to be difficult due to the possibility of off-target effects. Recent research has suggested that CRISPR/Cas9 nanocomplexes, which target BACE1, have the potential to decrease cognitive deficits and amyloid-associated pathologies in Alzheimer's disease. This demonstrates the significant potential of CRISPR/Cas9 gene editing for treating neurodegenerative diseases using either nonviral or viral vectors. However, there are several obstacles that need to be addressed prior to treating neurodegenerative disorders in humans by gene therapy, with safety concerns being a primary consideration. Gene silencing is a highly promising transgenic approach. RNA interference (RNAi) is a widely observed biological process where small interfering RNAs (siRNAs) play a crucial role in reducing protein synthesis by targeting and destroying the messenger RNAs (mRNAs) associated with them. Multiple clinical experiments have demonstrated that synthetic siRNAs can effectively inhibit specific proteins or genes in humans with generally favorable tolerability. The rationale for selecting a transgenic strategy is influenced by several factors, such as safety concerns, insertional mutagenesis, genotoxicity, and various disease conditions. Gene therapy for nervous system disorders faces several challenges, including the targeted delivery of therapeutic genes to specific areas of the brain, potential immunological reactions, and the requirement for sustained therapeutic effects over an extended period. Ongoing research and clinical studies are currently exploring innovative gene therapy approaches to address these challenges and improve potential treatments for various nervous system disorders (Chen et al. 2020).

14.6 Regulatory and Ethical Considerations

Gene therapy regulatory frameworks are critical for guaranteeing the safety, efficacy, and ethical conduct of gene therapy research and clinical applications. Regulatory authorities and government entities create these frameworks to give rules for the development, testing, and approval of gene therapy products. Due to the complexity of gene editing and gene therapy technologies, the public may find it challenging to comprehend their potential benefits and adverse effects. The safety, efficacy, and adverse effects of gene therapy should always be considered. The minimum requirement of gene therapy should be toward minimal adverse effects with safety and efficacy. Nonetheless, public and patient support is indispensable for the effective implementation of any novel technology. The integration of gene therapy into the treatment regimen for numerous hereditary, chronic, and genetic disorders is becoming increasingly significant. The treatment of various disorders can be achieved by employing gene therapy involving the somatic cell. Germline gene therapy involves the transmission of genetic modifications introduced into germ cells to subsequent generations. Therefore, they face social, legal, and ethical concerns. In order to prevent hazards to the general public and patients, appropriate biosafety and gene therapy protocols should be followed, in addition to the establishment of a regulatory framework and set of guidelines for gene therapy products. Gene therapy will achieve remarkable efficacy when regulatory concerns are accurately anticipated and adhered to in their entirety. Regulatory agencies such as the Food and Drug Administration (FDA) oversee the regulation of gene therapy products in the United States. The European Medicines Agency (EMA) is responsible for the evaluation and approval of gene therapy products in the European Union. Other regulatory bodies include the Pharmaceuticals and Medical Devices Agency (PMDA) in Japan, Health Canada, and the Therapeutic Goods Administration (TGA) in Australia (Pola and Padi 2023).

14.7 Conclusion and Summary

Gene therapy is a molecular medicine approach that manipulates gene expression to treat diseases or improve patient outcomes. It involves replacing mutated or malfunctioning genes with healthy ones, silencing hyperactive genes, and introducing modified genes into cells. Gene therapy has shown significant improvement in treating diseases like muscular dystrophy, hemophilia, cystic fibrosis, and cancers. Mechanisms include adding normal forms of genes, introducing genetically altered genes, introducing suicide genes, and introducing specific antigens. Gene therapy via viral vectors has faced numerous adverse effects, including insertional mutagenesis and immune reactions. However, these setbacks have also highlighted the potential of this therapeutic approach. Around 686 clinical trials resulted in the approval of Gendicine, Oncorine, Glybera, Strimvelis, Kymriah, Yescarta, Luxturna, and Zolgensma. Gene delivery systems are essential for facilitating the transportation of therapeutic genetic material into specific target cells. Adenoviruses are the most extensively used vector for gene transfer in human disease treatment, with advancements in production, genome design optimization, and novel biotechnologies contributing to their growth. Lentiviruses are also employed for the transportation of genes and are used in gene therapy applications, including immunodeficiency treatment and cancer treatment. Lentivirus vectors offer advantages such as efficient transduction, stable transgene expression, and compatibility with various cells. However, they can activate oncogenes, develop replication-competent viruses, and cause insertional mutagenesis. The liposomes, nanoparticles, and polymers can be employed as nonviral vectors, as they are less hazardous and less immunogenic. Nonbiodegradable polymers like polyethyleneimine (PEI) and poly(lactic-co-glycolic acid) (PLGA) can also be employed in gene therapy due to their flexibility and information-holding ability. Liposomes as vectors have several advantages, including inexpensive production, DNA protection, chromosomal-sized DNA fragment transport, and selective targeting of specific cells or tissues. However, they face challenges such as potential toxicity, stability issues during storage, and cargo capacity limitations compared to viral vectors. Nanoparticles are used in gene delivery to protect and transport genetic material to target cells. They are engineered with various materials and structures to enhance stability, improve cellular uptake, and ensure controlled release of the delivered genes. Gene therapy employing nanoparticles has also been developed for cancer and genetic disorders. CRISPR/Cas9 gene editing is an innovative method that allows accurate modifications to DNA within living organisms. It has applications in genome editing, transcription regulation, and gene therapy. However, challenges include off-target effects, repair efficiency, distribution methods, and ethical issues. RNA interference (RNAi) is an inherent cellular mechanism that controls gene expression by suppressing or reducing the activity of particular genes. Researchers are still refining and expanding their capabilities while tackling obstacles and ethical concerns related to their use. RNA interference (RNAi) molecules can be used in various therapeutic applications, including cancer treatment, antiviral treatment, genetic disorders, and neurodegenerative diseases. RNAi can silence specific genes, reduce tumor growth, and inhibit viral replication. It is also effective in treating neurodegenerative diseases, hypercholesterolemia, and rheumatoid arthritis. Gene therapy has made significant advancements in modern medicine. Ex vivo gene therapy techniques for treating genetic disorders are mostly focused on hematopoietic stem cells (HSCs). However, successful engraftment requires overcoming immunological barriers. Cancer gene therapy encompasses strategies that target tumors, modify genetic material, and augment the immune response against cancer cells. Effective strategies include expressing a gene, inserting a wild type of tumor suppressor gene, blocking oncogene expression, and enhancing tumor immunogenicity. However, gene therapy has a limited success rate among cancer patients and has limitations in targeting metastases. Combining gene therapy with chemotherapy, small molecule inhibitor therapy, radiation therapy, and immunotherapy could improve its effectiveness in treating various types of cancer. Gene therapy has also shown promise in treating neurodegenerative diseases, with early trials failing to achieve desired outcomes due to inadequate biodistribution. Advances in AAVs and nonviral delivery techniques have improved transgene expression and therapeutic safety. Regulatory frameworks are crucial for ensuring the safety, efficacy, and ethical conduct of gene therapy research and clinical applications. Regulatory agencies like the FDA, EMA, PMDA, Health Canada, and TGA oversee the regulation of gene therapy products. Public and patient support is essential for the effective implementation of any novel technology.

References

Aigner A (2007) Applications of RNA interference: current state and prospects for siRNA-based strategies in vivo. Appl Microbiol Biotechnol 76:9–21

Arabi F, Mansouri V, Ahmadbeigi N (2022) Gene therapy clinical trials, where do we go? An overview. Biomed Pharmacother 153:113324

Bobbin ML, Rossi JJ (2016) RNA interference (RNAi)-based therapeutics: delivering on the promise? Annu Rev Pharmacol Toxicol 56:103–122

Bulcha JT, Wang Y, Ma H, Tai PW, Gao G (2021) Viral vector platforms within the gene therapy landscape. Signal Transduct Target Ther 6(1):53

Chen J, Guo Z, Tian H, Chen X (2016) Production and clinical development of nanoparticles for gene delivery. Mol Ther Methods Clin Dev 3:16023

Chen W, Hu Y, Ju D (2020) Gene therapy for neurodegenerative disorders: advances, insights and prospects. Acta Pharm Sin B 10(8):1347–1359

Cockrell AS, Kafri T (2007) Gene delivery by lentivirus vectors. Mol Biotechnol 36:184–204

Cotrim AP, Baum BJ (2008) Gene therapy: some history, applications, problems, and prospects. Toxicol Pathol 36(1):97–103
Das SK, Menezes ME, Bhatia S, Wang XY, Emdad L, Sarkar D, Fisher PB (2015) Gene therapies for cancer: strategies, challenges and successes. J Cell Physiol 230(2):259–271
Ghosh S, Brown AM, Jenkins C, Campbell K (2020) Viral vector systems for gene therapy: a comprehensive literature review of progress and biosafety challenges. Applied Biosafety 25(1):7–18
Guo X, Huang L (2012) Recent advances in nonviral vectors for gene delivery. Acc Chem Res 45(7):971–979
Lundstrom K (2018) Viral vectors in gene therapy. Diseases 6(2):42
Lundstrom K (2019) RNA viruses as tools in gene therapy and vaccine development. Genes 10(3):189
Mendell JR, Al-Zaidy SA, Rodino-Klapac LR, Goodspeed K, Gray SJ, Kay CN, Boye SL, Boye SE, George LA, Salabarria S, Corti M (2021) Current clinical applications of in vivo gene therapy with AAVs. Mol Ther 29(2):464–488
Ng AP, Alexander WS (2017) Haematopoietic stem cells: past, present and future. Cell death discovery 3(1):1–4
Pola S, Padi D (2023) "Gene therapy": ethical and regulatory issues. In: Novel platforms for drug delivery applications. Woodhead Publishing, Cambridge, pp 607–618
Schlimgen R, Howard J, Wooley D, Thompson M, Baden LR, Yang OO, Christiani DC, Mostoslavsky G, Diamond DV, Duane EG, Byers K (2016) Risks associated with lentiviral vector exposures and prevention strategies. J Occup Environ Med 58(12):1159
Verma IM, Weitzman MD (2005) Gene therapy: twenty-first century medicine. Annu Rev Biochem 74:711–738
Wang H, La Russa M, Qi LS (2016) CRISPR/Cas9 in genome editing and beyond. Annu Rev Biochem 85:227–264
Wang D, Tai PW, Gao G (2019) Adeno-associated virus vector as a platform for gene therapy delivery. Nat Rev Drug Discov 18(5):358–378
Wirth T, Parker N, Ylä-Herttuala S (2013) History of gene therapy. Gene 525(2):162–169
Xu W, Jiang X, Huang L (2019) RNA interference technology. Compr Biotechnol 2019:560
Zhang F, Wen Y, Guo X (2014) CRISPR/Cas9 for genome editing: progress, implications and challenges. Hum Mol Genet 23(R1):R40–R46
Zheng CX, Wang SM, Bai YH, Luo TT, Wang JQ, Dai CQ, Guo BL, Luo SC, Wang DH, Yang YL, Wang YY (2018) Lentiviral vectors and adeno-associated virus vectors: useful tools for gene transfer in pain research. Anat Rec 301(5):825–836
Zu H, Gao D (2021) Non-viral vectors in gene therapy: recent development, challenges, and prospects. AAPS J 23(4):78

Chapter 15
Biotechnological Approaches in Infectious Diseases

Sabyasachi Banerjee, Sankhadip Bose, Amritesh Chandra Shukla, and Mirza R. Baig

Abstract Infectious conditions have a significant influence on the occurrence of illness and death worldwide. The management of human immunodeficiency virus (HIV), tuberculosis (Organization WH, Initiative ST, Treatment of tuberculosis: guidelines. World Health Organization, Geneva, 2010), and malaria infection poses significant difficulties, as seen by the continuous spread of these illnesses and their high fatality rates. The utilization of nano-sized carriers for the formulation of new and existing drugs offers a potential solution to various obstacles encountered in the treatment of diseases. These challenges include limited effectiveness of drugs at the desired target, insufficient drug concentration in areas where microbes hide and persist, and poor patient compliance due to drug-induced toxicities and prolonged treatment schedules. In addition, nanocarriers may be utilized for the formulation of vaccinations, which serve as a significant tool in our battle against infectious illnesses. This chapter gives a comprehensive overview of the present impacts of infectious illnesses, focusing on the primary factors that lead to illness and death. Furthermore, it highlights the potential of nanotechnology to improve the existing therapeutic approaches. It also provides a quick outline of the recent advances and future directions that might increase nanotechnology's worldwide impact.

Keywords Infectious diseases · Infection · Nanotechnology · HIV · TB · Malaria

---

S. Banerjee (✉)
Department of Pharmaceutical Chemistry, Gupta College of Technological Sciences, Asansol, West Bengal, India

S. Bose
School of Pharmacy, The Neotia University, Sarisa, West Bengal, India

A. C. Shukla
Department of Botany, University of Lucknow, Lucknow, India

M. R. Baig
Department of Clinical Pharmacy and Pharmacotherapeutics, Dubai Pharmacy College for Girls, Dubai, United Arab Emirates

S. Bose et al. (eds.), *Concepts in Pharmaceutical Biotechnology and Drug Development*, Interdisciplinary Biotechnological Advances,
https://doi.org/10.1007/978-981-97-1148-2_15

15.1 Introduction

Infections have a significant role in the overall burden of illness worldwide. Human immunodeficiency virus (HIV) infection, tuberculosis (World Health Organization, and Stop TB Initiative 2010), diarrhea, lower respiratory infections, and malaria are linked to high death rates. The mortality rates are most elevated in underdeveloped nations, where there may be restricted availability of resources such as anti-infectives and vaccinations (Roth et al. 2018). The mortality rates over time indicate that although the total number of fatalities is decreasing, there is still a significant difference in the number of deaths between countries with high socio-demographic index (SDI) and those with low SDI. Regrettably, clinical trials for infectious diseases (IDs) are falling behind in comparison to diseases like cardiovascular diseases and cancer (Roth et al. 2018). Consequently, it is essential to efficiently discover practical and effective solutions that facilitate improved treatment of IDs.

To efficiently handle IDs, several obstacles need to be addressed. The dearth of secure and efficacious medications is pivotal to our capability to address IDs (Kinch et al. 2014). Pathogen resistance may sometimes cause a lack of medication effectiveness, necessitating a more costly drug regimen for treatment (Munita and Arias 2016). The difficulties in treating infectious diseases are often exacerbated in nations with low SDI. Insufficient patient compliance with treatments and the need for continuous patient surveillance are significant barriers to achieving successful therapy (Sabaté 2003). Inadequate procurement methods, financial incapacity to afford medications, and the compromised stability of medicinal items under elevated humidity and temperature hinder the availability of effective therapies (Pheage 2017). To address these problems, it is essential to have a coordinated and emphasized approach at both scientific and regulatory levels.

Nanotechnology has the potential to revolutionize the identification and management of certain illnesses. These technologies, which use systems with a diameter around one-thousandth the thickness of a hair, have the potential to significantly influence the primary causes of illness and death worldwide. Extensive research has been conducted on nanosystems in recent decades, leading to the development of anesthetics, chemotherapeutics, nutritional supplements, imaging agents, and other products that have been authorized by the Food and Drug Administration (FDA) (Anselmo and Mitragotri 2016). Unsurprisingly, nanotechnology has undergone thorough evaluation in order to enhance the treatment of IDs. This chapter explores the potential of nanotechnology in treating IDs and provides an analysis of the advancements made in clinical and preclinical research, specifically focusing on TB, HIV, and malaria. Furthermore, the continuous influence of nanotechnology on the development of effective protein and mRNA vaccination against SARS-CoV-2 was acknowledged. Lastly, the obstacles involved in transitioning these technologies from the laboratory context to practical use in the clinical scenario were also addressed. This chapter excludes polymicrobial infectious diseases, alternative single-organism infectious diseases, the use of metal nanoparticles (NPs) (Huh and Kwon 2011) in disease management, and the topic of immunization (Irvine et al. 2015; Pardi et al. 2018).

15.2 Application of Nanotechnology in the Treatment of IDs

Treatment of IDs relies on the accessibility and appropriate administration of safe and effective drugs. The usage of nanotechnology has been extensively studied in preclinical research to enhance the effectiveness of ID medicines and make their administration easier. Researchers emphasize noteworthy preclinical research and address the obstacles to their use in clinical conditions.

15.2.1 Site-Specific Delivery to Infected Locations

There is a significant amount of interest in developing systems that allow for precise medicine delivery to specific infection locations. This is driven by instances where drugs have limited ability to reach infected tissues, such as antiretrovirals having low penetration into the lymph nodes and brain (Kirtane et al. 2016), as well as antituberculosis agents having low penetration into cavitary lesions (Sarathy et al. 2016). It also includes cases where drugs are distributed to sites that cause toxicity, like aminoglycosides in the ear (Huth et al. 2011), and drugs that kill beneficial bacteria (Tedijanto et al. 2018). The use of nanocarriers (NCs) for encapsulation has the potential to achieve therapeutic targeting; however, there is much opportunity for enhancement. Enhanced NC permeability at infection sites leads to improved accumulation on the target, as opposed to uninfected tissue. The surfaces of NCs may also be modified with ligands that have an affinity for infected tissues or microorganisms. The latter approach is referred to as ligand-mediated targeting or active targeting. Passive targeting refers to strategies that do not rely on specific ligands.

Targeted Medication Delivery Based on Ligands (Active) and Triggers

The levels of certain receptors/targets in affected regions may be increased. Placing ligands on the surface of NCs that adhere to such targets might increase their concentration in diseased areas. NCs that are conjugated with ligands may enhance their cellular absorption, hence assisting in treating intracellular infections. The mannose receptor is present in macrophages and dendritic cells that are found in tissues. Its function is to eliminate infections that have mannose-containing glycoproteins (Azad et al. 2014). NCs that have been modified with sugar moieties may exploit this uptake process and increase the absorption into macrophages that dwell in tissues. Alveolar macrophages in rats were investigated for drug exposure after the administration of ciprofloxacin-loaded liposomes that were functionalized with mannose and delivered as a pulmonary spray. Mannose-functionalized liposomes were 1.5 times more likely to be adopted by alveolar macrophages than unmodified liposomes. The targeted liposomes exhibited a 23-fold increase in bio-availability in the alveolar macrophages as compared to the free drug. Nevertheless, these trials were conducted in a model that was not infected, and the effectiveness of the treatment was not assessed.

NCs that specifically target epitopes shown on the pathogen's surface may enhance the delivery of drugs to infection sites. Pentamidine-loaded PLGA NPs were developed by Arias et al. with the purpose of combating *Trypanosoma brucei*, the causative agent of African sleeping sickness (Arias et al. 2015). A nanobody, or fragment of an antibody, was incorporated into the NPs to alter their properties. The 7.5% of the initial antibody mass that was fragmented was enough to pass through the microbial glycoprotein layer and reach the epitope. A tenfold decrease in dosage was achieved in vivo by nanoparticle-based therapy.

Cyclic peptides that attach to the surfaces of *Staphylococcus aureus* have been found using in vivo phage display screening. The regional dispersion of vancomycin-loaded NPs, which were modified with a fluorescein-tagged peptide, was assessed using confocal microscopy. The results revealed a selective buildup of these NPs in the lungs infected with *S. aureus* (Hussain et al. 2018). During a 20-day effectiveness trial, all of the mice that received the peptide-functionalized NPs were alive, but only around 40% of the animals in the free-vancomycin and nonfunctionalized NP groups survived. The studies indicate that the use of targeted NPs may enhance the effectiveness of therapy and perhaps lead to lower antibiotic dosages. Additionally, heat has been used to treat *S. aureus* infection by targeting the bacteria with antibodies. Kim et al. (2013) used biotinylated antiprotein A monoclonal antibodies to modify streptavidin-coated magnetic NPs. The quantification of the binding of antibody-functionalized NPs to *S. aureus* cells was performed using flow cytometry. The results demonstrated a binding that was 2.5 times greater than that of the NPs coated with IgG. Results from models of cutaneous infection in mice showed that *S. aureus* was almost eradicated twice more effectively by local injection of NPs functionalized with antiprotein A compared to the control group (Kim et al. 2013). Previous studies in the literature have mentioned the existence of NPs that target antibodies, but they did not assess their effectiveness in living organisms. Extensive investigation is now underway on the use of ligand-based targeting for delivering drugs to cancerous cells and particular tissues. However, further progress is needed to expand this technology further in the field of infectious diseases.

Drug molecules emitted during the circulation of NCs undergo nonselective dispersion. Therefore, it is beneficial to create binary systems that do not release any drugs into the bloodstream and instead release all of their contents when they come into contact with the pathogen. NCs composed of diblock copolymers of PEG and poly(phosphoesters) and triblock copolymers of poly(caprolactone), PEG, and poly(phosphoester) were employed for the targeted releasing of vancomycin in response to the presence of bacterial phosphatase (Xiong et al. 2012a) and lipase (Xiong et al. 2012b), respectively. These NCs rapidly disintegrate upon encountering these enzymes. The use of phosphatase-sensitive NPs in a zebrafish model of methicillin-resistant *S. aureus* infection resulted in a 20% increase in survival compared to the control group treated with free-drug. Stimuli-responsive NPs have also been used for the treatment of sepsis, with the objective of diminishing bacterial presence and inflammations in the lungs (Zhang et al. 2018). NPs were synthesized using a positively charged polyester (poly(β-amino ester)) to facilitate their breakdown by bacterial enzymes under acidic conditions, which are often present at infection sites. Infected tissues have been shown to have elevated surface expression of the protein intercellular adhesion molecule-1, which was used to coat the NPs. The NPs were loaded with both an anti-inflammatory and an antibiotic drug. In a murine model of lung *Pseudomonas aeruginosa* infection, the administration of specific NPs resulted in a nearly 50% increase in survival compared to the control group receiving free-drug therapy. An inherent constraint of these investigations is the absence of a comparison between stimuli-sensitive NPs and nondegradable NPs. For example, poly(β-amino esters) degrade rapidly in the absence of enzymes. This highlights the need for controls for nonbiodegradable NPs.

While nano-encapsulation may improve the concentration of drugs at infection sites, there are specific constraints when transitioning from rodent models to people. The extent to which improved permeability is exhibited in humans is uncertain since it is currently only established in rodent models. Moreover, several studies indicate that the level of targeting is closely associated with the intensity of infection (Bakker-Woudenberg et al. 1992, 1993). Therefore, it is uncertain whether targeting can effectively cure low levels of chronic infection. There should be some interesting comparisons made between the quantities of viruses and bacteria employed in rat illness models and those found in humans. If the quantities utilized in rodents are too elevated, then the data acquired up to this point may need re-evaluation. While medication concentration is heightened in infection sites in mouse models, a significant portion of the drug is also found in unintended areas. Drugs that exhibit toxicity at these unintended locations are not suitable candidates for nanoencapsulation. Additionally, it is necessary to ascertain if the rise in on-target accumulations is sufficient enough to support re-formulation. The drug targeting index, which evaluates the extent to which the medication is delivered to locations of high efficacy and low toxicity, is one metric that may be valuable in this context (Siegel et al. 2016). Even when no triggers are present, it is common for triggered-release systems to detect medicine leakage. The precise mechanism of release remains uncertain since even enzymes produced by the host organism might potentially serve as triggers. Furthermore, since NCs tend to be distributed throughout the body without targeting particular areas before releasing the medication, it is impossible to prevent drug release at unintended places. It is crucial to acknowledge that the intravenous method of delivering drugs may not be feasible in places with limited resources or in operations when drugs are given to a large number of people. Therefore, it is crucial to determine the suitability of these targeting techniques when the NCs are supplied by routes of administration that are relevant in a clinical context.

Passive Targeting

The pursuit of nanoencapsulation of anti-infectives has focused on two distinct objectives: delivering medications specifically to macrophages and targeting diseased tissues. The processes via which NCs deliver medications to these two locations, and hence the tactics for formulating them, are different. Macrophages are often targeted by viruses, such as HIV; fungi, such as *Aspergillus* species; and bacteria, such as *Mycobacterium tuberculosis*. Since NCs are mostly eliminated by these cells, they have been extensively used for delivering medicines specifically to macrophages. The administration of the antiviral drug azidothymidine in NPs made of poly(hexacyanoacrylate) resulted in enhanced accumulation of the drug in organs of the reticuloendothelial system (Arias et al. 2015), including the lungs, liver, and spleen (Löbenberg et al. 1998). After administering NPs to rats, it was shown that 60% of the medication dosage was present in the RES organs after 8 h. However, when treated simply with the soluble medicine, only 12% of the drug was found in the RES tissues.

NCs are used to improve drug deposition in regions of infection that are distinct from the RES organs. This phenomenon is derived from the discovery that blood arteries near sites of infection exhibit increased permeability, hence facilitating enhanced entrance of NCs. After injecting microbial peptidases under the skin of guinea pigs, Evan's blue dye was given intravenously to analyze the animals' vascular permeability at infection sites. As a good indication of vascular permeability, the dye accumulates in tissues thanks to its affinity for albumin in the circulation. While there was no development of the dye at the locations where saline was injected, there was an increased accumulation of the dye in the areas where peptidase was injected. The aforementioned effects were seen in enzymes obtained from *Aspergillus melleus*, *Pseudomonas aeruginosa* (Molla et al. 1989), *Candida albicans* (Kaminishi et al. 1990), and other sources. Furthermore, several investigations have shown a greater concentration of NCs near locations of infection in comparison to noninfected areas (Laverman et al. 1999). These investigations included infecting the calf muscle of rats with bacteria on one side and then administering radiolabeled liposomes to the animals via their veins. During a particular investigation including 48 subjects, it was shown that the number of liposomes in the abscess was about 40 times higher than in the muscle, as measured as a percentage of the injected dosage per gram of tissue.

The alteration in vascular permeability and medication targeting was somewhat contingent upon the infectious agent. Increased circulation duration may result in a greater number of traverses across both the intended and unintended tissues. Modeling studies indicate that increased penetration toward the target tissue may enhance targeting efficiency by prolonging circulation times (Siegel et al. 2016). Extended periods of circulation may be attained by diminishing the absorption of NCs by macrophages. Therefore, other ways are necessary to specifically target medicines to macrophages. One strategy is altering the outer layer of the NCs by including lipophobic polymers like PEG. This modification is believed to obstruct opsonization via steric hindrance. A study was conducted in a rat model of unilateral lung *Klebsiella pneumoniae* infection to investigate the impact of circulation half-life on the accumulation of infection at locations (Bakker-Woudenberg et al. 1992, 1993).

Infected rats were given liposomes tagged with Gallium-67, which had varying half-lives. Liposomes, irrespective of their duration in circulation, exhibited greater accumulation in the infected lung in comparison to the uninfected lung, hence confirming the previously mentioned increased permeability. Moreover, the build-up of liposomes exhibited a direct correlation with the severity of the illness. Significantly, in rats with the utmost extreme lung infection, the use of PEG-coated liposomes (with a half-life of 27 h) led to concentrations nearly four times greater than those achieved with uncoated liposomes (with a half-life of 19 h) (Bakker-Woudenberg et al. 1993). The advantage of PEG coating was diminished in conditions of low infection severity.

15.2.2 Systemic Anti-infective Sustained Delivery

Managing multiple IDs requires long-term therapy, which imposes a significant strain on both the healthcare system and the patient. Individuals infected with HIV require continuous therapy throughout their lives to achieve effective management of the viral load. To effectively manage TB, it is necessary to regularly ingest a combination of medications for a duration of many months to years, frequently numerous times each day (Organization WH, Initiative ST 2010). Extended and complex dosage schedules accompanied by a substantial number of pills lead to reduced patient compliance and, eventually, therapeutic failure. Occasionally, health-care personnel deliver drugs once daily for the sake of programming convenience, even if this may not result in achieving the most effective drug levels. Therefore, there is a need to create methods that decrease the frequency of dosage and simplify the dosing schedule. The pursuit of injectable nanocarriers that provide prolonged medication delivery has been ongoing. In general, these systems may be divided into two main categories—systems that regulate drug release using an excipient (such as a lipid or polymer) and those that depend on the gradual dissolving of drug crystals that are not easily soluble in the interstitial fluid.

Extensive investigation has been conducted on drug delivery systems composed of polyester materials that break down when exposed to esterases found in the body, such as poly(caprolactone) and poly(lactide-co-glycolide) (PLGA). Drug release is facilitated by the breakdown of the polymer and the diffusion of the drug (Jawahar and Meyyanathan 2012). This process may be adjusted by changing the lipophilicity of the monomer, the size of the particles (Makadia and Siegel 2011), and the length of the polymer chain. Additional polymers utilized for sustained release includedpoly(orthoesters),poly(anhydrides),poly(amides),andpoly(cyanoacrylates) (Manzari et al. 2021).

Liposomes, which are lipid vesicles capable of encapsulating both lipophilic and lipophobic medicines, may be used to provide sustained release of medications. NeXstar Pharmaceuticals invented MiKasomes, which is a liposomal version of amikacin (Fielding et al. 1998). Amikacin (Fielding et al. 1998) is a medication that has to be taken frequently and requires ongoing monitoring to ensure its effectiveness. In a rat model, MiKasomes were injected intravenously and disseminated to many tissues. Once there, they gradually released the medication, leading to a significant eightfold increase in the drug's half-life. The formulation development was terminated in 2000, despite initial favorable outcomes in clinical studies for *Mycobacterium* infections and urinary tract infections (Schiffelers et al. 2001).

One disadvantage of some polymeric and liposomal formulations is their reliance on significant amounts of excipients to regulate the release of drugs. This enlarges the volume of the injection and restricts the quantity of medication that may be delivered. In order to overcome this issue, researchers are now working on the development of aqueous dispersions containing nano-milled drug crystals (Trezza et al. 2015; Williams et al. 2015). After being injected subcutaneously or intramuscularly, the medication is absorbed into the bloodstream based on how quickly the drug crystals dissolve in the interstitial fluid (Williams et al. 2015). Therefore, the physicochemical characteristics of the medication and the dimensions of its crystals are crucial factors in influencing the kinetics of drug release. A frequently utilized method to extend the duration of medication absorption involves transforming the drug into prodrugs with increased lipophilicity or into salt forms with reduced solubility (McMillan et al. 2018).

A crucial factor to consider with sustained release systems is the possibility of extended exposure to concentrations of the drug that are below the therapeutic level. This prolonged exposure has the potential to result in the development of drug resistance (Wistrand-Yuen et al. 2018). The medication gradually decreases in quantity throughout the last stage of its release, resulting in a prolonged period of diminishing drug levels known as a "long pharmacologic tail" (Landovitz et al. 2016). Pulsatile release formulations, as opposed to prolonged release formulations, might potentially resolve this problem. Alternatively, less intrusive methods of removing the depot at the conclusion of the therapy time may be appealing.

15.2.3 Distribution of Drug Locally

Due to the fact that epithelial surfaces are often targeted by infections, delivering drugs directly to these surfaces has significant potential. When comparing systemic administration versus local delivery, it is probable that delivering the medicine directly to the spot of infection would increase the distribution of the drug to the intended target and reduce exposure to unintended targets. Naturally, its efficacy is restricted when it comes to systemic illnesses. Preclinical investigations, mostly conducted on rats, have found several benefits of employing nanoformulation for localized treatments. NCs enable the encapsulation of pharmaceuticals, resulting in prolonged drug release and thereby minimizing the need for frequent dosing. NCs may be designed to selectively release medications upon encountering certain stimuli, allowing for precise control over the timing and duration of drug delivery. NCs may improve the absorption of drugs into cells and boost their effectiveness against intracellular infections. Ultimately, the use of NCs may provide protection for sensitive pharmaceuticals, such as nucleic acids, against adverse physiological situations such as acidic environments or the presence of enzymes.

Pulmonary Delivery

Pulmonary delivery is a desirable method for treating respiratory infections because it allows for increased medication exposure directly at the target site. Formulations that are unable to permeate mucosal layers and biofilms have limited effectiveness in reaching the lungs owing to fast enzymatic inactivation, sequestration, and removal by coughing (Leal et al. 2017; Roy and Vij 2010). Mucous is produced by goblet cells located in the mucous membrane's epithelium in the lungs. Biofilms form when microorganisms gather on a surface and create a matrix outside their cells. This matrix consists of proteins, DNA, and large polysaccharides (such as alginate and *N*-acetyl glucosamine) (Han et al. 2017). Pulmonary delivery has been shown to produce higher local bio-availability of medicines compared to intravenous or oral treatment. When amorphous itraconazole NPs were delivered to the lungs of mice, the drug concentration was ten times greater compared to when the medication was administered orally in its commercial formulation (Sporanox) or in its unformulated state (Vaughn et al. 2006). In a mouse model of *Aspergillus fumigatus* infection, the researchers noticed that the group treated with pulmonary NPs had a higher median survival time (7.5 days) compared to the group treated with oral Sporanox (5 days) (Alvarez et al. 2007). The observed advantages cannot be solely ascribed to nanoformulation since the significant improvement is likely due to the change in delivery route from oral to pulmonary. Nevertheless, the low solubility of itraconazole in water may have prevented the direct delivery of the medication via the lungs without any formulation. In a study conducted by Wong et al., the researchers compared the efficacy of liposome-encapsulated ciprofloxacin, a medicine that dissolves well in water, with the efficacy of the free form of the antibiotic when administered to the lungs (Wong et al. 2003). The liposome-encapsulated medication had a half-life in mouse lungs that was almost twice as long as the free drug, with the liposome-encapsulated drug having a half-life of around 3 h compared to the free drug's half-life of around 1.5 h. All mice infected with *Francisella tularensis* in a mouse model died 14 days after infection, and there was no difference in mortality between the untreated and free-drug-treated groups. However, pulmonary liposome-treated animals did not have any mortality. Researchers were pleasantly surprised to find that liposomes were effective even when administered intravenously. The specific and nonspecific exposure to the targeted and unintended targets may have been better understood if the pharmacokinetics of pulmonary and intravenous liposomes had been compared head-to-head. Bacteria control their collaborative development and establish biofilms by the secretion and detection of signaling chemicals that induce pathogenicity. Compounds that interfere with this quorum-sensing process may be used to eliminate biofilms. However, mucous coating, especially in those with cystic fibrosis, makes it more difficult to deliver medications to bacterial biofilms. Nafee et al. (2014) encased quorum-sensing inhibitors in lipid NPs covered with polysorbate 80 to circumvent the resistance of mucus in the respiratory system. The PEG in this polysorbate 80 makes it a co-polymer surfactant (Nafee et al. 2014). Compared to the medicine that is available for free, the quorum-sensing inhibitor enclosed in NPs demonstrated superior control of virulence under laboratory conditions, as determined by measuring the quantities of a bacterial metabolite called pyocyanin. Confocal microscopy revealed that NPs coated with polysorbate 80 spread into porcine mucus. However, neither the free drug nor the NP form of the quorum-sensing inhibitor was tested for its ability to be transported. While these investigations provide a fascinating idea, it is crucial to assess its effectiveness in an in vivo model of cystic fibrosis. Another method for disrupting biofilms includes using polymeric NPs that release nitrous oxide. In order to provide a continuous and prolonged release of nitric oxide, which naturally has a brief duration of action, Duong et al. (2014) chemically bonded nitric oxide to star-shaped polymers. The scientists validated the continuous release of nitric oxide from the star-shaped polymer NPs in laboratory conditions using a bacterial reporter strain. In addition, it was shown that NPs were capable of reducing *Pseudomonas aeruginosa* biofilm development by more than 70% as compared to untreated controls (Duong et al. 2014). Nevertheless, in vivo validation was not conducted. *N*-Acetyl cysteine, a kind of mucolytic drug, may enhance the permeation of NCs by breaking down disulfide bonds in mucous. Patients with cystic fibrosis who took *N*-acetyl cysteine had a greater than fivefold increase in the diffusivity of DNA-loaded poly(lysine) polyplexes in their sputum (Suk et al. 2011). In order to confirm the efficacy of *N*-acetyl cysteine in living organisms, the researchers administered inflammatory lipopolysaccharide from *P. aeruginosa* to mice, resulting in increased production of mucus in the lungs. The transfection effectiveness of the polyplex in mice treated with lipopolysaccharide was about ten times lower compared to healthy animals. Remarkably, prior administration of *N*-acetyl cysteine restored the effectiveness of transfection in the inflammatory model. This research showcases the suppressive impact of mucous on the transportation of drugs to the lungs, as well as the advantages of merging mucolytics with nanotherapeutics.

Vaginal Delivery

Herpes simplex virus (HSV) and HIV may enter the body via the vaginal canal. The usage of microbicides administered locally as a preventive measure may decrease the likelihood of infection. To minimize the possibility of infection, it is advisable to have maximum coverage throughout the vaginal tract throughout the period of infection. Multiple obstacles hinder the achievement of full covering of the vaginal canal and the extended presence of the medication delivery system (das Neves et al. 2015). Initially, formulations have the potential to escape from the vaginal canal after the administration of a drug delivery device. Furthermore, the vaginal mucosa is coated with a layer of mucous, a thick biopolymer composed of glycoproteins known as mucins. This layer serves as a barrier that hinders the movement of medication particles. Furthermore, the mucous layer undergoes regular shedding, resulting in the potential loss of any particles trapped inside it (Leal et al. 2017). Furthermore, the vaginal surface has many folds or rugae, which diminish the accessibility of drugs. Ultimately, the acidic nature and enzymatic components of the vaginal fluid may break down vulnerable medications (das Neves et al. 2015). In order to improve the duration of vaginal engagement, a prevalent approach has attempted to design systems that can efficiently penetrate the mucous layer and reach the underlying mucosa. In a research conducted by Saltzman et al. (Cu et al. 2011), the vaginal absorption of dye-loaded PLGA NPs was examined. These NPs were either coated with PEG or avidin or left untreated. The largest amount of leaking from the vaginal canal occurred with uncoated NPs, with about 13% of the administered dosage being released within half an hour. The vaginal lavage, which permitted the accumulation of the mucous layer, included a high concentration of mucoadhesive NPs coated with avidin. PEG-coated NPs exhibited the greatest tissue absorption, albeit it was confined to the lower reproductive tract. The enhanced absorption of PEGylated NPs into tissues may be due to their increased ability to move through the mucous layer. This discovery was shown by various particle tracking studies conducted by the Hanes laboratory (Lai et al. 2007). To improve vaginal retention, one alternate approach is to deliver NPs that are embedded in films. While NP dispersions have the tendency to quickly escape from the vagina, films have the ability to retain the NPs in position for a certain period of time. Cunha-Reis et al. (2016) provided evidence that incorporating efavirenz, an anti-HIV medicine, into NPs and encapsulating them in a water-soluble film resulted in reduced leaking and increased drug concentrations in tissues compared to dispersing the NPs alone. Nevertheless, the enhanced levels of tissue concentrations were only detected during the first hour of injection, and not during subsequent periods. This research emphasizes two obstacles associated with NP-based vaginal medication delivery methods. Initially, there is a limited absorption of NPs in the upper reproductive canal, which raises doubts about the extent of protection provided by these systems. Furthermore, in experiments conducted on rodents, the levels of drugs in their bodies drop to a point where they cannot be detected around 24 h after administration. This indicates the ability to provide continuous drug delivery may only be feasible for a very short period of time. To address the problem of residence, NCs might be included in macro-drug delivery systems such vaginal rings, which have been shown in clinical studies to have prolonged residence durations. Although there are limits, the use of nanosystems for vaginal administration has shown therapeutic promise in experiments conducted on mice. Ensign et al. (2012). demonstrated that PEGylated PLGA NPs, when administered in a hypotonic solution, were quickly absorbed into the vaginal folds by advection. Acyclovir monophosphate NPs had a protective effect in about 54% of mice infected with HSV-2 when administered 30 min after infection. Conversely, the soluble medication provided protection to just 16% of the mice when exposed to the virus. Mice with lethal HSV-2 infection survived better when treated with a small interfering RNA (siRNA) combination consisting of lipofectamine and dsRNA that inhibits the expression of genes encoding viral envelope and DNA binding proteins (Palliser et al. 2006). PLGA NPs with cationic spermidine were utilized to achieve siRNA-mediated knockdown of the same target. These NPs have shown superior protective effects compared to lipofectamine lipoplexes (Steinbach et al. 2012). The researchers discovered that the PLGA NPs exhibited comparable knockdown effectiveness to lipoplexes while inducing less inflammation, as shown by the quantification of neutrophil infiltration using immunohistochemistry. The enhanced protective efficacy of PLGA

NPs was ascribed to their capacity to induce less inflammation (Kirtane et al. 2016). It is important to highlight that in all the trials mentioned, the interaction with the pathogen occurred within a short timeframe ranging from a few minutes to a few hours after treatment. Implementing such treatment regimens in resource-limited contexts is challenging, highlighting the need for long-acting local systems. However, the recent regulatory approval of Starpharma's VivaGel, an antimicrobial substance made of dendrimers, has shown its effectiveness against HSV, HIV, and bacterial vaginosis. VivaGel can be used as a gel or as a coating for condoms. This success highlights the immense potential for growth and the versatility of nanotechnological systems in this field.

Topical Delivery

Chronic and nonhealing wounds contribute to rising healthcare expenses and death rates (Sen et al. 2009). These wounds are a result of disruptions in the wound-healing process and persistent infection caused by opportunistic microorganisms. Chronic infections, in contrast to acute infections, are characterized by the presence of a concentrated group of microorganisms and have less severe symptoms. These infections are localized and occur repeatedly throughout time. In addition, whereas acute infections may be adequately controlled with drugs, chronic infections do not exhibit a positive response to traditional antibiotic treatment. Research revealed that biofilms were present in 60% of clinical samples taken from chronic infection sites, in contrast to just 6% observed in samples from acute infections (James et al. 2008). The findings indicate that biofilms have a significant impact on chronic infections and nonhealing wounds. Developing ways to reduce biofilm resistance might potentially be beneficial in treating such diseases. Topical medication administration and pulmonary delivery have similar obstacles and treatment goals. Both methods are affected by biofilm-forming opportunistic organisms including *P. aeruginosa* and *S. aureus*, which may cause infections in both the lungs and the skin. Consequently, several ideas of developing nanosystems that can penetrate biofilms for delivering drugs to the lungs may also be used to deliver drugs topically. The NPs were synthesized using sodium nitrite and reducing sugars to develop a sustained release of nitric oxide. The NPs exhibited sustained nitric oxide release for a duration exceeding 1 month. Nitric oxide-releasing NPs demonstrated a reduction in total bacterial burden in murine skin infection models of *S. aureus* and methicillin-resistant *S. aureus* biofilms, as compared to both unloaded NPs and the control group (Martinez et al. 2009). An effective approach to treating wound infections involves the use of silver solutions administered using cotton gauze. Nevertheless, because of the limited duration of effectiveness and abrupt exposure to elevated levels, the therapy must be administered with great frequency and may have harmful effects on the host cells (Mohiti-Asli et al. 2014). A suggested approach involves enclosing permeable silver microparticles inside a poly(lactide) nanofiber matrix. The hypothesis posited that a reduced rate of silver release from the scaffold might mitigate its toxicity to human cells while preserving its anti-infective characteristics. The scaffold facilitated the gradual release of silver, yet it did not decrease toxicity to host cells (Mohiti-Asli et al. 2014). Continued research in this domain might be very beneficial.

15.3 Eliminating Anti-infective Drug Resistance

The development of resistance to pathogens is a major and increasing challenge in the fight against IDs. Pathogens may acquire resistance by a number of mechanisms, including the production of drug-inactivating enzymes, an intracellular life cycle, biofilm formation, decreased metabolic activity, or the fact that they are obligate intracellular pathogens (Munita and Arias 2016). Ineffectiveness against intracellular infections may be caused by poor drug diffusion through the membrane of the host cells and its active efflux via drug transporters (Carryn et al. 2003). Since pathogens and drug molecules are not colocalized inside host cells, effectiveness is reduced, adding another layer of complication. Aminoglycosides, for instance, prefer to accumulate in the lysosomes, making them useless against cytosolic pathogens (Tulkens and Trouet 1978). Therefore, methods that boost drug accumulation within cells and direct medications to specific subcellular sites may be more effective. By enclosing anti-infectives in NCs, it may be possible to improve their absorption by cells and to modify their intracellular disposition. Improvements in gentamicin administration to cytoplasmic organisms like *Listeria monocytogenes* and *Salmonella typhimurium* were tried by Lutwyche et al. (1998). Fusogenic liposomes, which are stable at neutral pH but dissolve in the acidic environment of endolysosomes, were used to encapsulate gentamycin. The medication is released into the cytoplasm as the endosomal membrane is destabilized by lipid fusion. The scientists used two liposomal formulations, one having a fluorescence resonance energy transfer pair and the other without, to verify the acid-mediated disintegration of the liposomes. This mixture's low baseline fluorescence was due to the closeness of the fluorescent molecules. However, when the solution was made acidic, the liposomes ruptured and the lipids mingled, which dequenched the fluorescence. Gentamicin loaded into pH-sensitive liposomes was three times as effective as the free antibiotic in an in vitro model of infection with recombinant *S. typhimurium*. Although pH-insensitive liposomes were taken up by cells to a greater extent, their activity lagged behind that of the pH-sensitive formulation. The authors fluorescently labeled liposomes that were released and metabolized by cellular enzymes to investigate the process. Only pH-sensitive liposomes produced the marker intracellularly in mouse macrophages, as shown by confocal imaging. Bafilomycin, an endosome acidification inhibitor, suppressed marker release. The importance of intracellular targeting was shown, and the molecular specifics of the roles played by each component of the formulation were presented in this work.

Clemens et al. (2012) developed mesoporous silicon NPs with isoniazid to target intracellular *M. tuberculosis*. The NPs' pores were sealed with β-cyclodextrin, effectively preventing drug leakage under neutral pH conditions. However, as the pH becomes acidic, the plug dissolves, leading to the release of the medication. Sémiramoth et al. (2012) chemically linked penicillin G to hydrophobic squalene by a pH-sensitive ester bond. The cells quickly internalized NPs produced from this combination and then released the drug inside the endolysosomes. Consequently, there was enhanced efficacy compared to the unbound medication. The process of analyzing and understanding the ways by which bacteria take up NCs on their surface has been utilized to enhance the transport of substances within cells. Invasin, a protein produced by *Yersinia pseudotuberculosis*, attaches to integrins on the surface of host cells. This attachment leads to a rearrangement of the host cell's cytoskeleton and allows the bacteria to be taken up into membrane-bound structures. Labouta et al. (2015) conducted reverse engineering of this process by linking liposomes to the C-terminal portion of invasin, known as InvA497, on the surface. Confocal microscopy investigations conducted in Hep-2 cells demonstrated that InvA497-functionalized liposomes exhibited a significant increase in cell binding (>30-fold) compared to albumin-functionalized liposomes. Furthermore, this enhanced cell binding was eliminated in competition trials. InvA497-functionalized gentamycin-loaded liposomes exhibited a 30% decrease in bacterial load in a laboratory model of *Y. pseudotuberculosis* infection, but albumin-functionalized liposomes did not have any impact. This study has shown that the addition of InvA497 to the surface of polymeric NPs has positive effects on both the absorption of the NPs by cells in a laboratory context and their ability to kill microbes. This notion is incredibly creative and would benefit from in vivo examination. Moreover, if these systems are administered orally, then it is crucial to comprehend the stability of the peptides in the enzyme-abundant gastrointestinal environment. Understanding the immune response to systemically delivered NCs functionalized with bacterial peptides is crucial for determining their pharmacokinetics and tolerability.

Efficient intracellular transport is crucial for molecules like messenger RNA, which have garnered significant interest for their use in immunization. The majority of nucleic acid and protein delivery systems that are based on polymers are influenced by the pioneering research conducted by Langer and Folkman (1976). Their work demonstrated that macromolecules, including proteins, could be enclosed within small carriers made of polymers. This was initially considered improbable because of the usage of organic solvents in the synthesizing process and the perceived impermeability of polymers to macromolecules (Ditlev et al. 2018). After successfully demonstrating the proof-of-concept using polymers, researchers showed how RNA could be encapsulated in lipid carriers and its usefulness in immunization. NCs play a crucial role in facilitating the transportation of mRNA through the cellular membranes, allowing for endolysosomal escaping and the release of nucleic acid in the cytoplasm. For instance, self-replicating mRNA that codes for an antigen protein was combined with a polymer based on cationic dendrimers. Animals that received mRNA NPs exhibited immunological responses characterized by the production of antibodies and the activation of $CD8^+$ T-cells. Furthermore, these animals were fully shielded against fatal infections caused by *Toxoplasma gondii*, H1N1 influenza, and Ebola virus. Clinical trials have examined mRNA vaccines using lipid NCs, with participation from industries like Moderna, BioNTech, Acuitas Therapeutics, Pfizer, GlaxoSmithKline, Translate Bio, Sanofi, and others. Preliminary findings demonstrate that lipid-NP-formulated mRNA vaccines targeting SARS-CoV-2 and H7N9, H10N8 influenza viruses are well tolerated and elicit a strong humoral response (Jackson et al. 2020; Feldman et al. 2019). More precisely, mRNA-based vaccines have shown a level of effectiveness over 90% in combating SARS-CoV-2. The long-term viability of mRNA-based vaccinations over extended storage and transportation may pose significant challenges to their usage in resource-constrained environments. NPs containing peptide-based antigens have been produced for the purpose of vaccinating against SARS-CoV-2 (Tian et al. 2021). Nanoencapsulation has been utilized to target antibiotics to intracellular vesicles since bacteria and NCs have the ability to aggregate in these vesicles. Couvreur et al. demonstrated the co-localization of poly(cyanoacrylate) NPs with *S. typhimurium* in phagosomes of macrophages utilizing confocal and transmission electron microscopy. This led to an improved targeting of ampicillin inside the cells (Pinto-Alphandary et al. 1994). Toti et al. (2011) conducted a separate investigation and discovered that PLGA NPs, which contained a fluorescent dye, gathered together with chlamydial inclusion bodies in infected human lung epithelial cells (Toti et al. 2011). Experiments conducted in a controlled laboratory environment showed that antibiotics (rifampicin and azithromycin) were shown to be efficacious when administered promptly after infection, but their effectiveness diminished when administered after 24 or 48 h. Nevertheless, the activities of the NPs were partially recovered even when administered 24 and 48 h after infection. NPs possess the ability to hinder, interrupt, or disseminate bacteria found in biofilm infections (Han et al. 2017). The extracellular polymeric matrix in the biofilm acts as a diffusion barrier, providing protection to bacteria against elevated antibiotic concentrations, which often results in the development of persistent infections. Teirlinck et al. (2018) demonstrated that when gold NPs were exposed to laser irradiation, nanobubbles were developed, which caused the diffusion barrier to be disrupted. The local disturbance of the biofilms resulted in heightened permeability, thereby rendering them more susceptible to the antibiotic tobramycin.

15.4 Clinical and Preclinical Nanotechnology Advancements

This section focuses on the modern nanotechnologies used in the treatment of important infectious diseases such as tuberculosis, malaria, and HIV infection. The information indicates that nanotechnology has been extensively researched in clinical contexts and with large animals for the purpose of treating and preventing HIV infection. Conversely, nanotechnology was explored with less enthusiasm for TB and malaria. These IDs have mostly been tested in preclinical experiments involving animals. The data are listed in Table 15.1.

Table 15.1 Nanotechnology for the treatment of IDs in preclinical and clinical facilities (https://clinicaltrials.gov/)

Disease name	Formulation	Drug name	Testing phase	Limitation	Advancement
HIV	Injectable NPs	Rilpivirine	Phase III (NCT04542070, NCT02938520, NCT02951052)	Reactions at injection sites are rather frequent, refrigeration required	Reduced dose frequency
		Cabotegravir			
	Oral NPs	Lopinavir–efavirenz	Preclinical (rodent model)	Daily dosing is required	Omits potentially harmful excipients
	Injectable NPs	Dolutegravir		Needs the help of a caregiver	Reduced dose frequency
	Injectable lipid NPs	Lopinavir–ritonavir–tenofovir			Target macrophages
TB	Aerosolized micro-NPs	BCG vaccine		Will need patient training	Optimal drug transport to the lungs
	Oral polymer NPs	Rifampicin–isoniazid–pyrazinamide–ethambutol–streptomycin			Improved bio-availability, target alveolar Macrophages, avoid first-pass metabolism
	Oral lipid NPs	Rifampicin–isoniazid–pyrazinamide		Frequent dosing is required, devoid of ethambutol	Reduced dose frequency, improved bio-availability
Malaria	Injectable NPs	Artemether–lumefantrine		Needs the help of a caregiver	Superior bio-availability compared to oral forms, reduced dose frequency
	Microemulsion	Artemether–tafenoquine		Stabilizing microemulsions with the use of high surfactant concentrations	Improved bio-availability
	Injectable immune-liposomes	Chloroquine derivative (novel aminoalcohol derivative)		Needs the help of a caregiver	Target contaminated erythrocytes, enhance the lipophilic drug's solubility

15.5 Limitations and Future Prospects

The aforementioned examples demonstrate several potential therapeutic approaches in the development of nanomedicines for the treatment and prevention of infectious diseases. Nevertheless, in order to have the greatest influence, nanotechnology approaches must surmount several economic, commercial, and regulatory obstacles. Initially, the primary challenge is in minimizing the expenses associated with the development of novel medications and nanotechnology-driven systems. The discovery of novel pharmaceuticals for global health solutions is a costly endeavor.

Repurposing established medications inside NCs might potentially attain comparable objectives of heightened effectiveness and safety at a much-reduced cost. Nevertheless, a single nano-system that has been officially sanctioned (namely, an amphotericin B liposomal formulation) is not economically viable in poor nations (Sundar and Chakravarty 2010). The inclusion of excipients, such as lyoprotectants, in nanosystems may lead to an increase in the overall cost of various therapies. Furthermore, the production and preservation circumstances of some NPs might not be suitable for low-resource countries (Mitchell and Carlson 2018). Researchers must ensure the repeatability and scalability of these nanosystems while also taking into account their environmental impact.

Nanomedicines are expected to be complex structures consisting of numerous components arranged in certain spatial configurations. Minor alterations in procedure or structure may have a negative impact on the exact construction of nanomedicines (Desai 2012; Ioannidis et al. 2018). Furthermore, the presence of regional and national variations within and across regulatory bodies poses a significant problem, particularly when conducting multinational clinical trials. Ultimately, it is crucial to consider the level of acceptance among patients about these nanosystems. Legal action was taken against the FDA in 2006 by the International Centre for Technology Assessment and other consumer groups who claimed the agency was too slow to regulate products using nanomaterials. In light of recent developments, the FDA formed the Nano Task Force. An interconnected group of stakeholders must collaborate for the effective implementation of nanomedicines into clinical practice. This includes individuals from the academic field, investors from many industries, government representatives, and contract research and production organizations. Nanotechnology is a multidisciplinary area that necessitates the use of intellectual property and commercialization methods in order to expand and develop. However, the rapid application of nanotechnologies from the laboratory to the general population and the significant influence these systems have made on the SARS-CoV-2 epidemic provide considerable potential. This example implies that such achievements may be feasible in addressing other contagious illnesses as well.

15.6 Conclusions

Identification numbers have a significant influence on the occurrence of illness and death worldwide, with a particularly severe effect on nations with low SDI. Enhancing the accessibility and effectiveness of medications may enhance the patients' well-being and alleviate the impact of diseases. This chapter focuses on the use of nanotechnology-based methods to enhance oral medication delivery, reduce the frequency of drug administration, and target drugs to specific infection areas, hence enhancing the effectiveness of therapy. Ultimately, it examined instances of nanotechnologies now undergoing clinical trials for HIV therapy and in the preclinical stage of development for TB and malaria treatment. Nanotechnologies that are both innovative and cost-effective, and specifically designed to address the issues faced by low-SDI nations, are expected to have the most positive impact on patients.

References

Alvarez CA, Wiederhold NP, McConville JT, Peters JI, Najvar LK, Graybill JR et al (2007) Aerosolized nanostructured itraconazole as prophylaxis against invasive pulmonary aspergillosis. J Infect 55(1):68–74

Anselmo AC, Mitragotri S (2016) Nanoparticles in the clinic. Bioeng Transl Med 1(1):10–29

Arias JL, Unciti-Broceta JD, Maceira J, Del Castillo T, Hernández-Quero J, Magez S et al (2015) Nanobody conjugated PLGA nanoparticles for active targeting of African Trypanosomiasis. J Control Release 197:190–198

Azad AK, Rajaram MV, Schlesinger LS (2014) Exploitation of the macrophage mannose receptor (CD206) in infectious disease diagnostics and therapeutics. J Cytol Mol Biol 1(1):1000003

Bakker-Woudenberg IA, Lokerse AF, ten Kate MT, Storm G (1992) Enhanced localization of liposomes with prolonged blood circulation time in infected lung tissue. Biochim Biophys Acta Mol Basis Dis 1138(4):318–326

Bakker-Woudenberg I, Lokerse AF, Mt K, Mouton J, Woodle M, Storm G (1993) Liposomes with prolonged blood circulation and selective localization in Klebsiella pneumoniae-infected lung tissue. J Infect Dis 168(1):164–171

Carryn S, Chanteux H, Seral C, Mingeot-Leclercq M-P, Van Bambeke F, Tulkens PM (2003) Intracellular pharmacodynamics of antibiotics. Infect Dis Clin 17(3):615–634

Clemens DL, Lee B-Y, Xue M, Thomas CR, Meng H, Ferris D et al (2012) Targeted intracellular delivery of antituberculosis drugs to Mycobacterium tuberculosis-infected macrophages via functionalized mesoporous silica nanoparticles. Antimicrob Agents Chemother 56(5):2535–2545

Cu Y, Booth CJ, Saltzman WM (2011) In vivo distribution of surface-modified PLGA nanoparticles following intravaginal delivery. J Control Release 156(2):258–264

Cunha-Reis C, Machado A, Barreiros L, Araújo F, Nunes R, Seabra V et al (2016) Nanoparticles-in-film for the combined vaginal delivery of anti-HIV microbicide drugs. J Control Release 243:43–53

das Neves J, Nunes R, Machado A, Sarmento B (2015) Polymer-based nanocarriers for vaginal drug delivery. Adv Drug Deliv Rev 92:53–70

Desai N (2012) Challenges in development of nanoparticle-based therapeutics. AAPS J 14(2):282–295

Ditlev JA, Case LB, Rosen MK (2018) Who's in and who's out—compositional control of biomolecular condensates. J Mol Biol 430(23):4666–4684
Duong HT, Jung K, Kutty SK, Agustina S, Adnan NNM, Basuki JS et al (2014) Nanoparticle (star polymer) delivery of nitric oxide effectively negates Pseudomonas aeruginosa biofilm formation. Biomacromolecules 15(7):2583–2589
Ensign LM, Tang BC, Wang Y-Y, Tse TA, Hoen T, Cone R et al (2012) Mucus-penetrating nanoparticles for vaginal drug delivery protect against herpes simplex virus. Sci Transl Med 4(138):138ra79-ra 79
Feldman RA, Fuhr R, Smolenov I, Ribeiro AM, Panther L, Watson M et al (2019) mRNA vaccines against H10N8 and H7N9 influenza viruses of pandemic potential are immunogenic and well tolerated in healthy adults in phase 1 randomized clinical trials. Vaccine 37(25):3326–3334
Fielding RM, Lewis RO, Moon-McDermott L (1998) Altered tissue distribution and elimination of amikacin encapsulated in unilamellar, low-clearance liposomes (MiKasome®). Pharm Res 15:1775–1781
Han C, Romero N, Fischer S, Dookran J, Berger A, Doiron AL (2017) Recent developments in the use of nanoparticles for treatment of biofilms. Nanotechnol Rev 6(5):383–404
Huh AJ, Kwon YJ (2011) "Nanoantibiotics": a new paradigm for treating infectious diseases using nanomaterials in the antibiotics resistant era. J Control Release 156(2):128–145
Hussain S, Joo J, Kang J, Kim B, Braun GB, She Z-G et al (2018) Antibiotic-loaded nanoparticles targeted to the site of infection enhance antibacterial efficacy. Nature Bio Eng 2(2):95–103
Huth M, Ricci A, Cheng A (2011) Mechanisms of aminoglycoside ototoxicity and targets of hair cell protection. Int J Otol 2011:937861
Ioannidis JP, Kim BY, Trounson A (2018) How to design preclinical studies in nanomedicine and cell therapy to maximize the prospects of clinical translation. Nat Biol Eng 2(11):797–809
Irvine DJ, Hanson MC, Rakhra K, Tokatlian T (2015) Synthetic nanoparticles for vaccines and immunotherapy. Chem Rev 115(19):11109–11146
Jackson LA, Anderson EJ, Rouphael NG, Roberts PC, Makhene M, Coler RN et al (2020) An mRNA vaccine against SARS-CoV-2—preliminary report. N Engl J Med 383(20):1920–1931
James GA, Swogger E, Wolcott R, Ed P, Secor P, Sestrich J et al (2008) Biofilms in chronic wounds. Wound Repair Regen 16(1):37–44
Jawahar N, Meyyanathan S (2012) Polymeric nanoparticles for drug delivery and targeting: a comprehensive review. Int J Health Allied Sci 1(4):217
Kaminishi H, Tanaka M, Cho T, Maeda H, Hagihara Y (1990) Activation of the plasma kallikrein-kinin system by Candida albicans proteinase. Infect Immun 58(7):2139–2143
Kim M-H, Yamayoshi I, Mathew S, Lin H, Nayfach J, Simon SI (2013) Magnetic nanoparticle targeted hyperthermia of cutaneous Staphylococcus aureus infection. Ann Biomed Eng 41:598–609
Kinch MS, Patridge E, Plummer M, Hoyer D (2014) An analysis of FDA-approved drugs for infectious disease: antibacterial agents. Drug Discov Today 19(9):1283–1287
Kirtane AR, Langer R, Traverso G (2016) Past, present, and future drug delivery systems for antiretrovirals. J Pharm Sci 105(12):3471–3482
Labouta HI, Menina S, Kochut A, Gordon S, Geyer R, Dersch P et al (2015) Bacteriomimetic invasin-functionalized nanocarriers for intracellular delivery. J Control Release 220:414–424
Lai SK, O'Hanlon DE, Harrold S, Man ST, Wang Y-Y, Cone R et al (2007) Rapid transport of large polymeric nanoparticles in fresh undiluted human mucus. Proc Natl Acad Sci 104(5):1482–1487
Landovitz RJ, Kofron R, McCauley M (2016) The promise and pitfalls of long acting injectable agents for HIV prevention. Curr Opin HIV AIDS 11(1):122
Langer R, Folkman J (1976) Polymers for the sustained release of proteins and other macromolecules. Nature 263(5580):797–800
Laverman P, Boerman OC, Oyen WJ, Dams ETM, Storm G, Corstens FH (1999) Liposomes for scintigraphic detection of infection and inflammation. Adv Drug Deliv Rev 37(1–3):225–235
Leal J, Smyth HD, Ghosh D (2017) Physicochemical properties of mucus and their impact on transmucosal drug delivery. Int J Pharm 532(1):555–572

Löbenberg R, Araujo L, von Briesen H, Rodgers E, Kreuter J (1998) Body distribution of azidothymidine bound to hexyl-cyanoacrylate nanoparticles after IV injection to rats. J Control Release 50(1–3):21–30

Lutwyche P, Cordeiro C, Wiseman DJ, St-Louis M, Uh M, Hope MJ et al (1998) Intracellular delivery and antibacterial activity of gentamicin encapsulated in pH-sensitive liposomes. Antimicrob Agents Chemother 42(10):2511–2520

Makadia HK, Siegel SJ (2011) Poly lactic-co-glycolic acid (PLGA) as biodegradable controlled drug delivery carrier. Polymers 3(3):1377–1397

Manzari MT, Shamay Y, Kiguchi H, Rosen N, Scaltriti M, Heller DA (2021) Targeted drug delivery strategies for precision medicines. Nat Rev Mater 6(4):351–370

Martinez LR, Han G, Chacko M, Mihu MR, Jacobson M, Gialanella P et al (2009) Antimicrobial and healing efficacy of sustained release nitric oxide nanoparticles against Staphylococcus aureus skin infection. J Invest Dermatol 129(10):2463–2469

McMillan J, Szlachetka A, Slack L, Sillman B, Lamberty B, Morsey B et al (2018) Pharmacokinetics of a long-acting nanoformulated dolutegravir prodrug in rhesus macaques. Antimicrob Agents Chemother 62(1):10.1128/aac. 01316-17

Mitchell SL, Carlson EE (2018) Tiny things with enormous impact: nanotechnology in the fight against infectious disease. ACS infect dis 4(10):1432–1435

Mohiti-Asli M, Pourdeyhimi B, Loboa EG (2014) Skin tissue engineering for the infected wound site: biodegradable PLA nanofibers and a novel approach for silver ion release evaluated in a 3D coculture system of keratinocytes and Staphylococcus aureus. Tissue Eng Part C Methods 20(10):790–797

Molla A, Yamamoto T, Akaike T, Miyoshi S, Maeda H (1989) Activation of Hageman factor and prekallikrein and generation of kinin by various microbial proteinases. J Biol Chem 264(18):10589–10594

Munita JM, Arias CA. Mechanisms of antibiotic resistance. Virulence mechanisms of bacterial pathogens 2016: 4 (2) 481–511

Nafee N, Husari A, Maurer CK, Lu C, de Rossi C, Steinbach A et al (2014) Antibiotic-free nanotherapeutics: ultra-small, mucus-penetrating solid lipid nanoparticles enhance the pulmonary delivery and anti-virulence efficacy of novel quorum sensing inhibitors. J Control Release 192:131–140

Palliser D, Chowdhury D, Wang Q-Y, Lee SJ, Bronson RT, Knipe DM et al (2006) An siRNA-based microbicide protects mice from lethal herpes simplex virus 2 infection. Nature 439(7072):89–94

Pardi N, Hogan MJ, Porter FW, Weissman D (2018) mRNA vaccines—a new era in vaccinology. Nat Rev Drug Discov 17(4):261–279

Pheage T (2017) Dying from lack of medicines. Afr Renew 30(3):24–25

Pinto-Alphandary H, Balland O, Laurent M, Andremont A, Puisieux F, Couvreur P (1994) Intracellular visualization of ampicillin-loaded nanoparticles in peritoneal macrophages infected in vitro with Salmonella typhimurium. Pharm Res 11:38–46

Roth GA, Abate D, Abate KH, Abay SM, Abbafati C, Abbasi N et al (2018) Global, regional, and national age-sex-specific mortality for 282 causes of death in 195 countries and territories, 1980–2017: a systematic analysis for the Global Burden of Disease Study 2017. Lancet 392(10159):1736–1788

Roy I, Vij N (2010) Nanodelivery in airway diseases: challenges and therapeutic applications. Nanomedicine 6(2):237–244

Sabaté E (2003) Adherence to long-term therapies: evidence for action. World Health Organization

Sarathy JP, Zuccotto F, Hsinpin H, Sandberg L, Via LE, Marriner GA et al (2016) Prediction of drug penetration in tuberculosis lesions. ACS Infect Dis 2(8):552–563

Schiffelers R, Storm G, Bakker-Woudenberg I (2001) Liposome-encapsulated aminoglycosides in pre-clinical and clinical studies. J Antimicrob Chemother 48(3):333–344

Semiramoth N, Meo CD, Zouhiri F, Said-Hassane F, Valetti S, Gorges R et al (2012) Self-assembled squalenoylated penicillin bioconjugates: an original approach for the treatment of intracellular infections. ACS Nano 6(5):3820–3831

Sen CK, Gordillo GM, Roy S, Kirsner R, Lambert L, Hunt TK et al (2009) Human skin wounds: a major and snowballing threat to public health and the economy. Wound Repair Regen 17(6):763–771

Siegel RA, Kirtane AR, Panyam J (2016) Assessing the benefits of drug delivery by nanocarriers: a partico/pharmacokinetic framework. IEEE Trans Biomed Eng 64(9):2176–2185

Steinbach JM, Weller CE, Booth CJ, Saltzman WM (2012) Polymer nanoparticles encapsulating siRNA for treatment of HSV-2 genital infection. J Control Release 162(1):102–110

Suk JS, Boylan NJ, Trehan K, Tang BC, Schneider CS, Lin J-MG et al (2011) N-acetylcysteine enhances cystic fibrosis sputum penetration and airway gene transfer by highly compacted DNA nanoparticles. Mol Ther 19(11):1981–1989

Sundar S, Chakravarty J (2010) Liposomal amphotericin B and leishmaniasis: dose and response. J Global Infect Dis 2(2):159

Tedijanto C, Olesen SW, Grad YH, Lipsitch M (2018) Estimating the proportion of bystander selection for antibiotic resistance among potentially pathogenic bacterial flora. Proc Natl Acad Sci 115(51):E11988–E11E95

Teirlinck E, Xiong R, Brans T, Forier K, Fraire J, Van Acker H et al (2018) Laser-induced vapour nanobubbles improve drug diffusion and efficiency in bacterial biofilms. Nat Commun 9(1):4518

Tian J-H, Patel N, Haupt R, Zhou H, Weston S, Hammond H et al (2021) SARS-CoV-2 spike glycoprotein vaccine candidate NVX-CoV2373 immunogenicity in baboons and protection in mice. Nat Commun 12(1):372

Toti US, Guru BR, Hali M, McPharlin CM, Wykes SM, Panyam J et al (2011) Targeted delivery of antibiotics to intracellular chlamydial infections using PLGA nanoparticles. Biomaterials 32(27):6606–6613

Trezza C, Ford SL, Spreen W, Pan R, Piscitelli S (2015) Formulation and pharmacology of long-acting cabotegravir. Curr Opin HIV AIDS 10(4):239

Tulkens P, Trouet A (1978) The uptake and intracellular accumulation of aminoglycoside antibiotics in lysosomes of cultured rat fibroblasts. Biochem Pharmacol 27(4):415–424

Vaughn JM, McConville JT, Burgess D, Peters JI, Johnston KP, Talbert RL et al (2006) Single dose and multiple dose studies of itraconazole nanoparticles. Eur J Pharm Biopharm 63(2):95–102

Williams PE, Crauwels HM, Basstanie ED (2015) Formulation and pharmacology of long-acting rilpivirine. Curr Opin HIV AIDS 10(4):233–238

Wistrand-Yuen E, Knopp M, Hjort K, Koskiniemi S, Berg OG, Andersson DI (2018) Evolution of high-level resistance during low-level antibiotic exposure. Nat Commun 9(1):1599

Wong JP, Yang H, Blasetti KL, Schnell G, Conley J, Schofield LN (2003) Liposome delivery of ciprofloxacin against intracellular Francisella tularensis infection. J Control Release 92(3):265–273

World Health Organization, and Stop TB Initiative. (2010) Treatment of tuberculosis: guidelines. World Health Organization, Geneva

Xiong MH, Li YJ, Bao Y, Yang XZ, Hu B, Wang J (2012a) Bacteria-responsive multifunctional nanogel for targeted antibiotic delivery. Adv Mater 24(46):6175–6180

Xiong M-H, Bao Y, Yang X-Z, Wang Y-C, Sun B, Wang J (2012b) Lipase-sensitive polymeric triple-layered nanogel for "on-demand" drug delivery. J Am Chem Soc 134(9):4355–4362

Zhang CY, Gao J, Wang Z (2018) Bioresponsive nanoparticles targeted to infectious microenvironments for sepsis management. Adv Mater 30(43):1803618

Chapter 16
Forensic Drug Chemistry: Unravelling Evidence Through Scientific Analysis

Abu Md Ashif Ikbal, Rabin Debnath, Sabu Thomas, Debprasad Chattopadhyay, and Partha Palit

Abstract Forensic drug chemistry plays a pivotal role in modern criminal investigations by utilising scientific methods to identify, analyse, and interpret illicit substances at crime scenes. This interdisciplinary field merges principles from chemistry, pharmacology, and law enforcement to provide accurate and reliable evidence that assists in establishing connections between individuals, substances, and criminal activities. The primary objective of forensic drug chemistry is to determine the presence and composition of controlled substances, such as illegal drugs and prescription medications, in various samples recovered from crime scenes. The constant emergence of new designer drugs, synthetic compounds, and modified substances necessitates continuous research and method development to ensure accurate identification and classification. The results obtained from forensic analyses can determine the course of an investigation, influence legal decisions, and impact the lives of both victims and suspects. Moreover, forensic drug chemistry extends beyond identifying substances; it aids in uncovering patterns of drug trafficking, manufacturing, and distribution. In conclusion, forensic drug chemistry stands as an indispensable component of modern forensic science. Through its scientific rigour and multidisciplinary approach, this area provides crucial information

A. M. A. Ikbal · P. Palit (✉)
Department of Pharmaceutical Sciences, Drug Discovery Research Laboratory, Assam University (A Central University), Silchar, India

R. Debnath
ISF College of Pharmacy, Moga, Punjab, India

S. Thomas
School of Chemical Sciences, Mahatma Gandhi University, Kottayam, India
e-mail: sabu.thomas@mgu.ac.in

D. Chattopadhyay
Natural product, ICMR-National Institute of Traditional Medicine, Nehru Nagar, Belagavi, Karnataka, India

ICMR-National Institute of Traditional Medicine, Belagavi, India

School of Life Sciences, Swami Vivekananda University Barrackpore, Kolkata, India

S. Bose et al. (eds.), *Concepts in Pharmaceutical Biotechnology and Drug Development*, Interdisciplinary Biotechnological Advances,
https://doi.org/10.1007/978-981-97-1148-2_16

about the composition, origin, and the distribution of illicit substances. By deciphering the chemical complexities of crime scene evidence, these field not only facilitates criminal investigations but also contributes to maintaining the integrity and efficacy of the criminal justice system as a whole.

Keywords Crime scenes · Forensic drug chemistry · Forensic science

16.1 Introduction

A key area of forensic science is forensic drug chemistry, which is essential for criminal investigations, court cases, and public safety. For law enforcement organisations, courts, and regulatory authorities, it entails the examination and identification of pharmaceuticals, prohibited substances, and other chemical compounds. It provides help with the prosecution of drug-related crimes and to ensure public safety, the main goal of forensic drug chemistry is to offer precise and trustworthy evidence on the existence, composition, and purity of drugs (Sosa-Reyes et al. 2022). Forensic drug chemistry is a subfield of forensic science that focuses on the analysis and identification of restricted substances, pharmaceutical medications, and other chemical compounds for the purposes of criminal investigations, legal procedures, and regulatory compliance (Bechky 2020). The type, composition, purity, and quantity of substances are determined using a variety of techniques and methodologies, with a primary focus on those relevant to drug-related crimes, in this multidimensional profession. The goal of forensic drug chemistry is to offer precise and factually sound evidence that aids courts, regulatory bodies, and law enforcement agencies in determining the presence of illegal drugs, prescription drugs, and other chemical substances, ultimately promoting public safety and justice (Neal et al. 2022).

The main elements of forensic drug chemistry are:

1. Identification of the Substance:
    (a) Identification of restricted substances: Forensic drug researchers are mainly concerned with this task. Due to their potential for abuse and harm, many chemicals are subject to regulation by the government. They consist of prescription pharmaceuticals that are vulnerable to abuse as well as illegal substances like marijuana, cocaine, heroin, methamphetamine, and cocaine (Morrow et al. 2019).
    (b) Pharmaceutical Drugs: Forensic drug chemistry also includes the examination of pharmaceutical substances in order to establish their validity, spot and also look into cases of prescription medication abuse or diversion (Peltier-Rivest and Pacini 2019).
    (c) Other Chemical Compounds: In addition to identifying restricted and pharmaceutical substances, forensic drug chemists can be used to identify a wide range of chemicals, poisons, and compounds discovered at criminal scenes that may be connected to environmental contamination, poisoning, or other criminal activity (Steuer et al. 2019).

2. Methodologies and Analytical Techniques:
   (a) Gas chromatography-mass spectrometry (GC-MS): GC-MS is a successful analytical technique for distinguishing and identifying volatile and semi-volatile substances. It is frequently used to verify the chemical make-up of substances and tell apart distinct compounds (Dekebo et al. 2018).
   (b) Liquid chromatography-mass spectrometry (LC-MS): This adaptable technology is used to analyse different chemicals. It offers excellent sensitivity and specificity for forensic drug analysis, particularly when dealing with complicated combinations (Mullin et al. 2019).
   (c) Infrared Spectroscopy (IR): Using IR spectroscopy, one can characterise unknown substances and validate the existence of specific chemical bonds, as well as locate functional groups within a substance (Bakels et al. 2020).
   (d) Nuclear Magnetic Resonance (NMR) Spectroscopy: NMR offers insights into substances' molecular structures by providing extensive information about those structures (Liu et al. 2020a).

3. Quality Assurance and Chain of Custody.
   (a) Quality Control: It is crucial to guarantee the dependability and accuracy of forensic drug analyses. To ensure the greatest levels of analytical precision, laboratories follow strict quality control methods that include calibration, validation, and the utilisation of standard reference materials.
   (b) Chain of Custody: The chain of custody is managed with care in order to guarantee the validity of the evidence. Each individual who handles the evidence makes a note of their actions to verify that it is still acceptable in court (Hatzakis 2019).

4. Professional Testimony:
   (a) Forensic drug chemists are frequently called as expert witnesses in court cases. They offer their conclusions, methods, and interpretations in order to aid the court in understanding the scientific aspects of a case. Judges and juries need the defendant's testimony to be able to make informed decisions about the case (Wheate 2008).

5. Emerging Issues and Trends:
   (a) Designer drugs: The emergence of novel synthetic substances, frequently regarded as "designer drugs" or "legal highs," presents considerable difficulties for forensic drug scientists. For the reason that these chemicals are created to replicate the effects of prohibited narcotics while avoiding legal restrictions, analytical procedures must constantly adapt in order to advance.
   (b) Analytical Technology Advances: To stay up with the changing landscape of illegal drugs and novel chemical compounds, forensic drug scientists must constantly update their analytical procedures (Abbate et al. 2021).
   (c) Interdisciplinary Collaboration: For a thorough knowledge of drug-related situations, cooperation between forensic drug chemists, toxicologists, and other professionals is becoming increasingly important (Ott and Arroyo 2023).

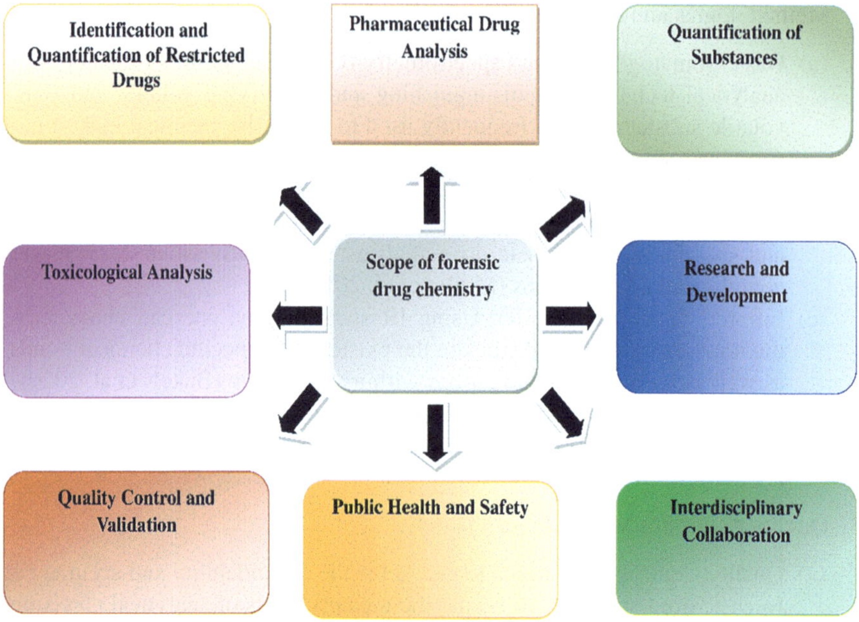

Fig. 16.1 Overview of the scope of forensic drug chemistry

(d) Opioid epidemic: The opioid crisis has increased public awareness. of the analysis, identification, and prevention of overdose-related deaths caused by opioids and their analogues (Biancuzzi et al. 2022) (Fig. 16.1).

16.2 Importance of Scientific Analysis in Criminal Investigations

It is impossible to overstate the importance of scientific analysis in forensic cases. A methodical, unbiased, and evidence-based approach to comprehending, reconstructing, and resolving crimes is provided by scientific analysis. It plays a major part in various important areas of criminal investigations, improving the judicial system's accuracy, dependability, and impartiality (McIntosh et al. 2021). The following are a few of the main justifications why scientific analysis is crucial in criminal investigations:

- Evidence Gathering and Storage:
- Physical evidence present at crime scenes should be collected, recorded, and preserved according to scientific analysis. By preventing contamination and degradation, this preserves the integrity of the evidence for later analysis.
- Crime Scene Reconstruction:

- By examining evidences such as bloodstains, ballistics, and DNA, scientific procedures assist detectives in reconstructing crime scenes. This makes it easier for detectives to piece together the timeline of what happened and pinpoint potential culprits (Gallagher and Thornton 2011).
- Identification of the Offenders:
- DNA analysis, fingerprint analysis, and facial recognition are forensic methodologies that can aid in identify and link suspects to a crime, resulting in arrests and legal action.
- Creating Deadlines:
- The timing of a death or an incident can be determined by scientific study, which is useful for establishing alibis or verifying witness testimony (Fenton et al. 2013).
- Identifying Tools and Weapons.
- Investigators can connect weapons or tools to a crime by using ballistic analysis, toolmark analysis, and firearms examination, which can assist them understand how a crime was perpetrated.
- Victims' Identification:
- By examining dental records, DNA, and other physical traits, forensic procedures can help identify victims in situations involving large-scale disasters or accidents (Delgado et al. 2021).
- Determining the toxicology and cause of death:
- Toxicological testing can identify the presence of medicines or poisons in the body of a person, assisting in diagnosing the cause of death and exposing suspicious activity.
- Crime Connection:
- Through the investigation of modus operandi, geographic profiling, or similar patterns of evidence, scientific research can assist in connecting crimes that at first glance appear unrelated (Santhy 2023).
- Restoration of Innocent Parties:
- The value of scientific rigour in investigations is illustrated by the exoneration of people falsely convicted of crimes using DNA analysis and other forensic procedures (Garrett 2020).

16.3 Interdisciplinary Foundations

The complicated area of forensic drug chemistry depends on the seamless integration of numerous disciplines to interpret the evidence and offer precise insights into drug-related situations. This chapter explores the crucial functions that chemistry, pharmacology, and law enforcement perform in this interdisciplinary field. Additionally, it looks at how accurate and dependable evidence interpretation depends on these fields' synergistic partnership.

16.3.1 *Role of Chemistry, Pharmacology, and Law Enforcement*

Chemistry in Forensic Drug Analysis

Chemistry provides the fundamental principles and methods required for substance identification and quantification, serving as the cornerstone of forensic drug analysis. Spectroscopy, chromatography, and mass spectrometry are examples of analytical chemistry techniques that are important for determining the chemical make-up of confiscated items (Lu et al. 2021). With the aid of these methods, forensic chemists can tell apart a variety of chemicals, such as illegal drugs, prescription pharmaceuticals, and adulterants.

Furthermore, organic chemistry plays an important role in revealing the molecular structures of complex compounds, aiding in the identification of designer pharmaceuticals and novel psychoactive chemicals (Bisen et al. 2023). To keep up with the continuously changing world of illicit substances, this knowledge is essential. Chemists can properly profile drugs using cutting-edge analytical techniques and spectroscopic technologies, enabling law enforcement agencies to make wise decisions (Hoffmann and Jackson 2015).

Pharmacology and Its Significance

Pharmacology adds value to chemical analysis by revealing important information regarding how medications affect the human body. To evaluate chemicals' possible hazards and effects on people, it is crucial to understand their pharmacological qualities. Pharmacological knowledge helps establish causality in cases involving drug-related occurrences by connecting particular drugs to observed physiological and psychological effects (Aronson 2011).

Additionally, pharmacology helps distinguish between drugs that, despite having seemingly similar chemical properties, have diverse pharmacological profiles. When multiple compounds with distinct effects are present, this difference is essential because it allows for correct classification and assessment of legal ramifications (Parvathaneni et al. 2019).

Law Enforcement's Role in Forensic Drug Chemistry

Law enforcement organisations are essential to the gathering, storage, and delivery of evidence to forensic labs. They are in charge of protecting crime sites and obtaining vital information for analysis. They are on the front lines of the war against the illicit drug trade. The reliability and admissibility of evidence in court depend on proper documentation and chain of custody protocols (Humphries et al. 2021).

In order to gain further information about the case's background, including the circumstances of the seizure, potential suspects, and associated criminal activities, forensic chemists must work with law enforcement. This information improves the precision and applicability of the studies performed, producing more fruitful investigation results.

16.3.2 Integration of Disciplines for Accurate Evidence Interpretation

Synergistic Collaboration

The seamless integration of chemistry, pharmacology, and law enforcement skills is essential for the proper interpretation of forensic drug evidence. It is crucial that professionals in various professions collaborate and communicate regularly. To provide a thorough understanding of the case, forensic chemists and pharmacologists must collaborate closely to correlate chemical discoveries with observable physiological effects (Bartley et al. 2010).

Additionally, constant interaction with law enforcement guarantees that forensic investigations are in line with the context of the investigation, allowing for the production of focused and useful results. This multidisciplinary approach reduces any case of misunderstanding and increases the overall dependability of the inferences made from the analysis.

Multidisciplinary Training and Education

Professionals in forensic drug chemistry must acquire multidisciplinary training to promote effective collaboration. These lessons ought to cover the fundamentals of chemistry, pharmacology, and law enforcement practices. Workshops, seminars, and training programmes that span disciplines can enhance knowledge sharing and a shared appreciation of each discipline's contributions (Joshi and Sisco 2023).

In addition, multidisciplinary case studies and mock scenarios can be helpful in imitating actual circumstances and enabling practitioners to hone their cooperation abilities. This training develops the forensic team's collective competence in addition to improving the proficiency of individual specialists.

The foundation of forensic drug chemistry is the interaction of chemistry, pharmacology, and law enforcement. Professionals in this sector can collaborate to solve challenging drug-related issues by acknowledging the essential contributions of each discipline. Forensic professionals may assure accurate evidence interpretation through teamwork and thorough training, ultimately aiding in the pursuit of justice and the protection of public safety (Spain et al. 2022).

16.4 Analytical Techniques in Forensic Drug Chemistry

16.4.1 Chromatography and Its Variants (GC, HPLC)

In forensic drug chemistry, chromatography, a potent analytical technique, is frequently employed to separate and identify complicated combinations of chemicals. It is essential to the analysis of narcotics and regulated substances because it helps forensic labs, law enforcement, and other organisations identify and quantify illegal drugs, medicines, and other chemical substances. Two types of chromatography that are widely applied in this field are gas chromatography (GC) and high-performance liquid chromatography (HPLC). The uneven distribution of components between stationary and mobile phases is the cornerstone of chromatography. In GC, the stationary phase is a high-boiling-point liquid immobilised on a solid support, whereas the mobile phase is a carrier gas (often helium or nitrogen). The stationary phase in HPLC is a solid support that is placed into a column, whereas the mobile phase is a liquid solvent that is pumped through the column. The degree of the interaction between the analytes and stationary phase, which is a prerequisite for both techniques, dictates how rapidly the analytes separate and elute (Dekebo et al. 2018).

GC: Gas Chromatography

The examination of volatile and thermally stable chemicals lends itself particularly well to GC, making it a useful technique in forensic drug chemistry. A sample is injected into the hot injector at the start of the procedure, where it vaporises and enters the GC column (Lu et al. 2021). On the basis of affinity of the constituents for flow rate the stationary phase and the stationary phase of the carrier gas, the constituents in the sample are separated, it is discovered by a detector, typically a mass spectrometer (MS) or a flame ionisation detector (FID). The resulting chromatogram shows the sample's components graphically, with each peak denoting a different compound. GC is commonly used in forensic drug chemistry to examine substances including cocaine, amphetamines, and cannabis. For law enforcement and legal proceedings, the capacity to precisely calculate drug concentrations and detect the presence of adulterants or cutting agents is essential (Alkhuder 2022).

Liquid Chromatography with High Performance (HPLC)

When it comes to separating numerous compounds, including non-volatile and thermally unstable molecules, HPLC is a flexible chromatographic technology that excels. Because of its versatility and sensitivity, it has established itself as a pillar in forensic drug chemistry. A liquid sample is delivered into an HPLC high-pressure pump, which forces it through a stationary phase-filled column. Based on how they interact with the stationary phase, the components are divided and elute at various rates (Kątny and Frankowski 2017). A variety of detectors, including UV-Vis, fluorescence, and mass spectrometry, are used to achieve detection. HPLC is useful for analysing pharmaceutical substances, opioids, and complicated mixes like narcotics or fake medicines that have been seized. The capacity to distinguish between potentially dangerous impurities and active pharmacological substances is essential for public safety (Colleoni et al. 2023).

Chromatography's benefits in forensic drug chemistry

- Sensitivity: The remarkable sensitivity of both GC and HPLC enables the detection of minute quantities of pharmaceuticals and their metabolites.
- Selectivity: Chromatography have a high degree of selectivity, making it possible to recognise chemicals that are closely related.
- Quantification: Chromatography techniques are excellent in accurately quantifying drug concentrations, which is essential in legal procedures.
- Versatility: HPLC and GC are indispensable in the field of forensic drug chemistry because they can be modified to analyse a wide variety of substances (Da Filicaia et al. 2023).

- Chromatography is essential to forensic drug chemistry, particularly gas chromatography (GC) and high-performance liquid chromatography (HPLC). These methods make it possible to accurately identify, quantify, and distinguish between narcotics and prohibited substances, which greatly aids law enforcement efforts and ensures public safety (Joshi and Sisco 2023).

16.4.2 Mass Spectrometry Substance Identification

One important method for qualitative chemical analysis is mass spectrometry (MS). Mass spectrometry has developed into an essential tool for proteome research in recent years. The capacity to analyse proteins has been completely transformed by desorption and ionisation methods like matrix-assisted laser desorption ionisation mass spectrometry (MALDI MS) and electrospray ionisation mass spectrometry (ESI MS). These advancements provide degrees of sensitivity and mass accuracy for the detection, identification, and structural characterisation of proteins that was not possible before (Li et al. 2022). Today, it is possible to determine peptides and proteins with low parts per million mass measurement accuracies and frequently quantify molecular weights beyond 200 kDa. The fast growth of protein and gene databases has tremendously improved the identification of proteins. Mass spectrometry is a potent analytical method that is frequently used in forensic drug chemistry to identify compounds. Forensic laboratories and law enforcement organisations rely on it constantly to analyse illegal narcotics, prescription medications, and other chemical substances because it offers accurate and trustworthy information regarding the composition, structure, and molecular weight of molecules (Locatelli et al. 2022).

Ions are separated in accordance to their mass-to-charge ratio (m/z) in mass spectrometry's fundamental operation. Here's how it functions:

- Ionisation: The first step in the procedure is to ionise the sample, which can be accomplished using a number of techniques, including electron impact (EI), chemical ionisation (CI), or electrospray ionisation (ESI). Ionisation is the process by which sample molecules are changed into ions by either removing or adding electrons, producing positively charged or negatively charged species.
- Mass separation: After the material has been ionised, the ions accelerate and then fed through a mass analyser to perform mass separation. Quadrupole, time-of-flight (TOF), ion trap, and magnetic sector analysers are examples of common mass analyser types. Each of these analysers separates ions according to their m/z ratio, making it possible to precisely estimate their mass.
- Detection: After being separated, the ions are next detected by a detector, which is typically an electron multiplier or a detector array. A mass spectrum is generated by the detector as it logs the intensity of ions at various m/z values (Lee et al. 2020).

Multiple benefits of mass spectrometry in forensic drug chemistry include:

- Since they are extremely sensitive instruments that can identify minute amounts of compounds, mass spectrometers are ideal for the analysis of illegal drugs and restricted substances.
- Mass spectrometry's remarkable allows for the distinction between molecules with identical masses but differing chemical structures. The identification of drug analogues and mixes depends on this specificity.
- Mass spectrometry offers structural data on chemicals, such as fragmentation patterns, which can be used to confirm the identify of unidentified molecules (Wang et al. 2022).

16.4.3 Spectroscopy Methods (FTIR, UV-Vis) in Drug Analysis

FTIR

In the area of forensic drug chemistry, the powerful analytical method known as Fourier transform infrared (FTIR) spectroscopy is frequently used in drug analysis. The identification, categorisation, and analysis of pharmaceuticals and their constituent parts are all made possible by this method (Alseekh et al. 2021). An in-depth explanation of how FTIR spectroscopy is used in forensic drug investigation is provided below:

The interaction between infrared radiation and matter is the foundation of FTIR spectroscopy. Infrared radiation is absorbed by molecules at particular wavelengths that correspond to their vibrational states. An absorption spectrum is a representation of the wavelengths of infrared light that are absorbed by a material. Since every pharmacological molecule has an own spectrum, FTIR is a fantastic tool for substance identification. Conventional spectroscopy, which records radiant power data as a function of frequency, is frequency domain spectroscopy. Radiant power data is recorded as a function of time in time domain spectroscopy, which is accomplished through Fourier transform (FT) (Suto et al. 2021). The frequency of electromagnetic radiation (EMR) from the source is changed to a correspondingly slower oscillating signal in the later case by the Michelson interferometer (MI), while the radiant power (v) is plotted against frequency (v1) (Hz) in the earlier instance. A computer is used to analytically split the slower oscillating signal into its constituent oscillations and calculate the oscillations of the appropriate frequencies of the detected radiation. This information is regularly compiled. Each resolved oscillation's amplitude depends on the radiation's intensity. A mathematical method for converting time domain spectrum to conventional frequency domain spectrum is called the Fourier Transform (FT) (Mateo and Talavera 2018).

Drug analysis applications:

- Drug Identification: The main application of FTIR in the non-destructive and quick identification of medications. In order to compare it to reference spectra stored in databases, analysts gather the infrared spectrum of an unknown material. Even in complicated mixes, exact identification is possible thanks to the distinctive spectral fingerprint (Alvarez-Ordóñez et al. 2012).
- Purity Evaluation: By measuring the strength of particular absorption peaks, FTIR can assess the purity of medications. A pharmacological sample's impurities or adulterants can be identified and measured using their distinctive spectral signatures.
- Structural Characterisation: FTIR can give important information about the functional groups and chemical structure of medicines. Understanding the composition and preparation of illegal substances is made easier by this information (Song et al. 2020).
- Forensic Examination: FTIR is utilised in forensic investigations to examine drug evidence collected from crime scenes. For example, it can assist in proving that illicit drugs are present in materials that have been confiscated or secret labs (He et al. 2021).

Advantages:

- Non-destructive: Since FTIR analysis does not involve destroying samples, it enables further testing or the preservation of evidence.
- Rapid: FTIR measurements are frequently speedy, making them appropriate for forensic laboratories with high throughput.
- Sample preparation is frequently done in a minimal manner, which lowers the chance of contamination or sample modification.
- High Specificity: The method has the potential to distinguish between chemically similar molecules and is extremely specific.

It is a pillar of forensic drug analysis due to its non-destructiveness and high specificity, aiding law enforcement in combating drug-related crimes and safeguarding public safety. When used in combination with other analytical methods, FTIR spectroscopy offers a thorough understanding of drug samples that have been confiscated, helping court trials and convictions in drug-related situations (Agatonovic-Kustrin et al. 2020).

UV-Vis Spectroscopy

The interaction between ultraviolet and visible light and matter is the foundation of UV-Vis spectroscopy. When electrons go through 2 h electronic transitions, molecules absorb this ultraviolet-visible electromagnetic energy. Different compounds have distinctive absorption spectra because of the difference between energies their ground and excited states, which is related to the wavelength of light that is absorbed (Alkhuder 2022).

Typically, absorbance (A) versus wavelength (λ) is how UV-Vis spectra are presented.

Applications for Drug Analysis

1. Quantitative evaluation.

   The quantitative investigation of pharmacological compounds in solution is frequently carried out using UV-Vis spectroscopy. It is based on Beer's Law, which asserts that a substance's absorbance is inversely correlated with both its concentration and the length of the sample cell (Estupiñán Méndez and Allscher 2022).
2. Identifying drugs and determining their purity:

   Identification: UV-Vis spectra, particularly when paired with reference standards, can be utilised as an efficient and instructive technique for drug identification. One chemical from another is distinguished by having distinctive absorption peaks.

   Examining the distinctive peaks in the UV-Vis spectrum might help you assess the purity of a substance. The spectral signatures of impurities or adulterants can be used to detect them (Ramohlola et al. 2020).
3. Analysis of Pharmaceutical Formulation:

   Pharmaceutical businesses must use UV-Vis spectroscopy to guarantee the reliability and uniformity of drug compositions. The concentration of active pharmaceutical ingredients (APIs) is evaluated, and formulation stability is tracked over time.
4. Forensic analysis:

   When examining crime scene evidence or dealing with confiscated drugs, UV-Vis spectroscopy is used in the analysis of forensic drug. It contributes to proving the existence of illegal drugs (Pratiwi and Nandiyanto 2022).
5. Drug Formulation Analysis:

   Pharmaceutical businesses must use UV-Vis spectroscopy to guarantee the accuracy and reliability of drug compositions. Active pharmaceutical ingredients (APIs) concentration and formulation stability are assessed using it.
6. Forensic investigation.

When working with captured drugs or examining evidence from crime scenes, UV-Vis spectroscopy is utilised in the analysis of forensic drug. It contributes to confirming the existence of illegal drugs.

Benefits of UV-Vis Spectroscopy

- High Sensitivity: UV-Vis spectroscopy is suitable for both quantitative and qualitative analysis since it is highly sensitive to even trace levels of analytes.
- Speed and effectiveness: UV-Vis measurements are frequently quick, enabling high-throughput analysis in forensic laboratories.
- Non-Destructive: Since this method does not cause any harm to the sample, it can still be used for analysis or as proof.
- Wide Range of Applications: UV-Vis spectroscopy is applicable to a wide range of sample types, such as liquids, solids, and even gases.
- UV-Vis spectroscopy is relatively inexpensive in regards of equipment and consumables when compared to several other analytical techniques (Zhu et al. 2021).

The importance in forensic drug chemistry (Fig. 16.1):

One cannot overestimate the value of UV-Vis spectroscopy in forensic drug chemistry. It performs several vital functions in the context of drug analysis, including:

- Rapid Substance Identification: UV-Vis spectroscopy offers a quick way to identify unknown chemicals discovered in seized items or at crime scenes. Forensic analysts can easily identify the existence of a material by comparing the UV-Vis spectrum of an unidentified substance to reference spectra.
- Drug Purity Assessment: In criminal investigations, figuring out the drug's purity is crucial. The purity or safety of a medicine can be impacted by impurities or adulterants. The strength and form of the absorption peaks can be examined using UV-Vis spectroscopy to determine the purity of confiscated medicines.
- Quantitative Analysis: UV-Vis spectroscopy provides an easy-to-use and trustworthy approach for law enforcement authorities to ascertain the precise concentration of a drug. This is crucial for determining the gravity of drug-related offences (Herrada-Manchón et al. 2020).

- FTIR, GC-MS, and LC-MS are a few complementary analytical techniques that are employed frequently alongside UV-Vis spectroscopy to provide a more thorough knowledge of samples of drugs that have been seized. The reliability and precision of forensic analyses is improved by combining various methodologies (Sharma et al. 2019).

16.4.4 Microcrystal Tests

Microcrystal tests are a collection of chemical procedures utilised in forensic drug chemistry to identify and categorise pharmaceuticals and illegal drugs. When a drug sample reacts with particular reagents, these tests include the production of distinctive crystals. Due to their simplicity, speed, and reproducibility, microcrystal examinations are frequently employed as a first step in forensic drug investigation (Tian et al. 2021).

Principle of Microcrystal Tests:

On the basis of the idea that specific chemical substances, when combined with a drug sample under controlled circumstances, can form distinctive and aesthetically distinct microcrystals, microcrystal tests are used to identify the presence of drugs. Under a microscope, one may see these microcrystals, which are exclusive to certain medication classes or unique compounds. The chemical components of the drug and the particular reagents employed in the test must interact in order for these crystals to form (Soursou et al. 2023).

The most common microcrystal tests and their uses

1. Scott test.

   Application: Barbiturates like pentobarbital and phenobarbital are distinguished from each other using the Scott Test.

   Principle: A drug sample is dissolved in a solution of sulphuric acid and chloral hydrate. The resultant microcrystals are particular to the kind of barbiturate that is there (Anzar et al. 2022).
2. The Marquis Test.

   Application: Numerous substances are detectable using the Marquis Test, including opioids (such as morphine, heroin) and amphetamines.

   Principle: The drug sample is mixed with a reagent that contains sulphuric acid and formaldehyde. The resulting microcrystals' distinctive colours make it easier to recognise the medication class (Viera-Herrera et al. 2020).
3. Mandelin Test.

   Application: Ecstasy (MDMA) and similar amphetamine compounds are most commonly recognised via the Mandelin Test.

   Principle: The drug sample is mixed with a reagent that contains concentrated sulphuric acid and ammonium metavanadate. The generated microcrystals exhibit distinctive hues that are unique to MDMA and related substance (Zhu and Hao 2020).
4. The Froehde Test.

   Application: Ecstasy (MDMA) is identified and distinguished from other compounds using the Froehde Test.

   Principle: The drug sample is mixed with a reagent that contains concentrated sulphuric acid and p-dimethylaminobenzaldehyde. For MDMA, the resultant microcrystals show unique colour shifts (Mani et al. 2020).
5. Keller-Kiliani test:

   Application: Barbiturates are identified and distinguished using the Keller-Kiliani Test.

   Principle: The drug sample is introduced to a reagent that is a combination of concentrated sulphuric acid and potassium dichromate. In order to identify the barbiturate present, specific microcrystals develop (Sánchez-Viesca and Gómez 2023).

The importance of forensic drug chemistry:

In forensic drug chemistry, microcrystal assays provide number uses and advantages.

- Rapid screening: Microcrystal examinations are rapid and can produce initial results in a matter of minutes, enabling forensic scientists to rank samples for additional examination.
- Low Cost: These tests are affordable and don't need expensive equipment, so even forensic laboratories with little resources can do them.
- Sensitivity: Microcrystal assays are helpful as screening tools because they can identify medicines even in low quantities.
- Evidence Preservation: Since microcrystal studies normally only require small amounts of a sample, a significant amount of the evidence can be kept for follow-up analyses or legal proceedings.
- Differentiation: These tests are able to distinguish between medications that belong to the same class or to locate particular chemicals in complex mixes (Olsen et al. 2018).

16.5 Sample Types and Evidence Collection

16.5.1 Biological Fluids (Blood, Urine, Saliva)

In drug-related cases, biological fluids are crucial sources of evidence. They offer vital information on a person's exposure to and interactions with different chemicals. To guarantee precise and trustworthy results, analysis of these fluids calls for careful procedures and considerations (de Campos et al. 2022a).

Blood

Blood is one of the biological samples that is most frequently used in forensic drug investigation. It offers a variety of information about a person's drug use, such as whether parent drugs and metabolites are present. To maintain the integrity of the blood samples, extreme caution must be used when collecting them. To avoid contamination or tampering, this entails utilising sterile equipment, adhering to the right extraction processes, and maintaining a chain of custody.

Following collection, the blood sample goes through a number of analytical procedures. It is done by using methods like mass spectrometry and liquid chromatography to separate and identify the molecules. Forensic chemists can assess the kind and quantity of chemicals in the bloodstream by comparing the results to established standards (Peters and Wissenbach 2023).

Urine

Another important biological sample utilised for the analysis of drug is urine. Compared to blood, it provides a wider window of detection, making it particularly helpful in situations involving chronic substance use. Utilising specialised containers and following stringent chain of custody guidelines are requirements when collecting urine samples. It is crucial to make sure the sample accurately reflects the individual's drug exposure.

Utilising methods like gas chromatography-mass spectrometry and immunoassays for urine sample analysis. Immunoassays enable a preliminary check for the existence of particular drug classes, while chromatography-mass spectrometry provides data that are highly precise and accurate for confirmation (Nicolaou et al. 2021).

Saliva

Recent years have seen a rise in popularity for saliva, a biological sample that is considerably less invasive. Instances requiring immediate testing, like roadside drug testing, can benefited from its simple and non-invasive collection. The traces of chemicals found in saliva samples can reveal information about recent drug usage.

Utilising specialised swabs created to absorb oral fluids painlessly while collecting saliva requires their use. The swab is after that delivered to the lab for examination. Utilising methods like liquid chromatography-tandem mass spectrometry (LC-MS/MS), it is likely to detect and quantify a variety of medications in saliva (Minho et al. 2023).

16.5.2 Powder and Plant Material Analysis

Powders and plant materials are two common types of substances found in drug-related instances. In order to examine the presence of prohibited substances and their purity levels in these items, analysis necessitates the use of a variety of procedures.

Powder Analysis

Among the most prevalent types of illegal narcotics are powdered compounds. It is possible to ascertain the chemical make-up and structural characteristics of powders by using methods like X-ray diffraction (XRD) and Fourier transform infrared (FTIR) spectroscopy. In order to make a preliminary identification, FTIR spectroscopy provides knowledge regarding the functional groups that are present. On the other side, XRD assists in the characterisation of compounds like designer medicines and adulterants by recognising crystalline structures (Sherma and Rabel 2019).

Aside from that, powdered substances can have their components separated and measured using chromatographic methods which involves gas chromatography and high-performance liquid chromatography (HPLC). For determining the purity levels and locating any potential pollutants, this is essential.

Plant Material Analysis

To identify and measure the active components in cases involving chemicals produced from plants (like cannabis), forensic chemists use a wide range of approaches. This frequently includes cannabinoids, which are what give marijuana its euphoric effects (Leghissa et al. 2018).

To identify and separate the individual compounds present in the plant material, methods like gas chromatography-mass spectrometry (GC-MS) and thin-layer chromatography (TLC) are used. To accurately classify the substance under controlled substance laws and assess its potency, this information is crucial.

16.5.3 Handling and Preserving Evidence for Accurate Analysis

In forensic drug chemistry, it is crucial to maintain the validity of the evidence. The accuracy and dependability of analyses can be compromised by improper handling or storage. To protect the chain of custody and avoid contamination or tampering, proper protocols must be followed (Sisodia 2022).

Collection Protocols

The reliability of forensic analyses depends on the establishment of standardised collection techniques. This entails using sterile tools, recording the procedure for collecting samples, and sealing samples in tamper-evident containers. The collection staff must also receive training to minimise the danger of cross-contamination.

Storage Conditions

In order to maintain the integrity of biological samples and substances, proper storage conditions are essential. Temperature, humidity, and light exposure are just a few variables that might have an impact on how stable evidence is. To prevent degradation, forensic laboratories need specialised storage facilities with controlled ambient conditions (Heinsvig et al. 2023a).

Chain of Custody

The reliability and admissibility of evidence in court depend on maintaining a clear and uninterrupted chain of custody. This entails recording each person who touches the evidence, along with the occasion, duration, and goal of each interaction. A well-kept chain of custody offers a timeline of the evidence's progression from collection to analysis.

Critical elements of forensic drug chemistry include an awareness of the various sample types and the application of appropriate evidence collection techniques. Forensic chemists can produce precise and trustworthy results by adhering to established procedures and making use of the proper analytical methods, thereby aiding in the pursuit of justice in drug-related situations (Roux et al. 2022).

16.6 Identification and Quantification of Restricted Substances

16.6.1 Principles of Substance Identification

An essential component of forensic drug chemistry is the identification of restricted drugs. It entails a methodical approach that draws on a variety of analytical tools, reference sources, and professional interpretation (Mercolini and Protti 2016). This section explores the guiding principles and techniques used throughout the procedure.

Analytical Techniques

To identify prohibited compounds, forensic chemists use a range of analytical methods. These methods consist of mass spectrometry, chromatography, and spectroscopy.

- Spectroscopy: This technique examines how matter and electromagnetic waves interact. Spectroscopy reveals details about the molecular make-up and structure of substances by examining the absorption or emission of light (Wang et al. 2021).
- Chromatography: Complex mixtures are separated out into their constituent parts using chromatographic techniques involving gas chromatography (GC) and high-performance liquid chromatography (HPLC). To isolate particular compounds for additional investigation, this is essential (Strynar et al. 2023).
- Mass spectrometry: By determining the mass-to-charge ratio of ions, mass spectrometry enables the measurement of molecular weights and structures. It is very helpful for verifying the origin of substances (Fisher et al. 2022).

Reference Materials and Libraries

- Reference resources are essential for identifying substances. These materials are made up of verified samples with well-known chemical compositions. In order to compare analytical data from unknown samples with those in reference libraries, forensic chemists use them as a baseline.
- Spectral Libraries: These databases hold spectra collected from well-known materials. Forensic chemists can make educated guesses by comparing the spectrum data of an unknown compound to those in the library (Elkins and Zeller 2021).
- Standard Compounds: These are unadulterated substances with well-known chemical make-ups. They act as concrete evidence for establishing a substance's identity.

Expert Interpretation

Forensic chemists use their knowledge and experience to interpret the findings of analytical procedures. This includes taking into account a number of variables, such as the analytical method utilised, the existence of impurities or adulterants, and any background information provided by law enforcement (Merone et al. 2021).

16.6.2 Quantitative Analysis for Determining Drug Concentrations

Forensic drug chemistry frequently includes the quantification of drug concentrations in addition to the identification of banned drugs. For both evaluating the potential effects on an individual and for legal actions, this information is essential (Donati and Amais 2019).

Calibration Curves and Standards

The creation of calibration curves is a prerequisite for quantitative analysis. To determine a correlation between the recorded signal and the concentration, it is necessary to examine known concentrations of the chemical. Forensic chemists can precisely determine the concentration by comparing the signal from an unknown sample to the calibration curve.

- Internal Standards: These are additives to both the standard solutions and the unidentified samples that have a known concentration. They aid in adjusting for differences in sample processing and instrument response (Fares et al. 2022).

Validation and Quality Control

Forensic laboratories adopt validation processes and quality control measures to guarantee the accuracy and dependability of quantitative analyses.

- Method validation: This procedure evaluates the analytical method's performance traits. It involves figuring out variables like linearity, selectivity, accuracy, and precision (Gliozzo 2020).
- Quality Control Samples: These are predetermined samples whose composition is known and which are frequently examined alongside unidentified samples. They act as controls to make sure the analytical process is operating within reasonable bounds (Lim et al. 2021).

16.6.3 Challenges in Detecting Trace Amounts

Forensic chemists face particular difficulties when detecting banned compounds in tiny levels. Specialised methods and increased sensitivity are necessary in these circumstances.

Sensitivity of Analytical Instruments

For detecting tiny levels, cutting-edge analytical tools with exceptional sensitivity are necessary. This also applies to sensors with improved signal-to-noise ratios and lower detection limitations (Sisco and Forbes 2021).

Sample Preparation Techniques

It is possible to concentrate and improve the detectability of trace compounds by using specialised sample preparation techniques, such as solid-phase extraction and derivatization.

Quality Assurance Measures

When working with trace levels, rigorous quality assurance procedures are very important. These procedures include routine instrument calibration and validation.

In conclusion, crucial components of forensic drug chemistry are the quantification and identification of prohibited drugs. Forensic chemists can deliver precise and trustworthy results by combining analytical methods, reference materials, and professional interpretation. Additionally, the quantification of drug concentrations adds an essential dimension to the investigation and makes it possible to fully comprehend the potential effects on people and legal actions. Despite the difficulties presented by trace quantities, specialised methods and strict quality control procedures enable forensic chemists to expertly and confidently traverse this challenging environment.

16.7 Evolution of Illicit Substances

16.7.1 Understanding Designer Drugs and Synthetic Compounds

With the rapid emergence of new chemicals, the landscape of illegal substances is always changing. This section explores the idea of designer drugs and synthetic substances, their history, and the difficulties they provide in forensic drug analysis (Huestis et al. 2017).

Definition and Characteristics

The term "designer drugs" (sometimes referred to as "novel psychoactive substances" or "NPS") refers to synthetic substances created to replicate the effects of prohibited substances. They are frequently produced by altering the chemical composition of already-approved medications, producing in compounds with comparable or even improved pharmacological effects. They are therefore particularly difficult to recognise and categorise (Jagadeesh et al. 2022).

Origins and Motivations

There are several different aspects that influence the development of designer medications. These can involve tries to get around current drug restrictions, take advantage of legal weaknesses, or satisfy certain market desires. In order to synthesise these molecules, underground labs and covert chemists are essential. They use their chemistry expertise to produce substances with modified structures and qualities (Nitescu 2019).

Diversity of Designer Drugs

Synthetic cathinones (often referred to as "bath salts"), synthetic cannabinoids, and new hallucinogens are only a few examples of the diverse chemicals that fall under the category of "designer drugs." Since each group has a different chemical makeup and pharmacological profile, identifying them can be challenging (Musselman and Hampton 2014).

16.7.2 Continuous Research and Method Development

Forensic drug chemists do constant research and method development to keep up with changing compounds in response to the quick rise of designer drugs. This section looks at how forensic science can change to meet new difficulties (Weber et al. 2023).

Research Initiatives

Research facilities and forensic labs are at the forefront of initiatives to comprehend and recognise synthetic medications. Researchers work together to investigate the chemistry, pharmacology, and analytical characteristics of novel compounds. This study contributes to the creation of analytical methods and reference collections for precise identification (Marinho et al. 2022).

Analytical Technique Advancements

It is frequently necessary to create new analytical techniques or adapt already existing ones in order to identify designer pharmaceuticals. Advanced spectroscopic methods, including as high-resolution mass spectrometry and nuclear magnetic resonance (NMR), provide improved capabilities for structural elucidation. Innovative sample preparation techniques are being created to extract and examine a larger variety of chemicals (Yang et al. 2023).

Reference Material Expansion

Reference collections are continually growing to include a wider variety of synthetic substances and designer medications. Forensic chemists can use these collections as priceless tools for comparative analysis to compare analytical data from unidentified samples to established standards.

16.7.3 Adapting to New Substances and Formulations

Forensic drug chemistry must constantly develop due to the dynamic nature of the black market for drugs. This section looks at the tactics used by forensic professionals to keep up with new drugs and formulations (de Boer et al. 2020).

Intelligence Gathering and Information Sharing

To stay up to date on new drugs, forensic laboratories work closely with regulatory organisations, international organisations, and law enforcement authorities. This exchange of knowledge enables specialists to foresee trends and proactively create analytical procedures for novel chemicals.

Targeted Screening and Non-targeted Analysis

To identify unidentified chemicals, forensic chemists combine targeted screening and non-targeted analysis techniques. While non-targeted analysis seeks to detect and identify a wider variety of drugs, targeted screening entails testing for particular groups of compounds. This two-pronged approach increases the possibility of finding compounds that have not been seen before (Jilala and Lwoga 2022).

Multidisciplinary Collaboration

Multidisciplinary collaboration is frequently necessary for the identification of new drugs. To comprehend the pharmacological effects and potential concerns connected with novel chemicals, forensic professionals collaborate closely with scientists, toxicologists, and pharmacologists. This coordinated effort guarantees a thorough method of drug identification (Labat et al. 2023).

In conclusion, forensic drug chemistry continues to face difficulties due to the evolution of illegal compounds. The advent of synthetic chemicals and designer pharmaceuticals demands ongoing study, method development, and field adaption. To keep abreast of new drugs and help to the continuous struggle against illicit drug use and trafficking, forensic professionals combine updated analytical techniques, larger reference libraries, and multidisciplinary teamwork.

16.8 Legal Regulations and Chain of Custody

16.8.1 Ensuring Evidence Admissibility in Court

To present evidence that is both legally admissible in court and scientifically sound is one of the main goals of forensic drug chemistry (Khan et al. 2021). This section explores the important elements that influence whether or not evidence is admissible in court.

Scientific Reliability and Validity

Evidence must satisfy the standards of scientific validity and reliability in order to be acceptable. This means that the analysis techniques must be reliable from a scientific standpoint and accepted in the field of forensic drug chemistry. Furthermore, the outcomes must be precise and repeatable.

To assure the accuracy of their tests, forensic chemists use tried-and-true analytical methods and follow set procedures. To stay current with developments in the industry, they also do ongoing research and development, which strengthens the validity of their conclusions.

Expert Testimony and Qualifications

An essential part of presenting evidence in court is the testimony of forensic experts. To be acknowledged as authorities in their profession, forensic chemists must meet the appropriate requirements, including relevant education, training, and experience. They can effectively and convincingly communicate their analyses, methodology, and outcomes to judges and juries thanks to their experience (Edmond et al. 2016).

They must also stay abreast of legal requirements and be aware of any modifications to laws or other standards that may affect the admissibility of their testimony.

16.8.2 Importance of Maintaining Proper Chain of Custody

A key component of forensic drug chemistry that ensures the validity and admissibility of evidence is the chain of custody. This section looks at how important it is to keep a chain of custody for all of the evidence you handle.

Definition and Purpose

The chronological recording of the custody, control, transfer, and analysis of physical evidence is known as the "chain of custody." It offers a reliable record of each person who has interacted with the evidence, along with information about the conditions in which that interaction occurred. Establishing the validity and integrity of the evidence in court is the main goal of the chain of custody (Roux et al. 2022).

Documentation and Record-Keeping

The collection, packaging, storage, analysis, and any transfers of the evidence between people or locations are all scrupulously recorded by forensic chemists. The chain of custody record comprises information about each entry, including the date, time, place, and names of everyone involved (Young and Goodman-Delahunty 2021).

This thorough record-keeping protects the evidence from tampering, contamination, or improper handling. It gives a precise and open account of how the evidence got from the crime scene to the forensic lab.

Tamper-Evident Packaging and Sealing

The evidence must be kept in tamper-evident packaging or containers to ensure its integrity. If these canisters have been opened or hacked, observable indicators will appear. Additionally, proper sealing guarantees that the contents are safe and uncontaminated during storage and transportation.

16.8.3 Role of Forensic Drug Chemists in Legal Proceedings

As expert witnesses and contributors to the case at hand, forensic drug chemists are crucial to the success of court cases. In this part, we examine the roles and contributions of forensic chemists in the courtroom (Meier-Augenstein 2019).

Expert Testimony and Presentation

When called upon to give expert testimony, forensic chemists describe to the court their studies, processes, and conclusions. To make difficult scientific knowledge accessible to judges and juries, they must successfully express it in a straightforward and understandable manner.

Their testimony helps the court to comprehend the importance of the evidence, its applicability to the case, and its ramifications for the legal proceedings. This professional advice frequently weighs heavily in the decision-making process.

Cross-Examination and Peer Review

Opposing counsel may subject forensic chemists to arduous cross-examination to evaluate the accuracy and dependability of their findings. The objective of this procedure is to make sure that the evidence is admissible and that the methodology is sound (Ashraf et al. 2022).

Furthermore, within the scientific community, forensic analyses and findings are subject to peer review. This outside examination further supports the legitimacy and dependability of the evidence.

In conclusion, forensic drug chemistry includes the chain of custody and legal requirements as essential elements. Adherence to strict scientific standards and keeping a transparent record of evidence handling are necessary to ensure the admissibility of evidence in court. In this procedure, forensic chemists are crucial because they offer expert witness and contribute their specialised knowledge to court cases. They are crucial in maintaining the integrity of the criminal justice system because of their knowledge and dedication to scientific integrity.

16.9 Forensic Drug Chemistry in Criminal Justice

16.9.1 Influence of Analysis Results on Investigations

The impact of forensic drug chemistry analysis findings on criminal investigations within the field of criminal justice is significant and wide-ranging. In situations involving illegal narcotics and regulated substances, forensic drug chemistry plays a crucial role in supplying law enforcement authorities, prosecutors, and the judiciary with crucial evidence and expertise (de Campos et al. 2022b). Given below is how forensic drug chemistry analysis outcomes affect cases and the criminal justice system:

- Proof of Drug Crimes:

- Drug-related crimes can be demonstrably shown by forensic drug chemistry analysis. Forensic chemists examine samples that have been captured by law enforcement and are thought to be illicit drugs to establish their authenticity and composition. These analyses' findings indicate the existence of controlled substances, which lays the groundwork for criminal cases.
- Substance Recognition:
- Identification of chemicals obtained during investigations is one of forensic drug chemistry's main tasks. If the substances are actually illegal narcotics, prescription pharmaceuticals, or other substances, the analysis results support that. To determine whether possession and distribution are lawful, this identification is essential (Rinaldi et al. 2020).
- Quantification of Substances:
- Forensic drug analysis measures the quantity or concentration of a substance in a sample that has been seized. This goes beyond simple identification. Given that drug offences frequently involve particular amount thresholds that might impact charges and punishment, this information is essential for determining the seriousness of an offence.
- Evaluation of composition and purity:
- The findings of the analysis provide light on the quality and make-up of the medications that were seized. To assess a substance's potential impact on public health and safety, its purity must be known. High-purity medicines may be riskier and result in harsher penalties (Woodman et al. 2020).
- Acknowledgement of Alleged Offences:
- Confirming the kind of chemicals that have been seized helps investigators develop a case. The findings of the analysis aid in establishing a link between the substances and the people engaged in drug-related offences including manufacturing, distribution, or possession.

- Comparison of Controlled Substances:
- Differentiating between distinct banned substances and designer drugs is also made possible by forensic drug chemistry. This distinction is essential since legal implications frequently range greatly depending on the particular substance (Moud 2022).
- Expert witness testimony:
- Those who study drugs for forensic purposes frequently witness in court as experts. In their evidence, they describe the outcomes of their analysis, the procedures they used, and the importance of their conclusions. Their knowledge helps the judge and jury to understand the scientific components of the evidence and gives the prosecution's case more credibility (Tettey et al. 2021).
- Influence on Charging Decisions.
- Prosecutors' judgements about charges may be affected by the findings of forensic drug analyses. These choices could be made about the severity of the charges (such as whether they are crime or whether defendants are given the option of plea bargains or diversionary programmes (Otgaar et al. 2022)).
- The Effect on the Sentence:
- The outcomes of the analysis could have a big impact on sentences in cases where offenders are found guilty. The amount and purity of the narcotics that were collected, as well as the defendant's prior record, can affect the length of sentences that the court imposes (Trozze et al. 2023).
- Assistance with Legal Action:
- The outcomes of the analyses serve as solid, fact-based evidence in the legal system. They support the fair administration of justice, safeguard the rights of both defendants and victims, and ensure that judicial proceedings are founded on facts (Brasileiro Dourado and Fischer 2021).
- Public safety and deterrence:
- In order to efficiently discover and stop drug-related crimes, law enforcement is given the tools they need by forensic drug chemistry, which helps to ensure public safety. Given that chemicals can be detected and analysed, this knowledge serves as a deterrent, possibly lowering drug-related offences (Hachem et al. 2023; Zapata et al. 2021; Tsochatzis et al. 2021).
- New Psychoactive Substances (NPS) Addressed:
- To detect and categorise the new psychoactive drugs (NPS) that are constantly being discovered, forensic drug chemistry is essential. This aids in keeping law enforcement and decision-makers abreast of changing drug usage patterns and addressing new public health issues (Graziano et al. 2019).

16.9.2 Impact on Legal Decisions and Outcomes

The criminal justice system's legal judgements and the resolutions of criminal cases are significantly influenced by the forensic drug chemistry of the drugs involved. Criminal investigations, prosecutions, and punishment are all directly impacted by the findings of forensic chemists' drug examinations. Here is a thorough examination of how forensic drug chemistry impacts court rulings and outcomes (Bishnoi and Sharma 2020):

1. Charges and Arrests:
   By establishing the nature of the seized compounds, forensic drug chemistry frequently starts the legal process. Police personnel gather samples for analysis when they believe illegal narcotics are present. The findings of these studies have an impact on whether or not someone are arrested and could decide the initial charges brought against them. The detection and measurement of restricted substances may result in accusations of manufacturing, distribution, trafficking, or possession, which has an impact on the outcome of any ensuing legal proceedings.
2. Proof of an Offence:
   Results of forensic drug analysis are vital pieces of evidence in situations involving drugs. To establish the elements of the crime, such as the nature, quantity, and purity of the chemicals collected, prosecutors rely on the conclusions of forensic scientists. This evidence bolsters the prosecution's case and plays a role in the judge's decision to press charges against the defendant (Welner et al. 2023).
3. Plea Agreement Talks:
   Results of forensic drug testing can affect how defendants and prosecutors negotiate plea agreements. When confronted with strong evidence of drug possession or distribution, defendants may be more willing to negotiate plea deals that result in lowered charges or sentencing recommendations. Such negotiations can lead to favourable legal outcomes for both parties and expedite the resolution of cases.
4. Trial Procedures:
   Forensic chemists frequently testify as experts in situations that end up in court. They give testimony regarding the analysis techniques they utilised, the outcomes they saw, and the importance of those outcomes. Their expert testimony enables juries and judges to understand the scientific components of the offered evidence. Legal judgements, including the verdict and sentence, can be affected by a witness's credibility and persuasiveness (Blais et al. 2022).
5. Choosing a sentence:
   The results of forensic drug tests have a direct bearing on sentence determination. When deciding on sentencing, courts take into account things including the quantity and type of narcotics involved, their purity, and the criminal history of the offender. These characteristics are often used to construct sentencing guidelines, and forensic drug chemistry findings are crucial in determining how to apply these guidelines in many countries.
6. Evidence Refusal or Reclusion:
   The admissibility of drug analysis results may occasionally be contested by defence lawyers on the grounds of mistakes, contamination, or procedural irreg-

ularities. The forensic evidence may be suppressed or excluded by a court if it finds that the analysis was improperly done or that the evidence was corrupted. Such court rulings may have a big effect on a case's outcome, possibly leading to charges being dismissed or convictions (Carrier 2021).

7. Impact on Probation and Parole.

   Forensic drug chemistry findings may affect a person's eligibility for probation and parole if they have been convicted of drug-related crimes. When defining the terms of probation or parole, including requirements for drug testing and substance misuse treatment, supervising authorities take into account the type and seriousness of the drug offence (Gibbs et al. 2019).

16.9.3 Communication of Complex Scientific Findings to Non-experts

A critical component of the criminal justice system is explaining intricate scientific findings from forensic drug chemistry to non-experts like juries, judges, attorneys, and the general public. For scientific evidence to be accepted, believed in, and correctly used in legal processes, it must be communicated effectively (Liu et al. 2020b). The following are some methods and factors to take into account when communicating complicated scientific material clearly:

1. Employ clear language:

   As much as possible, stay away from technical language. In ordinary, everyday language, explain scientific principles and procedures. Instead of saying "chromatography," you may say something like "a technique that separates and identifies substances in a mixture."
2. Metaphors and analogies:

   Use metaphors and analogies to clarify difficult ideas. To make separation easier for ordinary individuals to understand, we may compare chromatography to separating various candies into different categories in a bag.
3. Visual Supports.

   Use diagrams, charts, and graphs as visual aids to support your remarks. Scientific topics can be made more concrete and understandable through the use of visuals (Brazier 2020).
4. Explanation in Detail:

   Organise complicated procedures into step-by-step instructions. Describe each step of the forensic examination, emphasising the most important results and their importance.
5. Examples from Real Life.

   Give concrete instances and case studies that demonstrate the use of forensic drug chemistry in criminal investigations. Using relatable narratives can draw in non-experts and make the topic more approachable (Strynar et al. 2023).

16.10 Uncovering Patterns in Drug Trafficking

16.10.1 *Analysing Chemical Signatures and Isotopic Compositions*

Chemical signatures and isotopic compositions are useful tools that forensic drug scientists use to identify patterns in drug trafficking (Parihar et al. 2023). This section examines how these techniques aid in comprehending the provenance and manufacturing processes of illegal substances.

Chemical Signatures in Illicit Substances

Chemical signatures are distinctive or distinguishing chemical characteristics that are particular to particular geographic areas, production processes, or synthetic approaches. The raw ingredients, reagents, and synthesis techniques utilised in drug manufacture give birth to these signatures (Kim et al. 2023).

Traces of substances and impurities metals or non-metals that are present in trace concentrations might act as identifying markers. These elements may come from the synthesis's precursor compounds or from impurities added during manufacture.

Isomers and stereochemistry: A substance's isomeric forms can have different chemical characteristics. It is possible to learn more about a substance's production process by looking at its particular isomeric profile (Rivera et al. 2019).

Isotopic Ratios: Depending on the synthesis process and the geographic location, isotopic compositions, particularly stable isotopes of elements like carbon, hydrogen, and nitrogen, can change. In order to identify the source of precursor materials, isotopic analysis can be useful (Camin et al. 2016).

Mass Spectrometry and Isotopic Analysis

A potent method for figuring out a substance's isotopic composition uses mass spectrometry and isotopic analysis. This gives important details regarding its origin and manufacturing process.

High-Resolution Mass Spectrometry (HRMS): HRMS permits accurate measurement of ion mass, allowing for the determination of isotopic ratios. Forensic chemists can deduce important information regarding the history of the material by comparing these ratios to established benchmarks (Machethe and Mofokeng 2022).

Gas chromatography-Combustion-Isotope Ratio Mass Spectrometry (GC-C-IRMS): This specialised method examines stable isotopes in substances including carbon, hydrogen, and nitrogen. It offers extremely precise isotopic information that can shed light on the substance's production process and place of origin (Meola et al. 2021).

16.10.2 Tracing Origins and Production Methods

For effective law enforcement, it is essential to understand the sources and production processes of illegal substances. Experts in forensic drug chemistry are crucial in giving this important data.

Geographic Profiling

Geographic profiling is examining the chemical fingerprints and isotopic compositions of seized substances to ascertain their most likely geographical origin. This is especially true for substances produced from plants, like cocaine or cannabis, which are plant-based narcotics.

Cocaine profiling: Depending on the area where the coca plant was produced, cocaine, for instance, displays different molecular signatures. Forensic scientists can identify the most likely source region by further refining these characteristics using isotopic analysis (Scendoni et al. 2021).

Cannabis Strain Identification: Cannabis strains differ chemically, which may reveal information about their geographic origin. Validating these profiles uses isotopic analysis (Heinsvig et al. 2023b).

Synthetic Route Identification

Understanding the process of making synthetic medicines offers vital insights into the criminal networks at play. Forensic scientists can determine the most likely mode of manufacture by comparing the changes in chemical impurities and isotopic compositions produced by several synthetic pathways.

Synthesis of Methamphetamine and Amphetamine: The production of methamphetamine and amphetamine requires a number of different compounds and reagents as precursors. Experts in forensics can determine the method of synthesis by examining the impurities and isotopic signals (Kranenburg et al. 2022).

Synthetic cannabinoids: The creation of synthetic cannabinoids entails particular chemical procedures that leave behind distinctive traces. These signatures allow forensic scientists to connect confiscated chemicals to specific production processes (Marinho et al. 2022).

16.10.3 Contribution to Dismantling Drug Networks

Chemical fingerprints and isotopic analysis provide important information that aids law enforcement in destroying drug trafficking networks.

Targeted Investigations

Law enforcement authorities are better equipped to conduct focused investigations when they are aware of the likely source and production processes of seized drugs. By concentrating resources on particular areas or criminal networks, they can increase the possibility that their operations will be successful.

Intelligence-Led Operations

Law enforcement organisations plan and carry out actions using the intelligence collected through forensic analyses. It offers vital information about the organisation, supply networks, and major participants in drug trafficking networks (Piraianu et al. 2023).

Courtroom Evidence

Chemical signature and isotopic analysis data are significant sources of evidence for legal proceedings. It offers a rational scientific foundation for tying confiscated compounds to particular production processes and geographical origins. This makes the case against those who traffic in drugs stronger (Kim et al. 2023).

To sum up, chemical signature and isotopic composition analysis is essential for identifying trends in drug trafficking. Forensic drug chemists give essential knowledge on the origin and production processes of illegal drugs by using cutting-edge techniques and taking advantage of the particular features of substances. This information enables law enforcement organisations to carry out focused investigations, break up drug distribution networks, and provide solid proof in court. Through their knowledge, forensic chemists are vital to the ongoing fight against drug trafficking and the protection of local populations all over the world.

16.11 Future Trends and Challenges

16.11.1 Technological Advancements in Drug Analysis

The landscape of forensic chemistry, pharmaceutical research, and law enforcement operations to combat drug-related crimes is constantly changing due to technological developments in drug analysis. As we look to the future, a number of noteworthy trends and advancements are anticipated to have a big impact on drug analysis which are:

- High-Resolution Mass Spectrometry (HRMS): HRMS equipment, including Orbitrap and FT-ICR mass spectrometers, will get easier to use and more economical in the upcoming years. These tools provide hitherto unseen levels of mass precision and resolution, making it possible to precisely identify and quantify drugs even in intricate mixes. In order to remain ahead of new designer medications and to guarantee the accuracy of drug analysis results, HRMS will be crucial (Deschamps et al. 2023).

- Miniaturised and Portable Analytical equipment: The field of drug analysis is being revolutionised by the development of handheld and portable analytical equipment. These small, cutting-edge analytical gadgets will enable law enforcement personnel and emergency personnel to quickly and effectively identify drugs on the spot. These technologies will advance more and become more commonly used as a result.
- Artificial intelligence (AI) and machine learning algorithms are being progressively incorporated into analytical tools to automate data analysis, increase accuracy, and improve pattern identification. These instruments will speed up data processing, enabling forensic chemists and researchers to effectively make sense of enormous datasets. The identification of novel drug analogues and designer pharmaceuticals will also be aided by AI-driven algorithms, which will spot patterns and trends in analytical data (Deschamps et al. 2023).
- Hyphenated Techniques: Combining many analytical methods is now common practise in the field of drug detection. Examples of this are liquid chromatography-mass spectrometry (LC-MS) and gas chromatography-mass spectrometry (GC-MS). Hyphenated methods offer further data that make drug sample characterisation more thorough. Additional versions of these methods will provide quicker and more in-depth insights into complex medication combinations.
- Improvements in Sample Preparation: The drug analysis process will be made easier by improvements in sample preparation procedures such solid-phase microextraction (SPME) and microextraction techniques. These methods lessen the need for laborious sample handling, cut down on contamination risk, and improve sensitivity, especially when analysing minute quantities of medicines in diverse matrices (Yu et al. 2021).
- Spectroscopy using nuclear magnetic resonance (NMR): Development due to improvements in NMR instrumentation and methods, it is now a useful tool for drug study, providing structural insights and isotope-specific data (Yang et al. 2017).

16.11.2 Anticipating Emerging Substances and Techniques

To keep ahead of changing difficulties and guarantee community safety, drug analysis professionals must constantly anticipate new compounds and methodologies. In order to prepare for the introduction of novel substances and the development of methods for the manufacture and distribution of drugs, a range of strategies and cooperative efforts are used.

- One of the main strategies includes carefully tracking of drug usage patterns and gathering relevant data. Information on drug seizures, substance misuse trends, and newly emerging drug-related health issues are all included in this report. Networks and surveillance systems are set up to monitor drug trafficking routes, patterns, and market shifts. Experts can spot strange or unexpected trends and substances by regularly collecting and analysing this data (Brien et al. 2023).
- Another crucial stage in predicting new compounds in the creation of early warning systems. Collaboration between law enforcement organisations, medical specialists, forensic labs, and university scholars is crucial. These parties exchange knowledge and ideas, assisting in the development of alert systems that can quickly tell interested parties about new substances and trends (Coppey et al. 2020).
- In this procedure, forensic intelligence is crucial. To obtain information on drug trafficking organisations and new substances, specialised units within law enforcement organisations are frequently used. Analysis of drug samples captured in various operations allows forensic chemists and intelligence analysts to jointly identify new compounds.
- In order to confront the transnational aspect of the drug trade, international cooperation is crucial. Organisations like Interpol and Europol make it easier for people to share information and intelligence on new drugs and international drug trafficking. International alliances and agreements are encouraged in order to jointly combat the illicit drug trade and solve new drug issues (Hawgood et al. 2022).
- Programmes for education and training are essential for identifying the warning indications of developing medicines. Law enforcement, forensic analysts, and healthcare personnel are all given training so they may recognise developing chemicals and their effects and react accordingly. The hazards of designer drugs and novel psychoactive substances (NPS) are also the subject of public awareness campaigns (Shogar 2019).

16.11.3 Ethical Considerations in Forensic Drug Chemistry

In the field of forensic drug chemistry, ethics play a crucial role, especially as it develops in response to new trends and difficulties. The integrity of the criminal justice system, protection of individual rights, and public confidence in forensic drug analysis are all dependent on these ethical considerations. The following ethical factors will continue to be crucial in the future:

- Reliability and accuracy:
- Challenge: There is an increasing demand for drug analysis that is more accurate and reliable as analytical techniques develop.

- Ethical Concern: It is essential to ensure that analytical procedures are based on accurate data and that outcomes are reliable from a scientific standpoint. False results, whether favourable or negative, can have detrimental effects on both people and society (Dror 2020).
- Disclosure and Transparency.
- Challenge: It could be harder to communicate results to non-experts in a clear and understandable manner when using advanced analytical techniques.
- Ethical Concern: Honest and transparent communication of findings to all stakeholders, including judges, juries, and attorneys, is essential for enabling them to make judgements based on the facts presented (Shogar 2019).
- Aware Consent and Privacy:
- Challenge: Using cutting-edge analytical methods may call for access to personal information and biological materials.
- Ethical Concern: It is critical to uphold people's rights to privacy and seek their informed consent when gathering and analysing their samples. It is a persistent ethical challenge to strike a balance between the demand for proof and individual rights.
- Bias and Objectivity:
- Challenge: It can be difficult to establish findings that support prosecution or law enforcement purposes.
- Forensic chemists must keep their ethical obligation to maintain objectivity and impartiality. Analyses must be conducted objectively, free from other influences (Elkins and Fambegbe 2020).
- Continuous Education and Training:
- The difficulty is keeping up with the continually changing technologies and methods.
- Ethical Concern: To uphold the highest levels of competence and to keep current in their area, forensic chemists have an ethical duty to participate in ongoing training and education.
- Follow-through on Quality Assurance:
- The difficulty of ensuring the accuracy and dependability of forensic analyses increases as technology develops.
- Ethical Concern: To retain the credibility of their profession, ethical chemists must follow strict quality assurance and control requirement (Dror 2020).

16.12 Conclusion

An essential and vital place in modern criminal investigations is taken by forensic drug chemistry. Establishing links between people, substances, and criminal activity depends critically on the use of scientific procedures to find, examine, and interpret illicit compounds discovered at crime scenes. This multidisciplinary field, which draws on the fields of chemistry, pharmacology, and law enforcement, is motivated by the need to identify the presence and make-up of controlled substances across a variety of sample types recovered from crime scenes, from illegal drugs to prescription medications. In addition to identifying and quantifying drugs, forensic drug chemists use a variety of cutting-edge methods to learn more about their sources, origins, and potential adulterants. These methods include chromatography, mass spectrometry, spectroscopy, and microcrystal testing. Keeping up with the constantly changing environment of illicit substances, which is characterised by the continual introduction of new designer drugs, synthetic chemicals, and modified substances, is a persistent problem for these professionals. Continuous research and method development are essential to ensure the precision of classifications and identifications, and it's crucial to have a firm grasp of legal rules and chain of custody processes to uphold the admissibility of evidence in court cases. An indispensable tool in the field of criminal justice, forensic drug chemistry occupies a special place. The results of forensic investigations can have a significant impact on both the lives of victims and suspects, as well as the direction of an inquiry, legal judgements, and other repercussions. In addition, this field goes beyond merely classifying substances and is crucial in revealing trends of drug manufacturing, distribution, and trafficking. Forensic experts can pinpoint the production processes or geographic locations of batches of illegal substances by carefully examining chemical signatures, isotopic compositions, and impurity profiles. Law enforcement organisations can break up drug networks and capture important figures in the illegal drug trade thanks to the information that provides them with critical insight. The field of forensic drug chemistry is crucial to the advancement of contemporary forensic science. It is able to offer crucial insights into the make-up, origin, and distribution of illegal substances thanks to its persistent commitment to scientific rigour and interdisciplinary approach. Forensic drug chemistry helps criminal investigations by revealing the molecular details of crime scene evidence.

References

Abbate V, Guillou C, Schwenk M (2021) Psychoactive designer drugs: classes, mechanisms, and regulation. Regul Toxicol 1:1109–1125

Agatonovic-Kustrin S, Ristivojevic P, Gegechkori V, Litvinova TM, Morton D (2020) Essential oil quality and purity evaluation via ft-ir spectroscopy and pattern recognition techniques. Appl Sci 10(20):7294

Alkhuder K (2022) Attenuated total reflection-fourier transform infrared spectroscopy: a universal analytical technique with promising applications in forensic analyses. Int J Legal Med 136(6):1717–1736

Alseekh S, Aharoni A, Brotman Y, Contrepois K, D'Auria J, Ewald JC, Ewald J, Fraser PD, Giavalisco P, Hall RD, Heinemann M (2021) Mass spectrometry-based metabolomics: a guide for annotation, quantification and best reporting practices. Nat Methods 18(7):747–756

Alvarez-Ordóñez A, Prieto M, Alvarez-Ordóñez A, Prieto M (2012) Technical and methodological aspects of fourier transform infrared spectroscopy in food microbiology research. Springer, New York

Anzar N, Suleman S, Parvez S, Narang J (2022) A review on illicit drugs and biosensing advances for its rapid detection. Process Biochem 113:113–124

Aronson JK (2011) Adverse drug reactions: history, terminology, classification, causality, frequency, preventability. In: Stephens' detection and evaluation of adverse drug reactions: principles and practice, pp 1–19

Ashraf S, Sayeed SA, Ali R, Vohra F, Ahmed N, Alam MK (2022) Assessment of potential benefits of functional food characteristics of beetroot energy drink and flavored milk. Biomed Res Int 15:2022

Bakels S, Gaigeot MP, Rijs AM (2020) Gas-phase infrared spectroscopy of neutral peptides: insights from the far-IR and THz domain. Chem Rev 120(7):3233–3260

Bartley JK, Basu-Dutt S, Geisler VJ, Khan FA, Swamy-Mruthinti S (2010) Making chemistry relevant to science and engineering majors. In: Making chemistry relevant: strategies for including all students in a learner-sensitive classroom environment, pp 169–192

Bechky BA (2020) Evaluative spillovers from technological change: the effects of "DNA envy" on occupational practices in forensic science. Adm Sci Q 65(3):606–643

Biancuzzi H, Dal Mas F, Brescia V, Campostrini S, Cascella M, Cuomo A, Cobianchi L, Dorken-Gallastegi A, Gebran A, Kaafarani HM, Marinangeli F (2022) Opioid misuse: a review of the main issues, challenges, and strategies. Int J Environ Res Public Health 19(18):11754

Bisen AC, Sanap SN, Agrawal S, Biswas A, Bhatta RS (2023) Chemical metabolite synthesis and profiling: mimicking in vivo biotransformation reaction. Bioorg Chem 8:106722

Bishnoi N, Sharma US (2020) The crucial role of forensic science in solving modern-day crimes: an analysis of the criminal justice system. Ann Rom Soc Cell Biol 1:2257–2269

Blais E, Brisson J, Gagnon F, Lemay SA (2022) Diverting people who use drugs from the criminal justice system: a systematic review of police-based diversion measures. Int J Drug Policy 105:103697

Brasileiro Dourado LM, Fischer B (2021) Drug sentencing in Brazil: factors influencing sentencing outcomes in the criminal courts of Recife/PE. J Criminol Res Policy Pract 7(4):335–344

Brazier JL (2020) Use of isotope ratios in forensic analysis. In: Forensic applications of mass spectrometry. CRC Press, pp 259–289

Brien R, Volpe I, Grigg J, Lyons T, Hughes C, McKinnon G, Tzanetis S, Crawford S, Eade A, Lee N, Barratt MJ (2023) Co-designing drug alerts for health and community workers for an emerging early warning system in Victoria, Australia. Harm Reduct J 20(1):1–27

Camin F, Bontempo L, Perini M, Piasentier E (2016) Stable isotope ratio analysis for assessing the authenticity of food of animal origin. Compr Rev Food Sci Food Saf 15(5):868–877

Carrier MB (2021) The making of evident expertise: transforming chemical analytical methods into judicial evidence. NTM Z Gesch Wiss Tech Med 29(3):261–284

Colleoni E, Samaras VG, Guida P, Frassoldati A, Faravelli T, Roberts WL (2023) Unraveling the complexity of pyrolysates from residual fuels by Py-GCxGC-FID/SCD/TOF-MS with an innovative data processing method. J Anal Appl Pyrolysis 175:106204

Coppey F, Bécue A, Sacré PY, Ziemons EM, Hubert P, Esseiva P (2020) Providing illicit drugs results in five seconds using ultra-portable NIR technology: an opportunity for forensic laboratories to cope with the trend toward the decentralization of forensic capabilities. Forensic Sci Int 317:110498

Da Filicaia EG, Evershed RP, Peggie DA (2023) Review of recent advances on the use of mass spectrometry techniques for the study of organic materials in painted artworks. Anal Chim Acta 1246:340575

de Boer RA, Hulot JS, Tocchetti CG, Aboumsallem JP, Ameri P, Anker SD, Bauersachs J, Bertero E, Coats AJ, Čelutkienė J, Chioncel O (2020) Common mechanistic pathways in cancer and heart failure. A scientific roadmap on behalf of the translational research committee of the Heart Failure Association (HFA) of the European Society of Cardiology (ESC). Eur J Heart Fail 22(12):2272–2289

de Campos EG, da Costa BR, Dos Santos FS, Monedeiro F, Alves MN, Santos Junior WJ, De Martinis BS (2022a) Alternative matrices in forensic toxicology: a critical review. Forensic Toxicol 40(1):1–8
de Campos GE, Krotulski AJ, De Martinis BS, Costa JL (2022b) Identification of synthetic cathinones in seized materials: a review of analytical strategies applied in forensic chemistry. WIREs Forensic Sci 4(5):e1455
Dekebo A, Kwon SY, Kim DH, Jung C (2018) Volatiles analysis of honey by gas chromatography-mass spectrometry (GC-MS): comparison of SPME volatiles extraction methods. J Apic Res 33(2):117–128
Delgado Y, Price BS, Speaker PJ, Stoiloff SL (2021) Forensic intelligence: data analytics as the bridge between forensic science and investigation. Forensic Sci Int Synergy 3:100162
Deschamps E, Calabrese V, Schmitz I, Hubert-Roux M, Castagnos D, Afonso C (2023) Advances in ultra-high-resolution mass spectrometry for pharmaceutical analysis. Molecules 28(5):2061
Donati GL, Amais RS (2019) Fundamentals and new approaches to calibration in atomic spectrometry. J Anal At Spectrom 34(12):2353–2369
Dror IE (2020) Cognitive and human factors in expert decision making: six fallacies and the eight sources of bias. Anal Chem 92(12):7998–8004
Edmond G, Found B, Martire K, Ballantyne K, Hamer D, Searston R, Thompson M, Cunliffe E, Kemp R, San Roque M, Tangen J (2016) Model forensic science. Aust J Forensic Sci 48(5):496–537
Elkins KM, Fambegbe I (2020) Case studies and methods for teaching professional ethics for forensic science students. J Forensic Sci Edu 2(1):1
Elkins KM, Zeller C (2021) Approaches for teaching forensic body fluids analysis for remote learning. J Forensic Sci Edu 3(1):1
Estupiñán Méndez D, Allscher T (2022) Advantages of external reflection and transflection over ATR in the rapid material characterization of negatives and films via FTIR spectroscopy. Pol Sci 14(4):808
Fares MY, Hegazy MA, El-Sayed GM, Abdelrahman MM, Abdelwahab NS (2022) Quality by design approach for green HPLC method development for simultaneous analysis of two thalassemia drugs in biological fluid with pharmacokinetic study. RSC Adv 12(22):13896–13916
Fenton N, Neil M, Lagnado DA (2013) A general structure for legal arguments about evidence using Bayesian networks. Cogn Sci 37(1):61–102
Fisher CM, Peter KT, Newton SR, Schaub AJ, Sobus JR (2022) Approaches for assessing performance of high-resolution mass spectrometry–based non-targeted analysis methods. Anal Bioanal Chem 414(22):6455–6471
Gallagher MB, Thornton JI (2011) Trace evidence in crime reconstruction. Crime Reconstr 29:247–297
Garrett BL (2020) Wrongful convictions. Annu Rev Criminol 3:245–259
Gibbs BR, Lytle R, Wakefield W (2019) Outcome effects on recidivism among drug court participants. Crim Justice Behav 46(1):115–135
Gliozzo E (2020) Ceramics investigation: research questions and sampling criteria. Archaeol Anthropol Sci 12(8):202
Graziano S, Anzillotti L, Mannocchi G, Pichini S, Busardò FP (2019) Screening methods for rapid determination of new psychoactive substances (NPS) in conventional and non-conventional biological matrices. J Pharm Biomed Anal 163:170–179
Hachem M, Mizouni R, Alawadhi IM, Altamimi MJ (2023) Digital forensic intelligence for illicit drug analysis in forensic investigations. iScience 26(10):108023
Hatzakis E (2019) Nuclear magnetic resonance (NMR) spectroscopy in food science: a comprehensive. Compr Rev Food Sci Food Saf 18(1):189–220
Hawgood J, Woodward A, Quinnett P, De Leo D (2022) Gatekeeper training and minimum standards of competency: essentials for the suicide prevention workforce. J Crisis Interv Suicide Prev 43(6):516

He X, Wang J, Teng X, Fan L, Li X, Wei C, Zhang C (2021) On the rapid and non-destructive approach for barbiturates, benzodiazepines, and phenothiazines determination and differentiation using spectral combination analysis and chemometric methods. Microchem J 162:105853

Heinsvig PJ, Noble C, Dalsgaard PW, Mardal M (2023a) Forensic drug screening by liquid chromatography hyphenated with high-resolution mass spectrometry (LC-HRMS). Trends Analyt Chem 15:117023

Heinsvig PJ, Noble C, Dalsgaard PW, Mardal M (2023b) Forensic drug screening by liquid chromatography hyphenated with high-resolution mass spectrometry (LC-HRMS). Trends Anal Chem 15:117023

Herrada-Manchón H, Rodríguez-González D, Fernández MA, Suñé-Pou M, Pérez-Lozano P, García-Montoya E, Aguilar E (2020) 3D printed gummies: personalized drug dosage in a safe and appealing way. Int J Pharm 587:119687

Hoffmann WD, Jackson GP (2015) Forensic mass spectrometry. Annu Rev Anal Chem 8:419–440

Huestis MA, Brandt SD, Rana S, Auwärter V, Baumann MH (2017) Impact of novel psychoactive substances on clinical and forensic toxicology and global public health. Clin Chem 63(10):1564–1569

Humphries G, Nordvik R, Manifavas H, Cobley P, Sorell M (2021) Law enforcement educational challenges for mobile forensics. Forensic Sci Int Digit Investig 38:301129

Jagadeesh P, Mavinkere Rangappa S, Siengchin S, Puttegowda M, Thiagamani SM, Hemath Kumar M, Oladijo OP, Fiore V, Moure Cuadrado MM (2022) Sustainable recycling technologies for thermoplastic polymers and their composites: a review of the state of the art. Polym Compos 43(9):5831–5862

Jilala W, Lwoga N (2022) A brief history of forensic services in Tanzania: current challenges and mitigation efforts. Forensic Sci Int Synergy 4:100227

Joshi M, Sisco E (2023) Forensic seized drug analysis: current challenges and emerging analytical solutions. WIREs Forensic Sci 7:e1486

Kątny M, Frankowski M (2017) Impurities in drug products and active pharmaceutical ingredients. Crit Rev Anal Chem 47(3):187–193

Khan AA, Uddin M, Shaikh AA, Laghari AA, Rajput AE (2021) MF-ledger: blockchain hyperledger sawtooth-enabled novel and secure multimedia chain of custody forensic investigation architecture. IEEE Access 9:103637–103650

Kim K, Stoll S, Singh R, Lee WH, Hwang JH (2023) Recent advances in illicit drug detection sensor technology in water. Trends Anal Chem 18:117295

Kranenburg RF, Ou F, Sevo P, Petruzzella M, de Ridder R, van Klinken A, Hakkel KD, van Elst DM, van Veldhoven R, Pagliano F, van Asten AC (2022) On-site illicit-drug detection with an integrated near-infrared spectral sensor: a proof of concept. Talanta 245:123441

Labat DE, Goldfarb D, Evans JR, Schreiber Compo N, Koolmees CJ, Laporte G, Lothridge K (2023) A forensic science informational video can help mock jurors evaluate forensic expert testimony. Crim Justice Behav 50(11):1716–1736

Lee H, Abarghani A, Liu B, Shokouhimehr M, Ostadhassan M (2020) Molecular weight variations of kerogen during maturation with MALDI-TOF-MS. J Fuels 269:117452

Leghissa A, Hildenbrand ZL, Schug KA (2018) A review of methods for the chemical characterization of cannabis natural products. J Sep Sci 41(1):398–415

Li S, Wang X, Zhang Y, Wan Z, Wang Y (2022) Advances in the application of chromatographic techniques in forensic toxicology fund. Adv Anal Sci 3(1):12

Lim JW, Kim TY, Woo MA (2021) Trends in sensor development toward next-generation point-of-care testing for mercury. Biosens Bioelectron 183:113228

Liu Z, Liu D, Cai Y, Yao Y, Pan Z, Zhou Y (2020a) Application of nuclear magnetic resonance (NMR) in coalbed methane and shale reservoirs: a review. Int J Coal Geol 218:103261

Liu Y, Li A, Buchanan S, Liu W (2020b) Exposure characteristics for congeners, isomers, and enantiomers of perfluoroalkyl substances in mothers and infants. Environ Int 144:106012

Locatelli M, Covone S, Rosato E, Bonelli M, Savini F, Furton KG, Gazioglu I, D'Ovidio C, Kabir A, Tartaglia A (2022) Analysis of seven selected antidepressant drugs in post–mortem samples using fabric phase sorptive extraction followed by high performance liquid chromatography-photodiode array detection. Forensic Chem 31:100460

Lu X, Zhang Z, Gao R, Wang H, Xiao J (2021) Recent progress in the chemical attribution of chemical warfare agents and highly toxic organophosphorus pesticides. Forensic Toxicol 39(2):334–349

Machethe P, Mofokeng JT (2022) The impact of illicit drug networks on the effectiveness of law enforcement in South Africa. Tech Soc Sci J 27:338

Mani NS, Budak JZ, Lan KF, Bryson-Cahn C, Zelikoff A, Barker GE, Grant CW, Hart K, Barbee CJ, Sandoval MD, Dostal CL (2020) Prevalence of coronavirus disease 2019 infection and outcomes among symptomatic healthcare workers in Seattle, Washington. Clin Infect Dis 71(10):2702–2707

Marinho BA, Suhadolnik L, Likozar B, Huš M, Marinko Ž, Čeh M (2022) Photocatalytic, electrocatalytic and photoelectrocatalytic degradation of pharmaceuticals in aqueous media: analytical methods, mechanisms, simulations, catalysts and reactors. J Clean Prod 343:131061

Mateo C, Talavera JA (2018) Short-time fourier transform with the window size fixed in the frequency domain. Digit Signal Process 77:13–21

McIntosh LG, Janes S, O'Rourke S, Thomson LD (2021) Effectiveness of psychological and psychosocial interventions for forensic mental health inpatients: a meta-analysis. Aggress Violent Behav 58:101551

Meier-Augenstein W (2019) Forensic stable isotope signatures: comparing, geo-locating, detecting linkage. WIREs Forensic Sci 1(5):e1339

Meola S, Huhtala S, Broséus J, Jendly M, Jalava K, Aalberg L, Esseiva P (2021) Illicit drug profiling practices in Finland: an exploratory study about end users' perceptions. Forensic Sci Int 324:110848

Mercolini L, Protti M (2016) Biosampling strategies for emerging drugs of abuse: towards the future of toxicological and forensic analysis. J Pharm Biomed Anal 130:202–219

Merone GM, Tartaglia A, Rossi S, Santavenere F, Bassotti E, D'Ovidio C, Bonelli M, Rosato E, de Grazia U, Locatelli M, Savini F (2021) Fast quantitative LC-MS/MS determination of illicit substances in solid and liquid unknown seized samples. Anal Chem 93(49):16308–16313

Minho LA, de Lourdes CZ, Menezes HC (2023) Development of polypropylene membranes grafted with nanocellulose to analyze organic pollutants in environmental waters using miniaturized passive samplers based on liquid-phase microextraction. Microchem J 190:108641

Morrow JB, Ropero-Miller JD, Catlin ML, Winokur AD, Cadwallader AB, Staymates JL, Williams SR, McGrath JG, Logan BK, McCormick MM, Nolte KB (2019) The opioid epidemic: moving toward an integrated, holistic analytical response. J Anal Toxicol 43(1):1–9

Moud AA (2022) Fluorescence recovery after photobleaching in colloidal science: introduction and application. ACS Biomater Sci Eng 8(3):1028–1048

Mullin L, Katz DR, Riddell N, Plumb R, Burgess JA, Yeung LW, Jogsten IE (2019) Analysis of hexafluoropropylene oxide-dimer acid (HFPO-DA) by liquid chromatography-mass spectrometry (LC-MS): review of current approaches and environmental levels. Trends Analyt Chem 118:828–839

Musselman ME, Hampton JP (2014) "Not for human consumption": a review of emerging designer drugs. Pharmacotherapy 34(7):745–757

Neal TM, Martire KA, Johan JL, Mathers EM, Otto RK (2022) The law meets psychological expertise: eight best practices to improve forensic psychological assessment. Annu Rev Law Soc Sci 18:169–192

Nicolaou AG, Stavrou IJ, Louppis AP, Constantinou MS, Kapnissi-Christodoulou C (2021) Application of an ultra-performance liquid chromatography-tandem mass spectrometric method for the detection and quantification of cannabis in cerumen samples. J Chromatogr A 1642:462035

Nitescu V (2019) New psychoactive substances and the risks of consumption in children and adolescents. Rom J Pediatr 68(4):233–238

Olsen NJ, James JA, Arriens C, Ishimori ML, Wallace DJ, Kamen DL, Chong BF, Liao D, Chinchilli VM, Karp DR (2018) Study of anti-malarials in incomplete lupus erythematosus (SMILE): study protocol for a randomized controlled trial. Trials 19:1–9

Otgaar H, Howe ML, Dodier O (2022) What can expert witnesses reliably say about memory in the courtroom? Forensic Sci Int Mind Law 3:100106

Ott CE, Arroyo LE (2023) Transitioning surface-enhanced raman spectroscopy (SERS) into the forensic drug chemistry and toxicology laboratory: current and future perspectives. WIREs Forensic Sci 1:e1483

Parihar A, Choudhary NK, Sharma P, Khan R (2023) Carbon nanomaterials-based electrochemical aptasensor for point-of-care diagnostics of cancer biomarkers. Mater Today Chem 30:101499

Parvathaneni V, Kulkarni NS, Muth A, Gupta V (2019) Drug repurposing: a promising tool to accelerate the drug discovery process. Drug Discov Today 24(10):2076–2085

Peltier-Rivest D, Pacini C (2019) Detecting counterfeit pharmaceutical drugs: a multi-stakeholder forensic accounting strategy. J Financ Crime 26(4):1027–1047

Peters FT, Wissenbach D (2023) Current state-of-the-art approaches for mass spectrometry in clinical toxicology: an overview. Expert Opin Drug Metab Toxicol 19(8):487–500

Piraianu AI, Fulga A, Musat CL, Ciobotaru OR, Poalelungi DG, Stamate E, Ciobotaru O, Fulga I (2023) Enhancing the evidence with algorithms: how artificial intelligence is transforming forensic medicine. Diagnostics 13(18):2992

Pratiwi RA, Nandiyanto AB (2022) How to read and interpret UV-VIS spectrophotometric results in determining the structure of chemical compounds. Indones J Educ Sci Technol 2(1):1–20

Ramohlola KE, Iwuoha EI, Hato MJ, Modibane KD (2020) Instrumental techniques for characterization of molybdenum disulphide nanostructures. J Anal Chem 16:2020

Rinaldi R, Bersani G, Marinelli E, Zaami S (2020) The rise of new psychoactive substances and psychiatric implications: a wide-ranging, multifaceted challenge that needs far-reaching common legislative strategies. Hum Psychopharmacol 35(3):e2727

Rivera D, Mozayani A, Obi E, Drake JM (2019) The structural complexities of Cannabis sativa L. and profiling techniques for geographic source determination. J Forensic Res 10:1000443

Roux C, Bucht R, Crispino F, De Forest P, Lennard C, Margot P, Miranda MD, NicDaeid N, Ribaux O, Ross A, Willis S (2022) The Sydney declaration–revisiting the essence of forensic science through its fundamental principles. Forensic Sci Int 332:111182

Sánchez-Viesca F, Gómez R (2023) The chemistry of Mandelin's test for strychnine. Open Access Res J Chem Pharm 3(01):001–004

Santhy DK (2023) The realm of forensic toxicology in criminal justice system of India. Int J Law Policy Soc Rev 5(1):42–46

Scendoni R, Fedeli P, Cannovo N, Cingolani M (2021) The "magnificent seven errors" in forensic autopsy practice: the Italian context. Acad Forensic Pathol 11(4):208–214

Sharma V, Kumar R, Kaur P (2019) Forensic examination of textile fibers using UV-vis spectroscopy combined with multivariate analysis. J Appl Spectrosc 86:96–100

Sherma J, Rabel F (2019) Thin layer chromatography in the analysis of cannabis and its components and synthetic cannabinoids. J Liq Chromatogr 42(19–20):613–628

Shogar IA (2019) The nature of forensic ethics in bioanalytical science. Revel Sci 9(01):9–17

Sisco E, Forbes TP (2021) Forensic applications of DART-MS: a review of recent literature. Forensic Chem 22:100294

Sisodia U (2022) Chain of custody: scaling the investigation to the event. In: Crime scene management within forensic science: forensic techniques for criminal investigations. Springer, Singapore, pp 407–418

Song Y, Cong Y, Wang B, Zhang N (2020) Applications of fourier transform infrared spectroscopy to pharmaceutical preparations. Expert Opin Drug Deliv 17(4):551–571

Sosa-Reyes AM, Villavicencio-Queijeiro A, Suzuri-Hernández LJ (2022) Interdisciplinary approaches to the teaching of forensic science in the forensic science undergraduate program of the National Autonomous University of Mexico, before and after COVID-19. Sci Justice 62(6):676–690

Soursou V, Campo J, Picó Y (2023) A critical review of the novel analytical methods for the determination of microplastics in sand and sediment samples. TrAC Trends Anal Chem 17:117190
Spain RD, Hedge JW, Ohse D, White A (2022) The need for research-based tools for personnel selection and assessment in the forensic sciences. Forensic Sci Int Synergy 4:100213
Steuer AE, Brockbals L, Kraemer T (2019) Metabolomic strategies in biomarker research–new approach for indirect identification of drug consumption and sample manipulation in clinical and forensic toxicology? Front Chem 7:319
Strynar M, McCord J, Newton S, Washington J, Barzen-Hanson K, Trier X, Liu Y, Dimzon IK, Bugsel B, Zwiener C, Munoz G (2023) Practical application guide for the discovery of novel PFAS in environmental samples using high resolution mass spectrometry. J Expo Sci Environ Epidemiol 33(4):575–588
Suto H, Kataoka F, Kikuchi N, Knuteson RO, Butz A, Haun M, Buijs H, Shiomi K, Imai H, Kuze A (2021) Thermal and near-infrared sensor for carbon observation fourier transform spectrometer-2 (TANSO-FTS-2) on the greenhouse gases observing SATellite-2 (GOSAT-2) during its first year in orbit. Atmos Meas Tech 14(3):2013–2039
Tettey JN, Crean C, Rodrigues J, Yap TW, Lim JL, Lee HZ, Ong MC (2021) United Nations Office on drugs and crime: recommended methods for the identification and analysis of synthetic cannabinoid receptor agonists in seized materials. Forensic Sci Int Synergy 3:100129
Tian W, Chen G, Gui Y, Zhang G, Li Y (2021) Rapid quantification of total phenolics and ferulic acid in whole wheat using UV–vis spectrophotometry. Food Control 123:107691
Trozze A, Davies T, Kleinberg B (2023) Explaining prosecution outcomes for cryptocurrency-based financial crimes. J Money Laund Control 26(1):172–188
Tsochatzis ED, Giannopoulos G, Lopes JA, Guillou C (2021) Drugs analysis and citizens safety. Workflows for chemical analysis and structural elucidation of new psychoactive substances in the EU. Saf Sci 137:105126
Viera-Herrera C, Santamaría-Aguirre J, Vizuete K, Debut A, Whitehead DC, Alexis F (2020) Microcrystalline cellulose extracted from native plants as an excipient for solid dosage formulations in drug delivery. Nano 10(5):975
Wang Y, He T, Wang J, Wang L, Ren X, He S, Liu X, Dong Y, Ma J, Song R, Wei J (2021) High performance liquid chromatography fingerprint and headspace gas chromatography-mass spectrometry combined with chemometrics for the species authentication of Curcumae rhizoma. J Pharm Biomed Anal 202:114144
Wang H, Gao Y, He Q, Liao J, Zhou S, Liu Y, Guo C, Li X, Zhao X, Pan Y (2022) 2-Hydrazinoterephthalic acid as a novel negative-ion matrix-assisted laser desorption/ionization matrix for qualitative and quantitative matrix-assisted laser desorption/ionization–mass spectrometry analysis of N-Glycans in peach allergy research. J Agric Food Chem 71(1):952–962
Weber A, Hoplight B, Ogilvie R, Muro C, Khandasammy SR, Pérez-Almodóvar L, Sears S, Lednev IK (2023) Innovative vibrational spectroscopy research for forensic application. Anal Chem 95(1):167–205
Welner M, DeLisi M, Knous-Westfall HM, Salsberg D, Janusewski T (2023) Forensic assessment of criminal maturity in juvenile homicide offenders in the United States. Forensic Sci Int Mind Law 4:100112
Wheate R (2008) Australian forensic scientists: a view from the witness box. Aust J Forensic Sci 40(2):123–146
Woodman PA, Spiranovic C, Julian R, Ballantyne KN, Kelty SF (2020) The impact of chemical trace evidence on justice outcomes: exploring the additive value of forensic science disciplines. Forensic Sci Int 307:110121
Yang B, Liao GQ, Wen XF, Chen WH, Cheng S, Stolzenburg JU, Ganzer R, Neuhaus J (2017) Nuclear magnetic resonance spectroscopy as a new approach for improvement of early diagnosis and risk stratification of prostate cancer. J Zhejiang Univ Sci B 18(11):921
Yang Y, Yang L, Zheng M, Cao D, Liu G (2023) Data acquisition methods for non-targeted screening in environmental analysis. Trends Anal Chem 7:116966

Young G, Goodman-Delahunty J (2021) Revisiting Daubert: judicial gatekeeping and expert ethics in court. Psychol Inj Law 14:304–315
Yu HY, Myoung S, Ahn S (2021) Recent applications of benchtop nuclear magnetic resonance spectroscopy. Magnetochemistry 7(9):121
Zapata F, Matey JM, Montalvo G, García-Ruiz C (2021) Chemical classification of new psychoactive substances (NPS). Microchem J 163:105877
Zhu J, Hao G (2020) Design and test of a compact compliant gripper using the Scott–Russell mechanism. Arch Civ Mech 20:1–2
Zhu N, Li G, Zhou J, Zhang Y, Kang K, Ying B, Yi Q, Wu Y (2021) A light-up fluorescence resonance energy transfer magnetic aptamer-sensor for ultra-sensitive lung cancer exosome detection. J Mater Chem B 9(10):2483–2493

Chapter 17
Artificial Intelligence in Drug Discovery and Development

Geethaa Sahgal and Jeevandran Sundarasekar

Abstract Recently, machine learning (ML) and artificial intelligence (AI) have gained popularity in the healthcare sector. In the field of drug discovery and development, numerous strategies have been developed to create a human healthcare system that is beneficial. Many automated drug development design pipelines and validated ML algorithms are currently being imagined for various diseases. Preclinical research is a crucial stage in the drug development process before ML predictive models are used. These pipelines facilitate a better understanding of the diseases and related biological. These methods enable the exploration of previously unsolvable problems, such as property prediction, molecular design, and synthetic route optimisation, while also lowering the expense and duration of research associated with finding a new drug through in vitro or in vivo models. An overview of the application of AI and ML to drug development is given in this chapter. It talks about the various applications of these technologies as well as the issues that still need to be resolved. An examination of AI and ML's potential applications in drug discovery rounds out the chapter.

Keywords Artificial intelligence · Machine learning · Drug discovery and development · Future directions and challenges

17.1 Introduction

Drug discovery and development is the process of identifying and developing a novel medication with particular chemical properties to treat a particular disease. Precision and trial-and-error drugs are the two types of approaches currently used

G. Sahgal (✉)
Faculty of Pharmacy, AIMST University, Bedong, Kedah, Malaysia
e-mail: geethaa@aimst.edu.my

J. Sundarasekar
Faculty of Applied Science, AIMST University, Bedong, Kedah, Malaysia
e-mail: jeevan@aimst.edu.my

S. Bose et al. (eds.), *Concepts in Pharmaceutical Biotechnology and Drug Development*, Interdisciplinary Biotechnological Advances,
https://doi.org/10.1007/978-981-97-1148-2_17

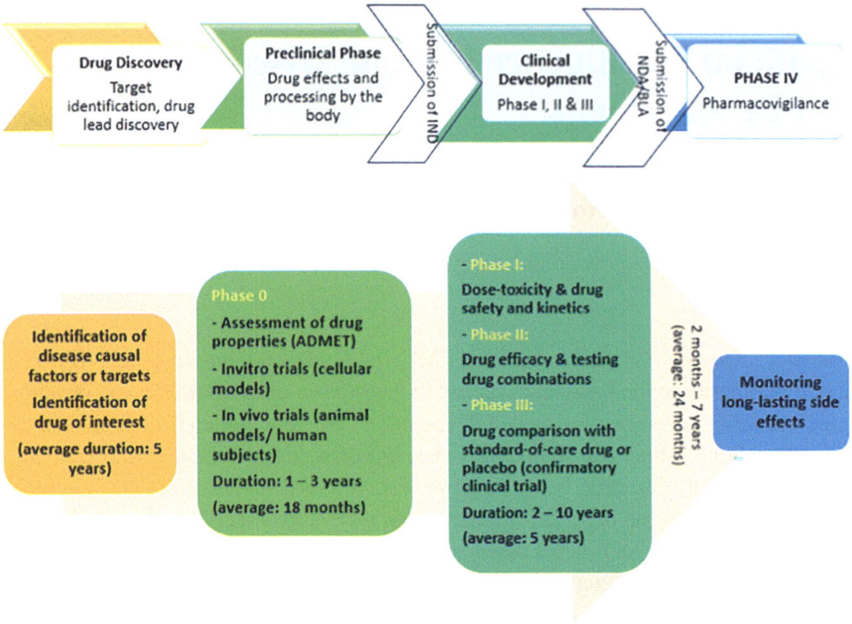

Fig. 17.1 The phases in the drug development process (Réda et al. 2020)

in drug discovery and development. Precision medicine, according to Collins and Varmus (2015), emphasises the unique variations in each person's genes, environment, and lifestyle (Collins and Varmus 2015). It can be more challenging to develop new methodologies for this approach because it differs from the traditional "one-size-fits-all" approach to treating diseases. On the other hand, precision medicine might lead to more successful and customised patient care (Carracedo-Reboredo et al. 2021). Trial-and-error medicine is another common strategy in drug discovery. This approach often necessitates testing multiple medications on patients before identifying the most effective one, which can make it challenging and expensive. Trial-and-error medicine can, nevertheless, remain a valuable instrument in the drug discovery process since it can be used to identify novel, promising therapeutic candidates. The production of pharmaceuticals has long been subject to a regulatory framework that safeguards the quality of completed goods. This is a complex and labour-intensive process that has traditionally relied on labour-intensive techniques like high-throughput screenings and trial-and-error in vitro and in vivo studies (Patel and Shah 2022). With this process, it can take over 10 years to get approval and treat a patient successfully. The whole process is described in Fig. 17.1. According to estimates from the Tufts Centre for the Study of Drug Development (CSDD), it would have cost roughly $2.6 billion as of 2014 to develop a new prescription medicine that is approved for sale (Deep Pharma 2020). Additionally, as shown in Fig. 17.2, the article stated that it takes overcoming multiple stages of obstacles before a single novel targeted drug is found (Réda et al. 2020).

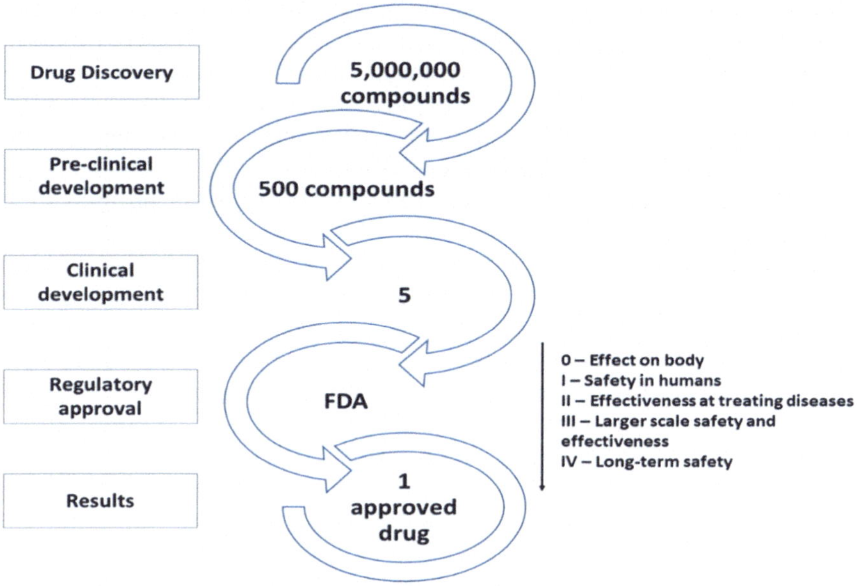

Fig. 17.2 Pharma efficiency: challenges (Deep Pharma 2020)

17.2 Revolution of Artificial Intelligence (AI) in Drug Discovery

In recent years, a large amount of biomedical and digitisation data has been produced by small individual laboratories and large international initiatives (Carracedo-Reboredo et al. 2021). The amount of data in the pharmaceutical industry is increasing dramatically, which makes it difficult to collect, analyse, and apply that knowledge to complex clinical issues (Paul et al. 2021; Ramesh et al. 2004). AI was therefore created to handle large amounts of data with better automation. AI is a technological system that mimics human intelligence in learning and problem-solving through a range of state-of-the-art tools and networks. AI is capable of interpreting input data, learning from it, and making decisions on its own to achieve specific goals. The employment of software and systems in the healthcare sector surely does not endanger people's ability to exist in person (Paul et al. 2021; Yang and Siau 2018; Wirtz et al. 2019). The advancement of this technology for use in the medical and pharmaceutical industries by experts will significantly improve patient care.

More data is being produced by the pharmaceutical industry than ever before, rendering outdated traditional data storage techniques. There is an enormous surge in drug candidates as we enter the big data (BD) era of contemporary drug discovery. Consequently, AI aids in the management and analysis of collective information through deep learning and relevant modelling studies (Paul et al. 2021; Solanki et al. 2022). The first attempts to use the drug discovery process in the pharmaceutical industry were funded by Merck and were known as the 2012 QSAR ML. This complex model significantly increases the predictivity of the developed drug candidates for the toxicity and the adsorption, distribution, metabolism, and excretion (ADME) data sets (Zhu 2019). This opened the door for the development of neural network techniques, which the pharmaceutical industry currently uses extensively for drug discovery and development.

The pharmaceutical industry employs 240 AI companies, 600 investors, and 90 corporations, as per the AI for Drug Discovery, Biomarker Development, and Advanced R&D Landscape, 2020 report (Deep Pharma 2020). However, as of 2022, 495 AI companies, 1120 investors, and 100 corporations are involved in the pharmaceutical industry, according to the AI for Drug Discovery Landscape Overview landscape Q1 teaser report. According to Deep Pharma Intelligence 2020, the United States (US), China, the United Kingdom (UK), Canada, the European Union (EU), Asia, and Australia contributed 54.4%, 2.5%, 14.2%, 6.7%, 13.4%, 8.4%, and 0.4% of AI for drug discovery in 2020 (Deep Pharma 2020). On the other hand, according to Deep Pharma Intelligence 2022, there will be increases in the US, China, EU, Asia, and the Middle East of 55.15%, 3.83%, 16.76%, and 9.89%, respectively (Deep Pharma 2022). When considering AI from a global perspective, the US leads the AI industry in terms of the number of businesses that use AI to drive R&D, research centres, institutes, and investments. Novartis is the major driver of increased AI activity in the pharmaceutical industry in the UK and EU. Additionally, AstraZeneca and BenevolentAI work together on innovative AI-generated targets for chronic kidney disease. China has made significant investments in AI totalling $5 billion. Tianjin, one of the largest cities in China, is investing $16 billion in the region's AIsector. In addition, Beijing will construct a $2.12 billion AI research project. China aims to lead the world in AI by 2030, according to the AI Strategic Plan, which was published in July 2017. According to Deep Pharma Intelligence's 2022 analysis, Saama Technologies, Inc., a prominent clinical data analytics company, is one of the driving forces behind the implementation of AI in the Asia-Pacific region. This demonstrates the rapid growth of AI in the pharmaceutical sector for global drug development (Deep Pharma 2022).

Figure 17.3 shows the market drug discovery timeline for AI. Just a small number of businesses decided to start the pilot collaborations and make modest investments in AI drug discovery and development during the first part of the timeline. Due to the failure of numerous pilot projects in the life sciences, deep learning for drug discovery and advanced R&D was heavily criticised during the second phase. Nevertheless, there is an intense rivalry to acquire the best AI startups, and technology testing gets underway. The AI industry's capitalisation continued to rise in the fifth year of the market timeline, seemingly justifying many of the early investors' stakes. The big pharma industry is engaged in fierce competition to prioritise AI and related technologies in drug development platforms. As a result, numerous AI startups partnered with major pharmaceutical companies, including AstraZeneca, Pfizer, Novartis, Takeda, Merck, and Jansen (Réda et al. 2020). The top pharma companies by number of pharma AI deals and the Top 10 AI and Technology Partners by number of significant pharma AI deals in 2021 and Q1 2022, respectively, are displayed in Figs. 17.4 and 17.5. The AI and Pharma partnership from February 2022 to the end of 2021 is outlined in Table 17.1 (Réda et al. 2020).

Fig. 17.3 AI for drug discovery market timeline (Deep Pharma 2022)

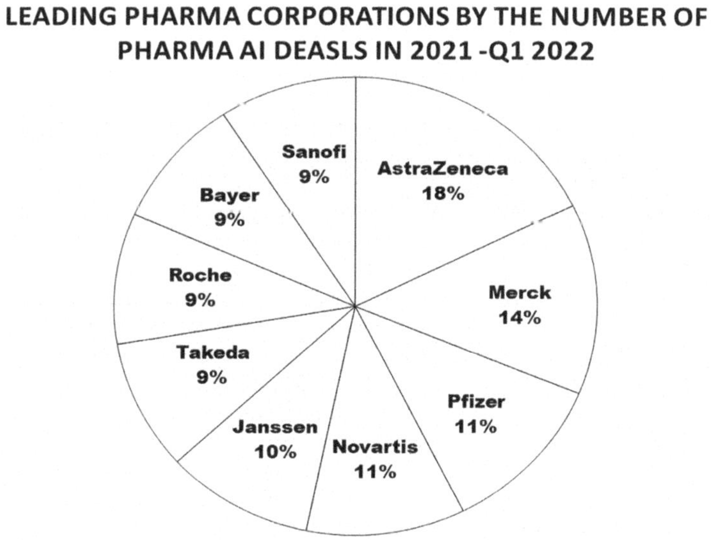

Fig. 17.4 The top pharmaceutical companies based on the quantity of deals in pharma AI between 2021 and Q1 2022 (Deep Pharma 2022)

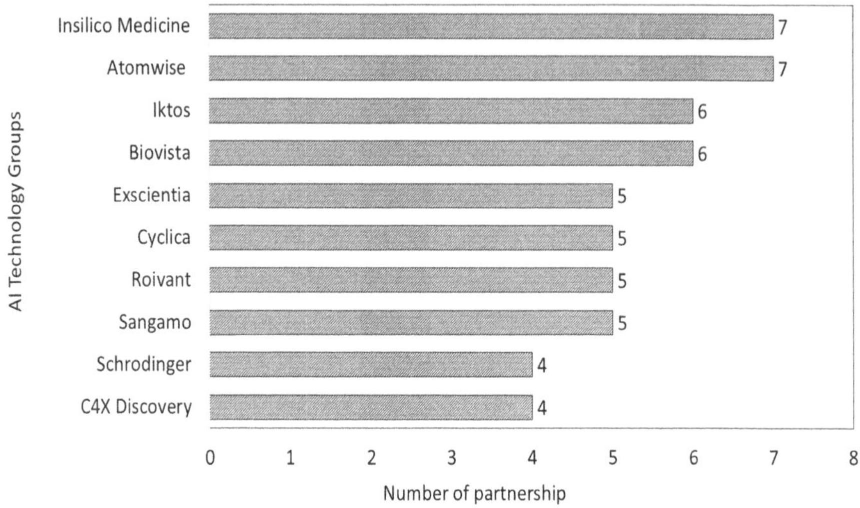

Fig. 17.5 Based on the quantity of significant pharmaceutical AI deals in 2021–Q1 2022, the to 10 AI and technology partners (Deep Pharma 2022)

Table 17.1 The collaboration between AI and pharma industry from the 2021 to February 2022 (Réda et al. 2020)

Month/ Year	Collaborators	Description
January 2021	Merck and Philips	To developments in customised reproductive treatment based on AI
February 2021	Roche and PatchAi	Together with PatchAi, Roche Italia is launching a virtual platform for cancer patients
March 2021	Pfizer and IKTOS	Pfizer's small-molecule initiatives will be relevant
April 2021	AstraZeneca, University of Florida, and NVIDIA	AI-powered de novo design software from Iktos
May 2021	Bristo Myers Squibb and Exscientia	AstraZeneca is collaborating on new AI research projects with NVIDIA and the University of Florida with the goal of improving patient care and drug discovery
June 2021	GSK and Progentec	GSK and the AI startup Progentec have announced an 18-month collaborative research agreement
June 2021	Teva and Insilico medicine	To make use of Insilico's ML technology, Teva and Insilico medicine partnered
July 2021	Eli Lilly and Verge genomics	Eli Lilly and Verge genomics collaborated to use AI in drug development to produce novel medications for the treatment of amyotrophic lateral sclerosis
August 2021	Insilico medicine and Westlake Pharma	Westlake Pharma and Insilico medicine have announced a collaboration to expedite the development of novel coronavirus drugs

Table 17.1 (continued)

Month/ Year	Collaborators	Description
September 2021	Summit pharmaceuticals international and CytoReason	The goal of the partnership between CytoReason and summit pharmaceuticals international (SPI) is to introduce CytoReason's ML platform to the Japanese clinical drug discovery market
October 2021	Poolbeg pharma and Eurofins genomics	Poolbeg pharma and Eurofins genomics launch an AI programme
December 2021	Roche and Genentech	Recursion's platform will be utilised by Genentech and Roche to find new drugs in the fields of oncology and neurobiology
January 2022	Merck and AbSci	In a $610 million deal, AbSci and Merck teamed to produce enzymes using AbSci's AI platform
January 2022	Sanofi and Exscientia	Exscientia and Sanofi worked together to create novel therapeutic candidates for immune-mediated illnesses and cancer
January 2022	Amgen and generate biomedicines	Amgen and generate biomedicines worked together to develop protein therapies for five different clinical targets. Amgen might be required to pay up to $1.9 billion in exchange for this partnership in order to obtain a cutting-edge AI-driven platform
February 2022	Bayer, Aalto and HUS	The cooperation between HUS, Aalto, and Bayer was extended to use AI to support clinical drug trials

17.3 The Computational Methods Used by AI Companies

The first stages towards the gradual advancement of AI in drug discovery are topological descriptors, molecular fingerprints, QSAR modelling, ML strategies (k-nearest neighbours, random forest, support vector machines, and genetic algorithms), and modelling techniques (simulated annealing and genetic algorithms) (Zhu 2019). Wen and colleagues report on a deep learning (DL) model designed to predict interactions between drugs and their biological target, based on 15,524 drug-target combinations extracted from the DrugBank database (Zhu 2019; Wen et al. 2017). ML, symbolic AI, deep learning (DL), evolutionary algorithms, reinforcement learning, convolutional neural networks (CNN), cheminformatics, bioinformatics, natural language processing (NLP), quantum computing, GANs, and federated learning are some of the computational techniques used by the most advanced AI companies in the pharmaceutical industries (Deep Pharma 2022). The technologies and computational techniques employed by AI companies are compiled in Table 17.2 (Deep Pharma 2022).

ML, a branch of AI, uses big datasets and algorithms to train and adapt to simulate how the human brain works. ML needs to be appropriately adjusted throughout the process to utilise any kind of data. For the algorithm to learn a model with available parameters through training data, a particular model must first define the necessary parameters (Dara et al. 2022). This training data model will be used to predict

Table 17.2 The most cutting-edge AI businesses employ these computational techniques and technologies (Deep Pharma 2022)

Company	Computational techniques employed	Technologies synopsis
Ardigen	Bioinformatics, deep learning, NLP	In addition to developing custom applications, Ardigen also works in the areas of BD integration, biological and clinical data analysis, and laboratory information management systems
Atomwise	ML, DL, CNN, chemoinformatics	In addition to developing custom applications, Ardigen also works in the areas of BD integration, biological and clinical data analysis, and laboratory information management systems
BenchSci	NLP, DL, ML	Decodes data from both open and closed sources about reagents, like antibodies, and provides published figures with useful information
BenevolentAI	ML, DL, symbolic AI, chemoinformatics	Decodes data from both open and closed sources about reagents, like antibodies, and provides published figures with useful information
Berg	ML, DL, bioinformatics	Create new biomarkers and treatment targets by analysing information derived from patient samples in both state of healthy and illness
Berkeley lights	Ml, bioinformatics	Automate cell selection, processing, and analysis. Enables scientists to: Automate the production of cellular therapies and hasten the development of cell lines
BioSymetrics	NLP, DL, ML	Manage raw drug data sets, phenotypic data, imaging data, and genomics data. Allows for the rapid integration of analytics and ML capabilities into ongoing commercial procedures by researchers
BIOZ	NLP, DL, ML	For the benefit of the life sciences community, bioz has developed a search engine that can sift across the vast majority of webpages containing intricate, unorganised scientific publications using ML and natural language processing
Bioxcel therapeutics	ML, DL, chemoinformatics	Leading the way in fusing pharmacological knowledge combined with AI and BD analytics is the biopharmaceutical business Bioxcel corporation
C4X discovery	ML, DL, chemoinformatics, bioinformatics	Using human genetic datasets, C4X's novel DNA-based target identification platform (Taxonomy3(R)) finds new targets that are specific to individual patients
CelerisTx	DL, bioinformatics	It is a deep learning company that validates lead drug candidates in automated lab settings and breaks down unbreakable targets using creative, computer-based techniques

Table 17.2 (continued)

Company	Computational techniques employed	Technologies synopsis
CytoReason	ML, DL, symbolic AI, chemoinformatics, bioinformatics	Cutting-edge ML technologies and unrivalled access to both proprietary and public data enable CytoReason to create proprietary biological models of disease, tissue, and drug
Data4Cure	ML, DL, NLP	A variety of heterogeneous data types can be integrated and subjected to advanced analysis by distinct system components, including both structured and unstructured molecular, phenotypic, and clinical data, thanks to the Data4Cure platform's modular architecture
Deep genomics	ML, DL, bioinformatics	Deep genomics is using AI to open up a new world of potentially life-saving genetic therapies
Desktop genetics	Bioinformatics, ML	Motivated by the practical applications of CRISPR technology, a team of genome editing specialists, bioinformaticians, and data scientists established desktop genetics. The biggest database on genome editing in the world was used to train DESKGEN AI, their primary technology
Envisagenics	ML, DL, high-performance computing	The SpliceCore platform from Envisagenics combines high-performance computing, RNA-splicing analytics, and proprietary ML algorithms to find disease-specific alternatively spliced RNA that may be used as a target for therapy
Euretos	ML, DL, bioinformatics	Through an intuitive application, Euretos offers easy access to the cloud-based platform for discovery and facilitates the safe integration of public and proprietary company data
Exscientia	ML, DL, bioinformatics, chemoinformatics	The business predicts a molecule's pharmacology, novelty, synthetic accessibility, and ADME using ML
Genialis	ML, DL	Genialis is a platform that combines AI-based techniques with computational biology to model and combine data at the nexus of clinical and translational medicine
GNS healthcare	ML, DL	Through the integration and conversion of numerous patient data formats into in silico patients, GNS healthcare AI technology reveals the intricate network of interactions underlying the progression of an illness and how well a treatment works
Healx	ML, NLP, symbolic AI, chemoinformatics, bioinformatics	By integrating and transforming a wide range of patient data types into in silico patients, GNS healthcare AI technology reveals the intricate network of interactions underlying the course of disease and the effectiveness of treatments

(continued)

Table 17.2 (continued)

Company	Computational techniques employed	Technologies synopsis
Iktos	ML, DL, cheminformatics	Iktos is creating a ligand-based de novo drug design technology based on deep learning (DL), with an emphasis on multi-parametric optimisation (MPO)
Insilico medicine	DL, GANs GANs + reinforcement learning, symbolic AI, ML, chemoinformatics, bioinformatics	All-inclusive DL pipeline. Biology: HTS analysis, DNNs for target ID, and signalling pathways. Chemistry: Generating novel molecules with GANs-RL
Kyndi	NLP, DL, ML	Leading AI software is offered by Kyndi, which can analyse lengthy texts and produce insights that can be put into action more quickly, intelligently, and understandably
MEdChemica	ML, chemoinformatics	The company's goal is to accelerate our clients' initiatives through its extensive knowledge in method development, lead optimisation, and lead generation
Nference	NLP, DL	Modern neural networks are used by nferX to automate real-time knowledge extraction from scientific, commercial, and regulatory literature
NuMedii	BD analytics, DL, ML	Analyse millions of raw human, biological, pharmacological, and clinical data points to find connections between medications and diseases at the systems level
Nuritas	DL, bioinformatics	Estimate the therapeutic potential of bioactive peptides derived from food. Permits the following: The economical development of highly focused treatments for particular diseases using natural food sources
Owkin	ML, federated learning	Owkin advances precision medicine by fusing his knowledge of biology and ML. Owkin's data connect service makes real-world data more accessible by therapeutic area
Peptone	DL (TensorFlow + Keras base)	The first protein database in the world with complete KerasTM and TensorflowTM integration designed especially for deep learning and AI applications
Phenomic AI	DL, reinforcement learning	Phenomics finds drugs that specifically target these resistant cells and makes predictions about which cells will withstand chemotherapy. Subsequently, the compounds will be developed and commercialised
ProteinQure	Quantum computing, reinforcement learning, chemoinformatics	To create protein medications, ProteinQure is integrating atomistic simulations, quantum computing, and reinforcement learning. Without crystal structures, they can investigate protein structures and create peptide-based medicines

Table 17.2 (continued)

Company	Computational techniques employed	Technologies synopsis
ReverieLabs	Evolutionary algorithms, ML	Structure-based ML predictive models for small-molecule potency and ADMET/PK characteristics
ReviveMed	ML, DL	With the help of ReviveMed's platform, metabolic data can be applied quickly, efficiently, and economically to identify novel disease mechanisms for drug discovery. Metabolomic biomarkers can also be used to determine which patients would benefit from treating the underlying disease mechanism
Structura biotechnology	ML (stochastic gradient descent and branch-and-bound maximum likelihood optimisation)	The cryoSPARC SystemTM uses machine learning (ML) to enable finding protein and molecular complex structures in high throughput using cryo-EM data
Xralpi	Quantum physics, ML	With XtalPi's ID4 platform, it is possible to accurately predict factors related to physiochemical and pharmaceutical aspects of solid-form selection, small-molecule candidates for drug design, and other important areas of drug development

the test data based on the required parameters. A few of the industries currently using ML (ML) in drug discovery and development are finding materials, consumables, chemical synthesis, data mining (ontology), biology research (target validation), lead discovery, preclinical development, biomarker discovery, clinical development, pharmacovigilance, drug repurposing, and virtual lab assistants (Dara et al. 2022). ML principles offer numerous opportunities for target identification, peptide synthesis, drug toxicity and physiochemical property evaluation, drug monitoring, drug efficacy and effectiveness, and drug repositioning when developing a new medication, all made possible by a chemical library that holds over 106 million compounds (Gupta et al. 2021).

The process of developing a targeted medicine must include lead optimisation, target identification and validation, hit recovery, and clinical trials (Vohora and Singh 2018). Physicians have been able to create meaningful targeted medications by using computer-assisted drug design (CADD), an artificial technology. This method produces the lead compounds with optimal properties in silico and provides theoretical molecular properties for the targeted medicine, such as distribution, absorption, metabolism, excretion, selectivity, and bioactivity (Dara et al. 2022). Preclinical drug development and discovery costs eventually decrease with the use of multi-objective ML techniques. The pharmaceutical industry has therefore been more willing to spend money on ML technologies, instruments for getting exact results, and targeted drugs for drug development.

The utilisation of BD facilitates the collection of pharmaceutical research findings, which are then transformed into studies involving the application of various algorithms and reusable techniques. The drug discovery process makes extensive use of a wide range of ML tools, such as support vector machine (SVM), random forest, Naive Bayesian classification (NBC), multiple linear regression (MLR), logistic regression (LR), linear discriminant analysis (LDA), probabilistic neural networks (PNN), multi-layer perceptron (MLP), and multiple linear regression (MLR) (Dara et al. 2022). Due to the lack of a standardised digital mechanisation method or universal software for computer monitoring of chemical reactions, a variety of AI-aided computational tools are available to aid in drug discovery, such as Alphafold, Chemputer, Conv_qsar_fast, Chemical VAE, DeepChem, DeepNeuralNet-QSAR, Deep Tox, Delta Vina, Hit Dextar, QML, SIEVE-Score, and many more (Sarkar et al. 2023).

17.4 Role of AI Tools in Drug Discovery

Drug design, drug repurposing, drug polypharmacology, and drug screening are the four main facets of drug discovery. The potential of AI to forecast drug characteristics may mean that repeated clinical trials involving human subjects are not necessary for ongoing research. This would be beneficial both in terms of cost and ethics. ML has two subsets: supervised and unsupervised learning. While the clustering model is an example of unsupervised learning, classification and regression analysis are examples of supervised learning (Dara et al. 2022). The clustering method finds and groups similar data points in larger datasets without regard to the exact result. The Hidden Markov model, Hierarchical clustering, K-means, and Neural Networks are the clustering models used in de novo molecular design, deep feature selection for biomarkers, feature reduction in single-cell data to identify cell types, and cell types of biomarkers from single-cell RNA data (Dara et al. 2022). Neighbour, SVMs, NLP Kernel Techniques, and NLP Bayesian classifiers are models that are used in classification. These models are used to study tissue-specific biomarkers from gene expression signatures, target disease drug associations, and target drug ability based on PK properties, protein structure or sequence, and protein structure (Dara et al. 2022). Models for regression analysis include linear regression, spare linear regression, random forests, and decision trees. These models can identify molecular features that predict drug responses to cancer, targets for diseases such as Huntington's disease, target drug ability from multi-dimensional data, and gene expression data that predict the success of clinical trials (Dara et al. 2022).

17.5 Research and Development (R&D) Approaches of AI in Drug Discovery

Biology and computer sciences applied mathematics differ in bioinformatics is gradually closing thanks to many emerging collaborations between academic institutions, research facilities, pharmaceutical labs, and AI and ML businesses. This would allow for the completion of drug development pipelines more quickly because they could be done computationally, automatically, and with a reduced chance of human error. The time it takes to find possible drug candidates could be reduced from several months to a year by using AI tools (Réda et al. 2020). It is advisable to proceed with caution and refrain from assuming that the failure rate problem can be entirely resolved through computational methods. However, it might also be able to shorten the duration and expense of drug development while simultaneously making therapies more patient-centred by integrating ML techniques into pipelines. This is due to the possibility that applying or enhancing precision medicine techniques may be made simpler by incorporating multi-view data. However, systematic approaches make it easier for studies to be replicated and reused, for data quality to be transparently and consistently controlled and shared, and for promising targets to be identified in silico. Furthermore, before conducting any preclinical or wet lab testing, these methods may provide numerical values for comparing and assessing the efficacy of possible compounds (Réda et al. 2020).

Nowadays, a large number of biopharmaceutical companies, including AbbVie, AstraZeneca, Amgen, BenevolentAI, Insilico Medicine, Inc., Sanofi, Recursion Pharmaceuticals, Owkin, Strados Labs, Antiverse, Standigm, and Deep Genomics, use AI in their research and development. The most inventive AI research and development in biopharma is compiled in Table 17.3. The mission of AbbVie, a multinational biopharmaceutical company focused on research, is to develop and commercialise innovative therapies for some of the most difficult and dangerous illnesses by leveraging its expertise, dedicated workforce, and innovative approaches. Alzheimer's, schizophrenia, and Parkinson's disease are AbbVie's main focus areas. AI has been used to increase patient adherence to medication, visually confirm medication ingestion, and use data to evaluate treatment efficacy. This business collaborates with BenchSci, Atomwise, AiCure, Calico Labs, and Mission Therapeutics. AstraZeneca's primary focus areas are the development, discovery, and commercialisation of prescription medications, mainly for the treatment of illnesses in the oncology, cardiovascular, renal and metabolism, and respiratory therapy areas. The company collaborates with Schrodinger, Mila, Melloddy, MLPDS Consortium, BenevolentAI, Roivant Sciences, and Tencent to create knowledge graphs that combine genomic, disease, drug, and safety data to develop an understanding of the disease. It also identifies new targets for novel medicines and provides quick and accurate image analysis (Réda et al. 2020). AI systems are trained to facilitate easier and more accurate sample analysis for pathologists. The analysis time could be reduced by more than 30% as a result of this. For one of their AI systems, they developed a method based on how some autonomous cars interpret their environment. In order to help inform decisions regarding immunotherapy-based bladder cancer treatment, the AI system was trained to evaluate tumour and immune cells for the PD-L1 biomarker (Réda et al. 2020).

A new breed of data-driven drug discovery firms that integrate automation and data science has also emerged as a result of the growth of robotics-focused businesses in the biomedical research sector. Arcotoris, Genomenon, and GATC Health represent three of the newest waves of data-driven pharmaceutical companies.

Table 17.3 Summary of the most innovative research and development for AI in biopharma (Source: Deep Pharma Intelligence, 2022) (Deep Pharma 2022)

Nu	Biopharma	Focus area	Involvement of AI	Collaboration companies	Findings
1	Amgen	Cardiovascular and osteoporosis	• To increase the precision of risk assessments; to offer individualised patient care; to assist adherence with precise and consistent real-time responses; to leverage data to identify the most efficient course of action	Owkin	Using natural language processing (NLP) tools and agile development methodologies, the team created a strong algorithm that can reasonably mimic the manual process
2	BenevolentAI	Kidney diseases, Parkinson's disease, cancer and fibrosis	• In molecular design; in patient stratification; in the collection of more varied data; in determining specific medication targets	Neuropore therapies, Novartis, and AstraZeneca	The company is the creator of the cutting-edge experimental and computational discovery platform known as the benevolent PlatformTM, which aids scientists in developing novel treatments for diseases and customising drugs for each patient. The three primary focal points for the benevolent PlatformTM are precision medicine, molecular design, and target identification
3	Insilico medicine, Inc	Cancer, immunity, kidney disease, fibrosis and inflammatory bowel disease (IBD)	• To identify drug molecules more quickly and cheaply • To identify novel biological targets for currently used medications • To use innovative chemistry to validate the targets. To predict clinical trial outcomes	BioTime, juvenescence AI limited, Bitfury and Fosun Pharma	Using a new AI system called GENTRL (generative Tensorial reinforcement learning), it has created a drug discovery process that can be completed in a matter of days rather than years (first synthesis and trials taking only 21 days instead of 3 years). As a result of Insilico's successful testing of the technology in the first experimental validation of this a type of AI technology for finding drugs in animals and cells, several brand-new compounds that can treat conditions like fibrosis were created

4	Sanofi	Central nervous system, heart diseases, cancer, diabetes, and pneumonia	• To diagnose illnesses early; to decentralise clinical trials; to enable remote (eHealth) communication between physicians and patients; to enhance marketing tactics	GSK and DNDi	(1) MouthLab is a solitary, non-invasive AI-powered device that measures more than ten distinct health indicators rapidly and precisely (2) LIFEdata is an easy-to-use, intuitive AI platform that automates customised dialogue experiences across all platforms (3) In real-world clinical trials, CART can provide continuous vital sign monitoring and be worn comfortably for everyday use (4) To help physicians make better clinical decisions, NeuroAdvise is a mobile application that provides clinical decision support (5) Mentalab is a wearable patch that measures ECG biosignals continuously to monitor and diagnose respiratory and cardiac conditions (6) ChatbotPack.com can assess and diagnose respiratory conditions and early warning indicators of neurological illnesses in the elderly (7) Wavy assistant continuously monitors heart health in real time using voice and AI technologies

(continued)

Table 17.3 (continued)

Nu	Biopharma	Focus area	Involvement of AI	Collaboration companies	Findings
5	Recursion pharmaceuticals	Genetic diseases, infectious diseases, inflammation and immunology	• The infrastructure layer is a highly synchronised network used for chemical and biological data design, execution, aggregation, and storage • Recursion data universe is a high-dimensional dataset covering various data modalities related to biology and chemistry • The software programme recursion map was created to handle and interpret data from the recursion data universe	Bayer and Roche	It focused on: (1) Sophisticated tools for biotechnology, such as CRISPR genome editing and synthetic biology (2) Reliable automation of complex laboratory research on a never-before-seen scale through the use of cutting-edge robotics (3) Neural network architectures for analysing and drawing conclusions from large, intricate internal datasets in an iterative manner (4) Increasing the elasticity of high-performance computing by utilising cloud solutions
6	Owkin	Oncology, immune system and cardiovascular diseases	• To better the drug development process by looking into the mechanisms that lead to varying drug efficacy from patient to patient • To determine which patient-specific medication candidates will yield the best results and enhance treatment	Roche	Owkin AI models: Owkin has 40 more models in development and has catalogued 30 live disease models. These models are constructed using interpreted AI, which sets them apart from conventional black box models. This enables the business to advance its research and pinpoint the biomarkers that underpin its predictions. To find new biological targets, optimise clinical trial designs with patient subgroups, and find patients qualified for specific treatments, new multimodal biomarkers must be discovered • Owkin loop, Owkin studio, and Owkin connect are the models

7	Deep genomics	Genetic diseases, CNS diseases, and metabolic diseases	• Target discovery. They look into every avenue, find mutations that cause disease, and figure out how to solve the genetic issue • Therapy design: To identify the most likely-to-be-effective targeted therapies, the AI evaluates hundreds of thousands to millions of possible options. After that, they are confirmed in the wet lab • To generate data on animal toxicity, cell viability, and genome-wide off-target effects	NIL	(1) To find the best therapeutic candidates, the deep genomics workbench mines, processes, and identifies novel targets from thousands of potential targets (2) Using their platform, the deep genomics team evaluated over 69 billion oligonucleotide molecules against a million targets in silico as part of project Saturn. The result was a library of 1000 compounds whose ability to manipulate cell biology was confirmed through experimentation
8	Strados lab	Respiratory management solution	• The respiratory management solution is an innovative approach that combines ML, patient centricity, and lung biomarkers • The life sciences sector stands to gain significant advantages from Strados labs' improvement of pulmonary care monitoring capabilities, which allow for the analysis of longitudinal lung acoustics to provide insight into patient drug response	• Decentralised trial compatibility: Clinpal, thread, icon, Iqvia, Curavit, Curebase, Medidata • Cough trial solution: Bayer, Bellus, GSK, Merck, Algernon • Digital acoustics biomarkers: Veeva, Medidata, Datatrak, PerkinELmer • Trial partner solution: Paraxel, Pfizer, Iqvia, icon, PPD, Amgen, Labcorp, Genentech, ERT, Bristol Myers squib	• The RESP biosensor is the only FDA-approved device that collects lung sound acoustics wirelessly and hands-free. This makes it perfect for clinical studies where the goal is to measure how patients respond to new medications by collecting coughs and other lung sounds in an efficient and objective way. The data collected can then be processed and analysed

(continued)

Table 17.3 (continued)

Nu	Biopharma	Focus area	Involvement of AI	Collaboration companies	Findings
9	Standigm	Cancer, Parkinson's disease, non-alcoholic steatohepatitis (NASH), mitochondrial disease	• Within 2 weeks, Standigm ASKTM offers unique targets that are precisely tailored to a customer's research context • Within 7 months, this workflow-optimised AI system can produce multiple first-in-class compounds	Hanmi, KIST, SAMJIN, SK chemicals	(1) Within 2 weeks, Standigm ASKTM offers unique targets that are precisely tailored to a customer's research context (2) Standigm's optimised workflow artificial intelligence system can generate multiple first-in-class compounds in less than 7 months (3) Standigm has developed customised partnership models, from providing AI solutions to licencing AI assets and platforms, that are perfect for every customer's needs (4) Standigm possesses a remarkable inventory of ready-made internal therapeutic assets that are both appealing and able to satisfy customers' pipeline requirements
10	Antiverse	New type antibody discovery	• Offers focused options so that, once there are too many options for antibody-antigen binding, testing can be concentrated on safely • Can assist in locating enough potential binders that traditional methods might overlook • Can create new binder variations that will be adequate for clients' needs	NIL	(1) In order to find more rare and diverse antibodies, it keeps diversity from being lost during amplification (2) To extract additional data from the workloads that are already in place, next-generation sequencing, or NGS, is employed (3) Utilising trained models and AI-enhanced drug discovery, the platform analyses data gathered from thousands of experiments. To choose the best candidates, these outputs are compared to known data

The biotech platform business Arctoris makes use of its completely automated drug discovery platform. It operates out of Singapore, Boston, and Oxford. An oncologist and a medicinal/synthetic chemist founded the company intending to utilise technology to its fullest potential and combine it with a wealth of industry knowledge to expedite the search and development of novel therapies. The company's fundamental tenet is that without richer, more consistent, and repeatable underlying data, drug discovery programmes cannot advance and hit the next milestone more quickly and with a higher probability of success. The most significant obstacle to using AI and ML for drug discovery is obtaining reliable, repeatable, well-structured, and heavily annotated data. Arctoris can generate massive amounts of ML-ready data thanks to robotics, which expedites the process from target to hit, lead, and candidate and improves decision-making. The combination of precision robotics, a specialised data science platform, and the seasoned knowledge of biotech and pharma veterans in drug discovery enables both quality and speed. Arctoris fully logs and analyses every experimental output, including temperature, humidity, CO_2, batch ID, provenance of the reagent, and a plethora of other metadata. Furthermore, the platform facilitates automated quality assurance and quality control, employing statistical methods to ensure the total validity and dependability of each outcome. Arctoris ensures that superior data will be generated in a faster manner, enabling better decisions to be made earlier in both human-powered and, more specifically, AI/ML-driven programmes. This is because AI models are trained using the best available data. Combining data science and robotics, Arctoris has produced a cutting-edge technology platform that powers drug discovery projects both inside the company and through partnerships with international biotech and pharmaceutical companies (Réda et al. 2020). Numerous factors are included in the industry standard for data generation and processing, including a sparse set of high-level results data, a highly fragmented file and storage system, vague reagent and cell line provenance, inconsistent application of techniques and protocols, variability, and human error. On the other hand, Arcotoris-enabled data generation and processing offers precise adherence to automated protocols, completely validated reagents and cell lines with comprehensive audit trails, repeatable results data in a standard format, extra rich research metadata collection, safe and helpful data storage and access, and sophisticated assay performance monitoring.

A genomics startup called Genomenon uses AI to organise the world's genomic knowledge in order to speed up the identification of genetic diseases and the creation of treatments. Offering a comprehensive understanding of the genetic drivers and clinical features of any genetic disease, the ProdigyTM Genomic Landscapes from Genomenon support the entire drug development process from discovery to commercialisation. Gene disorders, inherited illnesses, and somatic and hereditary cancers are the main therapeutic areas of Genomenon. Genomenon's ProdigyTM Genomic Landscapes, which use a special blend of professional, scientific review and internal genomic language processing (GLP), offer an evidence-based basis for every phase of the drug development process. A novel technique called GLP is used to systematically extract and standardise genomic and clinical data from scientific and medical literature. GLP is more sensitive than traditional techniques at identifying complex genomic information, which will lead to the identification of more variants and patients. Genomenon's database, which was created using GLP, now has more than 14.8 million variants, 8.8 million full-text articles, and 3 million supplemental datasets. Genomenon and Alexion, the Rare Disease division of AstraZeneca, collaborated to use AI technology to speed up the genetic diagnosis process for patients with uncommon diseases. With regard to ATP7B, a gene associated with Wilson disease, Genomenon's AI-powered approach discovered 3.7 times as many pathogenic/likely pathogenic variants (869 pathogenic variants) with evidence-supported results as the crowdsourced ClinVar database (235 pathogenic variants). In order to help doctors make more accurate diagnoses, this greatly increases the tools at their disposal.

The AI startup GATC Health speeds up the process of finding and developing new drugs. GATC Health, an AI startup, speeds up discovering and creating new medications. The company provides highly efficient services to pharmaceutical companies that reduce risk in the drug discovery process. GATC Health develops a cutting-edge AI-based drug development platform from beginning to end. The platform supports in silico clinical trial simulation, drug and therapeutic solution development, early disease detection, identification of disease biology, and in vitro and in vivo testing feedback loop simulation. To identify and validate new drugs, GATC's Platform replicates human biology using highly detailed disease-specific data and exclusive AI solutions. The company develops an innovative approach to drug discovery that boosts productivity and dramatically shortens the time to clinical development. Oncology, neurology, cardiology, immunology, rheumatology, and other diseases are among those that GATC Health targets. Through preclinical drug de-risking, drug compound discovery, and diagnostic biomarker discovery, GATC Health employs AI in research and development. To gain insight into the disease, AI tools are utilised to determine the causal relationship between the biomarkers and the illness. A set of distinctive medicinal compounds is produced with the help of AI-assisted compound discovery. The identified diagnostic biomarkers' causal and effect impacts were evaluated mathematically. The verified causal biomarkers and pathways are simulated and evaluated using AI-assisted database models in conjunction with human expertise. This will lead to a final set of treatment targets. Six to nine months is GATC Health's drug discovery timeline.

17.6 Challenges of Using AI in Drug Discovery

Despite the potential benefits of AI in drug discovery, there are many challenges and limitations to consider. One of the primary challenges is the availability of relevant data (Vamathevan et al. 2019). AI-based techniques typically require large data sets to be trained (Tsuji et al. 2021). The precision and reliability of the results may be impacted by the fact that there is frequently a limit to the amount of data that is available, or the data may be inconsistent or of low quality

(Blanco-Gonzalez et al. 2023). A problem with AI-based solutions is that they can raise ethical concerns about bias and fairness. For example, if the training data is biassed or unrepresentative, the predictions produced by an ML algorithm may be unfair or inaccurate (Silvia and Carr 2019). One of the most important things to consider is making sure AI is applied fairly and ethically when creating new therapeutic compounds. Numerous strategies and techniques can be employed to address the obstacles AI encounters in the field of chemical medicine. One tactic is data augmentation, which entails producing artificial data to improve datasets that already exist. According to Grabner et al. (2021), contemporary AI methods are insufficient to replace traditional experimental protocols or the expertise and experience of human investigators (Grebner et al. 2021). AI can only forecast what it can infer from available data; human scientists will then need to confirm and analyse the results (Gilpin et al. 2018). On the other hand, merging AI with traditional experimental methods can also help the process of discovering new medications. Combining AI's predictive capability with human researchers' expertise and knowledge can expedite the development of new drugs and optimise the drug discovery process (Wang et al. 2019). The biopharma industry still faces a significant challenge from the global lack of talent in AI. However, it is expected that in the coming years, an increasing number of specialised university courses and programmes centred on data science and AI applications will at least partially address this problem (Réda et al. 2020).

17.7 Summary of the Potential AI for Revolutionising Drug Discovery

AI holds the potential to revolutionise the drug discovery process by enhancing speed and accuracy, producing more personalised and effective treatments, and speeding up drug development. However, for AI to be successfully applied in drug discovery, high-quality data must be available, ethical concerns must be resolved, and the limitations of AI-based techniques must be understood. Promising strategies for overcoming the difficulties and constraints of AI in the context of drug discovery are provided by recent developments in the field, such as the use of explainable AI, data augmentation, and integration of AI with conventional experimental methods. This is an intriguing and promising area of research that has the potential to drastically alter the drug discovery process, especially in light of the increasing interest and attention from academics, pharmaceutical companies, and regulatory bodies, as well as the possible advantages of AI. AI in the pharmaceutical sector will enable the creation of drugs that are individually customised to meet the requirements of every patient, including the right dosage, release parameters, and other components. By highlighting and advancing the significance of mechanisation, using the current AI-assisted algorithms should shorten the time it takes for the goods to reach the market, improve their quality, and increase the overall security of the manufacturing scheme. Better resource and cost efficiency should follow from it as well.

References

Blanco-Gonzalez A, Cabezon A, Seco-Gonzalez A, Conde-Torres D, Antelo-Riveiro P, Pineiro A, Garcia-Fandino R (2023) The role of ai in drug discovery: challenges, opportunities, and strategies. Pharmaceuticals 16(6):891. https://doi.org/10.3390/ph16060891

Carracedo-Reboredo P, Liñares-Blanco J, Rodríguez-Fernández N, Cedrón F, Novoa FJ, Carballal A, Maojo V, Pazos A, Fernandez-Lozano C (2021) A review on machine learning approaches and trends in drug discovery. Comput Struct Biotechnol J 19:4538–4558. https://doi.org/10.1016/j.csbj.2021.08.011

Collins FS, Varmus H (2015) A new initiative on precision medicine. N Engl J Med 372(9):793–795. https://doi.org/10.1056/nejmp1500523

Dara S, Dhamercherla S, Jadav SS, Babu CM, Ahsan MJ (2022) Machine learning in drug discovery: a review. Artif Intell Rev 55(3):1947–1999. https://doi.org/10.1007/s10462-021-10058-4

Deep Pharma (2020) Artificial intelligence for drug discovery landscape overview Q1. https://www.deep-pharma.tech/ai-in-drug-discovery-2020. Accessed 20 Sept 2023

Deep Pharma (2022) Artificial intelligence for drug discovery landscape overview Q1. https://www.deep-pharma.tech/ai-in-drug-discovery-2022-q1. Accessed 20 Sept 2023

Gilpin LH, Bau D, Yuan BZ, Bajwa A, Specter M, Kagal L (2018) Explaining explanations: an overview of interpretability of machine learning. In: 2018 IEEE 5th international conference on data science and advanced analytics (DSAA). IEEE, pp 80–89. https://doi.org/10.1109/dsaa.2018.00018

Grebner C, Matter H, Kofink D, Wenzel J, Schmidt F, Hessler G (2021) Application of deep neural network models in drug discovery programs. ChemMedChem 16(24):3772–3786. https://doi.org/10.1002/cmdc.202100418

Gupta R, Srivastava D, Sahu M, Tiwari S, Ambasta RK, Kumar P (2021) Artificial intelligence to deep learning: machine intelligence approach for drug discovery. Mol Divers 25:1315–1360. https://doi.org/10.1007/s11030-021-10217-3

Patel V, Shah M (2022) Artificial intelligence and machine learning in drug discovery and development. Intell Med 2(3):134–140. https://doi.org/10.1016/j.imed.2021.10.001

Paul D, Sanap G, Shenoy S, Kalyane D, Kalia K, Tekade RK (2021) Artificial intelligence in drug discovery and development. Drug Discov Today 26(1):80. https://doi.org/10.1016/j.drudis.2020.10.010

Ramesh AN, Kambhampati C, Monson JR, Drew PJ (2004) Artificial intelligence in medicine. Ann R Coll Surg Engl 86(5):334. https://doi.org/10.1308/147870804290

Réda C, Kaufmann E, Delahaye-Duriez A (2020) Machine learning applications in drug development. Comput Struct Biotechnol J 18:241–252. https://doi.org/10.1016/j.csbj.2019.12.006

Sarkar C, Das B, Rawat VS, Wahlang JB, Nongpiur A, Tiewsoh I, Lyngdoh NM, Das D, Bidarolli M, Sony HT (2023) Artificial intelligence and machine learning technology driven modern drug discovery and development. Int J Mol Sci 24(3):2026. https://doi.org/10.3390/ijms24032026

Silvia H, Carr NK (2019) When worlds collide: protecting physical world interests against virtual world malfeasance. Mich Tech L Rev 26:279. https://doi.org/10.36645/mtlr.26.2.when

Solanki P, Baldaniya D, Jogani D, Chaudhary B, Shah M, Kshirsagar A (2022) Artificial intelligence: new age of transformation in petroleum upstream. Pet Res 7(1):106–114. https://doi.org/10.1016/j.ptlrs.2021.07.002

Tsuji S, Hase T, Yachie-Kinoshita A, Nishino T, Ghosh S, Kikuchi M, Shimokawa K, Aburatani H, Kitano H, Tanaka H (2021) Artificial intelligence-based computational framework for drug-target prioritization and inference of novel repositionable drugs for Alzheimer's disease. Alzheimers Res Ther 13(1):1–5. https://doi.org/10.1186/s13195-021-00826-3

Vamathevan J, Clark D, Czodrowski P, Dunham I, Ferran E, Lee G, Li B, Madabhushi A, Shah P, Spitzer M, Zhao S. Applications of machine learning in drug discovery and development. Nat Rev Drug Discov 2019;18(6):463–477. https://doi.org/10.1038/s41573-019-0024-5

Vohora D, Singh G (2018) Pharmaceutical medicine and translational clinical research. Academic Press, an imprint of Elsevier, London

Wang L, Ding J, Pan L, Cao D, Jiang H, Ding X (2019) Artificial intelligence facilitates drug design in the big data era. Chemom Intell Lab Syst 194:103850. https://doi.org/10.1016/j.chemolab.2019.103850

Wen M, Zhang Z, Niu S, Sha H, Yang R, Yun Y, Lu H (2017) Deep-learning-based drug–target interaction prediction. J Proteome Res 16(4):1401–1409. https://doi.org/10.1021/acs.jproteome.6b00618

Wirtz BW, Weyerer JC, Geyer C (2019) Artificial intelligence and the public sector—applications and challenges. Int J Public Adm 42(7):596–615. https://doi.org/10.1080/01900692.2018.1498103

Yang Y, Siau KL (2018) A qualitative research on marketing and sales in the artificial intelligence age. MWAIS 2018 Proc 41:1. http://aisel.aisnet.org/mwais2018/41

Zhu H (2019) Big data and artificial intelligence modeling for drug discovery. Annu Rev Pharmacol Toxicol 60(1):573–589. https://doi.org/10.1146/annurev-pharmtox-010919-023324

Chapter 18
AI: Catalyst for Drug Discovery and Development

Khushbu Nailwal, Sumit Durgapal, Khushboo Dasauni, and Tapan Kumar Nailwal

Abstract This abstract explores the multifaceted roles of AI in expediting the drug development process. AI-driven algorithms analyze vast datasets, unraveling complex biological interactions and identifying potential drug targets with unprecedented speed and precision. In silico drug screening, powered by machine learning models, accelerates the identification of promising compounds, minimizing the time and resources traditionally required. Furthermore, AI facilitates personalized medicine by analyzing individual patient data to tailor treatments based on genetic and molecular profiles. In clinical trials, AI optimizes patient recruitment, enhances trial design, and expedites data analysis, leading to more efficient and cost-effective drug development. While the adoption of AI presents unprecedented opportunities, challenges such as data security, interpretability, and ethical considerations must be navigated. This chapter provides a comprehensive overview of the current state of AI in drug discovery, emphasizing its potential to reshape the pharmaceutical industry and improve global healthcare outcomes.

Keywords AI · Drug discovery · Drug development · Machine learning

18.1 Introduction

In the span of the past 200 years, advancements in modern medicine have significantly enhanced the lives of innumerable patients. Illnesses and ailments that were once considered untreatable or lethal have been vanquished with the aid of therapeutic drugs developed to extend and enhance the quality of life. The rise in new diseases and the resurgence of once-controlled illnesses, often in altered forms, are

K. Nailwal · S. Durgapal
Department of Pharmaceutical Sciences, Kumaun University, Nainital, India

K. Dasauni · T. K. Nailwal (✉)
Department of Biotechnology, Kumaun University, Nainital, India

S. Bose et al. (eds.), *Concepts in Pharmaceutical Biotechnology and Drug Development*, Interdisciplinary Biotechnological Advances,
https://doi.org/10.1007/978-981-97-1148-2_18

attributed to the changes in the environment, lifestyle changes, food contamination, the widespread use of antibiotics and chemicals in farming, and the growing problem of drug-resistant pathogens. So, accelerated drug discovery and development are required across diverse therapeutic areas of the healthcare system to address these growing unmet medical needs (Fig. 18.1).

Drug discovery is the process by which new medications or pharmaceutical compounds are identified, designed, and developed for the treatment of diseases and medical conditions. The drug discovery and development process consists of several steps, including identifying drug targets, validating those targets, transforming initial hits into leads, refining the leads, determining preclinical molecules, evaluating them preclinically, conducting clinical trials, and obtaining regulatory approval (Sarkar et al. 2023). Pharmaceutical companies primarily aim to bring new drugs to market by navigating the intricate stages of drug discovery and development. However, this process of creating a new drug is both time-consuming and expensive, requiring a diverse range of technologies and expertise. Generally, it takes approximately 15 years and an average investment of $2.8 billion to discover and develop a new drug (Qureshi et al. 2023). The challenges posed by conventional approaches in drug discovery, such as limited effectiveness and significant expenses, have become significant obstacles. Consequently, it has become necessary to explore novel methods to address the time-consuming and costly nature of this endeavor (Chen et al. 2022). With the revolutionizing research and developments in the areas of high-performance hardware, cloud prowess, and availability of a high volume of classified genomic, proteomic, pharmacological, clinical trial, and omics data, artificial intelligence and machine learning had paved the way for drug discovery and development (Singh et al. 2023). Machine learning (ML) algorithms, especially deep learning (DL) algorithms, have energized the drug discovery process. Recently, deep learning methods, such as artificial neural networks (ANNs) with

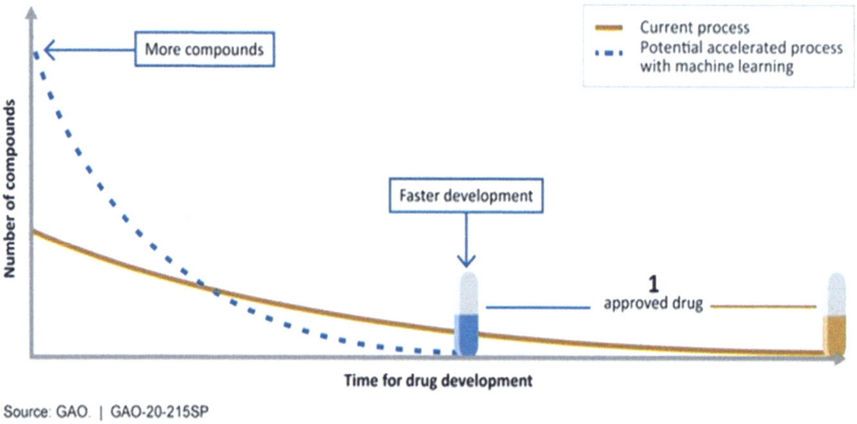

Fig. 18.1 Number of compounds vs time for drug development; the number of compounds represents the chemical compounds being scanned to identify a drug candidate

multiple hidden layers, have experienced a resurgence. They are gaining attention for their ability to automatically extract meaningful information from input data and capture complex relationships between inputs and outputs. These capabilities enhance traditional machine learning techniques, which rely on manually engineered molecular descriptors. The initial skepticism surrounding the usefulness of AI in pharmaceutical discovery is gradually fading away, opening up exciting possibilities for the field of medicinal chemistry. The integration of AI and advanced experimental insights is anticipated to transform the quest for innovative and enhanced pharmaceuticals. It promises to make the process faster, more cost-effective, and increasingly compelling. Deep learning-enabled methods are beginning to tackle critical challenges in drug discovery. Moreover, numerous technological advancements, such as "message-passing paradigms," "spatial-symmetry-preserving networks," "hybrid de novo programming," and other innovative machine learning approaches, are poised to become widely adopted (Sarkar et al. 2023). These methods will play a pivotal role in unraveling some of the most profound and intriguing questions in the field.

This chapter establishes a fundamental groundwork for a variety of machine learning principles and techniques. It subsequently delves into their application at various points within the drug discovery and development process.

18.2 Machine Learning in Drug Discovery and Development

Machine learning is an area of artificial intelligence (Fig. 18.2) involving the development of algorithms and statistical models that allow computer systems to infer the trends and patterns in existing data without being explicitly programmed; these models and data can subsequently be employed for making forecasts on fresh information (Vilar and Costanzi 2012).

18.2.1 Types of Machine Learning

Categorized on the basis of the learning approach, we have the following classification of ML algorithms (Fig. 18.3).

Supervised Learning

This type is based on learning from labeled training data, which includes input-output pairs. The algorithm learns the relationship between the input features and the output labels, with the goal of predicting the output for new, unseen data. Supervised learning is used for tasks such as classification, regression, and multi-label classification. In drug discovery, QSAR models are used to predict the

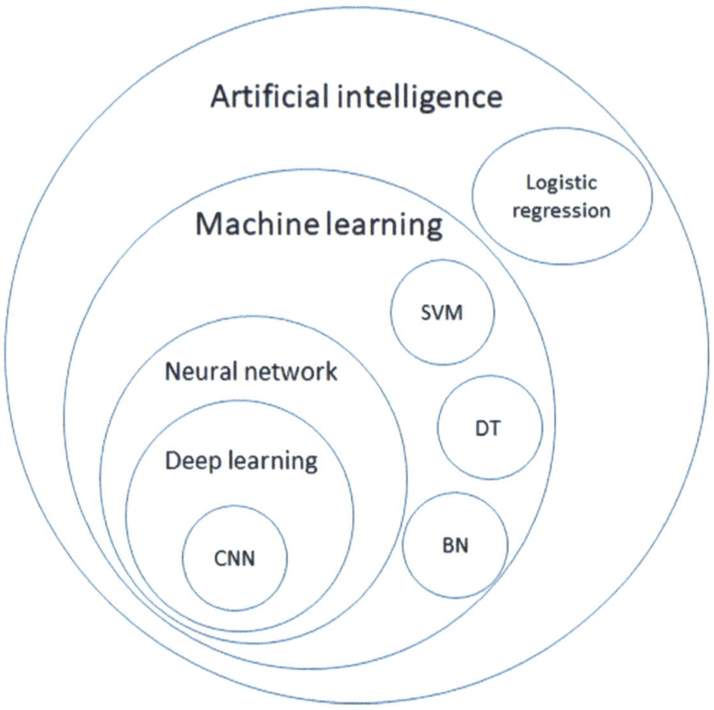

Fig. 18.2 Visual representation of AI as a superset

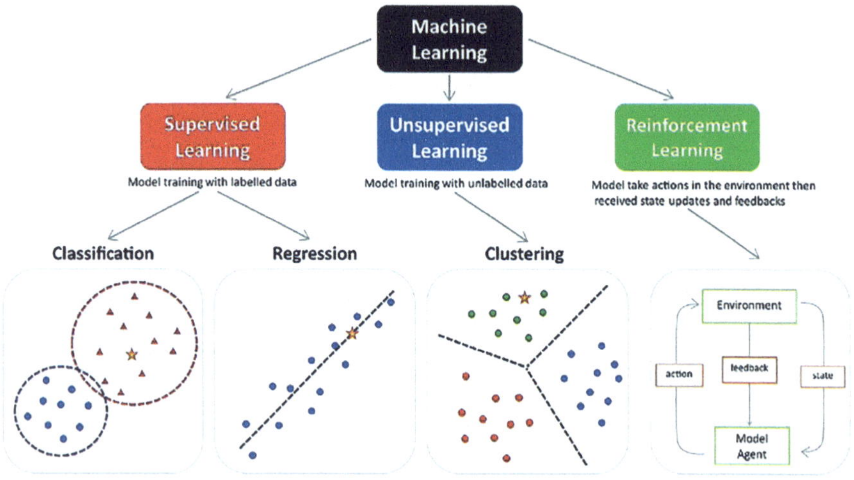

Fig. 18.3 Main types of ML

biological activity of potential drug candidates based on their chemical structures (Vilar and Costanzi 2012). Using a dataset of known chemical structures and their corresponding biological activities, the supervised learning algorithm learns to correlate molecular features with the desired biological effects. This helps in the identification of promising drug candidates for further testing and optimization (Vora et al. 2023).

Unsupervised Learning

In this type, the algorithm is provided with an unlabeled dataset and must find patterns, relationships, or structures within the data without prior knowledge of the desired output. Unsupervised learning is suitable for tasks such as clustering, anomaly detection, and dimensionality reduction. Molecular clustering is an unsupervised learning technique used to group similar compounds together based on their structural or physicochemical properties (Kovács et al. 2005).

Semi-supervised Learning

This category incorporates both annotated and unannotated data during the training process. The algorithm can use the labeled data to guide its learning process while also discovering patterns in the unlabeled data. Semi-supervised learning is useful when labeled data are limited or expensive to obtain. In the drug discovery process, it is used to group similar compounds together based on their structural or physicochemical properties. This can help researchers identify the relationships between various compounds and their biological activities, leading to the discovery of new chemical scaffolds and drug candidates. For example, hierarchical clustering or self-organizing maps (SOM) can be applied to analyze large compound libraries and identify groups of molecules with similar properties (Takács et al. 2021).

Reinforcement Learning

Reinforcement learning algorithms learn through interaction and feedback, aiming to maximize cumulative rewards. They are used in robotics, gaming, and resource allocation and can improve prediction accuracy in drug discovery tasks. In de novo drug design, reinforcement learning guides the generation of novel chemical structures with desired properties based on feedback from a scoring function. This can lead to discovering new drug candidates more efficiently (Popova et al. 2018).

18.2.2 Machine Learning Life Cycle (Fig. 18.4)

The machine learning process starts with the formulation of a problem statement and then follows a set of predefined steps to tackle the problem like *"A pharmaceutical company wants to identify potential drug candidates for a specific disease, optimizing for both efficacy and safety."* ML life-cycle generally contains the following seven major stages:

Gathering Data

This stage involves the identification and gathering of relevant data from different data sources, which is then combined together to make a dataset, which would become the base for the training and testing of the ML model, e.g., *for the above problem statement collecting molecular data, biological data, and experiment data* (e.g., *assay results in clinical trial data).*

Data Preparation

The dataset prepared above needs to be prepared for the further steps. This step generally involves data exploration. The statistical characteristics, format, and quality of the data are understood.

Data Wrangling

In this step, we convert the raw data into a usable format by performing cleaning and preprocessing by handling the missing values, normalizing features, and encoding categorical values, e.g., *creating informative features such as molecular fingerprints, protein-drug interaction scores, and biological pathway indicators.*

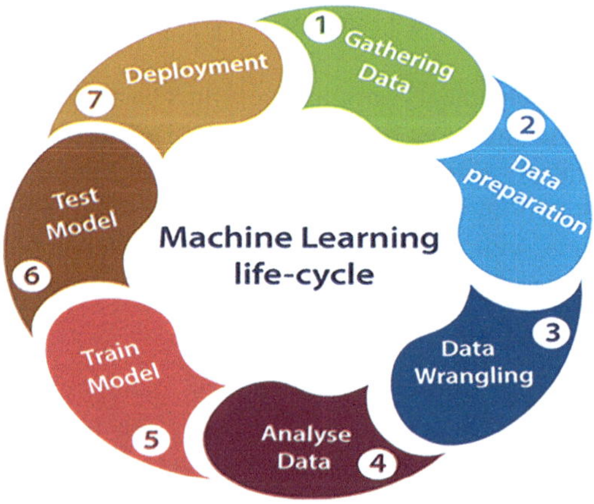

Fig. 18.4 Machine learning life cycle

Analyze Data

With the available formatted data, different statistical techniques are deployed in this stage to explore, summarize, and gain insights and then build a machine learning model. It deploys mathematical techniques such as dimensionality reduction that selectively pick the features from the raw dataset to serve requirements and further engineering new features that would assist in the model training. Apart from this, it involves key activities such as using descriptive statistics, data visualization, and exploratory data analysis, e.g., *choosing appropriate machine learning models, such as deep neural networks for predicting drug-target interactions or ensemble methods for compound screening.*

Train Model

Filtered relevant data are passed on to the ML model, where it learns patterns and relationships between input and the target variables, e.g., *train the selected model using labeled data, where the labels represent known drug properties or outcomes* (e.g., *drug binding affinity and toxicity*).

Test Model

Following the training phase, the model's performance is assessed using metrics tailored to the particular task, such as accuracy, precision, recall, and F1 score for classification or mean-squared error for regression. Cross-validation or holdout validation is often used to assess how well the model generalizes to new, unseen data, e.g., *assess the model's performance using relevant evaluation metrics (*e.g., *ROC-AUC for binary classification tasks) on a validation dataset.*

Deployment

Once the model performs satisfactorily during evaluation, it can be deployed into a real-world environment where it will make predictions or decisions on new, incoming data, e.g., *deploy the trained model into a drug discovery platform or system where it can predict the properties of new candidate compounds.*

The deployed machine learning model, available through the above process, is used for making predictions or inferences on new data. This task may involve classification, making recommendations, forecasting future values, identifying anomalies, and solving various other problems. As newly classified data or data with variations emerge, the existing model requires tuning. Therefore, the model is periodically retrained with updated data to adapt to changing conditions and enhance its predictive capabilities.

18.2.3 *Machine Learning Algorithms*

Machine learning algorithms have been identified to play a very important role in drug discovery and development. These algorithms are aimed at utilizing leveraging data-driven approaches to accelerate the identification of potential drug candidates, optimize their properties, and enhance the efficacy of the drug development procedure. Below are discussed the popular ML models with utility proven in the drug discovery process.

Random Forests and Decision Trees

Decision trees serve as models for both classification and regression tasks. They operate by posing a series of sequential questions, leading us down specific paths within the tree based on our responses (Song and Ying 2015). This model essentially follows a set of "if this, then that" conditions to ultimately arrive at a particular outcome. The concept of "tree depth" is essential here, indicating how many questions are asked before reaching a final classification.

Decision trees offer several advantages, including (Fig. 18.5):

Interpretability: They are highly interpretable and allow for intuitive visualizations, simplifying the understanding of complex decisions.

Overfitting: Decision trees can be prone to overfitting, especially when they are very deep, introducing errors due to bias and variance. Their internal workings are transparent, facilitating result reproducibility.

Versatility: Decision trees are versatile, as they can accommodate both numerical and categorical data, rendering them suitable for a wide range of applications.

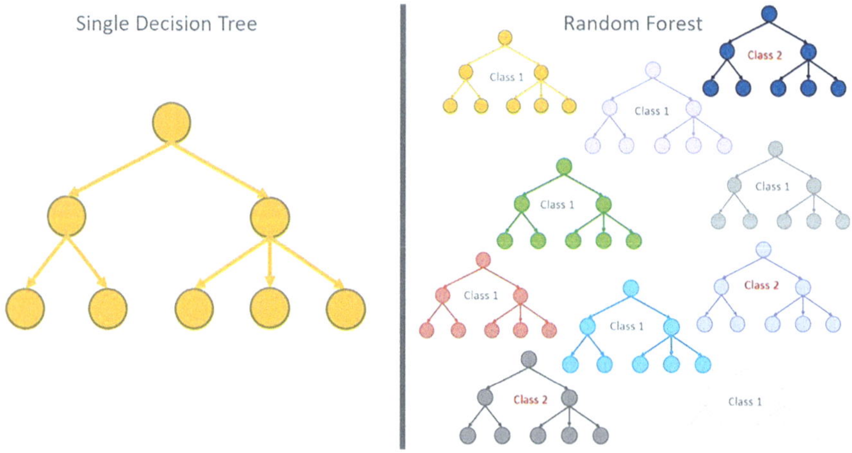

Fig. 18.5 Random forest and decision trees

Scalability: They perform well with large datasets, making them efficient for data analysis.

Speed: Decision trees are known for their exceptional speed, enabling rapid decision-making and analysis. These advantages collectively make decision trees a valuable tool in data analysis and machine learning. However, they also have some disadvantages:

Greedy Algorithm: Hunt's algorithm, used during training, is greedy and may not always produce globally optimum results.

Overfitting: Decision trees can be prone to overfitting, especially when they are very deep, introducing errors due to bias and variance.

Random forests are the remedy to overcome the above-mentioned disadvantages. A random forest comprises multiple decision trees, and their individual results are combined to produce a final outcome. What makes random forests powerful is their capability to reduce overfitting while keeping bias-related errors in check. Random forests have found utility in many research studies that deal with drug discovery. In the research conducted by Kapsiani and Howlin (2021), they investigated the utilization of the DrugAge database for the prediction of compounds with anti-aging properties (Kapsiani and Howlin 2021). To be more specific, they utilized the random forest algorithm to forecast whether a compound would extend the lifespan of *C. elegans* (Kapsiani and Howlin 2021). Ahn et al. (2022) and their team developed a random forest model based on Kullback-Leibler divergence (KLD) feature vectors. They successfully used this model to predict drug-target interactions (DTIs) for 17 representative targets (Ahn et al. 2022).

Support Vector Machines (SVMs)

Support vector machines (SVMs) are a type of supervised machine learning algorithm that has found diverse applications in various fields, including drug discovery and management (Fig. 18.6). In drug discovery and management, SVM finds utility in various tasks, including target identification, virtual screening, quantitative structure-activity relationship (QSAR) modeling, and drug repositioning (Vilar and Costanzi 2012). SVMs are based on the idea of finding the best hyperplane that separates the data points of different classes or predicts continuous output values. In the case of classification, SVM aims to maximize the margin between two classes by finding the optimal hyperplane, which is a line or plane (in higher-dimensional space) that separates the data points belonging to different classes. Support vectors are the data points positioned closest to the hyperplane, and they play a crucial role in defining the location and orientation of the hyperplane. A wider margin between the support vectors and the hyperplane indicates that the SVM model generalizes better to new, unseen data (Rokach 2009).

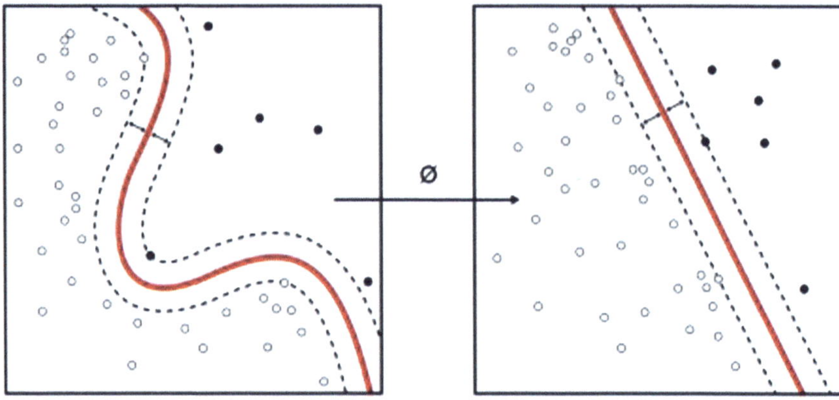

Fig. 18.6 Support vector machine

Ensemble Learning

This machine learning approach combines multiple individual models, known as base models, to construct a more robust and precise predictive model (Fig. 18.7). It combines the predictions of multiple models that would surpass the performance of any single model. It is based on the concept of "wisdom of crowd" where multiple perspectives contribute to better decision-making. There are multiple ensemble techniques such as voting in which each model's prediction is counted and the class with maximum predictions in favor is selected as the output of the model. In the bagging technique, different base models are trained on different data subsets and then averaged on their predictions; random forest is an example of bagging technique. The boosting technique tends to focus more on difficult examples, thus improving accuracy by giving more weight to examples that are misclassified by the previous models. Another method known as stacking combines the predictions generated by multiple base models by training an additional model, called a meta-model, which leverages the predictions made by the base models (Verikas et al. 1999; Zhou et al. 2002; Freund and Schapire 1997).

Reinforcement Learning (RL)

Reinforcement learning (RL) is a subfield of machine learning that focuses on learning through interaction with an environment. It has demonstrated successful applications in diverse fields, including game-playing, robotics, and recommendation systems (Han et al. 2023). In recent years, there has been growing interest in applying RL in drug discovery and development. RL has the potential to accelerate drug discovery by optimizing the selection and design of molecules for testing. Through interaction with a virtual environment that simulates biological systems, RL algorithms can learn to predict which molecules are most likely to be

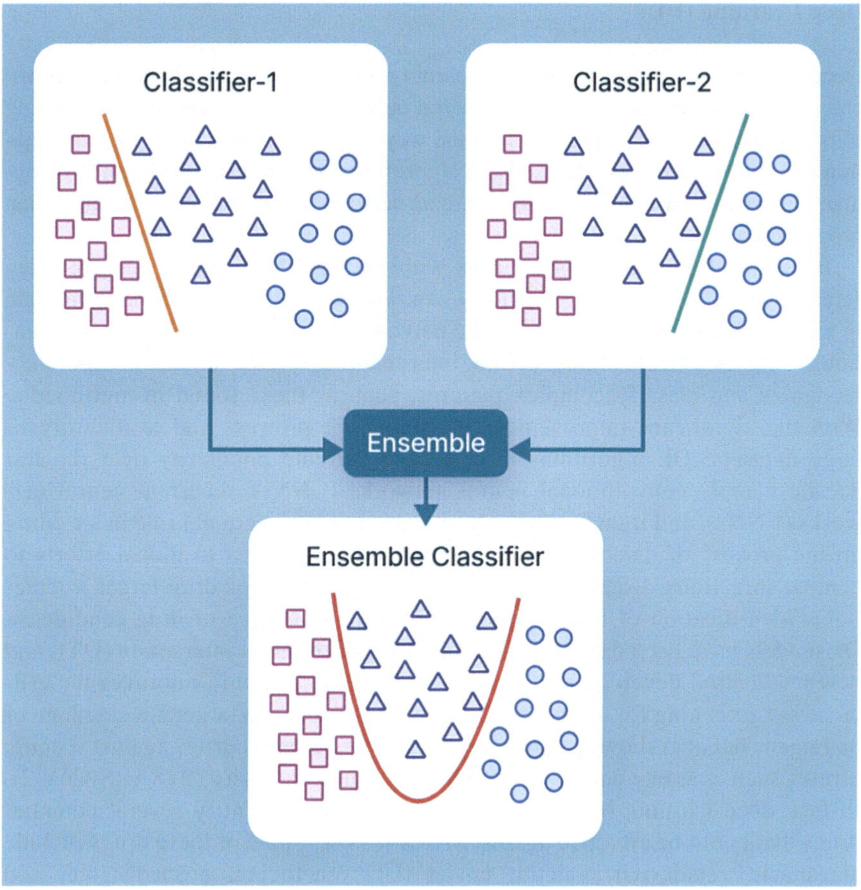

Fig. 18.7 Ensemble learning: classifier

effective in treating a particular disease. This can help reduce the time and cost associated with traditional trial-and-error methods of drug discovery. RL can also be used to optimize the dosing of drugs in clinical trials. By learning from patient data, RL algorithms can determine the optimal dosage for a particular patient, taking into account factors such as age, weight, and medical history (Liu et al. 2020).

One of the challenges of applying RL in drug discovery and development is the availability of high-quality data. The effectiveness of RL algorithms is contingent on the quality and quantity of data accessible for training. In drug discovery, data may be limited due to the complexity of biological systems and the high cost of experiments. Despite these challenges, RL has the potential to revolutionize drug discovery and development by reducing the time and cost associated with traditional methods and by improving the effectiveness of treatments for patients. There has not been any significant work in the area of utilizing RL for this purpose.

Deep Learning (DL)

Deep learning is a subset of machine learning, and it focuses on the usage of artificial neural networks (ANNs). These neural networks aim to mimic the functioning of the human brain. It kind of mimics the way the human brain processes information and learns from experience. These algorithms can automatically learn to recognize patterns in data, make decisions, and perform tasks without explicit human programming.

It uses multiple layers of neurons within the ANN. These layers are often referred to as "deep" networks. They can process and transform the input data in a hierarchical manner, allowing the network to learn features and representations at varying levels of abstraction, thus enabling the DL models to accurately recognize and classify complex patterns, such as those found in multimedia. With the significant improvement in computing prowess and availability of large datasets, DL algorithms have gained extreme popularity over the last decade namely convolutional neural networks (CNNs), recurrent neural networks (RNNs), and transformers. Deep learning plays a crucial role in speeding up the process of discovering new drugs and contributes to global efforts to combat infectious diseases by enabling the prediction of drug-target interactions, identification of lead compounds, and optimization of drug candidates. DL models have been used in the prediction of drug-target interaction (DTI) and new medication development (Kim et al. 2021). It not only improves the efficiency of screening for antimicrobial compounds that can target a wide range of pathogens but also shows promise in identifying potential drugs against specific viruses such as severe acute respiratory syndrome coronavirus 2 (SARS-CoV-2). In fact, deep learning has been successfully used to identify several potential drugs that could be effective against SARS-CoV-2. Some of these drugs include atazanavir, remdesivir, kaletra, enalaprilat, venetoclax, posaconazole, and daclatasvir (Zhang et al. 2021) (Fig. 18.8).

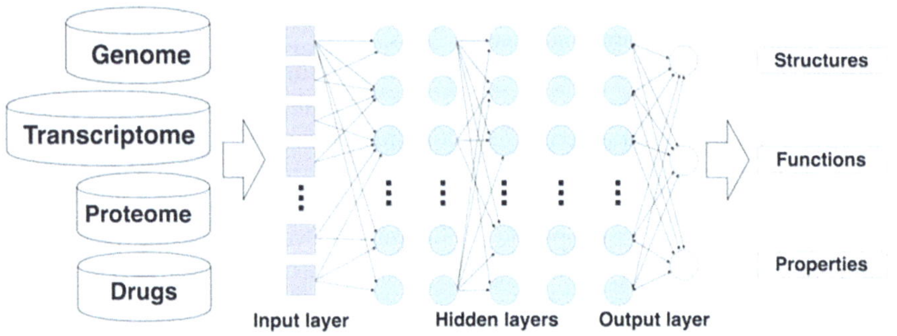

Fig. 18.8 Simple deep neural network architecture for drug discovery

18.3 Transformation of Drug Discovery with AI

Conventionally, the drug comprises several sequential stages: target identification, target validation, lead identification, lead optimization, preclinical testing, and clinical trials. The high cost, time, and limited success rate associated with conventional drug discovery methods for these methods have led researchers to seek innovative approaches to improve the process. The transition from traditional paper-based systems to electronic laboratory notebooks and regulatory submissions has streamlined data management, making it easier to store, access, and share information among researchers, clinicians, and regulatory authorities. With this, the massive ocean of datasets available to us for the different stages of drug development has eased the entry of artificial intelligence. Researchers can access vast repositories of scientific literature, clinical trial data, and databases of repurposable drugs. These resources help identify potential drug targets, explore new therapeutic approaches, and accelerate the discovery and development of novel treatments. Technological advancements have also revolutionized omics profiling in drug development. While genotyping and whole-genome sequencing have been instrumental in understanding genetic variations, new technologies such as microfluidics and antibody tagging have made single-cell technologies widely accessible. This enables researchers to study the transcriptome (e.g., using RNA-seq) (Heath et al. 2015) and the proteome (e.g., via mass cytometry) (Spitzer and Nolan 2016) at a single-cell resolution. Integrating multiple omics modalities (McGinnis et al. 2019) provides a more comprehensive view of the molecular mechanisms underlying diseases and drug responses. In addition to these advancements, digital technologies have bolstered the success rate of clinical trials. Electronic patient data capture systems allow real-time data collection, reducing errors and enhancing data quality. Moreover, artificial intelligence and machine learning algorithms can be deployed for pattern recognition in large datasets, predict treatment outcomes, and optimize trial designs. The transformative power of digital technologies in the drug development process is undeniable. From data collection and analysis to precision medicine and clinical trial optimization, these technologies have revolutionized the way drugs are discovered, developed, and delivered to patients, ultimately improving patient outcomes and advancing health care as a whole.

18.4 AI in Target Identification and Biomarker Discovery

Identification of therapeutic targets and biomarkers represents a fundamental step toward the search for novel therapies for many diseases starting from cancer and neurodegenerative disorders up to infectious diseases (Bodaghi et al. 2023). Biomarkers are endpoints in drug development that provide insights into disease states, progression, and treatment response. Targets are specific molecular pathways within the human body that can be modulated to achieve a therapeutic endpoint. Historically, this has been an empirical, experimental, and domain-reliant process that has consumed much time, often with undue uncertainty bedeviling its outcome. The introduction of artificial intelligence (AI) has catalyzed a revolution in areas of target identification and biomarker discovery. This is where AI as an important partner comes into action by using advanced machine learning algorithms, and its large-scale analysis potentialities to push drug development and precision medicine forward in a fast way (Hathout and Metwally 2023). Such a transformation lies in the ability of AI to process colossal volumes of biological data, pinpoint complex trends, and suggest potential targets or biomarkers at the earliest stages. We will go further and investigate how AI has transformed the process of target identification and biomarker discovery and explore different techniques used, their applicability, and benefits. This is an epic trip that may yield breakthrough revelations leading to rapid advancement in therapeutic innovation and taking medicine closer to individualized remedies for a multitude of diseases. Multiple state-of-the-art works have been done by researchers in biomarker discovery. Zhang et al. embarked on the task of developing a machine learning (ML) algorithm to accurately quantify tumor-stroma ratio (TSR) in hematoxylin-and-eosin (H&E)-stained whole-slide images (WSI). Subsequently, they delved into an investigation of its prognostic implications for patients diagnosed with muscle-invasive bladder cancer (MIBC). Their approach involved the utilization of an optimized cell classifier that had been previously constructed, leveraging the QuPath open-source software and a machine learning algorithm to achieve a quantitative assessment of TSR (Zheng et al. 2023). Daamen et al. (2023) have developed a novel iterative machine learning pipeline that employs gene enrichment profiles derived from blood transcriptome data. This pipeline stratifies COVID-19 patients according to disease severity and distinguishes severe COVID-19 cases from other patients experiencing acute hypoxic respiratory failure (Daamen et al. 2023). Nasimina et al. have successfully engineered a drug sensitivity prediction model tailored specifically for the receptor tyrosine kinase inhibitor sorafenib. Remarkably, their model demonstrated an impressive prediction accuracy rate exceeding 80% when it came to identifying AXL dependency in patients afflicted with leukemia. This achievement not only underscores the significance of their groundbreaking research but also holds promise for the personalized treatment of leukemia by enabling more precise and effective therapeutic choices for individuals with this devastating disease (Nasimian et al. 2023). Karaglani and colleagues have implemented an in silico pipeline that has been instrumental in scrutinizing high-throughput methylome datasets. Their primary objective was to pinpoint distinctive methylation patterns within three prominent pathological conditions that pose significant health burdens, namely breast cancer (BrCa), osteoarthritis (OA), and diabetes mellitus (DM). This innovative approach promises to enhance our understanding of the epigenetic underpinnings of these conditions, offering potential insights for improved diagnosis and treatment strategies (Karaglani et al. 2022).

18.5 Chemoinformatics and QSAR Modeling

Chemoinformatics is the use of computational methods and tools to analyze and interpret chemical data. It involves the storage, retrieval, and analysis of chemical structures and properties, as well as the prediction of chemical behavior and properties. Quantitative structure-activity relationship (QSAR) modeling is a specific application of chemoinformatics that involves the development of mathematical models to predict the biological activity or properties of chemical compounds. QSAR models are based on the relationship between the structural features of a compound (such as its molecular descriptors) and its biological activity or property. Artificial intelligence (AI) and machine learning (ML) play a crucial role in chemoinformatics and QSAR modeling (Hathout and Metwally 2023). In the field of chemoinformatics, artificial intelligence (AI) and machine learning (ML) techniques are employed to examine and model extensive datasets containing chemical structures and properties. ML algorithms can learn from these data to recognize patterns and correlations between molecular descriptors and biological activities or attributes. These models can subsequently be utilized to predict the activity or property values of novel chemical compounds. AI and ML approaches have significantly enhanced QSAR modeling, facilitating the creation of more precise and dependable models. ML algorithms, such as support vector machines, random forests, and neural networks, are commonly used in QSAR modeling to build predictive models from large datasets. These models can capture complex relationships between molecular descriptors and activity or property values, allowing for more accurate predictions. Additionally, AI and ML techniques can also be used for feature selection, where they automatically identify the most relevant molecular descriptors to include in the model. This helps in reducing dimensionality and improving model performance. Moreover, AI and ML can aid in virtual screening, which involves the rapid screening of large chemical databases to identify compounds with desired properties.

18.6 High-Throughput Screening (HTS) and Automation

High-throughput screening (HTS) is an essential method in drug discovery and chemical biology that enables the swift assessment of numerous chemical compounds for their biological activities. In HTS, automated systems and robotics are employed to conduct experiments on a large scale, allowing for the screening of thousands or even millions of compounds within a short timeframe. The high-speed nature of this process generates an immense volume of data, which must be analyzed and interpreted.

The integration of artificial intelligence (AI) in HTS and automation has significantly transformed the drug discovery process. AI algorithms can efficiently process the copious amounts of data produced by HTS, identifying patterns, relationships, and potential candidates for further exploration. Machine learning algorithms can learn from these data and make predictions about the properties or activities of novel compounds. Additionally, they can recommend new compounds or chemical modifications to enhance the desired biological activity. AI can also optimize the HTS process itself by assisting in experiment design and helping researchers choose the most relevant compounds for screening. It can automate data analysis and decision-making steps, reducing human error and increasing efficiency. AI-powered robotics and automation systems can manage complex tasks such as compound handling, assay setup, and data recording, allowing high-throughput screening to be conducted at an unparalleled scale. The integration of AI with HTS and automation has accelerated the drug discovery process, allowing researchers to screen a vast chemical space and identify potential drug candidates more efficiently. It has also led to the discovery of novel compounds and targets that may have been overlooked using traditional methods. As AI continues to advance, it holds tremendous potential in enhancing the speed and precision of high-throughput screening (HTS) and automation, ultimately contributing to the identification of novel drugs and therapeutic strategies.

18.7 Molecular Design and Optimization

Molecular design and optimization for drug discovery and development are a meticulous and transformative process. It begins with the identification of key biological targets, such as proteins or receptors, pivotal in disease pathways, which serve as the foundation for therapeutic interventions. Researchers then embark on the journey of identifying small molecules, or "hits," through advanced techniques such as high-throughput screening and virtual screening, ultimately seeking compounds that bind to the target and exhibit promising initial activities. Conventional approaches for molecular design and optimization have generally involved slow, expensive, and arduous trial-and-error experimental efforts. The biggest challenge for researchers was determining the most viable compounds, uncovering intricate molecular interactions, and projecting the safety and efficiency of novel chemicals. The integration of artificial intelligence (AI) and machine learning (ML) techniques has indeed revolutionized the field of molecular design and optimization. These advanced technologies have significantly expedited the process of designing and fine-tuning molecules for various applications, such as drug discovery and materials science (Karaglani et al. 2022; Karthikeyan and Priyakumar 2021; Fuhr and Sumpter 2022). In recent years, artificial intelligence (AI) and machine learning (ML) have gained widespread adoption to accelerate molecular design across various fields, including drug discovery and development. These cutting-edge technologies have shown remarkable potential for enhancing high-throughput virtual screening and implementing global optimization methods for inverse design of materials (Selvaraj et al. 2021). AI-driven predictive models have greatly benefited the field by accurately forecasting molecular properties, behaviors, and interactions. This aids in the selection of candidate molecules with the most promising characteristics, reducing the need for extensive and costly laboratory experiments. Rational drug design has also been empowered by AI and ML. Researchers can now use these technologies to design molecules that specifically target disease-related proteins or biological pathways. This targeted approach enhances the effectiveness of drug candidates and minimizes side effects, improving patient outcomes (Brown et al. 2020). Beyond pharmaceuticals, AI and ML are instrumental in discovering novel materials with unique properties. In fields such as materials science, where the design of advanced materials for electronics, energy storage, or environmental applications is paramount, AI and ML algorithms can help identify materials with desired characteristics. AI algorithms can also optimize chemical reactions, leading to increased yields and reduced waste. This is not only environmentally responsible but also cost-effective in industries such as chemical manufacturing. By fine-tuning reaction conditions and parameters, AI can help create more efficient and sustainable processes. Furthermore, machine learning models can adapt and improve in real time as new data become available (Karaglani et al. 2022). This iterative process ensures that the molecular design and optimization methods continually evolve and become more accurate. In conclusion, the integration of AI and ML techniques has revolutionized molecular design and optimization. These technologies have enabled efficient exploration of chemical space, predictive modeling, rational drug design, material discovery, optimization of chemical reactions, and real-time learning. With the help of AI and ML, researchers can accelerate the discovery and design process, leading to the development of more effective drugs and advanced materials (Fig. 18.9).

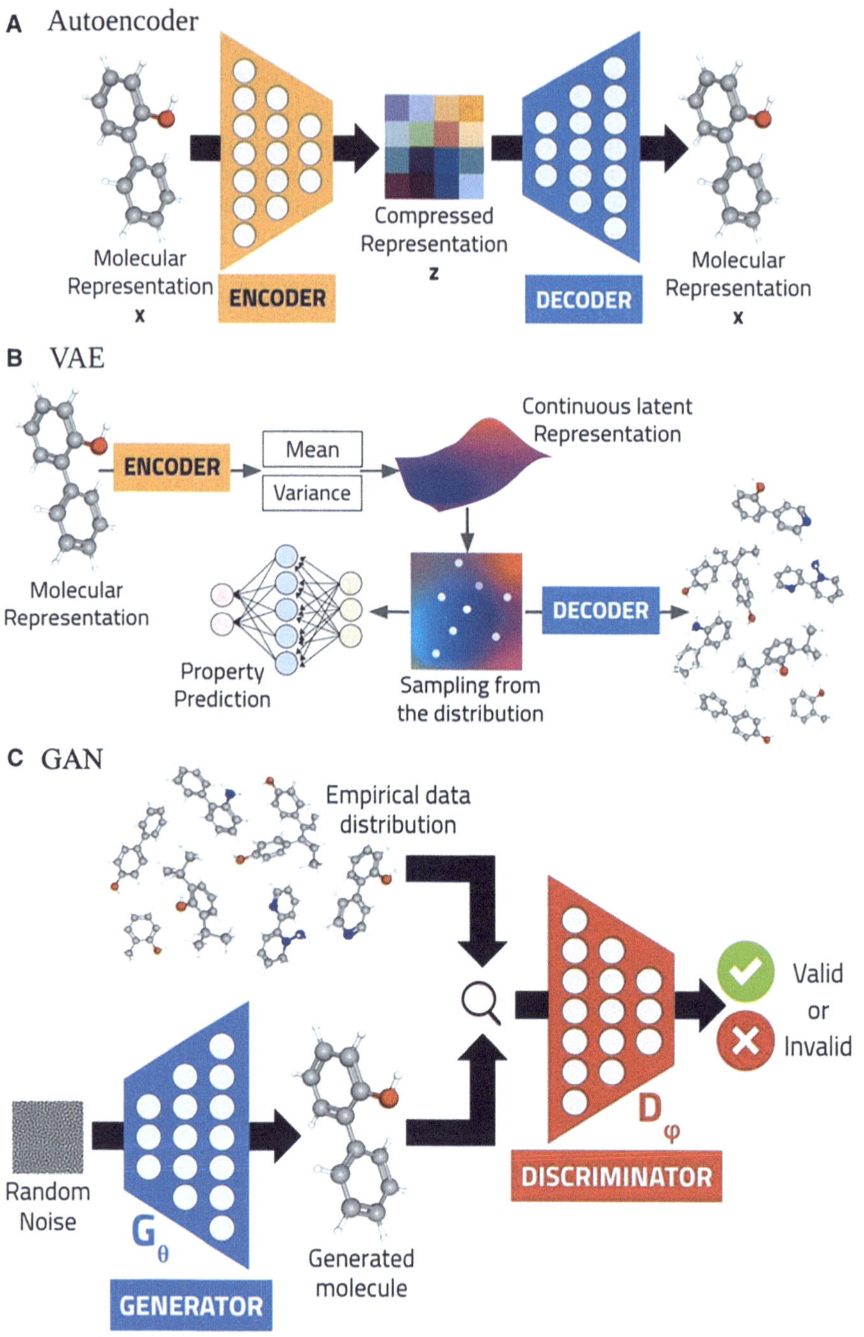

Fig. 18.9 Generative models typically used in molecular generation: AE (**a**), VAE (**b**), and GAN (**c**). (Karthikeyan and Priyakumar 2021)

18.8 AI in Clinical Trials

Clinical trials involve human volunteers participating in research studies to evaluate the safety and effectiveness of new treatments, medications, or medical devices. Clinical trials play a critical role in drug discovery by providing the evidence needed to determine whether a drug is safe, effective, and appropriate for use in humans (Araujo et al. 2020). Clinical trials are conducted in stages, with each stage serving a different purpose in the drug development process. Phase I trials test the safety and tolerance of the intervention in a subset of subjects, while phase II trials test efficacy in a larger group. Phase III trials involve a large group of subjects and compare interventions to existing treatments. Finally, once the intervention is approved, a phase IV trial is conducted to evaluate long-term safety and efficacy. Bringing a new drug to market is a lengthy and expensive process, taking an average of 10–15 years and costing around $1.5–2.0 billion. Clinical trials account for half of this time and investment, while the other 50% covers preclinical compound discovery, testing, and regulatory processes. Despite increased R&D investment from pharma and biotech companies, the number of new drugs approved per billion dollars spent has decreased by half every 9 years. With the drug discovery and development process getting overly complex and difficult, there has been a growing interest in using artificial intelligence (AI) in clinical trials (Scannell et al. 2012). AI has stepped into all phases of the clinical trial cycle. Many pharmaceutical manufacturers have been leveraging AI for multiple reasons, which are given in the following sections.

18.8.1 Faster and More Efficient Patient Recruitment

AI can be used to identify the potential trial participants based on demographics, medical history, and other criteria, enabling companies to recruit patients more quickly and efficiently. Several machine learning models based on natural language processing, deep learning, and rule-based techniques have been deployed to achieve this. NLP can analyze electronic health records (EHRs) and other clinical documents to identify patients who meet specific criteria for a trial (Kersloot et al. 2020). Deep learning algorithms can be deployed to analyze the medical imagery and genome dataset to identify eligible patients. Much simpler ML techniques such as random forests find utility in finding suitable volunteers by building rule-based systems.

18.8.2 Patient Stratification

Patient stratification is an important aspect of clinical trials that allows us to ensure diversity and representation of the patient population. It helps identify patient subgroups that may respond differently to the treatment being tested and minimizes the risk of adverse events. Stratification based on relevant characteristics ensures meaningful and applicable trial results. AI can help identify patient subgroups based on genetic, environmental, and other factors, enabling researchers to tailor treatments for specific patient populations.

18.8.3 Improved Trial Design

AI can be used in trial design with its capability to analyze large amounts of patient data accurately and quickly, identify patterns for improved trial design and decision-making, and determine the best patient population, optimal treatment dosage, and potential safety concerns. AI can also create predictive models to estimate trial success probabilities, identify problems beforehand, and identify potential biomarkers for personalized treatment plans (Harrer et al. 2019). The use of AI in trial design can improve clinical trial efficiency and accuracy and ultimately lead to faster drug development and better patient outcomes.

18.8.4 Real-Time Monitoring

Monitoring patient safety and efficacy data in real time is a crucial aspect of clinical trials. It enables researchers to collect and analyze patient data continuously and make informed decisions about various trial parameters. Real-time monitoring of patient data includes tracking patient safety and efficacy outcomes, assessing adverse events, and monitoring treatment dosages. By collecting patient data in real time, researchers can identify potential safety issues early on and take appropriate measures to mitigate risk. They can also adjust treatment dosages and management strategies based on the data to ensure that patients receive the optimal treatment. Additionally, real-time monitoring enables researchers to identify trends and patterns in patient data, which can help them make statistics-backed decisions about trial design and protocol modifications. Overall, real-time monitoring of patient safety and efficacy data is critical for ensuring patient safety and the success of clinical trials. By continuously monitoring patient data, researchers can optimize trial outcomes, improve patient care, and provide valuable insights that can inform future clinical research.

Now, we briefly discuss the role AI plays in all the stages involved in clinical trial (Shah et al. 2019):

Phase 1: AI holds significant promise in improving the efficiency and accuracy of phase 1 clinical trials. By analyzing large quantities of patient data, AI algorithms can identify potential safety concerns or adverse events more quickly and accurately than traditional methods (Askin et al. 2023). Machine learning techniques can also be used to optimize dosing and scheduling, helping to ensure that patients receive the correct amount of the drug for maximum effectiveness. This can help speed up the drug development pipeline and cut the cost of involvement in clinical trials.

Phase 2: AI has the potential to greatly enhance the efficiency and accuracy of phase 2 clinical trials. Advanced algorithms can be used to rapidly analyze large volumes of patient data, identifying potential safety concerns or adverse events. Machine learning techniques can also be utilized to optimize the dosing and scheduling of the drug, allowing for more precise and effective treatment. As a result, phase 2 trials can provide more detailed and reliable information about the safety and efficacy of the drug. This information can then inform the design of subsequent phase 3 trials, enabling more effective and targeted treatment options for patients.

Phase 3: Phase 3 clinical trials are conducted with thousands of patients to evaluate the safety and effectiveness of a new drug or treatment. They confirm the results of earlier trials and provide more detailed information about the drug's safety and efficacy. Phase 3 trials are designed to reduce bias and ensure accuracy, and the results are used to determine whether the drug can be approved by regulatory agencies. AI can be used in clinical trials to handle and analyze large amounts of data quickly and accurately, helping to speed up the drug development process.

Phase 4: Phase 4 clinical trials monitor the long-term safety and efficacy of drugs in a larger patient population after regulatory approval. These trials involve thousands of patients and may last for years, using a variety of study designs. The results inform clinical practice guidelines and may identify new indications or optimize dosing. Machine learning and AI analyze real-world data to identify potential safety concerns, monitor drug effectiveness, and improve patient outcomes.

18.9 Challenges and Considerations

The integration of artificial intelligence (AI) into drug development and design presents an exciting frontier with the potential to revolutionize the pharmaceutical industry. However, this transformative journey is not without its array of challenges and multifaceted considerations that demand a comprehensive approach to unlock AI's full potential. Foremost among these challenges is the need for vast and high-quality data. The complexity of drug development and the quest for precise predictions require AI models to be built upon a foundation of extensive clinical and molecular information. Regrettably, the limited availability of such data represents a significant stumbling block, necessitating innovative solutions for data acquisition and management (Patel and Shah 2022). Additionally, the "black box" nature of many AI algorithms poses an interpretability conundrum. In the context of health care and drug development, transparency is paramount to engender trust. Consequently, there is a pressing need to develop AI models that provide clear and intelligible explanations for their decisions, bridging the gap between complex algorithms and human understanding. Addressing regulatory compliance is another pivotal consideration. The absence of comprehensive guidelines from esteemed agencies, such as the FDA, introduces uncertainty in establishing a clear regulatory pathway for AI-driven drug discovery (Blanco-González et al. 2023). Collaborative efforts between industry stakeholders and regulatory bodies are imperative to devise frameworks that accommodate the unique challenges and opportunities presented by AI in drug development. Ethical concerns loom large in this landscape, encompassing issues such as potential bias in AI algorithms and privacy. The establishment and adherence to rigorous ethical guidelines for data collection, model training, and decision-making processes are vital to ensure the responsible and ethical development of AI models. The reproducibility and validation of AI models are indispensable in establishing their reliability and generalizability. Fostering a culture of transparency and open collaboration by rigorously validating procedures and openly sharing data and code can significantly contribute to achieving this goal. Simultaneously, the substantial costs and resource intensity associated with building and maintaining AI systems must be weighed against their potential benefits and returns on investment. Exploring partnerships and collaborations and harnessing the capabilities of cloud-based services can alleviate the financial burden tied to AI implementation. Integrating AI tools seamlessly with existing drug development pipelines and methodologies represents another formidable challenge. Approaches for integrating AI-driven insights with existing workflows require prioritizing data security using encryption methods and strict access controls, which guarantee the safeguarding of confidential patients and research. Addressing the shortage of AI and domain-specific experts in health care and life sciences is pivotal, and it can be achieved through investments in training and education programs. Fostering collaborations between AI experts and domain specialists fosters a synergy that accelerates innovation and problem-solving in the field. Finally, establishing clear guidelines for intellectual property rights and adapting existing legal frameworks to account for AI-generated inventions are indispensable steps. This approach not only fosters innovation but also safeguards the rights of all parties involved, promoting a balanced and conducive environment for advancement. In conclusion, the effective integration of AI into drug development and design necessitates a holistic and multifaceted approach. It encompasses addressing data quality, enhancing algorithm interpretability, navigating regulatory compliance, mitigating ethical concerns, ensuring reproducibility, optimizing cost-effectiveness, seamless integration with traditional methods, bolstering data security, fostering expertise, and establishing equitable guidelines for intellectual property rights. By interweaving these diverse aspects, the potential of AI in drug development can be maximized, propelling innovation and enhancing patient outcomes, leading to a brighter and more promising future in health care.

18.10 Conclusion

In this chapter, we have delved deeply into the influence of artificial intelligence on the processes of drug discovery and development. It began with the recognition of AI's prominent role in the pharmaceutical industry, signaling a transformative era in the domain of drug development. It navigated through the various facets of AI's influence; it became evident that machine learning, in particular, has emerged as a powerful tool. It not only handles large datasets but also automates processes that were historically labor-intensive and time-consuming. AI is not merely a technological advancement; it signifies a fundamental shift in the approach to drug discovery. The integration of human expertise with AI algorithms has unlocked the potential to unveil novel treatments concealed within the intricate complexity of biological systems. Furthermore, it highlights the concrete advantages brought by AI. It improves the precision of identifying potential drug targets, expedites the decision-making processes related to promising drug candidates, and plays a pivotal role in designing medicines with specific characteristics, ultimately enhancing their effectiveness and safety profiles. However, it is essential to acknowledge the challenges accompanying the integration of AI into drug discovery. These challenges encompass concerns about data quality, the interpretability of AI-generated results, and ethical considerations inherent to the use of AI in such a critical field. In summary, this chapter unequivocally demonstrates AI's central role as a catalyst for transforming all aspects of drug discovery and development. As we conclude this chapter, it becomes evident that AI holds the potential to revolutionize the pharmaceutical industry, promising the acceleration of life-saving drug development. While challenges persist along the journey, the undeniable prospects of reshaping our approach to drug development through AI should motivate researchers, practitioners, and all stakeholders to fully embrace this transformative change. By collectively embracing AI's capabilities, we can expedite the discovery of new therapies, ultimately leading to improved health outcomes for individuals around the world.

References

Ahn S, Lee SE, Kim M (2022) Random-forest model for drug–target interaction prediction via Kullback–Leibler divergence. J Cheminform. https://doi.org/10.1186/s13321-022-00644-1

Araujo MP, Al-Yaseen W, Innes NP (2020) A road map for designing and reporting clinical trials in paediatric dentistry Int J Paed Dent 31:14–22

Askin S, Burkhalter D, Calado G, El Dakrouni S (2023) Artificial intelligence applied to clinical trials: opportunities and challenges. Health Technol 13:203–213

Blanco-González A, Cabezón A, Seco-González A, Conde-Torres D, Antelo-Riveiro P, Piñeiro Á, Garcia-Fandino R (2023) The role of AI in drug discovery: challenges, opportunities, and strategies. Pharmaceuticals 16:891

Bodaghi A, Fattahi N, Ramazani A (2023) Biomarkers: promising and valuable tools towards diagnosis, prognosis and treatment of Covid-19 and other diseases. Heliyon 9:e13323

Brown N, Ertl P, Lewis R, Luksch T, Reker D, Schneider N (2020) Artificial intelligence in chemistry and drug design. J Comput Aided Mol Des 34:709–715

Chen S, Li Z, Zhang S, Zhou Y, Xiao X, Cui P, Xu B, Zhao Q, Kong S, Dai Y (2022) Emerging biotechnology applications in natural product and synthetic pharmaceutical analyses. Acta Pharm Sin B 12:4075–4097

Daamen AR, Bachali P, Grammer AC, Lipsky PE (2023) Classification of COVID-19 patients into clinically relevant subsets by a novel machine learning pipeline using transcriptomic features. Int J Mol Sci 24:4905

Freund Y, Schapire RE (1997) A decision-theoretic generalization of on-line learning and an application to boosting. J Comput Syst Sci 55:119–139
Fuhr AS, Sumpter BG (2022) Deep generative models for materials discovery and machine learning-accelerated innovation. Front Mater 9:865270. https://doi.org/10.3389/fmats.2022.865270
Han D, Mulyana B, Stankovic V, Cheng S (2023) A survey on deep reinforcement learning algorithms for robotic manipulation. Sensors 23:3762
Harrer S, Shah P, Antony B, Hu J (2019) Artificial intelligence for clinical trial design. Trends Pharmacol Sci 40:577–591
Hathout RM, Metwally AA (2023) Tracing the nose-to-brain nanoparticulate drug delivery using bio/chemoinformatics tools. In: A handbook of artificial intelligence in drug delivery. Academic Press, pp 241–257
Heath JR, Ribas A, Mischel PS (2015) Single-cell analysis tools for drug discovery and development. Nat Rev Drug Discov 15:204–216
Kapsiani S, Howlin BJ (2021) Random forest classification for predicting lifespan-extending chemical compounds. Sci Rep 11:13812. https://doi.org/10.1038/s41598-021-93070-6
Karaglani M, Panagopoulou M, Baltsavia I, Apalaki P, Theodosiou T, Iliopoulos I, Tsamardinos I, Chatzaki E (2022) Tissue-specific methylation biosignatures for monitoring diseases: an in silico approach. Int J Mol Sci 23:2959
Karthikeyan A, Priyakumar UD (2021) Artificial intelligence: machine learning for chemical sciences. J Chem Sci 134. https://doi.org/10.1007/s12039-021-01995-2
Kersloot MG, van Putten FJP, Abu-Hanna A, Cornet R, Arts DL (2020) Natural language processing algorithms for mapping clinical text fragments onto ontology concepts: a systematic review and recommendations for future studies. J Biomed Semant. https://doi.org/10.1186/s13326-020-00231-z
Kim J, Park S, Min D, Kim W (2021) Comprehensive survey of recent drug discovery using deep learning. IJMS 22:9983
Kovács F, Legány C, Babos A (2005) Cluster validity measurement techniques. In: 6th International symposium of Hungarian researchers on computational intelligence, p 35
Liu S, See KC, Ngiam KY, Celi LA, Sun X, Feng M (2020) Reinforcement learning for clinical decision support in critical care: comprehensive review. J Med Internet Res 22:e18477
McGinnis CS, Patterson DM, Winkler J et al (2019) MULTI-seq: sample multiplexing for single-cell RNA sequencing using lipid-tagged indices. Nat Methods 16:619–626
Nasimian A, Al Ashiri L, Ahmed M, Duan H, Zhang X, Rönnstrand L, Kazi JU (2023) A receptor tyrosine kinase inhibitor sensitivity prediction model identifies AXL dependency in leukemia. Int J Mol Sci 24:3830
Patel V, Shah M (2022) Artificial intelligence and machine learning in drug discovery and development. Intell Med 2(3):134–140
Popova M, Isayev O, Tropsha A (2018) Deep reinforcement learning for de novo drug design. Sci Adv 4(7):eaap7885. https://doi.org/10.1126/sciadv.aap7885
Qureshi R, Irfan M, Gondal TM, Khan S, Wu J, Hadi MU, Heymach J, Le X, Yan H, Alam T (2023) AI in drug discovery and its clinical relevance. Heliyon 9:e17575
Rokach L (2009) Using Fuzzy logic in data mining. In: Data mining and knowledge discovery handbook. Springer, pp 505–520
Sarkar C, Das B, Rawat VS, Wahlang JB, Nongpiur A, Tiewsoh I, Lyngdoh NM, Das D, Bidarolli M, Sony HT (2023) Artificial intelligence and machine learning technology driven modern drug discovery and development. Int J Mol Sci 24:2026
Scannell JW, Blanckley A, Boldon H, Warrington B (2012) Diagnosing the decline in pharmaceutical R&D efficiency. Nat Rev Drug Discov 11:191–200
Selvaraj C, Chandra I, Singh SK (2021) Artificial intelligence and machine learning approaches for drug design: challenges and opportunities for the pharmaceutical industries. Mol Divers 26:1893–1913

Shah P, Kendall F, Khozin S, Goosen R, Hu J, Laramie J, Ringel M, Schork N (2019) Artificial intelligence and machine learning in clinical development: a translational perspective. NPJ Digit Med 2:69. https://doi.org/10.1038/s41746-019-0148-3

Singh S, Kumar R, Payra S, Singh SK (2023) Artificial intelligence and machine learning in pharmacological research: bridging the gap between data and drug discovery. Cureus 15(8). https://doi.org/10.7759/cureus.44359

Song Y-Y, Ying L (2015) Decision tree methods: applications for classification and prediction. Shanghai Arch Psychiatry 27(2):130–135

Spitzer MH, Nolan GP (2016) Mass cytometry: single cells, many features. Cell 165:780–791

Takács G, Sándor M, Szalai Z, Kiss R, Balogh GT (2021) Analysis of the uncharted, druglike property space by self-organizing maps. Mol Divers 26:2427–2441

Verikas A, Lipnickas A, Malmqvist K, Bacauskiene M, Gelzinis A (1999) Soft combination of neural classifiers: a comparative study. Pattern Recogn Lett 20:429–444

Vilar S, Costanzi S (2012) Predicting the biological activities through QSAR analysis and docking-based scoring. Methods Mol Biol 914:271–284

Vora LK, Gholap AD, Jetha K, Thakur RRS, Solanki HK, Chavda VP (2023) Artificial intelligence in pharmaceutical technology and drug delivery design. Pharmaceutics 15:1916

Zhang Y, Ye T, Xi H, Juhas M, Li J (2021) Deep learning driven drug discovery: tackling severe acute respiratory syndrome coronavirus 2, vol 12. Front Microbiol, p 739684

Zheng Q, Jiang Z, Ni X et al (2023) Machine learning quantified tumor-stroma ratio is an independent prognosticator in muscle-invasive bladder cancer. Int J Mol Sci 24:2746

Zhou Z-H, Wu J, Tang W (2002) Ensembling neural networks: many could be better than all. Artif Intell 137:239–263

Chapter 19
Artificial Intelligence: A Major Landmark in the Novel Drug Discovery Pathway for the Remarkable Advancement in the Healthcare System

Rabin Debnath, Abu Md Ashif Ikbal, Ankita Choudhury, Subhash C. Mandal, and Partha Palit

Abstract The healthcare industry is entering a new era of efficiency and personalisation because of the integration of artificial intelligence (AI), which has completely transformed medical research and development. With the use of machine learning and deep learning algorithms, artificial intelligence has had a particularly significant impact on drug discovery, simplifying the design and optimisation of novel molecules. Artificial intelligence (AI) optimises research designs for clinical trials, reducing procedures and increasing data analysis efficiency. Dynamic treatments and customised patient care are guaranteed by AI-powered real-time monitoring. Multidisciplinary innovation in medicine development is fuelled by the cooperative efforts of data analysts, physicians, and subject matter experts. AI optimises pharmaceutical resource potential and supports sustainable practices by accelerating drug discovery procedures. Personalised medicine is made possible by biomarker discovery, which is powered by AI and allows for the identification of critical indicators for disease states and treatment responses. Artificial intelligence (AI) in clinical trials leads to faster and more dependable therapeutic outcomes by streamlining designs, identifying patient subpopulations, and improving data processing efficiency. Artificial intelligence-driven surveillance enables instantaneous monitoring,

---

R. Debnath
ISF College of Pharmacy, Ghall Kalan, Punjab, India

A. M. A. Ikbal · P. Palit (✉)
Department of Pharmaceutical Sciences, Drug Discovery Research Laboratory, Assam University (A Central University), Silchar, India

A. Choudhury
Silchar Medical College and Hospital, College of Pharmacy, Silchar, India

S. C. Mandal
Pharmacognosy & Phytotherapy Research Laboratory, Division of Pharmacognosy, Department of Pharmaceutical Technology, Faculty of Engineering & Technology, Jadavpur University, Kolkata, India

S. Bose et al. (eds.), *Concepts in Pharmaceutical Biotechnology and Drug Development*, Interdisciplinary Biotechnological Advances,
https://doi.org/10.1007/978-981-97-1148-2_19

guaranteeing adaptable and customised patient interventions. A flexible and patient-centred healthcare system is made possible by AI systems' continuous learning capabilities, which adapt based on actual patient data. Overall, AI's transdisciplinary influence drives medication research and discovery towards more precise, effective, and personalised healthcare solutions, with the potential to yield ground-breaking discoveries that will benefit people all over the world.

Keywords Artificial intelligence (AI) · Drug discovery · Personalised medicine · Clinical trials · Healthcare efficiency

19.1 Introduction

19.1.1 Natural Language Processing, Deep Learning, and Machine Learning

Drug discovery technologies are being revolutionised by the strong tools of natural language processing (NLP), deep learning, and machine learning, which are bringing new speed and efficiency to the process. An enormous amount of biological data is being analysed, intricate patterns are being determined, and the process of finding new drug candidates is being accelerated considerably with the help of artificial intelligence (AI) (Gupta et al. 2021). With new approaches to persistent problems, the combination of artificial intelligence and drug discovery has enormous potential for the pharmaceutical sector. The interaction of computers with human language is the focus of the AI subfield of natural language processing (Jiménez-Luna et al. 2021). Natural language processing (NLP) plays a key role in drug development by helping to extract useful information from a wide range of sources, including patents, trials, and scholarly publications. It helps scholars find undiscovered connections and remain up to date on new advancements. The time and effort needed for literature mining can be greatly reduced by using natural language processing (NLP) techniques to navigate through large datasets and find pertinent molecular relationships, disease pathways, and possible therapeutic targets (Wang and Lin 2023). A kind of machine learning called deep learning has shown a remarkable ability to handle complex biological data. Conventional approaches are insufficient for thorough analysis due to the intricacy of molecular interactions and the volume of omics data. Deep learning models especially neural networks are very good at finding hidden correlations, identifying complex patterns in datasets, and forecasting the behaviours of molecules (Mahmud et al. 2021). These powers are used in drug development for tasks including molecular structure optimisation, drug-drug interaction prediction, and virtual screening. Because deep learning models can identify minute details in biological data, researchers will be better equipped to decide on the safety and possible effectiveness of drug candidates. Drug discovery can be accomplished with a variety of tools and approaches provided by machine learning, the larger parent field of deep learning (Tiwari and Singh 2022). Machine learning algorithms are useful in identifying possible drug candidates and streamlining drug development procedures through various applications such as predictive modelling and clustering analysis. In one noteworthy application, machine learning models are used to identify possible therapeutic targets linked to certain diseases by analysing a variety of biological parameters (Hussain et al. 2021). This focused strategy reduces resource waste and maximises performance by optimising the drug discovery process. The use of AI in drug discovery has sped up the entire drug development lifecycle, resulting in the creation of novel platforms. In order to speed up the diagnosis and molecular understanding of diseases, AI-driven platforms, for example, can make it easier to identify biomarkers unique to each disease. Using this information in conjunction with sophisticated analytics makes it possible to identify potential drug targets and create more accurate and successful therapeutic interventions (Dara et al. 2022). Artificial intelligence has been shown to be quite helpful in improving patient recruitment and clinical trial design. The effectiveness and success rates of clinical trials can be increased by using machine learning algorithms to evaluate a variety of datasets and identify patient groups that are most likely to respond favourably to a specific treatment. This focused strategy lowers the time and expense of medication research while simultaneously raising the possibility of successful medicines being introduced to the market (Harrer et al. 2019).

Pharmaceutical research is changing as a result of the integration of deep learning, machine learning, and natural language processing in AI-powered drug discovery technologies. Not only are these technologies speeding up procedures, but they are also radically altering how scientists approach drug development (Liu et al. 2021). AI's uses in drug development are probably going to get even more advanced as it develops, opening the door for a new era of precision medicine and tailored therapeutic interventions. AI and drug discovery working together could speed up the creation of ground-breaking therapies, ultimately leading to better patient outcomes and a revolution in the pharmaceutical sector (Gennatas and Chen 2021). The combination of deep learning, machine learning, and natural language processing (NLP) in drug discovery technologies is not only revolutionising research procedures but also solving major issues that have long plagued the pharmaceutical sector. The massive amount of biomedical literature, which is growing rapidly every year, has been one of the major challenges in drug development. NLP is revolutionary because it can understand large text libraries and extract valuable information from them. It helps scientists stay up to date on the most recent discoveries in science, determine possible therapeutic targets, and evaluate hypotheses more quickly (Öztürk et al. 2020).

NLP is essential not only for literature mining but also for knowledge representation and extraction from unstructured data. NLP makes it easier to integrate textual data by converting it into structured representations. NLP is important for knowledge representation and extraction from unstructured data, in addition to literature mining. Natural language processing (NLP) enables the integration of disparate data sources by converting textual content into structured representations (Baviskar et al. 2021). This integration enables researchers to find connections between genes, proteins, and pathways that may have gone unnoticed using more conventional techniques, which is crucial for a comprehensive knowledge of complex biological systems. Proteomics, genomics, and other omics domains possess complex and high-dimensional data, which is where deep learning, especially with neural networks, shines. Deep learning models are well suited for applications such as medication repurposing and chemical structure prediction because they can learn hierarchical representations of data. Deep learning models have the capability to analyse the three-dimensional structures of compounds and predict their biological activities. This makes it possible to identify medications that are currently on the market and potentially repurpose them for new uses (Sidak et al. 2022).

19.1.2 Examining Enormous Databases to Identify Potential Drug Candidates

One of the most important areas of drug discovery technologies is the analysis of massive databases to find possible drug candidates. Artificial intelligence (AI) is a key player in this revolutionary process (Paul et al. 2021). Modern biological data are so large and complicated that navigating them and extracting useful information require sophisticated computational techniques. AI, along with its subfields such as deep learning and machine learning, is becoming more and more the driving force that drives the identification and validation of possible therapeutic options. The time-consuming and resource-intensive process of searching through enormous chemical libraries for possible compounds has been one of the main obstacles in traditional drug discovery (Górriz et al. 2020). AI solutions improve and automate the screening process in order to overcome this difficulty. Machine learning algorithms have the ability to forecast the possibility of a novel chemical interacting with certain biological targets after being trained on large datasets of documented drug-target interactions. The early phases of drug development can be greatly accelerated by using this prediction power to help researchers choose which compounds they should investigate further. Large-scale biological dataset analysis benefits greatly from deep learning's capacity to identify intricate patterns in data (Abbasi et al. 2021). Deep learning models, a subset of artificial neural networks, excel at extracting intricate patterns from vast datasets for tasks like image recognition and natural language processing. In biology, they adeptly learn from raw proteomic, metabolomic, and genomic data to predict protein interactions, metabolic outcomes, gene functions, and disease associations. This skill is crucial for locating possible medication candidates with particular target characteristics, allowing for a more focused and effective drug development procedure. The extraction of information from scientific publications, patents, and clinical trial reports also heavily relies on natural language processing, or NLP (Sen et al. 2021). Natural language processing (NLP) helps identify possible therapeutic targets, comprehend disease pathways, and validate current theories by drawing insights from this enormous body of knowledge. Scientists may make well-informed decisions on which compounds to pursue by utilising NLP's integration with other AI technologies, which guarantees a thorough study of all accessible data. Through the use of these technologies, AI-driven drug discovery platforms develop a comprehensive method for identifying potential candidates (Demner-Fushman et al. 2021). To give a comprehensive image of possible medication candidates, these platforms combine information from multiple sources, such as clinical data, experimental results, and literature. These tools help scientists decide which compounds have the best chance of making it through the drug development pipeline by automating the study of various data types. In addition, artificial intelligence technologies support the idea of drug repurposing, the process of finding new therapeutic indications for already-approved medications. Machine learning algorithms can reveal previously unknown relationships between drugs and disorders by examining big datasets that contain clinical outcomes, molecular profiles, and drug interactions. This could result in the identification of innovative uses for already-approved pharmaceuticals. The hazards involved in creating completely new medications are decreased by this method, which also speeds up the drug development process by utilising the safety profiles of existing molecules (Tripathi et al. 2022).

The use of AI technologies is revolutionising the process of searching through massive databases for possible medication candidates. Together, these technologies enable researchers to more effectively traverse the huge terrain of biological data. Machine learning predicts interactions, deep learning reveals intricate correlations in large datasets, and natural language processing extracts knowledge from extensive textual sources. AI is expected to play a bigger part in drug development as technology develops, perhaps leading to more focused, effective, and creative methods for finding the next wave of therapeutic interventions. AI and drug development are working together to create a synergistic effect that will lead to faster, more accurate, and more informed identification of possible therapeutic candidates in the future (Husnain et al. 2023).

19.1.3 Quickly Identifying Drug-Target Interactions and Biomarkers to Expedite Preclinical Research

To accelerate preclinical research, it is necessary to quickly identify drug-target interactions and biomarkers. Artificial intelligence (AI) is one such advanced technology that can be used to precisely and quickly identify biomarkers linked to particular biological processes or disease states, as well as possible interactions between drug compounds and molecular targets. The quick identification of drug-target interactions and biomarkers is an essential phase in the changing landscape of drug discovery that greatly affects preclinical research efficiency (Boniolo et al. 2021). The process of identifying potential candidates and biomarkers is being revolutionised with unprecedented speed and precision through the utilisation of artificial intelligence (AI), specifically through the application of advanced technologies such as machine learning, deep learning, and natural language processing (NLP). As a result of their training on large datasets with data regarding established drug-target interactions, machine learning models have become extremely effective at forecasting novel interactions. These models have the ability to examine intricate patterns in biological data and identify possible connections between medicinal substances and particular molecular targets (Torfi et al. 2020). AI-driven algorithms can quickly go through massive databases to identify possible candidates for more research, greatly cutting down on the amount of time needed to complete the task. By finding and validating potential therapeutic candidates and relevant biomarkers more quickly, the aim is to accelerate the preclinical research stage of drug discovery.

- Drug Interactions with Targets:
  - Traditional Challenges: Traditionally, selecting the appropriate molecular targets for medicinal substances has required rigorous screening procedures and substantial experimentation.
  - Artificial Intelligence Solutions: By analysing vast datasets, which include details on established drug-target interactions, machine learning and deep learning algorithms can more accurately forecast possible interactions. These algorithms identify trends in biological data and identify pharmacological candidates that have a higher probability of interacting with particular molecular targets (Selvaraj et al. 2021).
- Identification of Biomarkers:
  - Traditional Challenges: Determining biomarkers linked to certain biological processes or disorders frequently necessitates extensive and time-consuming investigation.
  - AI Remedies: Potential biomarkers can be quickly identified using machine learning models that have been trained on a variety of datasets, including genomes, proteomics, and clinical data. These models identify indicators that might be connected to specific circumstances by examining patterns and correlations in the data. This expedites the process of identifying biomarkers, hence facilitating prompt confirmation in preclinical investigations (Sun et al. 2023).
- Accelerated Preclinical Research:
  - Conventional Timelines: Before moving on to clinical trials, preclinical research usually entails a number of tests and validations to guarantee the efficacy and safety of possible medication candidates.
  - AI Impact: AI accelerates preclinical research by rapidly discovering biomarkers and drug-target interactions. The time and resources needed for preliminary investigations can be decreased by researchers concentrating their efforts on candidates and biomarkers with better prediction success rates (Zhavoronkov et al. 2019).
- Potential for Personalised Medicine:
  - Conventional Approaches: Creating treatments based on a one-size-fits-all concept may cause patients' responses to differ from one another.
  - AI Developments: Quick biomarker identification enables more individualised treatment. AI systems have the capacity to evaluate patient-specific data

in order to find distinct biomarkers, opening the door for customised treatments based on personal traits. This lowers the possibility of side effects while simultaneously increasing therapeutic efficacy (Schork 2019).

- AI-Powered Systems: Integration of Technologies: AI-driven platforms integrate machine learning, deep learning, and natural language processing (NLP) to conduct comprehensive analyses of a variety of data sources.
- Automation and Efficiency: These platforms automate the analysis of large datasets, giving researchers a comprehensive understanding of possible drug-target interactions and biomarkers. Automation improves efficiency and facilitates data-driven decision-making (Evangelista 2020).
- To sum up, using AI to rapidly identify drug-target interactions and biomarkers is a game-changing strategy in drug discovery as it speeds up preclinical research, improves the identification of promising candidates, and has the potential to usher in a new era of personalised medicine by customising treatments for each patient.

19.2 Using AI to Repurpose Drugs

The field of pharmaceutical research and development has seen a significant transformation in recent years due to the growing influence of artificial intelligence (AI) on drug discovery. One important part of this revolution is repurposing existing drugs for novel medicinal purposes. This section examines the several applications of artificial intelligence (AI) in the investigation of already licenced drugs and their potential across a range of therapeutic areas (Mak and Pichika 2019). The investigation consists of three primary parts: III.A. examining already-approved medications for potential novel therapeutic applications; III.B. analysing pharmacological interactions and biological networks in silico; and III.C. examining medications that are well designed and optimised for a variety of functions.

19.2.1 Investigating Previously Authorised Drugs for Potential Novel Therapeutic Uses

"Repurposing" medicines refers to investigating uses for well-known pharmacological compounds outside of their original intended purposes (Pushpakom et al. 2019). This strategy relies heavily on AI since it can sift through massive datasets and identify candidates who could benefit from repositioning. Using complex algorithms and machine learning models, the system evaluates data from a range of sources, including genetic databases, clinical trial data, and biological literature (Vo et al. 2019).

By utilising artificial intelligence (AI) to thoroughly evaluate the safety and efficacy characteristics of presently marketed drugs, researchers can identify promising candidates for novel therapeutic indications. This tactic reduces the amount of time needed for drug development while also utilising the wealth of information found in approved medications. For example, a drug that initially showed promise in treating one condition may have properties that make it suitable for treating a completely different medical condition (Vatansever et al. 2021).

Furthermore, using AI to repurpose medications increases the chance of successful clinical studies. Predictive modelling and data analytics allow researchers to identify pharmacological candidates with the highest chance of success, which helps them prioritise their efforts and save overall drug development costs.

When evaluating already-approved pharmaceuticals for novel therapeutic uses, a detailed analysis of their recognised pharmacological properties is required. Artificial intelligence algorithms search through massive databases containing information on drug interactions, side effects, and outcomes of clinical trials. By identifying patterns and connections within this massive information, artificial intelligence (AI) enables the discovery of new candidates that may demonstrate efficacy against diseases outside of their initial indications. This data-driven approach expedites the discovery of new therapeutic applications and provides a workable replacement for traditional drug development (Bobo et al. 2016).

In the area of previously approved medication investigations, artificial intelligence's ability to analyse real patient data and electronic health records contributes to our growing knowledge of drug reactions. By taking into consideration, variables such as patient demographics, comorbidities, and treatment outcomes, AI assists in identifying certain patient populations that might benefit most from the repurposed drugs (Ahmed et al. 2022). By ensuring that repositioned pharmaceuticals are not only effective but also tailored to the individual needs of the individuals they are meant to treat, this patient-centred approach increases the accuracy of treatment programmes.

19.2.2 Biological Networks and Pharmacological Interactions Investigated In Silico

The phrase "in silico" describes computational methods based on computers that have been shown to be crucial for the repurposing of medications. AI-driven biological network and pharmacological interaction analysis can help us better understand the complex relationships that exist between drugs, biological processes, and diseases (Russo et al. 2020).

Biological networks are challenging to manually understand because they entail intricate relationships between genes, proteins, and other components. AI system's deep learning models in particular are highly adept at navigating these intricate networks and identifying potential sites for pharmaceutical repurposing interventions. Knowing more about the underlying biology allows researchers to more precisely and effectively move drugs to target specific nodes in the network.

Moreover, AI enables the speed and volume of pharmacological interaction analysis that are not achievable with traditional methods. It is feasible to identify candidates that may exhibit the requisite pharmacological activity by virtually screening drug libraries against specific targets relevant to disease states. This in silico screening significantly expedites the search for new treatments by selecting drugs based on a reason.

In silico research of biological networks and pharmacological interactions is the application of artificial intelligence (AI) to model and simulate complex biological processes (Davahli et al. 2021). Artificial intelligence algorithms have exceptional abilities to decipher the intricate web of connections between biological pathways, offering valuable perspectives on potential targets for pharmaceutical intervention. Researchers may theoretically screen treatment possibilities against specific molecular targets, allowing them to rank compounds with the best possibility of success. By accelerating the discovery of prospective treatment candidates and providing a sound basis for experimental validation, this computational technique maximises the use of resources in the drug development pipeline.

In in silico studies of biological networks and pharmacological interactions, AI-driven analyses also identify potential side effects and drug-drug interactions. Artificial intelligence (AI) uses a model of biological systems' behaviour to identify potential safety risks in pharmaceuticals (Selvaraj et al. 2021). This makes it possible for researchers to foresee potential issues that may arise during clinical trials. This proactive strategy streamlines the medication development process, improves the overall safety and efficacy of repurposed pharmaceuticals, and increases the likelihood of positive outcomes in a range of patient populations.

19.2.3 Drugs that Are Effectively Designed and Optimised for a Variety of Uses

Optimising drug design is essential for successful repurposing, and artificial intelligence (AI) brings a fresh dimension to this endeavour. Through sophisticated computational methodologies, artificial intelligence (AI) aids in the design of pharmaceuticals with increased therapeutic properties for new purposes in addition to their original effectiveness for their original indications (Fischer et al. 2019).

Machine learning algorithms have the capability to predict the most effective chemical modifications to alter the specificity or boost the efficiency of medicine, thanks to training from vast databases of molecular structures and the biological activities associated with them. A technique known as "de novo drug design" makes it possible to create analogues or derivatives of existing drugs that have superior pharmacological properties (Yang et al. 2019).

Not only are molecular structures but also dosing schedules and treatment approaches susceptible to AI-driven optimisation. By assessing clinical results and actual patient data, AI is able to offer personalised treatment plans that enhance therapeutic effectiveness and reduce side effects. This systematic approach increases the overall effectiveness of drug repurposing by tailoring therapies to the unique characteristics of each patient.

Using AI-driven design, drug optimisation for a variety of therapeutic purposes is a challenging task. Artificial intelligence (AI) systems use molecular structures to predict chemical changes that could increase or decrease a drug's effectiveness (Tiwari et al. 2023). Another use of predictive modelling is dosage optimisation, where AI prescribes customised dose regimens based on factors such as therapy response, genetic variations, and patient demographics. By tailoring drug design to specific patient characteristics, AI ushers in a new era of precision medicine in the area of medication repurposing. This maximises therapeutic efficacy and reduces the possibility of adverse reactions (Zhou et al. 2020).

AI affects the creation of combination therapies and the optimisation of drugs for a range of uses. By analysing the synergistic effects of many drugs in a therapeutic environment, AI develops strategies to mix pharmaceuticals in ways that maximise benefits and reduce bad effects. This technique not only broadens the therapeutic window for drugs that are already on the market, but it also opens doors for innovative therapeutic strategies. AI also simplifies the process of looking into medication repurposing in the context of recently identified illnesses, rapidly altering existing medications to satisfy evolving healthcare requirements and providing timely answers to novel health hazards (Sharma et al. 2022).

In summary, the use of AI in pharmaceutical repurposing represents a major breakthrough in the hunt for novel therapeutic strategies. Enhancing health care and expediting drug discovery are interdependent. Methodical review of approved drugs, biological network analysis, and AI-optimised drug design are required. As technology progresses, how AI uses previously approved drugs to uncover new therapeutic uses will likely have a significant impact on how medicine evolves in the future.

19.3 Clinical Trials and AI's Impact

Artificial intelligence (AI) has brought forth a new era in drug research and significantly impacted clinical trials. This section examines the ways in which artificial intelligence (AI) is profoundly altering the field of clinical trials. It concentrates on four main points: IV.A. better clinical trial design; IV.B. more patient acquisition; IV.C. patient reaction predicting to boost trial success rates; and IV.D. cutting down on the length of time and expense associated with medication development.

19.3.1 Improved Design of Clinical Trials

The effectiveness of clinical trials is largely dependent on their design, and artificial intelligence (AI) has emerged as a potent tool for expediting this process. Machine learning algorithms, trained on extensive datasets encompassing a range of patient groups and treatment results, facilitate the identification of relevant biomarkers and patient classification techniques. This paves the way for more targeted and efficient clinical trial designs (Sarker 2022).

AI also makes it possible for researchers to construct adaptive trials that can dynamically change based on real-time data, giving them the information they need to make informed judgements and modifications as the trial goes on (Malandraki-Miller and Riley 2021). This adaptability ensures that the study keeps up to date with the quickly evolving area of medicine and increases the trial's sensitivity to new discoveries. Last but not least, applying AI to enhance clinical trial design expedites the drug development process and raises the likelihood of discovering effective medicines for specific patient populations (Fitzgerald et al. 2021).

In an effort to improve clinical trial design, AI's sophisticated algorithms consider genetic variants in addition to patient demographics and treatment outcomes. By evaluating genetic data, AI can identify biomarkers that may serve as indicators of treatment response, enabling the development of more precise and individualised solutions (Liu et al. 2020). This personalised medicine approach brings in a new era of targeted therapy and increases the efficacy of clinical trials by tailoring medications to the unique genetic profile of each participant.

AI also makes it easier to incorporate empirical data into clinical trial design. The integration of data from wearables, patient-reported outcomes, and electronic health records facilitates a more comprehensive understanding of a medication's performance in real-world scenarios. This all-encompassing approach ensures that clinical trials are not only rigorous but also reflective of the various healthcare settings in which medications will be utilised, which improves the generalisability of study findings (Majeed and Hwang 2021).

19.3.2 Enhanced Patient Acquisition

Patient recruitment is an essential part of clinical trials, and using traditional methods to identify and enrol suitable patients can be challenging. AI breaks down this barrier with data-driven strategies to enhance patient acquisition (Locke et al. 2021). Natural language processing (NLP) algorithms are able to search through a vast amount of patient data, electronic health records, and medical literature in order to identify potential volunteers who meet the rigorous criteria for a particular study (Venugopal 2019).

Furthermore, AI can assist in selecting suitable sites for clinical trials by looking at demographic and geographic data. This makes the trial's participant pool more diverse and representative, which improves the findings' generalisability. Artificial intelligence (AI) streamlines the patient recruitment process, which not only helps clinical trials start more smoothly but also increases their validity and overall success.

In the area of better patient acquisition, AI locates potential participants and facilitates patient involvement. Chatbots and virtual assistants driven by artificial intelligence (AI) can provide information on trials, assist with inquiries, and answer questions, encouraging a more patient-centric approach. Apart from improving participant understanding and adherence, this engagement raises the public's positive perception of clinical trials (Qi et al. 2021).

Another part of patient recruitment that AI helps with is optimising the diversity of trial participants. Machine learning algorithms can identify variations in participant demographics and suggest targeted remedies to ensure equitable representation. This commitment to diversity is both ethically sound and enhances the trial results' generalisability to different demographic groups (Woodman and Mangoni 2023).

19.3.3 Forecasting Patient Reactions to Boost Trial Success Rates

One of AI's analytical capabilities, predicting patient responses to experimental medicines, is crucial to the success of clinical trials. Utilising machine learning algorithms to analyse historical data and identify patterns in patient responses, it is feasible to develop prediction models that forecast potential responses from different patient cohorts to novel medications (Naveed 2023).

This forecasting power allows researchers to tailor inclusion criteria, stratify patient groups, and adjust dosages based on expected responses. By incorporating AI-driven predictive analytics into trial design, researchers can increase patient safety, decrease the likelihood of unfavourable outcomes, and increase the overall success rates of clinical trials. This proactive approach not only safeguards participants but also upholds the ethical conduct of clinical research (de Thé et al. 2023).

In the area of patient reaction forecasting, AI not only anticipates therapy response but also recognises patients who might not respond to treatment. By finding minute patterns in treatment outcome data that indicate positive reactions, artificial intelligence (AI) can assist in selecting subgroups that stand to gain the most from experimental medicine. Focusing resources on patient populations who are more likely to respond favourably enhances the effectiveness of clinical studies and ultimately raises success rates.

Furthermore, virtual patient cohorts created by AI-powered simulations enable researchers to test various hypotheses and enhance trial designs before beginning actual research. This virtual testing reduces the likelihood of unforeseen problems and contributes to the development of more robust trial designs (Bates et al. 2020). Because AI can anticipate patient reactions, researchers may make data-driven decisions and adopt a more flexible and adaptive approach to the dynamic field of clinical research.

19.3.4 Reducing the Time and Expense of Drug Development

Clinical trials have benefited greatly from artificial intelligence's reduction in the time and expense associated with drug development. Data analysis and protocol development are two of the many process phases that AI streamlines, leading to more efficient and cost-effective trials.

By utilising predictive modelling to assist in identifying potential roadblocks and impediments in the drug development pipeline, artificial intelligence (AI) helps researchers proactively address issues that could delay trials. Real-time patient data monitoring enables prompt identification of unanticipated trends or undesirable events, hence minimising disruptions and facilitating appropriate actions (Majeed and Hwang 2021).

Additionally, AI-driven automation of data extraction and analysis speeds up the process of generating insights from clinical trial data (Benzidia et al. 2021). By reducing the need for manual data processing and speeding up decision-making, this reduces the likelihood of human error. The ultimate result is a more streamlined and cost-effective drug development process that expedites the release of innovative treatments into the healthcare sector.

AI contributes to the improvement of supply chain logistics in the endeavour to reduce the time and expenses related to medication development (Alanazi 2023). Predictive algorithms have the ability to forecast and accelerate trial-related material needs and procurement process optimisation, leading to efficient resource allocation. Effective supply chain management helps to minimise costs and minimise delays during the entire drug development process.

Moreover, AI affects data analysis in a way that extends beyond the trial phase and involves the use of post-trial data. Researchers can continue to obtain meaningful knowledge from completed studies that will assist in the design of post-marketing surveillance and treatment guidelines in the future by utilising AI to assess long-term follow-up data. This ongoing analysis helps to improve treatment regimens on an ongoing basis and ensures that the full value of clinical trial data is realised beyond the first study period (Xu et al. 2021).

In conclusion, artificial intelligence can benefit clinical trials in many ways, such as improved trial design, improved patient acquisition, the capacity to forecast patient reactions, and a reduction in the time and expense involved in drug development. As technology advances, combining AI with clinical trials could hasten medication development and usher in a new era of patient-centric, data-driven healthcare innovation.

19.4 How AI Is Revolutionising the Search for and Development of Drugs

19.4.1 Integration of AI Technologies with Domain Expertise

The pharmaceutical business is experiencing a paradigm shift as a result of the revolutionary advancement of drug discovery and development through the integration of artificial intelligence (AI) technologies with domain expertise. Through this synergy, which combines data-driven insights with specialised knowledge, the computational power of AI and the sophisticated comprehension provided by domain specialists are harnessed to speed and optimise the drug discovery process (Zhu 2020):

- Driven by Data in Drug Discovery: AI in Data Analysis: AI, especially deep learning and machine learning, is very good at analysing large, complicated datasets. In the field of drug discovery, where an abundance of biological and chemical data might be difficult, this is very helpful (Bonate et al. 2023).
- Integration of Domain Expertise: Domain specialists provide their in-depth knowledge of biological systems, pharmacology, and disease causes. This knowledge aids in the development of algorithms and models, guaranteeing the biological relevance of AI-driven investigations (Vamathevan et al. 2019).
- Identification and Validation of the Target: AI Algorithms: By examining biological data, machine learning algorithms are able to forecast possible therapeutic targets. These algorithms find proteins or genes linked to particular diseases by learning from a variety of datasets, such as proteomics and genomics data.
- Expert Interpretation: When it comes to evaluating predictions produced by AI, domain specialists are essential. Their expertise aids in verifying if the anticipated targets coincide with our understanding of the disease's underlying biological mechanisms (Brown et al. 2020).
- Drug Development and Enhancement: Artificial Intelligence in Molecular Design: Deep learning in particular helps in molecular design by forecasting potential interactions between various chemicals and target proteins. This expedites the process of identifying possible therapeutic candidates.
- Domain Expert Input: Taking into account variables such as toxicity, formulation, and bioavailability, domain experts offer valuable insights into the practical elements of medication design. By doing this, candidates produced by AI are guaranteed to meet actual pharmaceutical needs (Saeed and El Naqa 2022).
- Increasing the Effectiveness of Clinical Trials: AI for Patient Stratification: To find subpopulations that might react better to particular therapies, machine learning algorithms examine patient data. This aids in the creation of more focused and effective clinical trials (Al-Taie et al. 2021).
- Clinical Expertise: Domain experts work together with researchers to develop clinical trials, taking into account patient safety, trial endpoints, and ethical issues. Their knowledge guarantees that AI suggestions are consistent with the requirements of clinical research.
- Repurposing Drugs: AI Data Analysis: AI systems are able to examine current datasets in order to find possible new applications for currently approved medications. This technique is referred to as drug repurposing (Ashri 2019).
- Assessment by Domain Experts: Candidates for repurposing are assessed for safety and viability by domain experts. Their knowledge is essential for identifying possible hazards and advantages and for directing the formulation of new drugs.
- AI-Powered Networks and Partnerships:

  - Integrated Platforms: To deliver comprehensive solutions, AI-driven drug discovery platforms combine deep learning, machine learning, and domain-specific expertise.

– Interdisciplinary Collaboration: A comprehensive strategy is fostered through collaboration between domain experts, data scientists, and AI experts. Multidisciplinary teams are better equipped to handle the challenges of drug discovery (Alowais et al. 2023).

The combination of domain knowledge and AI technologies is revolutionising the drug discovery process. AI's computing power and domain specialists' subtle insights work together to make the process more focused and efficient, based on a thorough understanding of biological processes. This cooperative strategy might shorten the timeframes for medication discovery while also raising the chances of successfully introducing new and efficient medicines to the market.

19.4.2 Quicker Progress of the Treatment

Artificial intelligence (AI) is revolutionising the search and development of medications, and one significant result of this is the acceleration of treatment development. The pharmaceutical sector is undergoing a transformation as a result of artificial intelligence (AI) technologies, which have the capacity to handle enormous datasets and produce insights at an unprecedented velocity (Ahmed et al. 2020). Artificial intelligence speeds up the process of identifying possible drug targets in the search for better treatments. In order to forecast targets for additional study, machine learning systems quickly analyse complicated biological data. Drug development's initial phase is considerably shortened by this quick target selection, setting the stage for quicker advancement. Another area seeing rapid progress is AI-driven medication design. This simplifies the design process by using deep learning models to predict how different chemicals would interact with target proteins (Jiang n.d.).

Drug candidates are optimised more quickly, which decreases the need for laborious iterations and raises the possibility of quickly discovering effective treatment solutions. The faster pace of therapeutic development is further aided by AI-facilitated virtual screening and medication repurposing. Large chemical libraries are quickly processed through machine learning methods, which help find possible candidates for new indications. This method not only saves time but also makes the most of already-approved medications by perhaps repurposing them for other medical purposes. Artificial intelligence's analytical powers are applied to clinical trials to improve trial design. Machine learning finds distinct patient subpopulations for targeted interventions and optimises trial parameters by analysing a variety of datasets. Clinical trials are conducted more efficiently with this strategy, which facilitates faster data production and decision-making (Serov and Vinogradov 2022).

AI helps uncover biomarkers rapidly by analysing a variety of omics data, which is a critical step in customising medicines. Quickly identifying the biomarkers linked to diseases opens the door to more individualised and efficient treatment plans. The precision medicine age is one where this acceleration is very significant. Early disease identification and intervention are aided by the inclusion of AI-powered diagnostics. The prompt modification of treatment procedures, which guarantees timely interventions and may enhance patient outcomes, is made possible by real-time patient data monitoring. Artificial intelligence (AI) systems' capacity for continuous learning improves treatment plans by promoting flexible methods derived from continuing data analysis. AI additionally makes it easier for researchers, doctors, and data scientists to collaborate across disciplines. With the help of this cooperative ecosystem, data sharing is accelerated, resulting in more effective research methods and a group effort to comprehend diseases and create focused remedies more quickly (Jiang n.d.).

Conclusively, the transformative influence of artificial intelligence on drug discovery and development results in a more rapid progression from preliminary investigations to the delivery of efficacious therapies. AI substantially accelerates the development of new treatments by speeding up target identification, drug design, clinical trials, and biomarker discovery. The healthcare industry could be drastically altered by this paradigm shift, which promises to give more individualised and quicker answers (Mbatha et al. 2023).

19.4.3 Better Results for Patients

A new age of better outcomes for patients has been brought about by artificial intelligence's (AI) revolutionary impact on drug discovery and development. Artificial intelligence is making therapies more individualised and effective by processing enormous volumes of data, spotting trends, and optimising different phases of drug discovery and development. AI is helping patients have better outcomes in the following ways (Lipinski et al. 2019):

- Strategies for Targeted Treatment: Machine learning algorithms are used in target identification to more accurately anticipate possible drug targets by analysing a variety of biological data.
- Results That Are Focused on the Patient: Targeted identification makes it possible to create medications that are unique to certain molecular pathways, improving the chances of success and reducing side effects.
- Design and Optimisation of Drugs:
  - Deep Learning Models: AI-powered models calculate the interactions between various chemicals and target proteins, accelerating the development of new drugs.
  - Optimised Formulations: Patients can receive novel therapies sooner thanks to shorter development times caused by speedier drug design and optimisation (Jyothi n.d.).
- Enhanced Customisation Using Biomarkers:
  - AI in Biomarker Discovery: Machine learning examines omics data to find disease-related biomarkers.

  - Tailored Therapies: Quick biomarker identification enables more individualised therapy plans, guaranteeing that therapies are in line with unique patient traits and illness profiles (Schwager et al. 2021).

- Effective Clinical Trial Architecture:

AI helps to improve the design of clinical trials by identifying patient subpopulations that are more likely to respond favourably.

- Faster Clinical Validation: Effective trial designs expedite the validation of drug candidates, enabling patients to receive potentially beneficial treatments more quickly.

- Early Identification and Management of Diseases:
  - AI-Powered Diagnostics: To diagnose diseases early, AI apps examine patient data.
  - Better Results: By addressing diseases at earlier, more manageable stages, early diagnosis allows for prompt therapies that may improve patient outcomes.

- Adaptive Approaches to Therapy:

AI systems acquire knowledge through continuous learning from real-world data and ongoing clinical trials. By optimising therapies based on the most recent insights and patient responses, this ongoing learning enables adaptable treatment techniques (Greene et al. 2010).

- Reduced Adverse Reactions and Safety Issues:
  - Real-Time Monitoring: AI systems keep an eye on patient data in real time, which makes it easier to quickly identify negative consequences.
  - Patient Safety: Early detection of safety issues guarantees that patients receive care that has the fewest possible hazards, improving overall safety and well-being (Choudhury and Asan 2020).

19.4.4 Dealing with Difficult Medical Problems

When it comes to solving challenging medical issues, artificial intelligence (AI) has a particularly big impact on medication discovery and development. These difficulties are frequently caused by the complexity, scarcity, or resistance to traditional therapies connected to specific illnesses. AI's revolutionary potential helps create new strategies and accelerate resolutions for these difficult medical problems. AI analytics thrives in the field of complicated diseases by interpreting complex biological data. This makes it possible to identify particular molecular targets, which is an essential first step in creating more accurate and potent treatments that are suited to the complexity of these disorders (Yang et al. 2019). AI is proving to be a valuable tool in drug creation, especially for uncommon disorders with few available treatments. Accelerating drug design procedures satisfies a pressing need and offers possible benefits. The practice of drug repurposing, which looks for novel answers to challenging medical issues, is one notable example of AI's impact. Artificial intelligence (AI) provides novel approaches for treatment development by evaluating pre-existing datasets to find promising drug candidates for novel therapeutic indications. This strategy offers innovative applications for difficult conditions while optimising the effectiveness of currently available drugs (Fröhlich et al. 2018).

The use of AI in patient stratification for complex diseases is critical. Personalised therapies are made possible by the use of machine learning algorithms to identify specific populations that respond differently to different treatments. This method ensures individualised therapy based on the unique qualities of each patient, acknowledging the diversity of patients dealing with challenging medical issues (Pun et al. 2023).

AI-powered analysis of varied omics data greatly aids in the development of biomarkers, an essential component of early diagnosis and focused therapies. The ability to quickly identify biomarkers linked to difficult medical diseases improves diagnostic capacity and aids in the creation of focused therapies. Another area where AI helps overcome obstacles related to illnesses that are difficult to cure is clinical trial optimisation. AI increases the likelihood of successfully discovering cures for diseases that typically present formidable obstacles by optimising trial designs. AI systems, which learn from actual patient data, are inherently capable of continuous learning and adaptability. Because of its flexibility, treatment plans can be continuously improved, guaranteeing that methods change in response to new information and patient feedback. AI-powered surveillance enables real-time monitoring, which is essential in cases of extreme or quickly worsening situations (Kalid et al. 2018).

Real-time monitoring of patient data makes timely interventions possible, which is essential when delays might have a major negative influence on patient outcomes. AI enables healthcare practitioners to make well-informed decisions by providing comprehensive insights through the holistic analysis of varied information. By taking into account a variety of factors that can impact complex medical situations, this integrated approach helps to provide more efficient and individualised patient treatment (Sebastian and Peter 2022).

To sum up, artificial intelligence's revolutionary impact on medication research and discovery presents fresh promise for solving challenging medical issues. AI stimulates creativity, expedites drug development, and makes more individualised and efficient therapies possible for both uncommon and complicated diseases. AI technologies have the ability to overcome medical obstacles that were previously intractable, which could lead to better outcomes for those with challenging health issues [80].

19.5 Conclusion

19.5.1 Summary of AI's Enormous Potential for Medication Development and Discovery

An overview of AI's immense potential for drug discovery and development of an overview of artificial intelligence's (AI) immense potential for drug research and discovery, in conclusion, provides a revolutionary vision of the future of health care. AI has become a transformative force that surpasses conventional methods and greatly quickens the rate of medication development. With its potentially revolutionary potential, it holds great promise for revolutionising our understanding, discovery, and delivery of drugs. A paradigm shift in the field of medicine development has been brought about by AI's ability to quickly analyse large and complicated datasets, forecast prospective therapeutic targets, and optimise drug design processes. Combining machine learning with deep learning algorithms makes it easier to identify medication candidates more accurately and efficiently, which accelerates the process of turning scientific discoveries into practical therapies. Artificial intelligence has made major advances in drug repurposing by sifting through massive databases and finding previously undiscovered relationships between approved medications and cutting-edge therapeutic indications. This offers a long-term strategy for utilising current drugs to address novel medical problems in addition to shortening the development timeframe. As AI makes it possible to identify biomarkers and classify patient populations, patient-centric care becomes increasingly important. With this individualised approach, treatments will be catered to the unique qualities of each patient, maximising effectiveness and reducing side effects. AI's influence on clinical trial design also improves efficiency, ensuring that patients receive promising treatment candidates sooner. Artificial intelligence systems exhibit perpetual learning and adaptability that facilitate instantaneous monitoring and intervention. Not only does this flexibility guarantee patient security, but it also makes continuous therapy advancements possible.

19.5.2 Opening the Door to a New Age of Individualised Care

Artificial intelligence (AI) represents a significant turning point in the new medication discovery process, ushering in a period of extraordinary progress in the healthcare system and ushering in a phase of tailored treatment. Drug development is changing, and an era when therapies are customised to meet the specific needs of each patient is beginning, thanks to the revolutionary junction of cutting-edge technology and medical science. With the introduction of AI, the one-size-fits-all methodology of drug discovery has given way to a more individualised strategy, marking a paradigm change. Artificial intelligence's capacity to evaluate large and complex datasets makes it possible to gain a better knowledge of the unique traits of each patient, which opens the door to tailored treatments based on genetic, biochemical, and lifestyle factors. The foundation of AI in health care is made up of machine learning and deep learning algorithms, which are essential for understanding the intricacies of illnesses. These algorithms are capable of quickly sorting through massive databases, identifying possible therapeutic targets and improving the structure of new molecules. Drug discovery proceeds at a faster rate as a result, advancing the development of effective medications that are also adapted to the complex molecular underpinnings of various diseases.

Ultimately, the use of AI in the drug discovery process creates new opportunities for the healthcare industry. That is a significant turning point that ushers in a new era of personalised health care, when patients receive therapies that are not only more customised to their specific needs but also more successful. With AI technology developing further, there is a chance that personalised medicine will lead to groundbreaking discoveries that could revolutionise the global healthcare system.

References

Abbasi K, Razzaghi P, Poso A, Ghanbari-Ara S, Masoudi-Nejad A (2021) Deep learning in drug target interaction prediction: current and future perspectives. Curr Med Chem 28(11):2100–2113

Ahmed Z, Mohamed K, Zeeshan S, Dong X (2020) Artificial intelligence with multi-functional machine learning platform development for better healthcare and precision medicine. Database 2020:baaa010

Ahmed F, Soomro AM, Salih AR, Samantasinghar A, Asif A, Kang IS, Choi KH (2022) A comprehensive review of artificial intelligence and network based approaches to drug repurposing in Covid-19. Biomed Pharmacother 153:113350

Alanazi A (2023) Clinicians' views on using artificial intelligence in healthcare: opportunities, challenges, and beyond. Cureus 15(9):e45255

Alowais SA, Alghamdi SS, Alsuhebany N, Alqahtani T, Alshaya AI, Almohareb SN, Aldairem A, Alrashed M, Bin Saleh K, Badreldin HA, Al Yami MS (2023) Revolutionizing healthcare: the role of artificial intelligence in clinical practice. BMC Med Educ 23(1):689

Al-Taie Z, Liu D, Mitchem JB, Papageorgiou C, Kaifi JT, Warren WC, Shyu CR (2021) Explainable artificial intelligence in high-throughput drug repositioning for subgroup stratifications with interventionable potential. J Biomed Inform 118:103792

Ashri R (2019) The AI-powered workplace: how artificial intelligence, data, and messaging platforms are defining the future of work. Apress

Bates DW, Auerbach A, Schulam P, Wright A, Saria S (2020) Reporting and implementing interventions involving machine learning and artificial intelligence. Ann Intern Med 172(11_Supplement):S137–S144

Baviskar D, Ahirrao S, Potdar V, Kotecha K (2021) Efficient automated processing of the unstructured documents using artificial intelligence: a systematic literature review and future directions. IEEE Access 9:72894–72936

Benzidia S, Makaoui N, Bentahar O (2021) The impact of big data analytics and artificial intelligence on green supply chain process integration and hospital environmental performance. Technol Forecast Soc Chang 165:120557

Bobo D, Robinson KJ, Islam J, Thurecht KJ, Corrie SR (2016) Nanoparticle-based medicines: a review of FDA-approved materials and clinical trials to date. Pharm Res 33:2373–2387
Bonate PL, Barrett JS, Ait-Oudhia S, Brundage R, Corrigan B, Duffull S, Gastonguay M, Karlsson MO, Kijima S, Krause A, Lovern M (2023) Training the next generation of pharmacometric modelers: a multisector perspective. J Pharmacokinet Pharmacodyn 51(1):5–31
Boniolo F, Dorigatti E, Ohnmacht AJ, Saur D, Schubert B, Menden MP (2021) Artificial intelligence in early drug discovery enabling precision medicine. Expert Opin Drug Discov 16(9):991–1007
Brown N, Ertl P, Lewis R, Luksch T, Reker D, Schneider N (2020) Artificial intelligence in chemistry and drug design. J Comput Aided Mol Des 34:709–715
Choudhury A, Asan O (2020) Role of artificial intelligence in patient safety outcomes: systematic literature review. JMIR Med Inform 8(7):e18599
Dara S, Dhamercherla S, Jadav SS, Babu CM, Ahsan MJ (2022) Machine learning in drug discovery: a review. Artif Intell Rev 55(3):1947–1999
Davahli MR, Karwowski W, Fiok K, Wan T, Parsaei HR (2021) Controlling safety of artificial intelligence-based systems in healthcare. Symmetry 13(1):102
de Thé FX, Baudier C, Pereira RA, Lefebvre C, Moingeon P (2023) Patrimony Working Group. Transforming drug discovery with a high-throughput AI-powered platform: a 5-year experience with patrimony. Drug Discov Today 28(11):103772
Demner-Fushman D, Elhadad N, Friedman C (2021) Natural language processing for health-related texts. In: Biomedical informatics: computer applications in health care and biomedicine. Springer International Publishing, Cham, pp 241–272
Evangelista PN (2020) Artificial intelligence in fashion: how consumers and the fashion system are being impacted by AI-powered technologies
Fischer T, Gazzola S, Riedl R (2019) Approaching target selectivity by de novo drug design. Expert Opin Drug Discov 14(8):791–803
Fitzgerald J, Higgins D, Vargas CM, Watson W, Mooney C, Rahman A, Aspell N, Connolly A, Gonzalez CA, Gallagher W (2021) Future of biomarker evaluation in the realm of artificial intelligence algorithms: application in improved therapeutic stratification of patients with breast and prostate cancer. J Clin Pathol 74(7):429–434
Fröhlich H, Balling R, Beerenwinkel N, Kohlbacher O, Kumar S, Lengauer T, Maathuis MH, Moreau Y, Murphy SA, Przytycka TM, Rebhan M (2018) From hype to reality: data science enabling personalized medicine. BMC Med 16(1):1–5
Gennatas ED, Chen JH (2021) Artificial intelligence in medicine: past, present, and future. In: Artificial intelligence in medicine. Academic Press, pp 3–18
Górriz JM, Ramírez J, Ortíz A, Martinez-Murcia FJ, Segovia F, Suckling J, Leming M, Zhang YD, Álvarez-Sánchez JR, Bologna G, Bonomini P (2020) Artificial intelligence within the interplay between natural and artificial computation: advances in data science, trends and applications. Neurocomputing 410:237–270
Greene SK, Kulldorff M, Lewis EM, Li R, Yin R, Weintraub ES, Fireman BH, Lieu TA, Nordin JD, Glanz JM, Baxter R (2010) Near real-time surveillance for influenza vaccine safety: proof-of-concept in the Vaccine Safety Datalink Project. Am J Epidemiol 171(2):177–188
Gupta R, Srivastava D, Sahu M, Tiwari S, Ambasta RK, Kumar P (2021) Artificial intelligence to deep learning: machine intelligence approach for drug discovery. Mol Divers 25:1315–1360
Harrer S, Shah P, Antony B, Hu J (2019) Artificial intelligence for clinical trial design. Trends Pharmacol Sci 40(8):577–591
Husnain A, Rasool S, Saeed A, Hussain HK (2023) Revolutionizing pharmaceutical research: harnessing machine learning for a paradigm shift in drug discovery. Int J Multidiscip Sci Arts 2(2):149–157
Hussain W, Rasool N, Khan YD (2021) Insights into machine learning-based approaches for virtual screening in drug discovery: existing strategies and streamlining through FP-CADD. Curr Drug Discov Technol 18(4):463–472

Jiang J. (n.d.) Overcoming challenges of solid dosage formulation development by using emerging technologies (Doctoral dissertation)
Jiménez-Luna J, Grisoni F, Weskamp N, Schneider G (2021) Artificial intelligence in drug discovery: recent advances and future perspectives. Expert Opin Drug Discov 16(9):949–959
Jyothi NM (n.d.) AI-enabled genomic biomarkers: the future of pharmaceutical industry and personalized medicine
Kalid N, Zaidan AA, Zaidan BB, Salman OH, Hashim M, Muzammil HJ (2018) Based real time remote health monitoring systems: a review on patients prioritization and related "big data" using body sensors information and communication technology. J Med Syst 42:1–30
Lipinski CF, Maltarollo VG, Oliveira PR, Da Silva AB, Honorio KM (2019) Advances and perspectives in applying deep learning for drug design and discovery. Front Robot AI 6:108
Liu X, Rivera SC, Moher D, Calvert MJ, Denniston AK, Ashrafian H, Beam AL, Chan AW, Collins GS, Deeks AD, ElZarrad MK (2020) Reporting guidelines for clinical trial reports for interventions involving artificial intelligence: the CONSORT-AI extension. Lancet Digit Health 2(10):e537–e548
Liu Z, Roberts RA, Lal-Nag M, Chen X, Huang R, Tong W (2021) AI-based language models powering drug discovery and development. Drug Discov Today 26(11):2593–2607
Locke S, Bashall A, Al-Adely S, Moore J, Wilson A, Kitchen GB (2021) Natural language processing in medicine: a review. Trends Anaesth Crit Care 38:4–9
Mahmud M, Kaiser MS, McGinnity TM, Hussain A (2021) Deep learning in mining biological data. Cogn Comput 13:1–33
Majeed A, Hwang SO (2021) Data-driven analytics leveraging artificial intelligence in the era of COVID-19: an insightful review of recent developments. Symmetry 14(1):16
Mak KK, Pichika MR (2019) Artificial intelligence in drug development: present status and future prospects. Drug Discov Today 24(3):773–780
Malandraki-Miller S, Riley PR (2021) Use of artificial intelligence to enhance phenotypic drug discovery. Drug Discov Today 26(4):887–901
Mbatha SZ, Mulaudzi TV, Mbita Z, Adeola HA, Batra J, Blenman KR, Dlamini Z (2023) Artificial intelligence–enhanced drug discovery and the achievement of next-generation human-centered health system. In: Society 5.0 and next generation healthcare: patient-focused and technology-assisted precision therapies. Springer Nature Switzerland, Cham, pp 155–177
Naveed A (2023) Transforming clinical trials with informatics and AI/ML: a data-driven approach. Int J Comput Sci Technol 7(1):485–503
Öztürk H, Özgür A, Schwaller P, Laino T, Ozkirimli E (2020) Exploring chemical space using natural language processing methodologies for drug discovery. Drug Discov Today 25(4):689–705
Paul D, Sanap G, Shenoy S, Kalyane D, Kalia K, Tekade RK (2021) Artificial intelligence in drug discovery and development. Drug Discov Today 26(1):80
Pun FW, Ozerov IV, Zhavoronkov A (2023) AI-powered therapeutic target discovery. Trends Pharmacol Sci 44:561
Pushpakom S, Iorio F, Eyers PA, Escott KJ, Hopper S, Wells A, Doig A, Guilliams T, Latimer J, McNamee C, Norris A (2019) Drug repurposing: progress, challenges and recommendations. Nat Rev Drug Discov 18(1):41–58
Qi M, Cahan O, Foreman MA, Gruen DM, Das AK, Bennett KP (2021) Quantifying representativeness in randomized clinical trials using machine learning fairness metrics. JAMIA Open 4(3):ooab077
Russo G, Reche P, Pennisi M, Pappalardo F (2020) The combination of artificial intelligence and systems biology for intelligent vaccine design. Expert Opin Drug Discov 15(11):1267–1281
Saeed H, El Naqa I (2022) Artificial intelligence in clinical trials. In: Machine and deep learning in oncology, medical physics and radiology. Springer International Publishing, Cham, pp 453–501
Sarker IH (2022) Ai-based modeling: techniques, applications and research issues towards automation, intelligent and smart systems. SN Comput Sci 3(2):158

Schork NJ (2019) Artificial intelligence and personalized medicine. In: Precision medicine in Cancer therapy, pp 265–283

Schwager E, Jansson K, Rahman A, Schiffer S, Chang Y, Boverman G, Gross B, Xu-Wilson M, Boehme P, Truebel H, Frassica JJ (2021) Utilizing machine learning to improve clinical trial design for acute respiratory distress syndrome. NPJ Digit Med 4(1):133

Sebastian AM, Peter D (2022) Artificial intelligence in cancer research: trends, challenges and future directions. Life 12(12):1991

Selvaraj C, Chandra I, Singh SK (2021) Artificial intelligence and machine learning approaches for drug design: challenges and opportunities for the pharmaceutical industries. Mol Divers 26:1893–1913

Sen P, Lamichhane S, Mathema VB, McGlinchey A, Dickens AM, Khoomrung S, Orešič M (2021) Deep learning meets metabolomics: a methodological perspective. Brief Bioinform 22(2):1531–1542

Serov N, Vinogradov V (2022) Artificial intelligence to bring nanomedicine to life. Adv Drug Deliv Rev 184:114194

Sharma A, Virmani T, Pathak V, Sharma A, Pathak K, Kumar G, Pathak D (2022) Artificial intelligence-based data-driven strategy to accelerate research, development, and clinical trials of COVID vaccine. Biomed Res Int 2022:7205241

Sidak D, Schwarzerová J, Weckwerth W, Waldherr S (2022) Interpretable machine learning methods for predictions in systems biology from omics data. Front Mol Biosci 9:926623

Sun J, Dong QX, Wang SW, Zheng YB, Liu XX, Lu TS, Yuan K, Shi J, Hu B, Lu L, Han Y (2023) Artificial intelligence in psychiatry research, diagnosis, and therapy. Asian J Psychiatr 87:103705

Tiwari A, Singh S (2022) Computational approaches in drug designing. In: Bioinformatics. Academic Press, pp 207–217

Tiwari PC, Pal R, Chaudhary MJ, Nath R (2023) Artificial intelligence revolutionizing drug development: exploring opportunities and challenges. Drug Dev Res 84:1652

Torfi A, Shirvani RA, Keneshloo Y, Tavaf N, Fox EA (2020) Natural language processing advancements by deep learning: a survey. arXiv preprint arXiv 2003.01200

Tripathi L, Kumar P, Swain K, Pattnaik S (2022) Drug repurposing based on machine learning. In: Drug design using machine learning. Wiley, pp 143–164

Vamathevan J, Clark D, Czodrowski P, Dunham I, Ferran E, Lee G, Li B, Madabhushi A, Shah P, Spitzer M, Zhao S (2019) Applications of machine learning in drug discovery and development. Nat Rev Drug Discov 18(6):463–477

Vatansever S, Schlessinger A, Wacker D, Kaniskan HÜ, Jin J, Zhou MM, Zhang B (2021) Artificial intelligence and machine learning-aided drug discovery in central nervous system diseases: state-of-the-arts and future directions. Med Res Rev 41(3):1427–1473

Venugopal M (2019) Evolution of digital technologies and use of virtual assistants in drug development. In: Intelligent systems for healthcare management and delivery. IGI Global, pp 1–20

Vo AH, Van Vleet TR, Gupta RR, Liguori MJ, Rao MS (2019) An overview of machine learning and big data for drug toxicity evaluation. Chem Res Toxicol 33(1):20–37

Wang YH, Lin GY (2023) Exploring AI-healthcare innovation: natural language processing-based patents analysis for technology-driven roadmapping. Kybernetes 52(4):1173–1189

Woodman RJ, Mangoni AA (2023) A comprehensive review of machine learning algorithms and their application in geriatric medicine: present and future. Aging Clin Exp Res 35(11):2363–2397

Xu Y, Liu X, Cao X, Huang C, Liu E, Qian S, Liu X, Wu Y, Dong F, Qiu CW, Qiu J (2021) Artificial intelligence: a powerful paradigm for scientific research. Innovations 2(4):100179

Yang X, Wang Y, Byrne R, Schneider G, Yang S (2019) Concepts of artificial intelligence for computer-assisted drug discovery. Chem Rev 119(18):10520–10594

Zhavoronkov A, Mamoshina P, Vanhaelen Q, Scheibye-Knudsen M, Moskalev A, Aliper A (2019) Artificial intelligence for aging and longevity research: recent advances and perspectives. Ageing Res Rev 49:49–66
Zhou Y, Wang F, Tang J, Nussinov R, Cheng F (2020) Artificial intelligence in COVID-19 drug repurposing. Lancet Digit Health 2(12):e667–e676
Zhu H (2020) Big data and artificial intelligence modeling for drug discovery. Annu Rev Pharmacol Toxicol 60:573–589

Chapter 20
Phytopharmaceutical Biotechnology: Integration of Botany, Pharmacology and Plant Biotechnology to Deliver the Best Therapeutic Potential of Herbs

Suma Krishnaswamy

Abstract It is common knowledge that plants are a source of valuable preventive, curative and therapeutic substances. This knowledge has been handed down by generations of herbalists, ethnomedicine practitioners, healers and indigenous health ecosystems verbally or with practical demonstrations or occasionally documented. Traditional medicine, using plants, plant parts or by-products derived through some form of processing of plants, has evolved across civilisations and cultures to become an integral part of the human lifestyle. In recent years, the urgent need for standardising herbal medicine through scientific research and the development of modern phytopharmaceuticals have been recognised. Regulatory frameworks have been defined, and compliance is not only necessary but also mandatory. "Phytopharmaceutical drug" includes purified and standard fractions with a defined minimum of four qualitatively and quantitatively assessed bioactive or phytochemical compounds for internal or external use of human beings or animals for diagnosis, treatment, mitigation or prevention of any disease or disorder but excludes parenteral. While traditional approximations and ratios were historically used for plant-based medicines, modern science and regulatory practices aim to provide a more rigorous and evidence-based approach to ensure the consistency, reproducibility, safety and efficacy of these remedies. This chapter, through specific case studies, attempts to highlight the role of plant sciences and plant biotechnology in the accurate identification of plant species, recognising the potential therapeutic benefit, isolation of bioactives, validation of processes and intermediates, best practices to standardise the safety, drug delivery mechanisms, dosage forms and efficacy assessment.

Keywords Phytopharmaceuticals · Plant stem cells · Plant biotechnology · Green synthesis · Bioactive efficacy

S. Krishnaswamy (✉)
Phytazen Biosciences Private Limited, Bangalore, Karnataka, India
URL: https://phytazenbiosciences.com/

S. Bose et al. (eds.), *Concepts in Pharmaceutical Biotechnology and Drug Development*, Interdisciplinary Biotechnological Advances,
https://doi.org/10.1007/978-981-97-1148-2_20

20.1 Introduction

Rapid technological progress, while proving invaluable in easing several lifestyle processes, poses a huge challenge to populations remaining healthy, active, agile and fit. Extrinsic factors such as stress, lack of exercise, unhealthy imbalanced diet, irregular sleeping and feeding patterns and substance abuse interact with intrinsic factors such as physiology, genetics and environmental pollution, leading to lifestyle diseases or noncommunicable diseases (NCDs). WHO report (Noncommunicable diseases (who.int)) indicates that globally about 70% of the fatalities in 2019 were due to leading non-infectious diseases such as cardiovascular diseases, cancers, chronic respiratory diseases and diabetes mellitus. Nontransmissible diseases are responsible for nearly 74% (almost 41 million people) of all deaths globally every year, especially in low- and middle-income countries. Cardiovascular diseases cause the most deaths (about 17.9 million people annually), followed by cancers (9.3 million), chronic respiratory diseases (4.1 million) and diabetes (2.0 million including diabetes-induced kidney disease deaths or CKD). These statistics will indicate the limited efficacy of the existing pharmaceutical products and provide an opportunity for natural/botanical drug development to fill the lacuna. An ideal health solution may be preventive health care by consuming nutritionally and therapeutically rich natural products using the concept of "Food as medicine" (Downer et al. 2020). "Medically tailored meals" customised by a nutritionist, while reducing hospital visits and medical expenditure, have also been shown to successfully treat patients with NCDs such as diabetes, heart failure and chronic liver disease.

According to a global market research organisation, the growth of the global herbal medicine market is estimated to reach almost $347.50 billion by 2029 (https://www.fortunebusinessinsights.com/herbal-medicine-market-106320). In 2021, the European market value was around USD 69.20 billion, while the US market stood at around $23 billion. The use of herbal products has been increasing in popular demand, especially in developed countries where, according to WHO reports, almost 50–60% of the population use plant-based remedies because of their immunomodulatory and long-term benefits, lacking synthetic drugs. Nutritional health supplements such as plant-based proteins and "healthy ageing" formulas, designed to reduce oxidative damage, improve mitochondrial efficiency, improve inflammation control, improve collagen levels and so on, are popular preventive healthcare options. The second highest market share is held by herbal derivatives or recipes with the twin promise of better quality of life and healthy ageing especially those that impact visible age indicators in hair, face and skin.

Naturally sourced and formulated products are perceived to be safe generally compared to synthetic drugs, with fewer and short-term side effects (Veeresham 2012). There are growing awareness and acceptance of all products that are natural globally. While traditional herbal medicine is reputed to provide valuable therapeutic benefits, it is essential to acknowledge that the lack of precise dosing in some cases poses risks, including the potential for underdosing or overdosing (Ekor 2014). A literature survey of herbal components records adverse reactions due to mild to severe toxicities, sometimes leading to life-threatening conditions. For instance, *Ephedra sinica*, belonging to the family Ephedraceae, is a source of many sympathomimetic bioactive alkaloids including the ephedrine, pseudoephedrine, methyl ephedrine and norephedrine, which collectively enhance mental acuity, improve sexual performances, increase circulation and decrease weight due to elevated sympathetic nervous system activity and thermogenesis. However, *Ephedra's* value as an effective phytotherapeutic is diminished due to serious adverse responses such as hepatic injury and cardiovascular responses, impacting vital organs such as the heart and brain (National Institute of Diabetes and Digestive and Kidney Diseases 2012).

While the herbal pharmaceutical market is expanding rapidly and appears poised for exponential growth in the coming years, there are serious concerns over potentially harmful constituents in herbal products and their efficacy, consistency and safety. There is a dire need for the standardisation of therapeutic or curative herbal substances through scientific research, critical inquiry and regulatory harmonisation (Mandal and Mandal 2011). Recognition of this lacuna marked the genesis of a new class of plant products called phytopharmaceuticals.

The Greek terms "phyton" for plant and "pharmakon" for medicine were combined to arrive at the term "phytopharmaceutical" to describe plant-derived medicines (Bhusnure et al. 2019). A "phytopharmaceutical drug" emphasises the purification and standardisation of fractions from crude plant extracts, with a defined minimum of four qualitatively and quantitatively assessed bioactive or phytochemical compounds intended for internal or external use of human beings or animals for diagnosis, treatment, mitigation or prevention of any disease or disorder but does not include administration by parenteral route. While traditional approximations and ratios were historically used for plant-based medicines, modern science and regulatory practices aim to provide a more rigorous and evidence-based approach to ensure the consistency, reproducibility, safety and potency of these remedies. Good manufacturing practices to ensure high levels of quality (GMP), clinically proven activity, optimal dosage and safeness of standardised or quantified medicinal plant extracts are key factors that determine a good phytopharmaceutical product.

Regulatory frameworks have been precise and well-defined, and compliance is not only necessary but mandatory for safe and successful phytotherapy. The new regulation for scientific evaluation and data generation for phytopharmaceuticals in India conforms to regulations in the USA, China, Europe and other countries (Narayana 2013), ensuring that the standards are of international quality, opening up the entire spectrum of phytoactive market, spurring economic growth.

20.2 Phytopharmaceutical Drug Development Trajectory

The last decade has witnessed the launch of many phytopharmaceutical drugs backed by exhaustive evidence-based studies, clinical safety and efficacy data. The genesis of a new phytomedicine often begins with a market need and a research concept or idea. An exhaustive literature survey is taken up to augment the existing knowledge and lay firm foundations for the rationale and methodology of the new drug formulation, form of presentation, dose and dosage. The parameters of quality control and quality assurance are drafted, experimented and documented for future compliance. New drug/formulation developed through these steps are then subjected to intense phytochemical, biological and physiological testing and protocol-based preclinical studies. Phytomedicinal formulations emerging successfully out of the preclinical research are then subjected to well-structured human clinical trial protocols and CTRI-registered (Clinical Trial Registry of India) clinical studies, which are conducted on the basis of approvals from ethics committee and regulatory authorities. Once the clinical study presents positive results in tandem with the claims made and the necessary licences are obtained, the new phytopharmaceutical drug or formulation may be launched in the market to cater to the inadequately addressed medical needs, unfulfilled by standard mainstream medicines (Fig. 20.1).

A brief indicative list highlighting some of the standardised drugs and the wide range of phytocompounds present is given in Table 20.1.

Phytopharmaceuticals can be used for various purposes including:

- **Preventive Health Care:** Many phytopharmaceuticals are used as nutritional supplements to promote general health and well-being. They may contain antioxidants, vitamins, proteins and minerals that help strengthen the immune system and protect against various diseases.

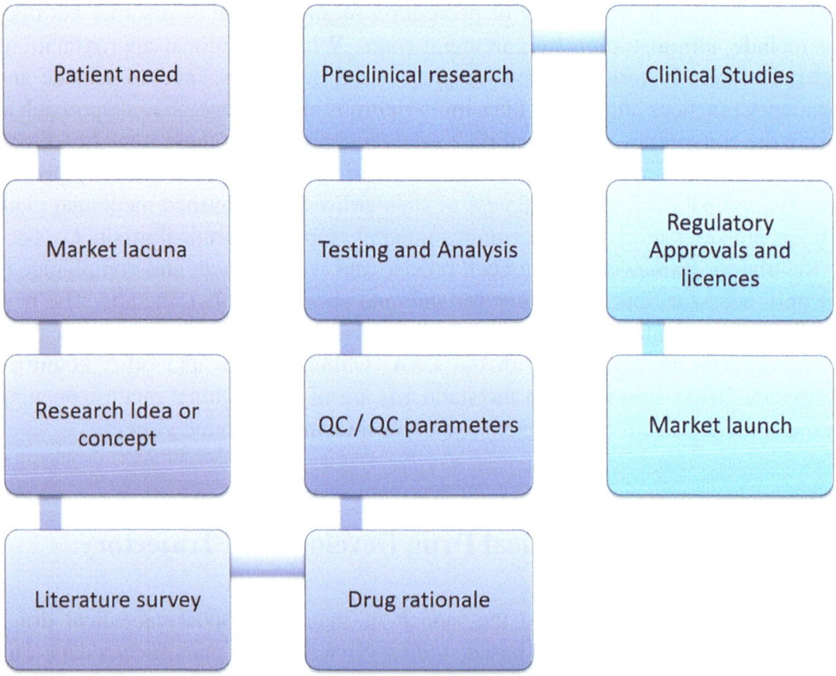

Fig. 20.1 Phytopharmaceutical drug discovery journey: commencement to conclusion

Table 20.1 Phytocompounds derived from plant species: a few examples

S/N	Plant species	Class of compound	Phytopharmaceutical	Use
1	*Andrographis paniculata*	Diterpenoid lactones	Andrographolide (Okhuarobo et al. 2014)	Anti-inflammatory, antimicrobial, Hepatoprotective
2	*Artemisia annua*	Sesquiterpene lactone	Arteether (Hien 1993)	Malaria
3	*Atropa belladonna*	Alkaloids	Atropine, scopolamine, tiotropium (Passos and Mironidou-Tzouveleki 2016)	Chronic obstructive pulmonary disease (COPD)
4	*Bacopa monnieri*	Saponins	Asiatic acid, brahmic acid (Banerjee et al. 2021)	Neurodegeneration
5	*Callistemon citrinus*	Triketones	Nitisinone (Mitchell et al. 2001)	Antityrosinaemia
6	*Camptotheca acuminata*	Quinoline alkaloids	Camptothecin (Martino et al. 2017)	Cancer
7	*Capsicum annuum*	Phenolic compound	Capsaicin (Fattori et al. 2016)	Analgesic
8	*Centella asiatica*	Pentacyclic triterpenoid Saponins	Asiaticosides (Diniz et al. 2023)	Neuromodulatory
9	*Cinchona*	Alkaloids	Quinine (Raza et al. 2021)	Malaria
10	*Ephedra sinica*	Amino alkaloid	Ephedrine (Tang et al. 2023)	Bronchodilator, asthma, COPD
11	*Galanthus nivalis*	Alkaloids	Galantamine (Lilienfeld 2002; Olazarán and García 2002)	Alzheimer's disease
12	*Gymnema sylvestre*	Triterpene saponins	Gymnemic acid (Liu et al. 1992)	Diabetes
13	*Moringa oleifera*	Sterol glycoside	β-Sitosterol (Park et al. 2011)	Antihyperglycaemic
14	*Nothapodytes foetida*	Alkaloids	Camptothecin (Namdeo and Sharma 2012)	Tumours
15	*Papaver somniferum*	Morphinan alkaloids	Apomorphine (Butnariu et al. 2022)	Parkinson's disease
16	*Podophyllum*	Lignans	Podophyllotoxin (Anand et al. 2022)	Hodgkin's lymphoma, monocytoid leukaemia
17	*Salacia chinensis*	Thiosugar sulfonium	Salacinol (Matsuda et al. 2005)	Diabetes
18	*Taxus baccata*	Tetracyclic diterpene	Taxol/paclitaxel (Gallego-Jara et al. 2020)	Cancer

- **Treatment of Specific Ailments:** Some phytopharmaceuticals are used to treat specific medical conditions. For example, the folklore use of willow bark extract, which contains salicylic acid, led to the development of aspirin, a widely used pain reliever and anti-inflammatory drug.

- **Complementary and Alternative Medicine (CAM):** Phytopharmaceuticals are used as alternative or complementary treatments alongside conventional medicine across the world in increasing proportions. They are perceived as relatively safe with fewer side effects as compared to synthetic drugs.
- **Traditional Medicine:** In traditional medicine systems such as Ayurveda, TCM and traditional African medicine, plants and their extracts are integral components of treatment regimens. They are used to balance the body's energy, treat diseases and maintain overall health.
- **Pharmaceutical Research:** Modern pharmaceutical research has also explored the potential of phytopharmaceuticals. Many drugs used today, such as levodopa (*Mucuna pruriens*), paclitaxel (derived from the Pacific yew tree *Taxus baccata*) and quinine (from the cinchona tree), have originated from plant compounds.
- **Cosmetic Ingredients and Formulations:** The spotlight has been on the potential of phytopharmaceuticals for anti-ageing, skincare and beauty products. Plant-based sources may be utilised in the form of wholesome extracts (most common example being *Aloe vera* juice/gel), purified fractions from extracts (anthraquinones from *Aloe vera)*, selective extracts (proteins from wheat germ oil) or single natural molecules (co-enzyme Q10, lutein from spinach or carrot).

20.3 Concerns of the Current Phytopharmaceutical Industry

20.3.1 Safety

Safety pharmacology is a branch of pharmacology that aims to decipher the potential harmful consequences that may ensue following use of products developed through biotechnological processes or new chemical entities meant for human use (Pugsley MK. 2008). It includes regulations that require a statement of risk, pertaining to a drug that may potentially exert a lethal effect on the human body. High sensitivities in bioassays and accuracy in the prediction of possible adverse events during clinical studies are the main goals of safety pharmacology studies. However, such stringent levels of testing may not be required for topical formulations for skin or ocular applications. Earlier, the pharmaceutical sector relied heavily on animal testing to establish the safety parameters of a drug. The candidate drug/drug substance was initially tested in at least two animal species: a rodent (mouse or rat) and canines or primates before it could be tested on human volunteers. The latest amendment to the "New Drugs and Clinical Trial Rules (2023)" strongly urges a reduction in dependency on animal testing and the adoption of ethical methods that reduce the suffering of test animals. Emphasis is on the adoption of innovative testing methods such as in silico studies, in vitro testing and methods relevant to human systems but excludes trials on animals, including 3D organoids, organs-on-chip, advanced technologies, computational methods and development of new-age alternative test methods to prove the safety of a new drug (https://www.cdsco.gov.in/opencms/opencms/en/Acts-and-rules/New-Drugs/ NDCTR_G.S.R. 175(E) dt_09.03.2023_ Amendment in First Schedule regarding use of induced Pluripotent Stem Cells additional testing methods).

20.3.2 *Efficacy*

Phytopharmaceutical compounds are complex mixtures of various lead compounds, their intermediates, by-products and precursors. The choice of a drug substance as a lead compound will depend on its therapeutic value at lower concentrations, safety at higher limits and interactions with other drug components in a given formula. The effectivity and harmlessness of any pharmaceutical product are determined by its constituents and the response time taken to achieve the desired outcome and extent of recovery. In a clinical trial, efficacy is determined as primary and secondary outcomes. During a phase 2, multicentred, randomised controlled open-label, clinical trial evaluating potency and lack of toxicity of antiviral phytopharmaceutical drug AQCH (abbreviation for aqueous extract of *Cocculus hirsutus*) obtained from *Cocculus hirsutus, treatment for COVID-19, the measurable* outcomes were recorded as follows (Table 20.2) (Joglekar et al. 2022).

However, this kind of in-depth study to ascertain efficacy and safety is usually lacking in traditional herbal medicines.

Table 20.2 Evaluation of safety and efficacy of a phytopharmaceutical drug

List of criteria for evaluation of safety and efficacy of a phytopharmaceutical drug derived from *Cocculus hirsutus* for treatment of COVID-19		
Primary outcomes	Secondary outcomes	Safety outcomes
Clinical improvement by 14th day	Number of patients showing clinical betterment by 7th and 28th days	Treatment-emergent adverse events
Reduction in disease severity	Time to clinical improvement	Serious adverse events
	Time to normalisation of fever without the use of antipyretics in the past 24 h	
	Time to alleviate cough	
	Time to viral clearance (first negative respiratory tract SARS-Cov-2 RT-PCR test)	
	Duration of supplemental oxygen therapy	
	Proportion of patients showing deterioration of clinical condition	
	Duration of hospitalisation	
	Number of deaths	

20.3.3 Consistency

The quality of phytopharmaceuticals depends on multiple factors such as accuracy in identification of plant, time and season of harvesting, improper harvesting (mixture of immature plant parts or plant parts that do not contain active ingredient resulting in dilution of bioactive concentration), presence of foreign matter such as soil, pebbles, seeds of weeds and sticks resulting in dilution and adulteration of the herb, improper drying and processing methods, pests, diseases and contaminants such as fungus and insect infestation. The consistent yield of biomass and phytopharmaceutical content in the biomass may be achieved only by the cultivation of specific herbs and standardising the agronomy protocols. For instance, the total alkaloid content in *Holarrhena antidysenterica* is highest (3.2–3.9%) only during the two flowering seasons: March–May and September–October. *Adhatoda vasica* , grown in different agroclimatic zones during different seasons, still yields the highest content of total alkaloids at 2.5% only between July and October. Thus, unseasonal harvesting may result in poor-quality raw materials with low or no efficacy. Cataloguing the different quality parameters for each herb may be valuable for establishing consistent herb quality leading to the quality of phytopharmaceuticals (Agrawal 2012).

20.3.4 Toxicity

Toxicity pharmacology refers to the preliminary tests for the evaluation of adverse reactions to a new phytopharmaceutical drug. Toxicity may be due to some factors that irritate the skin in the case of dermal products, due to the presence of heavy metals, drug interactions and unstable formulations leading to degradation of original drug and so on. The presence of alarming levels of heavy metals in herbs indicates the extent of pollution in soils, air and water. Heavy metal stress may possibly stimulate different, variable or enhanced bioactive production in some herbs. Growing specific herbs on heavy metal-contaminated soils may prove beneficial provided they are tolerant to high concentrations of soil heavy metals and growth and secondary metabolite production are optimised (Maleki et al. 2017). Two of the traditionally used neuromodulators, *Centella asiatica* and *Bacopa monnieri,* are known to accumulate heavy metals such as lead, nickel, cadmium or arsenic from the soil (Biswas et al. 2020). This process, known as bioremediation, involves the uptake of metal ions by the plant's roots and their translocation to different parts. It was found that both *Centella asiatica* and *Bacopa monnieri* store cadmium and lead in the roots with very little translocation to the aerial parts (Chandra et al. 2010). If the whole plant of either *Centella* or *Bacopa* is taken for processing as is the norm in the market, it is very likely that the resultant extract may have higher than accepted levels of cadmium or lead. Heavy metal contamination is a concern for both the phytopharmaceutical industry and consumers alike. This risk can be mitigated only with proper cultivation and harvesting practices, the adoption of heavy metal trapping strategies and strict quality control measures during production. Plants that are known hyperaccumulators may be cultivated through soil-less cultivation methods such as hydroponics and aquaponics that use inert growth substrates such as cocopeat and light expanded clay aggregate (LECA) balls, to reduce the risk of heavy metal contamination in herbs.

Toxicity could be due to drug–drug interactions between phytochemicals that are part of a polyherbal formulation. The physical parameters and solvents used to extract the bioactives may also impact the toxicity potential of a phytopharmaceutical drug or ingredient.

Detailed toxicity studies of an investigational product, while evaluating the safety profile, also provide important information about the absorption, distribution, metabolism and excretion (ADME) of the compound in the body (Govindaraghavan and Sucher 2015). The preclinical toxicity testing on various biological systems reveals the species-, organ- and dose-specific toxic effects of the candidate compound being tested. The toxicity of substances can be observed by a. studying the accidental exposures to a substance, b. cells/cell line-based in vitro study and c. in vivo testing on experimental animals. The pre-clinical toxicity testing helps to derive the important "No Observed Adverse Effect Level" without which further clinical evaluation of investigational product cannot continue.

Though traditional herbal medicine has a long history of providing therapeutic benefits, it is an undeniable fact that the risk potential for underdosing or overdosing has been quite high.

Plant-based medicines have indeed been traditionally derived based on approximations or ratios that were often rooted in empirical knowledge and observations rather than precise scientific measurements. These approximations were used to determine the intended dosage of the active compounds or drugs within the plant material. Here are a few reasons why this approach was common:

- Speculative Knowledge: Traditional herbal medicine systems, such as ethnomedicine, Ayurveda, Siddha, traditional Chinese medicine (TCM) and other indigenous healing practices, have evolved through centuries based on trial and error. Practitioners relied on their observations of the effects of various plant-based remedies on patients to determine dosages. These observations formed the basis for approximate dosing guidelines. Classical Ayurveda has clear-cut boundaries for harvesting, processing and mixing drugs in specific methods documented in the Samhitas, Nighantus and other treatises, which are considered best practices. Ethical manufacturers usually conform to these recommendations with no or few deviations.
- Variability in Plant Material: Plants are complex biological entities, and the concentration of active compounds can vary significantly based on numerous factors such as the plant's age, growing conditions, seasons, agroclimatic differences and even the time of harvest. This variability poses a great challenge to establish unequivocal standards.

- Lack of Analytical Tools: In the past, there were limited analytical tools and techniques available to quantify the exact content of active compounds within plant materials. Without these tools, it was difficult to determine precise dosages.
- Cultural and Historical Practices: Traditional medicine often incorporated cultural and historical practices that were passed down verbally through generations. There were few instances of documentation on palm leaves, but largely, oral transfer of formulation details was the norm. Any deviations or misrepresentation of the original formula or practices may have been lost through time. These practices included specific preparation methods, combinations of herbs and dosing regimens that were considered effective based on tradition or individual experiences.

In recent years, however, there has been growing interest to standardise herbal medicine by scientific validation of plant bioactives through exhaustive analysis, pre-clinical study and clinical trials, stringent quality control to standardise various parameters leading to the consistency of potency, dosage and safety and governed by strict regulatory guidelines demanding accountability from the grower/collector, researcher, manufacturer and medical practitioners alike. It was against this backdrop that a need for an integrated approach to standardisation of herbal therapeutics was envisaged. The Indian Pharmacopoeia, which is the official book of standards for drugs, has also included phytopharmaceuticals and specifies quality standards of identification, purity and potency of the drugs imported, manufactured for sale, stocked or exhibited for sale or distributed in India (Phytopharmaceuticals—Indian Pharmacopoeia Commission (ipc.gov.in)). The specific herbal monograph should state the title and synonym, if any, definition, limits of active ingredient/marker compounds, description, category, identification, relevant chemical tests, design of assay of the marker constituents, adulterants and contaminants, specific tests, stability and storage conditions, etc.

20.4 Phytopharmaceutical Biotechnology

Phytopharmaceutical biotechnology, an interdisciplinary field, integrates the botanical knowledge of plants, their efficacies and the use of plant biotechnology methods to devise novel strategies to deliver the best therapeutic potential of herbs and other plant-derived compounds, best practices for the authentication, cultivation, extraction and utilisation of medicinal plants. Phytopharmaceutical biotechnology forms a vital bridge between traditional herbal medicine and modern science. It enables the development of evidence-based herbal remedies that can be integrated into mainstream healthcare systems, providing patients with safe and effective treatment options. Additionally, it contributes to the conservation and sustainable use of natural biodiversity. When built over the solid foundation of time-tested and proven tenets of ethnomedicine, phytopharmaceutical biotechnology can help develop innovative strategies for the optimisation of the processes that lead to evidence-based, safe, effective sets of novel bioactive compounds. The main aspects of the critical study of phytopharmaceutical biotechnology include:

1. Botany
2. Standardisation
3. Plant biotechnology
4. Phytochemistry
5. Pharmacology
6. Formulation development
7. Validation and proof of concept
8. Preclinical Studies
9. Clinical trials

20.4.1 Botany

A deep study and analysis of various features of an herb/medicinal plants such as morphology, anatomy, histochemistry, ecological preferences and associations and unique adaptations help identify morphological markers unique to each plant species. These markers serve as differentiators or identifiers, aiding varietal authentication and raw material validation. For instance, in an exhaustive botanical investigation of *Andrographis serpyllifolia (*Rottl. Ex. Vahl) Wight, an Acanthaceae member, presence of peltate and glandular trichomes on the leaves, presence of cystoliths (calcium carbonate crystals) and presence of numerous white hairs at the base of anthers were identified as markers unique to this species as all three features were preserved in crude drug powder as well (Figs. 20.2 and 20.3) (Krishnaswamy and Kushalappa 2017).

Pharmacobotanical evaluation, a term used in the field of pharmacognosy, explores the quality of herbs, their derivatives and their potential therapeutic properties. Pharmacobotanical evaluation involves the comprehensive assessment and analysis of botanical materials, such as plants, plant parts (leaves, roots, stems, etc.) and plant extracts, to determine their pharmacological and medicinal potential. Further, pharmacobotanical analysis aids in highlighting the plant part most likely to be therapeutic by virtue of its structure and the presence of organs/organelles meant for producing and storing bioactives. These organs or organelles may be cultured through plant biotechnology processes to scale up targeted production of bioactive. This evaluation typically includes several key aspects:

- Botanical Identification: Accurate botanical identification of the live plant and the features of dry material is essential. Taxonomical studies, chemotaxonomy and DNA fingerprinting are conducted to ensure that the correct plant species or variety is being used for medicinal purposes.
- Morphological Characterisation: The physical characteristics of the plant, including its appearance, size, shape, colour and any distinctive features, are documented.
- Microscopic Evaluation: Microscopic analysis involves examining the plant material under a microscope to study its anatomical features and cellular struc-

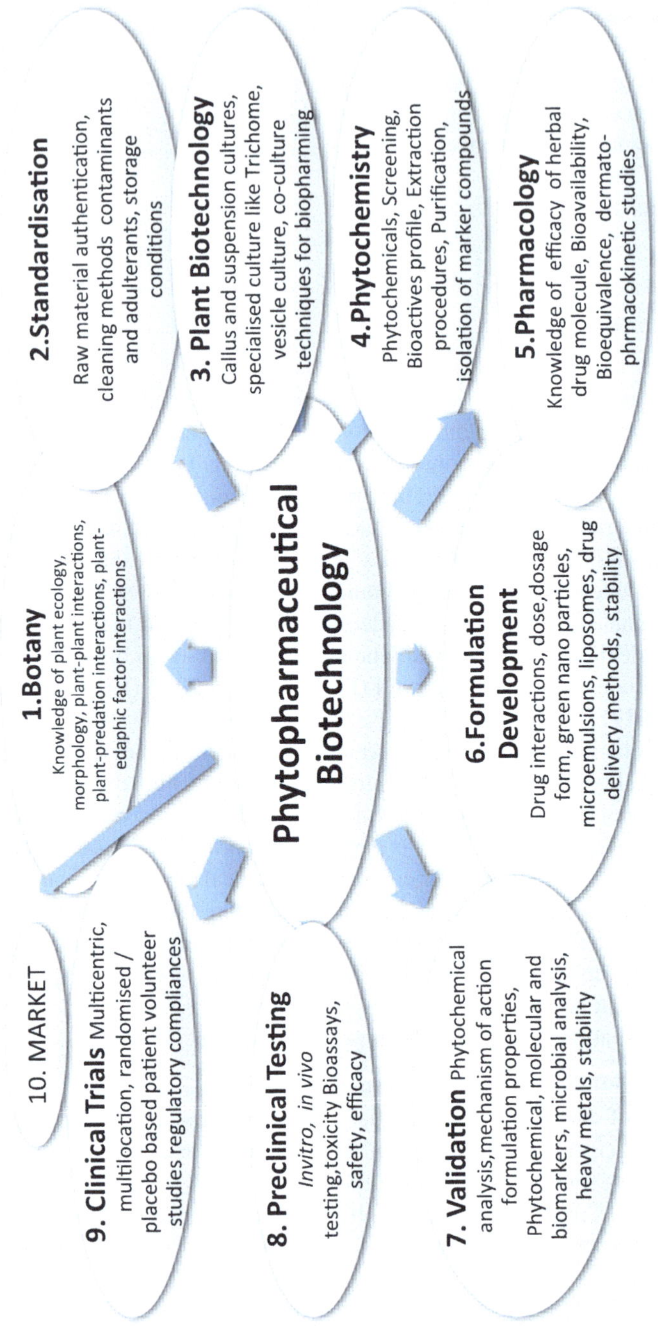

Fig. 20.2 Stages in the development of phytopharmaceutical drug/formulation

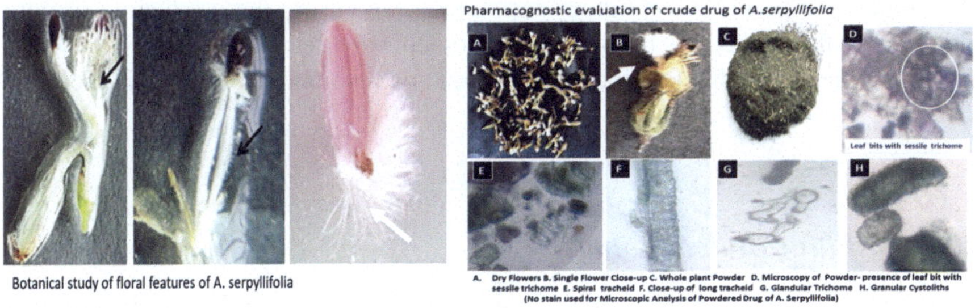

Fig. 20.3 Biomarkers (persistent hairs on anther in fresh and dried flower) in dried herbage of *Andrographis serpyllifolia* (Reused from Krishnaswamy and Kushalappa, 2017, distributed under Creative Commons Attribution 4.0 International Licence, which permits unrestricted use, distribution and reproduction in any medium. http://creativecommons.org/licenses/by/4.0) (Krishnaswamy and Kushalappa 2017)

ture, which can reveal important information about the plant's identity, cellular contents and quality.

- Toxicological Assessment: Safety is an important aspect of pharmacobotanical evaluation. Toxicological studies help determine whether the plant material or its extracts have any harmful effects on humans or animals.

20.4.2 Standardisation

Establishing quality control parameters and standards for a safe, effective and stable phytopharmaceutical ingredient or formulation is necessary at multiple levels, starting from planting material to the conduct of clinical trials and post-market surveillance. Standardisation helps to set non-negotiable guidelines ensuring high levels of compliance to maintain standards, processes and documentation, which are absolutely necessary for achieving batch-to-batch consistency, efficacy and stability.

The preliminary standards of cultivation or collection of herbs need to be set at the field level, with good cultivation practices or good collection practices at the source. The field staff responsible for cultivation or collection need to be trained to be able to distinguish salient features of clones of selected varieties and their substitutes or adulterants. Authentication of the botanical specimen and periodical botanical analysis through DNA fingerprinting to ensure genetic purity are necessary. If the herb is cultivated using the conventional method, adopting soil test crop response (STCR) along with integrated pest management techniques that avoid the use of harmful chemical pesticides may be preferred. STCR is a nutrient management method of determining the optimal fertiliser dose for an herb based on the soil analysis values and the target yield. STCR is used to improve productivity, soil health, nutrient availability and nutrient use efficiency. The use of hydroponics for the production of herbal biomass may be expensive as compared to the conventional cultivation model but is valuable in producing heavy metal-free herbage. However, the adoption of plant biotechnology techniques can ensure the production of pure plant material provided the tissue culture protocols are managed well under aseptic conditions that avoid microbial contamination and resultant endotoxins in the plant extract.

20.4.3 Plant Biotechnology

Plant biotechnology takes advantage of modern techniques to improve the production, quality and consistency of herbs and their bioactive principles. This can include genetic modification, tissue culture, biopharming and other advanced methods. Plant biotechnology helps in developing high-yield and disease-resistant plant varieties, enhancing the production of bioactive compounds and ensuring product uniformity. Plant tissue culture has the potential to develop alternate phytochemical pathways generating de novo synthesis of new molecules that may be potentially valuable as phytopharmaceuticals. Thus, plant tissue culture may aid herbal drug discovery by creating a vast pool of bioactive intermediates. Plant stem cell extracts are derived from callus cultures of specific plants, and their pathways may be modulated to over-produce certain phytochemicals. Specific plant organ culture is a specialised technique where minute plant parts storing a minuscule quantity of phytochemicals can be mass propagated to provide a valuable source of phytomedicine. Some plants have specialised hair-like structures called glandular trichomes on their leaves, flowers or other aerial parts. These trichomes bear secretory glands that can produce and store phytochemicals, such as essential oils or secondary metabolites, which serve various ecological roles, including defence against herbivores and pathogens.

Phytopharmaceutical Biotechnology

Phytopharmaceutical biotechnology is an interdisciplinary field that combines elements of botany, pharmacology and plant biotechnology to utilise the therapeutic potential of herbs and other plant-derived compounds. This approach seeks to develop innovative strategies for the propagation, cultivation, extraction and utilisation of medicinal plants to deliver safe and effective herbal medicines and natural remedies. The combination of phytopharmaceuticals and biotechnology has aided in the development of new drugs, dosage forms and treatments. By using contemporary methods of biotechnology, plants with unique chemical compositions are propagated and genetically upgraded for the mining of pharmaceuticals (Singh et al. 2022). Some key aspects of phytopharmaceutical biotechnology are to modify secondary product metabolic pathways by genetic modification of plants and the design of robust in vitro plant regeneration systems. Continued advancements in genetic tools and techniques, such as gene cloning, advancement in the identification of secondary plant metabolites and high-throughput monitoring technologies, have made it easier to implement strategies for manipulating a plant's secondary products.

Plant Tissue Culture

Plant stem cells are propagated by adopting the technique of plant tissue culture, resulting in aseptic cultures of a whole new plant, tissue or specific types of single cells from which plant metabolites may be harvested. Tissue material obtained for culture from a selected elite plant is called an explant. Regeneration of new cells occurs along the cut surfaces and edges of the explant. This is similar to a wound-healing reaction. Normal plant cells undergo dedifferentiation to develop into a colourless/coloured cell mass called callus or plant stem cells. Totipotent meristematic tissues are the source of plant stem cells. The callus cells mimic the intact mother plants' meristematic cells in terms of cell division rates, physiology and cell contents. Plant stem cells may be grown as individual cells or small cell clusters in a liquid culture for higher yield. In certain plant cells that accumulate compounds such as anthocyanins and carotenoids, the callus cells may be coloured, blue, purple, red, pink, black or orange depending on the pH, light or dark conditions provided during the culture regime.

Researchers globally are focused on the optimised development of high-value stem cell extracts with promising potential in the cosmetic, biomedical and therapeutic domains (Aggarwal et al. 2020). Stem cells derived from plants such as grapes (*Vitis vinifera*), lilacs (*Syringa vulgaris*) and others possess strong antioxidant and anti-inflammatory properties and are invaluable in cosmetic applications (Barthel and Aberdam 2005). Plant stem cells, their extracts and formulations have a wide range of cosmetic applications including skin whitening, de-tanning, moisturising and cleansing. Despite all the promising developments, plant stem cell technology remains hugely unexplored due to the staggering number of varieties of plant species and the exhaustive array of their phytochemical components.

In an experiment to enhance the production of metabolite β-pinene in callus culture developed in light conditions, using petals of *Rosa damascene* (Olgunsoy et al. 2017), it was found that deep pink-coloured callus cells were formed, accumulating about 236 times more metabolite than was found in intact petals. The influence of light on anthocyanin production is visible in the rapidly proliferating callus where the rose petal stem cells grown in the dark appear creamish yellow, whereas those grown in light developed pigment-bearing stem cells. This exponentially increased bioactive production was achieved using phenylalanine as a precursor in light conditions.

The stem cell of a rare, bitter apple variety, *Malus Domestica* ("Uttwiler Spätlauber"), with excellent keeping quality over extended periods, perhaps due to the presence of certain unique metabolites, was derived through plant cell culture technology (Trehan et al. 2017). Large-scale cultivation of apple callus (stem) cells was undertaken to overproduce the metabolites. It was found that human young epidermal cells treated with this extract had the potential to grow into multi-layers resulting in skin renewal. Untreated older epidermal cells could not develop into layers under normal laboratory conditions but resumed epithelial growth on treatment with 2% emulsion of apple stem cell extract consisting of phytocomponents such as carotenoids and flavonoids (Prhal et al. 2014).

Another product was developed by using biotechnological production methods in cloudberry (*Rubus chamaemorus*) cells (Martinussen et al. 2004). Bioreactors were used to develop entrenched callus and suspension cultures of *Rubus chamaemorus*. Optimal growth of callus was achieved on Murashige and Skoog medium, supplemented with kinetin and α-naphthalene acetic acid as plant growth hormones at specific concentrations. This process was standardised and proved to be a sustainable technique for consistent and continuous sources of fresh cells/ extracts from cell fractionation for isolating biologically potent compounds, freeze-dried cell products, aromatics or colouring agents. Stem cells cultured from tomato (*Lycopersicon esculentum*) cells possess tremendous potential to protect skin from adverse effects due to heavy metal toxicity (Tito et al. 2011).

A cosmetically active hydrophilic ingredient, extracted from liquid cultures of *L. esculentum,* consisted of comparatively higher concentrations of flavonoids and phenolic acids such as quercetin, rutin, coumaric, protocatechuic and chlorogenic acids. The tomato stem cell extract also contained higher phytochelatins, an antioxidant and chelating agent responsible for capturing metal ions and protecting cellular materials and organelles from potential toxicity. Extracts obtained by this method were very effective at enhancing healthy skin growth and complexion, contributing a valuable phytopharmaceutical in the area of skincare cosmetics.

The short-wave (UVC 200–280 nm), mid-wave (UVB 280–315 nm) and long-wave (UVA 315–400 nm) UV rays can cause cancer, erythema or photoageing on constant exposure. Ginger extract with skin-preserving bioactives (*Zingiber officinale*) is derived through biotechnology-based cell multiplication of improved cosmetic cell lines. Reduced pores and mattifying effect, contributing to almost 50% enhancement of skin quality, were achieved in female subjects in a clinical study (http://www.naolys.com/media/refine_ginger_en.pdf. Accessed 15 Jul 2018). This effect was further augmented by a consequent reduction in skin sebum and oily shininess. Increased synthesis of elastin fibres in the skin consequently reduced the rate of sebum production. The protective and potent anti-collagenase and hyaluronidase activity of an anti-ageing component obtained from stem cell extracts of edelweiss (*Leontopodium alpinum*) was reported to be rich in leontopodic acids A and B, which are responsible for exhibiting a strong and potent antioxidant effect on the skin (Cho et al. 2020).

Specialised organ culture is another phytopharmaceutical biotechnology-driven route to elicit the overproduction of pure bioactives from plant cell cultures, with fewer intermediates, which may otherwise appear as impurities. The juice-bearing vesicles from the fruits of citrus species have been cultured using tissue culture techniques to harvest vitamin C, naringin and hesperidin (Einset 1978). Citrus vesicle stem cells may also accumulate high concentrations of kinetin, a cytokine that strongly inhibits ageing processes in human cells. The innovations and technological advancement in the field of plant biotechnology enable novel product designs such as nourishing and hydrating, easily absorbed skin nutrients. Such highly efficacious plant stem cell-based cosmetics are in great demand.

The process flow for the production of phytopharmaceuticals using biotechnology methods is given in Fig. 20.4.

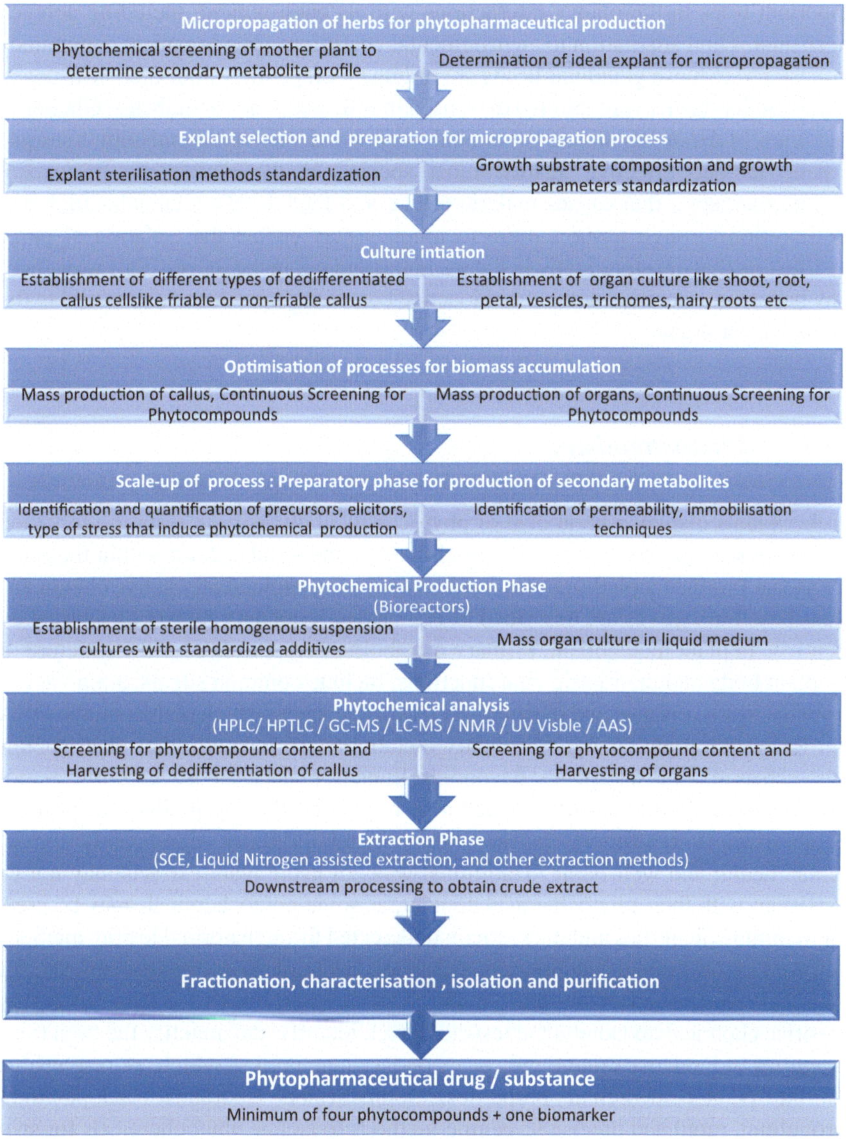

Fig. 20.4 Flowchart depicting the production of phytopharmaceutical or secondary metabolites through plant biotechnology methods

Another under-explored but highly promising aspect is the culture of glandular trichomes (Atito et al. 2018), which stores volatile compounds such as terpenes, essential oils, flavours and fragrances for targeted production. Glandular trichomes are specialised hair-like structures on the surfaces of various plants, including some species of flowering plants, herbs and even certain trees. These trichomes are involved in the production and secretion of various chemicals, often playing important roles in the plant's interactions with its environment, including defence mechanisms and adaptation to environmental conditions. The botanical compounds sequestered in these glandular heads are of huge therapeutic value. *Cleome droserifolia* plant has been overexploited to be used in folk medicine for diabetes. Glandular trichomes of *Artemisia annua,* the annual wormwood, contain artemisinin, a sesquiterpene lactone effective against most species of the pathogenic protozoan, *Plasmodium* spp., that causes malaria (Duke and Paul 1993). Glandular trichome proliferation through tissue culture techniques may enable the isolation of purer volatile oils that get stored only in the trichome and not in the leaves. This way, there may be less interference with other bioactive compounds produced and stored in the whole leaf or organ.

20.4.4 Phytochemistry

Higher plants display a great deal of phytochemical diversity in response to varied environmental, physiological and agroclimatic stimuli and species within the same genus can produce a plethora of related or dissimilar biologically active chemicals. Plant-based bioactives are described as primary or secondary metabolites based on their role in plant metabolism. Primary metabolites, responsible for the structure of the plant body and their basic vital functions, include common sugars, amino acids, proteins, lipids, purines and pyrimidines of nucleic acids, which constitute the DNA and RNA, chlorophylls, etc. Secondary metabolites include all other plant chemicals such as alkaloids, terpenes, flavonoids, lignans, plant steroids, curcumins, saponins, phenolics and glucosides, which may be produced as by-products of primary metabolism. Both primary and secondary metabolites can significantly impact human health and well-being. Traditional ways of using plant extracts may not be consistent with their phytochemical content, as some phytochemicals may be present in minute quantities and may remain undetected through normal testing methods followed by traditional practitioners, but in the realm of phytopharmaceuticals, the chemical constitution of the plant material is analysed through a range of highly sensitive chemical and analytical tests to detect, identify and quantify the even trace quantities of bioactive compounds or secondary metabolites, which are responsible for the plant's medicinal properties. Sophisticated laboratories adopt high-throughput, rapid and precise screening methods to isolate phytochemicals for specific targeted pharmacological activity. A thorough evaluation of the plant material/crude drug quality, including factors such as moisture content, ash content and other quality parameters, is crucial to ensure consistency and effectiveness in medicinal preparations.

According to regulatory stipulations, a phytopharmaceutical formulation mandatorily comprises a minimum of four biologically active phytochemical compounds in addition to a biomarker.

Phytopharmaceuticals and Biomarkers

It is important to distinguish the difference between phytochemical compounds and biomarkers. Phytopharmaceuticals and biomarkers are distinct but associated with each other in the fields of biology, pharma and medicine.

Phytopharmaceuticals are also known as plant-derived biologically active substances with curative or therapeutic properties. The bioactives or phytochemicals are usually secondary metabolites or by-products of primary metabolic processes taking place in a plant species and may be found stored in any or all regions of the plants such as leaves, stem, flowers and roots and are extracted from the plant parts using specific methods and conditions.

The usual phytochemicals used for therapy can be categorised as phenolics, tannins, saponins, alkaloids, polysaccharides, terpenes, gums, etc.

Biomarkers, however, are characteristics or a means of evaluating efficacies or outcomes of the use of phytopharmaceuticals, biological processes or conditions response to treatment. Biomarkers could be of different types:

1. Molecular or DNA-Based: genetic markers
2. Chemical Moieties: unique to a plant species, genus or family
3. Biochemical: proteins, enzymes
4. Response-Based: reduced blood sugar levels in the treatment of diabetes with a polyherbal formulation, reduction in itching and scaling in the treatment of psoriasis using a polyherbal skin cream

Biomarkers are necessary to assess the efficacy and safety of numerous interacting phytochemical molecules found in polyherbal formulations. In this context, a biomarker can help determine whether the formulation has the desired therapeutic effects and whether it can be safely used. Multiple biomarkers may also be necessary in ensuring safety and efficacy of complex herb mixtures. For instance, a polyherbal treatment for healing of non-healing wounds should reduce inflammation, arrest infection, remove bad odour and improve granulation to achieve even minimal healing. Here, reduced inflammation, lack of bad odour and clear red zones, indicating restored blood circulation, may be considered biomarkers.

In certain cases, phytopharmaceutical compounds themselves may be considered biomarkers. For instance, polysaccharides are complex carbohydrates that consist of many monosaccharide units linked together by glycosidic bonds. Polysaccharides have various biological functions, such as energy storage, structural support and cell recognition. Polysaccharides can also be considered biomarkers in a polyherbal phytopharmaceutical formulation, which is a medicine derived from natural sources and containing multiple herbs. Polysaccharides can reveal the source and authenticity of the herbs, determine the optimal dosage and extraction method and aid in the evaluation of the stability and shelf-life and the pharmacological and therapeutic effects of the formulation (Mohammed et al. 2021). Some examples of polysaccharides that have been used as biomarkers in polyherbal formulations are:

- Mangiferin: A xanthone glycoside is found in various plants, especially in mango. Mangiferin has anti-inflammatory, antioxidant, antidiabetic, anticancer and neuroprotective properties. Mangiferin can be used as a biomarker to standardise Mahasudarshan Churna, an Ayurvedic polyherbal medicine that is used for fever, cold, malaria and other conditions (Kaur et al. 2020).
- Andrographolide: The chief bioactive component of *Andrographis paniculata*, a herb used in ethnomedicine for respiratory infections, inflammation and liver disorders, is a diterpenoid lactone, andrographolide. It exerts immunomodulatory, antiviral, antibacterial, antimalarial and hepatoprotective effects. Andrographolide can be used as a biomarker to standardise PartySmart capsule, a polyherbal formulation that is used for preventing hangovers by preventing acetaldehyde build-up in the liver and protecting the liver (Khanvilkar et al. 2023).
- Curcumin: The principal phytochemical curcumin, displaying marked anti-inflammatory, antioxidant, anticancer, antidiabetic and neuroprotective properties in turmeric, is a polyphenol. Curcumin can be designated as a biomarker to indicate standardisation in the PartySmart capsule along with andrographolide (Venkataranganna et al. 2008; Gopumadhavan et al. 2008).

Thus, the choice of biomarker needs to necessarily align with the intended therapeutic effect or expected outcome desired by the treatment regime or medicine.

20.4.5 Pharmacology

Pharmacological effect or efficacy is the study of how drugs and compounds interact with the body and produce therapeutic effects. In phytopharmaceutical biotechnology, pharmacological investigation focuses on unravelling the mechanisms of action and safety profiles of plant-derived compounds. Pharmacological studies help determine the efficacy of herbal extracts and compounds in treating specific diseases or conditions. This can involve testing for activities such as anti-inflammatory, antimicrobial, antioxidant, neuroprotective, immunomodulatory or analgesic effects.

20.4.6 Formulation Development

Extraction

Phytopharmaceutical biotechnology also focuses on developing efficient methods for extracting and formulating plant-derived compounds. This includes optimising processes and extraction techniques to maximise compound yield, recovery and stability. Quality control standardisation is a crucial step, needing strict vigilance and enforcement. Novel methods of extraction of wholesome extracts and fractionated molecules have been developed. Some examples of these innovative extraction technologies are counter-current extraction, oil-as-solvent extraction (using bio-renewable, bio-based solvents such as citrus peel and rapeseed oil), ultra-sonication-assisted solvent extraction (Azmir et al. 2013).

Subcritical water hydrolysis (SWH), also known as subcritical water extraction (SWE), is a novel modern extraction process that exploits the unique properties of water below critical points of 100–374 °C and 1–22.1 MPa temperature and pressure, respectively, where the distinction between liquid and gaseous phases is obliterated and only one phase with strong catalytic character prevails. At subcritical ranges of temperature and pressure, low dielectric levels and varying densities cause a marked shift in the behaviour and properties of water. Under these conditions, non-polar substances such as proteins and amino acids can be recovered from biomass with tough cell walls (e.g. leftover biomass from coffee and tea processing), without breakdown of their structural integrity. This highly compressed water provides a clean, green alternative for extraction of biological/bioactive compounds through "valorisation" of biological raw material to valuable nutrients (Maheshwari et al. 2020; Di Domenico Ziero et al. 2020).

Further purification and isolation of phytochemicals are accomplished through preparative column chromatography and fermentation through biotechnological techniques using microbial strains such as *Bifidobacterium bifidum*, *Lactobacillus acidophilus* and others to achieve greater efficacy of the phytopharmaceuticals. *Codonopsis lanceolata* (*C. lanceolata*), Campanulaceae, consists of many phytocompounds such as saponins, alkaloids, tannins and polysaccharides (Ushijima et al. 2008). Using steam and fermentation in sequential processes, the extract (SFCE—steam fermented Codonopsis extract) was derived and used for the treatment of scopolamine-treated memory-impaired mice. Steamed *C. lanceolata* followed by fermentation recovered more phenolic acid including gallic acid and vanillic acid than original *C. lanceolate* (Weon et al. 2013). SFCE-treated cells showed significant neuroprotection at 500 μg/ml. The combination of steam and fermentation to obtain *C. lanceolata* extract enhanced cognitive activity related to memory processes and neuroprotective effects as compared to the conventionally derived *C. lanceolata* extract (Shinde et al. 2022). These results highlight the role biotechnological processes play in bringing out valuable synergistic efficiencies in phytopharmaceutical formulations.

The plant stem cell extracts are then used as single drugs or combined with multiple other extracts to derive a polyherbal formulation. The preferred solvents for extraction are water and ethyl alcohol as they are considered generally accepted as safe (GRAS). The phytopharmaceutical drug may be dispensed as such or run through other biotechnological techniques that augment the efficacies multifold.

Drug Delivery Systems

Biologically active materials such as plant parts, fungi, algae, bacteria or their extracts are incorporated into metal ions to develop green microscopic nanoparticles (NP) in a process known as "green synthesis". Advances in the field of nanotechnology enable the synthesis of a wide class of microscopic materials in the nanoscale range (ranging from 1 to 100 nanometres) as opposed to bulk materials, which range more than 100 nm in size and are visible to the naked eye (Alsaiari et al. 2023). The green synthesis process, while being environmentally friendly and degradable, ensures the delivery of optimally loaded doses of the drug onto metallic nanoparticles. These drug-impregnated nanoparticles have better penetrability, faster action and greater efficacies as a result of synergies (Letchumanan et al. 2021). Developed using metal/metal oxides of gold (Au), silver (Ag) and copper (Cu), nanoparticle-mediated delivery methods possess various advantages, such as long-term persistence, preservation of enzyme structure, precision targeting of cells and upregulation of the immune system (Mohammed et al. 2018). Copper being relatively cheaper is most preferred for green synthesis. Furthermore, it is a natural trace component in the human body and is necessary for important enzymes such as tyrosinase, superoxide dismutase and cytochrome oxidase, which mediate vital metabolic processes. On oxidation, Cu NPs form inorganic NPs of copper oxides (CuOs). Both Cu and CuO NPs possess anticancer, antimicrobial and antioxidant properties (Huang et al. 2020).

Formulation studies involve creating herbal medicines in various forms, such as capsules, tablets, creams and tinctures, to enhance their bioavailability and ease of use. Plant biotechnology methods support efficient impregnation of plant extracts in novel drug delivery systems such as liposomes, phytosomes, green synthesised nanoparticles, polymeric nanoparticles (PNPs), nanocapsules and nanospheres, polymeric nanogels and nanosuspensions, carbon nanotubes (CNTS) and nanofibers, polymeric nanomicelles and polymeric nanoliposomes. For example, green extract nanoparticles may be dispersed in a semi-solid cream to deliver antioxidants of high efficacy. Among the recently developed technology-driven drug delivery carriers, polymeric nanoparticles (PNPs) with high drug loading capacity, long half-life and high capacity to prevent drug deterioration are very effective. The PNPs offer easy surface penetration possibilities for ligands to pass through the blood-brain barrier (BBB) in Alzheimer's disease, dementia, Parkinson's disease and other CNS-related maladies.

20.4.7 Validation

Phytopharmaceuticals are manufactured through various processes of pharmacology and biotechnology resulting in specific dosage forms with different drug delivery mechanisms. Validation is vital across the entire gamut of the phytopharmaceutical value chain, encompassing research, manufacturing, testing and distribution activities from analytical methods used for quality control, checking infrastructure/equipment facilities and processes adopted for manufacturing. Validation protocols that set testing guidelines are developed for the kind of tests or assays to be carried out, frequency of testing and expected and realised outcomes. Process validation includes the identification of critical process parameters in a given process and the generation of data that aligns with the pre-set specifications.

20.4.8 Pre-clinical Studies

The safety and efficacy of phytopharmaceutical drugs are evaluated during preclinical studies before they are tested on humans. These studies include in vitro and in vivo experiments to determine the pharmacological properties of the drug, its toxicity profile and its mechanism of action. The preclinical studies also help to identify potential drug interactions and prediction of adverse effects that may occur when the drug is administered to humans. Preclinical research generally falls into one of three categories: in silico (computer modelling or simulations), in vitro (cell culture or partial biological systems) and in vivo (animal studies or complete nonhuman biological systems). Preclinical research can be classified into two distinct categories depending on the aim and purpose of the study, such as "hypothesis-generating" (exploration) and "hypothesis-testing" (confirmation) research (Venkatraman and Mehta 2023).

The pharmaceutical industry relies heavily on preclinical PK/PD modelling to select, predict and optimise effective human doses and/or dose regimens (Schuck et al. 2015). Knowledge of pathophysiology is vital in target validation, biomarker identification, patient grouping or stratification and impacts preclinical/clinical study success rates. Quantitative systems pharmacology (QSP) is another tool that pharmaceutical companies are adopting for evaluating all stages of drug discovery/ development. Though preclinical QSP modelling is still in nascent stages, it may be designed to optimise processes of lead generation, proof-of-concept validation, clinical efficacy forecasting, drug-dose optimisation, biomarker identification and elucidation and compound selection (van der Graaf and Benson 2011).

Laboratory experimentation is followed by the first phase of pre-clinical study, which aims to establish proof of safety of the new drug and is conducted on a limited number of individuals. The procedures to determine the effectiveness of the drug in the first phase are similar except for the increased number of individuals deployed in the third phase. Qualified doctors administer the new drug to a small group of 10–20 recruited patients, in small doses initially. The gradual increase in the dosage known as dose escalation enables the identification of the dosage that produces optimal results without causing harsh side effects. Drug metabolism and absorption in the human body are studied. Vital data collected on the dose, dosage timing, method of drug administration and the suitability of the treatment set the stage for the next phase. Usually, side effects are not observed in the first administration of the drug to a patient. The first phase of preclinical trials usually goes on for about a year.

20.4.9 Clinical Trials and Regulatory Compliance

The third phase of preclinical trials collates the potential outcomes of administering a new therapy or drug targeted to alleviate a specific ailment, to a small number of patients and compares them with the responses to the standard of care for similar conditions. Once the pre-clinical raw data are obtained and analysed, the phytopharmaceutical product moves on to phase 1–4 clinical trials. Phytopharmaceutical biotechnology leans heavily on the three pillars namely, quality, safety and consistency of herbal products. This includes implementing good manufacturing practices (GMP) to establish quality control standards, conducting quality assessments and validation of the therapeutic potential of herbal medicines. Human clinical trials are conducted to assess their safety and efficacy. Regulatory compliance ensures that phytopharmaceutical products meet the requirements of health authorities and are safe for consumers. Compliance with specific regulations and guidelines related to herbal medicines and dietary supplements is necessary, depending on the region and the intended use of the plant material.

20.5 Conclusion

Plant-based natural products including extracts, enriched fractions, essential oils, phytomolecules, flavours and fragrances are popular among the pharmaceutical, nutraceutical and cosmetic industries. Phytopharmaceuticals include enriched fractions of plant drugs consisting of at least four specific chemical markers with one biomarker. This new classification provides the way forward for the plant-based enriched fractions to be used as a drug, which do not belong to any part of Ayurvedic, Siddha or Unani literature. It has been very important to know the chemical composition and precise proportion of pharmacologically active ingredients in the formulation. The provisions stipulated for synthetic drugs are not appropriate or relevant for botanical-based products. Regulatory guidelines have been framed to bridge science-based conventional medicine and traditional medicine with anecdotal promise through a new class of evidence-based, safe and efficacious phytopharmaceutical drugs. This class of drugs in a new avatar may encourage a deeper investigation into the vast potential of plant-based therapeutics to provide quick-acting healthcare solutions to unmet medical needs under the standard of care.

The current trend is the domestication of medicinal plants through various innovative methods of cultivation, production and biotechnological interventions for in vitro over-production through plant tissue culture and genetic improvement of herbs. Discouraging the use of wild-crafted plants will force the industry to resort to the cultivation of herbs, taking away the huge stress on rapidly depleting biodiversity and natural populations. It will then be possible to obtain uniform and high-quality raw materials, which are fundamental to the efficacy and safety of herbal drugs. Furthermore, the development of skincare products based on plant stem cell extracts is an emerging trend at present due to the vast potential of plant stem cells that may be developed from different plant species. Presently, various forms of plant stem cells and the products derived from their extracts are commercially accessible to the cosmetic industry.

However, plant stem cells as a sustainable source of curative phytopharmaceuticals will be highly valuable in treating cases of antimicrobial resistance (AMR) and orphan diseases apart from NCDs. Plant stem cells have been found to amass plant constituents and therapeutically relevant plant products such as phytohormones (kinetin) and other bioactives, in sufficient quantities to be commercially viable. Optimising the biotechnological processes to enhance efficacy and yield resulting in augmented production of phytopharmaceuticals may make them economically affordable. The rich natural biodiversity offers an excellent source of plant stem cells that may be utilised by the phytopharmaceutical industry. Destructive harvesting should be penalised heavily, and the biodiversity should only serve as the gene pool for initiating experiments in phytopharmaceutical biotechnology and not as the final product of commerce. Earth's biodiversity, comprising millions of unique biological specimens, has evolved over nearly 3.5 billion years. Unfortunately, man, armed with technology, needs only a few decades to decimate them. After all, every living organism in the earth's ecosystem has a right to exist, irrespective of their contribution towards the betterment of the human race. Responsible utilisation of natural resources is not only necessary but is critical for the sustainable development of the phytopharmaceutical industry.

References

Aggarwal S, Sardana C, Ozturk M, Sarwat M (2020) Plant stem cells and their applications: special emphasis on their marketed products. 3 Biotech 10(7):291. https://doi.org/10.1007/s13205-020-02247-9. Epub 2020 Jun 5. PMID: 32550110; PMCID: PMC7275108

Agrawal SS, Paridhavi M (2012) Standardisation of herbal drugs. In: Herbal drug technology, 2nd edn. Universities Press (India) Private Ltd, pp 643–663

Alsaiari NS, Alzahrani FM, Amari A, Osman H, Harharah HN, Elboughdiri N, Tahoon MA (2023) Plant and microbial approaches as green methods for the synthesis of nanomaterials: synthesis, applications, and future perspectives. Molecules 28(1):463. https://doi.org/10.3390/molecules28010463

Anand PU, Biswas P, Kumar V, Ray D, Ray P, Loake VIP, Kandimalla R, Chaudhary A, Singh B, Routhu NK, Chen Z-S, Proćków J, Dey A (2022) Podophyllum hexandrum and its active constituents: novel radioprotectants. Biomed Pharmacother 146:112555, ISSN 0753-3322. https://doi.org/10.1016/j.biopha.2021.112555

Atito E, Heneidak S, Magdi A (2018) El-Sayed in vitro culture initiation of cleome droserifolia glandular trichomes for potential biotechnological applications. J Biol Stud 1(2):90–97. https://onlinejbs.com/

Azmir J et al (2013) Techniques for extraction of bioactive compounds from plant materials: a review. J Food Eng 117:426–436

Banerjee S, Anand U, Ghosh S, Ray D, Ray P, Nandy S et al (2021) Bacosides from Bacopa monnieri extract: an overview of the effects on neurological disorders. Phytother Res 35:5668. https://doi.org/10.1002/ptr.7203

Gallego-Jara J, Lozano-Terol G, Sola-Martínez RA, Cánovas-Díaz M, de Diego Puente T. A (2020) Compressive Review about Taxol®: History and Future Challenges. Molecules. 25(24):5986. https://doi.org/10.3390/molecules25245986

Barthel R, Aberdam D (2005) Epidermal stem cells. J Eur Acad Dermatol Venereol 19:405–413

Bhusnure OG, Shinde MC, Vijayendra Swamy SM, Gholve SB, Giram PS, Birajdar MJ (2019) Phytopharmaceuticals: an emerging platform for innovation and development of new drugs from botanicals. J Drug Deliv Ther 9(3-s):1046–1057. https://doi.org/10.22270/jddt.v9i3-s.2940

Biswas T, Parveen O, Pandey VP, Mathur A, Dwivedi UN (2020) Heavy metal accumulation efficiency, growth and centelloside production in the medicinal herb *Centella asiatica* (L.) urban under different soil concentrations of cadmium and lead. Ind Crops Prod 157:112948, ISSN 0926-6690. https://doi.org/10.1016/j.indcrop.2020.112948

Butnariu M, Quispe C, Herrera-Bravo J, Pentea M, Sarac I, Küşümler AS, Özçelik B, Painuli S, Semwal P, Imran M, Gondal TA, Emamzadeh-Yazdi S, Lapava N, Yousaf Z, Kumar M, Eid AH, Al-Dhaheri Y, Suleria HAR, del Mar M, Contreras JS-R, Cho WC (2022) *Papaver* plants: current insights on phytochemical and nutritional composition along with biotechnological applications. Oxidative Med Cell Longev 2022:2041769. https://doi.org/10.1155/2022/2041769

Chandra R et al (2010) Bio-accumulation of heavy metals in *Bacopa Monnieri* (L.) Pennell growing under different habitat. Int J Ecol Dev 15(W10). Issn 0972-9984 (Print); Issn 0973-7308 (Online) Copyright © 2010 IJED (CESER Publications)

Cho WK, Kim H-I, Kim S-Y, Seo HH, Song J, Kim J, Shin DS, Jo Y, Choi H, Lee JH et al (2020) Anti-aging effects of *Leontopodium alpinum* (Edelweiss) callus culture extract through transcriptome profiling. Genes 11(2):230. https://doi.org/10.3390/genes11020230

Di Domenico Ziero H, Buller LS, Mudhoo A, Ampese LC, Mussatto SI, Carneiro TÃ¢n F (2020) An overview of subcritical and supercritical water treatment of different biomasses for protein and amino acids production and recovery. J Env Chem Eng 8(5):104406. https://doi.org/10.1016/j.jece.2020.104406

Diniz LRL, Calado LL, Duarte ABS, de Sousa DP (2023) *Centella asiatica* and its metabolite asiatic acid: wound healing effects and therapeutic potential. Meta 13(2):276. https://doi.org/10.3390/metabo13020276. PMID: 36837896; PMCID: PMC9966672

Downer S, Berkowitz SA, Harlan TS, Olstad DL, Mozaffarian D (2020) Food is medicine: actions to integrate food and nutrition into healthcare. BMJ 29;369:m2482. https://doi.org/10.1136/bmj.m2482. PMID: 32601089; PMCID: PMC7322667.

Duke SO, Paul RN (1993) Development and fine structure of the glandular trichomes of *Artemisia annua* L. Int J Plant Sci 154:107–118

Einset JW (1978) Citrus tissue culture: stimulation of fruit explant cultures with orange juice. Plant Physiol 62(6):885–888. http://www.jstor.org/stable/4265555

Ekor M (2014) The growing use of herbal medicines: issues relating to adverse reactions and challenges in monitoring safety. Front Pharmacol 4:177

Fattori V, Hohmann MS, Rossaneis AC, Pinho-Ribeiro FA, Verri WA (2016) Capsaicin: current understanding of its mechanisms and therapy of pain and other pre-clinical and clinical uses. Molecules 21(7):844. https://doi.org/10.3390/molecules21070844. PMID: 27367653; PMCID: PMC6273101

Gopumadhavan S, Rafiq M, Azeemuddin M, Mitra SK (2008) Ameliorative effect of PartySmart in rat model of alcoholic liver disease. Indian J Exp Biol 46(2):132–137

Govindaraghavan S, Sucher NJ (2015) Quality assessment of medicinal herbs and their extracts: criteria and prerequisites for consistent safety and efficacy of herbal medicines. Epilepsy Behav 52(Part B):363–371. https://doi.org/10.1016/j.yebeh.2015.03.004. ISSN 1525-5050. https://toolbox.eupati.eu/resources/general-toxicity-studies/

Hien, D. (1993) Qinghaosu. Lancet 341(8845):603–608. https://doi.org/10.1016/0140-6736(93)90362-k

Huang W, Percie du Sert N, Vollert J, Rice ASC (2020) General principles of preclinical study design. Handb Exp Pharmacol 257:55–69. https://doi.org/10.1007/164_2019_277. PMID: 31707471; PMCID: PMC7610693

Joglekar S, Iyer SS, Parate R, Khobragade AA, Patil R, Bhushan S, Halnor DM, Rajadhyaksha GC, Parmar K, Dhawan S, Mehta S, Joshi SR (2022) Efficacy and safety of a phytopharmaceutical drug derived from *Cocculus hirsutus* in adults with moderate COVID-19: a phase 2, open-label, multicenter, randomized controlled trial. Infect Dis Ther 11(2):807–826. https://doi.org/10.1007/s40121-022-00604-0. Epub 2022 Feb 18. PMID: 35179709; PMCID: PMC8855350

Kaur P, Gupta RC, Dey A et al (2020) Validation and quantification of major biomarkers in 'Mahasudarshan Churna'- an ayurvedic polyherbal formulation through high-performance thin-layer chromatography. BMC Complement Med Ther 20:184. https://doi.org/10.1186/s12906-020-02970-z

Khanvilkar V, Hande P, Mandle S (2023) Marker-based standardization of polyherbal capsule formulation by high-performance thin-layer chromatography (HPTLC): densitometric method using andrographolide and catechin. J Planar Chromatogr–Mod TLC 36:55. https://doi.org/10.1007/s00764-023-00223-6

Krishnaswamy S, Kushalappa BA (2017) Correlative evaluation of the impact of adaptive plant morphology on bioactive accumulation based on micro-morphological studies in Andrographis serpyllifolia (Rottler ex Vahl) Wight. Not Sci Biol 9(2):263–273. https://doi.org/10.15835/nsb9210079

Letchumanan D, Sok SP, Ibrahim S, Nagoor NH, Arshad NM (2021) Plant-based biosynthesis of copper/copper oxide nanoparticles: an update on their applications in biomedicine, mechanisms, and toxicity. Biomol Ther 11:564

Lilienfeld S (2002) Galantamine—a novel cholinergic drug with a unique dual mode of action for the treatment of patients with Alzheimer's disease. CNS Drug Rev 8(2):159–176

Liu HM, Kiuchi F, Tsuda Y (1992) Isolation and structure elucidation of Gymnemic acids, antisweet principles of *Gymnema sylvestre*. Chem Pharm Bull 40:1366–1375

Maheshwari R, Kuche K, Advankar A, Soni NA, Sharma P, Tekade M, Tekade RK (2020) Industrial scale-up strategy and manufacturing technologies for botanical drug products. In: Lokhande J, Pathak YV (eds) Botanical drug products: recent developments and market trends. CRC Press, pp 87–116

Maleki M, Ghorbanpour M, Kariman K (2017) Physiological and antioxidative responses of medicinal plants exposed to heavy metals stress, plant. Gene 11:247–254. https://doi.org/10.1016/j.plgene.2017.04.006

Mandal SC, Mandal M (2011) Quality, safety, and efficacy of herbal products through regulatory harmonization. Ther Innov Regul Sci 45:45–53

Martino E, Della Volpe S, Terribile E, Benetti E, Sakaj M, Centamore A, Sala A, Collina S (2017) The long story of camptothecin: from traditional medicine to drugs. Bioorg Med Chem Lett 27(4):701–707. https://doi.org/10.1016/j.bmcl.2016.12.085. Epub 2016 Dec 31 PMID: 28073672.

Martinussen I, Nilsen G, Svenson L et al (2004) *In vitro* propagation of cloudberry (*Rubus chamaemorus*). Plant Cell Tissue Organ Cult 78:43–49. https://doi.org/10.1023/B:TICU.0000020392.85854.28

Matsuda H, Yoshikawa M, Morikawa T, Tanabe G, Muraoka O (2005) Antidiabetogenic constituents from *Salacia* species. J Trad Med 22(Suppl. 1):145–153

Mitchell G, Bartlett DW, Fraser TE, Hawkes TR, Holt DC, Townson JK, Wichert RA (2001) Mesotrione: a new selective herbicide for use in maize. Pest Manag Sci 57(2):120–128. https://doi.org/10.1002/1526-4998(200102)57:2<120::AID-PS254>3.0.CO;2-E. PMID 11455642

Mohammed WM, Mubark TH, Al-Haddad R (2018) Effect of CuO nanoparticles on antimicrobial activity prepared by sol-gel method. Int J Appl Eng Res 13:10559–10562

Mohammed ASA, Naveed M, Jost N (2021) Polysaccharides; classification, chemical properties, and future perspective applications in fields of pharmacology and biological medicine (a review of current applications and upcoming potentialities). J Polym Environ 29(8):2359–2371. https://doi.org/10.1007/s10924-021-02052-2. Epub 2021 Jan 27. PMID: 33526994; PMCID: PMC7838237

Namdeo AG, Sharma A (2012) HPLC analysis of camptothecin content in various parts of Nothapodytes foetida collected on different periods. Asian Pac J Trop Biomed 2(5):389–393. https://doi.org/10.1016/S2221-1691(12)60062-8. PMID: 23569936; PMCID: PMC3609305

Narayana DA, Katiyar C (2013) Draft amendment to drugs and cosmetics rules to license science-based botanicals, phytopharmaceuticals as drugs in India. J Ayurveda Integr Med 4:245–246

National Institute of Diabetes and Digestive and Kidney Diseases (2012) Liver Tox: clinical and research information on drug-induced liver injury. Bethesda (MD) . Ephedra. [Updated 2018 Feb 10]. https://www.ncbi.nlm.nih.gov/books/NBK548711/

Okhuarobo A, Falodun JE, Erharuyi O, Imieje V, Falodun A, Langer P (2014) Harnessing the medicinal properties of *Andrographis paniculata* for diseases and beyond: a review of its phytochemistry and pharmacology. Asian Pac J Trop Dis 4(3):213–222. https://doi.org/10.1016/S2222-1808(14)60509-0. PMCID: PMC4032030

Olazarán J, García G (2002) Galantamine: a novel cholinergic agent for Alzheimer's disease. Neurologia 17(8):429–436

Olgunsoy P et al (2017) Metabolite production and antibacterial activities of callus cultures from Rosa damascena Mill. petals. Marmara Pharm J 21(3):590–597

Park EJ, Cheenpracha S, Chang CL, Kondratyuk PT, Pezzuto MJ (2011) Inhibition of lipopolysaccharide-induced cyclooxygenase-2 expression and inducible nitric oxide synthase by 4-[(2′-oacetyl-α-L-rhamnosyloxy)benzyl] isothiocyanate from *Moringa oleifera. Nutr. Cancer* 63:971–982. https://doi.org/10.1080/01635581.2011.589960

Passos ID, Mironidou-Tzouveleki M (2016) Hallucinogenic plants in the Mediterranean countries. In: Neuropathology of drug addictions and substance misuse, vol 2. Elsevier, pp 761–772. https://doi.org/10.1016/b978-0-12-800212-4.00071-6

Prhal J, Milić J, Krajišnik D, Vuleta GM (2014) Properties and use of plant stem cells in cosmetic products, vol 64, p 26

Pugsley MK, Authier S, Curtis MJ (2008) Principles of safety pharmacology. Br J Pharmacol. 154(7):1382–99. https://doi.org/10.1038/bjp.2008.280. Epub 2008 Jul 7. PMID: 18604233; PMCID: PMC2492105.

Raza MA, Ur Rehman F, Anwar S, Zahra A, Rehman A, Rashid E, Kalsoom M, Ilahi H (2021) The medicinal and aromatic activities of cinchona: a review. Asian J Adv Res 8:42–45

Schuck E et al (2015) Preclinical pharmacokinetic/pharmacodynamic modeling and simulation in the pharmaceutical industry: an IQ consortium survey examining the current landscape. AAPS J 17:462–473

Shinde MU, Patwekar M, Patwekar F, Bajaber MA, Medikeri A, Mohammad FS, Mukim M, Soni S, Mallick J, Jawaid T (2022) Nanomaterials: a potential hope for life sciences from bench to bedside. J Nanomater 2022:5968131

Singh S, Datta R, Johri P, Trivedi M (2022) Phytopharmaceuticals and biotechnology of herbal plants, 1st edn. CRC Press. https://doi.org/10.1201/b22917

Tang S, Ren J, Kong L, Yan G, Liu C, Han Y, Sun H, Wang XJ (2023) Ephedrae Herba: a review of its phytochemistry, pharmacology, clinical application, and alkaloid toxicity. Molecules 28(2):663. https://doi.org/10.3390/molecules28020663. PMID: 36677722; PMCID: PMC9863261

Tito A, Carola A, Bimonte M, Barbulova A, Arciello S, De LD, Monoli I, Hill J, Gibertoni S, Colucci G, Apone F (2011) A tomato stem cell extract, containing antioxidant compounds and metal chelating factors, protects skin cells from heavy metal-induced damages. Int J Cosmet Sci 33:543–552

Trehan S, Michniak-Kohn B, Beri K (2017) Plant stem cells in cosmetics: current trends and future directions. Future Sci 3:4

Ushijima M, Komoto N, Sugizono Y, Mizuno I, Sumihiro M, Ichikawa M, Hayama M, Kawahara N, Nakane T, Shirota O, Sekita S, Kuroyanagi M (2008) Triterpene glycosides from the roots of *Codonopsis lanceolata*. Chem Pharm Bull 56:308–314. https://doi.org/10.1248/cpb.56.308

van der Graaf PH, Benson N (2011) Systems pharmacology: bridging systems biology and pharmacokinetics-pharmacodynamics (PKPD) in drug discovery and development. Pharm Res 28:1460–1464

Veeresham C (2012) Natural products derived from plants as a source of drugs. J Adv Pharm Technol Res 3(4):200–201. https://doi.org/10.4103/2231-4040.104709. PMID: 23378939; PMCID: PMC3560124

Venkataranganna MV, Gopumadhavan S, Sundaram R, Peer G, Mitra SK (2008) Pharmacodynamics & toxicological profile of PartySmart, a herbal preparation for alcohol hangover in Wistar rats. Indian J Med Res 127(5):460–466

Venkatraman S, Mehta TI (2023) Overview of preclinical research. Elsevier BV

Weon JB, Yun B-R, Lee J et al (2013) Effect of *Codonopsis lanceolata* with steamed and fermented process on scopolamine-induced memory impairment in mice. Biomol Ther 21(5):405–410. https://doi.org/10.4062/biomolther.2013.055

SPRINGER NATURE

GPSR Compliance

The European Union's (EU) General Product Safety Regulation (GPSR) is a set of rules that requires consumer products to be safe and our obligations to ensure this.

If you have any concerns about our products, you can contact us on ProductSafety@springernature.com

In case Publisher is established outside the EU, the EU authorized representative is:

Springer Nature Customer Service Center GmbH
Europaplatz 3
69115 Heidelberg, Germany

The manufacturer's authorised representative in the EU is Springer Nature Customer Service Centre GmbH, Europaplatz 3, 69115 Heidelberg, Germany. If you have any concerns regarding our products, please contact ProductSafety@springernature.com

Printed and bound by CPI Group (UK) Ltd, Croydon, CR0 4YY

01/10/2026

02225863-0016